Molecular Aspects of Fish and Marine Biology – Vol 2

Fish Development and Genetics

The Zebrafish and Medaka Models

Molecular Aspects of Fish and Marine Biology

Series Editors: **Choy Leong HEW & Zhiyuan GONG**
(National University of Singapore)

Published

Vol. 1: Fish Antifreeze Proteins
eds. C. L. Hew & K. V. Ewart

Vol. 2: Fish Development and Genetics: The Zebrafish and Medaka Models
eds. Z. Y. Gong & V. Korzh

Vol. 3: Current Trends in the Study of Bacterial and Viral Fish and Shrimp Diseases
ed. K. Y. Leung

Molecular Aspects of Fish and Marine Biology – Vol 2

Fish Development and Genetics

The Zebrafish and Medaka Models

editors

Gong Zhiyuan
National University of Singapore, Singapore

Vladimir Korzh
Institute of Molecular and Cell Biology, Singapore

World Scientific

NEW JERSEY • LONDON • SINGAPORE • BEIJING • SHANGHAI • HONG KONG • TAIPEI • CHENNAI

Published by

World Scientific Publishing Co. Pte. Ltd.
5 Toh Tuck Link, Singapore 596224
USA office: 27 Warren Street, Suite 401–402, Hackensack, NJ 07601
UK office: 57 Shelton Street, Covent Garden, London WC2H 9HE

British Library Cataloguing-in-Publication Data
A catalogue record for this book is available from the British Library.

Molecular Aspects of Fish and Marine Biology — Vol. 2
FISH DEVELOPMENT AND GENETICS
The Zebrafish and Medaka Models

ISBN-13 978-981-238-821-6
ISBN-10 981-238-821-4

Typeset by Stallion Press
Email: enquiries@stallionpress.com

Contents

Preface

This is an interesting time in fish and marine biology. Following the publication of the whole genome sequence of the first fish species, *Fugu ubripes*, in 2002, now the genome sequencing projects are near completion for the two popular experimental fish models, the zebrafish (*Danio rerio*) and medaka (*Oryzias latipes*). Both models, particularly the zebrafish model, have been well established as experimental systems in developmental and genetic analyses, as well as increasingly in medical research. In the past few years, we have witnessed an explosion of information from these researches. In order to keep track of the rapid development of this active field, we have devoted the entirety of volume two of the *Molecular Aspects of Fish and Marine Biology* series to the two small aquarium fish models. Thus, the name of this volume, *Fish Development and Genetics, The Zebrafish and Medaka Models.*

In this volume, there are 19 chapters. We have a broad coverage of zebrafish development from early embryogenesis to organogenesis. The topics include maternal factors (Chapter 1), gastrulation (Chapter 2), organizer and notochord (Chapter 3), floor plate (Chapter 4), central nervous system (Chapters 5 and 6), olfactory sensory system (Chapter 7), somites and segmentation (Chapters 8 and 9), muscle development (Chapter 10), skeletogenesis (Chapter 11) and endoderm formation (Chapter 12). We also have a few chapters on popular genetic tools in developmental analyses, including morpholino gene knockdown (Chapter 13), transgenic technology (Chapter 14), fish cloning (Chapter 15), transposons (Chapter 16), and evolution of the zebrafish genome (Chapter 17). In addition, two chapters focus on medaka genome mapping (Chapter 18) and medaka embryonic stem cells (Chapter 19).

These chapters summarize the current state-of-the-arts studies in the two fish models (particularly in the zebrafish) and focus in particular on the molecular aspects of development. We hope that this book will be a valuable reference for students to learn basic aspects of the two fish models as well as for researchers to look for resources in zebrafish and medaka research.

Finally, we wish to thank all contributors for their time and efforts to make the volume successful. We would also like to thank Ms. Serene Ong, editor of World Scientific, for her hard works and efficient efforts to ensure that this book to be published in the shortest possible time, despite the delay of submission of a few of the chapters.

Zhiyuan Gong
Vladimir Korzh
August 2004

Chapter 1

The Role of Maternal Factors in Early Zebrafish Development

Francisco Pelegri
Laboratory of Genetics
University of Wisconsin, Madison

The earliest events in zebrafish development are driven by maternal factors deposited in the egg during oogenesis that become activated upon fertilization and initiate cascades of events that drive early development. This review summarizes the forward and reverse genetic methods used to identify and analyze genes coding for such maternal factors. I also discuss current knowledge on the cellular processes involved in two important developmental transitions: the redistribution and activation of maternal factors at fertilization, and the transition from maternal to zygotic genetic programs. In addition, I summarize current knowledge on the function of maternal factors, both before and after zygotic gene activation, in embryonic processes involved in general cellular functions, axis formation, and cell fate specification.

1. Maternal-Effect Genes

The earliest embryonic processes are carried out by maternal factors produced during oogenesis by the mother and stored in the mature egg. Upon egg activation and fertilization, these factors become activated and they, in turn, initiate cascades of events necessary for early development. Only when the zygotic genome becomes activated at the midblastula transition (MBT),[1,2] which in the zebrafish begins at the 13th cell cycle,[3,4] does the embryo begin to utilize products derived from its own genes. Therefore, all processes that occur before zygotic gene activation must rely solely on maternal factors stored in the egg. Maternal products present in the oocyte can, in principle, be utilized for developmental functions that occur even after zygotic gene activation, and indeed this expectation has been confirmed experimentally for some genes. In this chapter, I will first review the functional definition of maternal factors and describe current tools used to identify such factors and analyze their function. Subsequently, I will review our current knowledge on the redistribution of these factors upon egg activation and fertilization, and their function during early embryonic development.

1.1. Definition

The term "maternal factor" refers to any product produced during oogenesis, which is stored in the egg, in the form of mRNA, protein or another type of molecule, and which functions during early embryogenesis. Maternal genes are those activated during oogenesis to produce such maternal products. In other words, maternal genes are essential in a female for the development of its progeny. Zygotic genes, on the other hand, refer to those genes active in an organism when their function is important for the organism itself (as opposed to its offspring).

The functional definition of a maternal gene has been traditionally determined through genetic analysis, so that a gene is determined to act maternally when mutations in the female germ line that produces the egg result in an embryonic phenotype in the progeny derived from those eggs. Reciprocal crosses, where the mutations are present in the male germ line and not in the female germ line, result in normal progeny. For simplicity, both individuals with an allelic combination that causes a phenotypic effect in their offspring, and the progeny embryos themselves are referred to as genotypically mutant.

1.2. Classification of Maternal Effect Genes

Maternal effect genes can be classified as having either strictly maternal or maternal/zygotic effects.

1.2.1. *Strictly maternal-effect genes*

Strictly maternal genes are expressed during oogenesis and this maternal expression is both required and sufficient to carry out all the function of the gene in the early embryo. A landmark characteristic of strictly maternal-effect genes is that the embryonic phenotype produced by mutations in them is only dependent on the genotype of the female germ line from which the eggs are derived and is independent of the genotypic constitution of the embryo itself (Table 1). In other words, all functional product is supplied maternally and either there is no early embryonic expression or, if there is, its functional contribution is negligible. Because of this, progeny from mutant mothers show the same phenotype regardless of the genotype of their fathers. This category of genes includes genes such as *futile cycle, janus, nebel* and *ichabod* (Table 2; Fig. 1).

It is important to note, however, that mutations resulting in strict maternal effects may be special alleles of genes that are required both maternally and zygotically. This situation can be envisioned for a hypomorphic allele when the zygotic function requires a lower threshold of wild-type function than the maternal function. It is also possible that mutations may specifically affect the maternal expression of a gene that is also used zygotically, for example, if it affects a maternal-specific transcript. Because of these possibilities, only after substantial analysis, such as the determination of the nature of the allele by molecular genetic analysis as well as the expression pattern of the gene, is it possible to conclude that a gene is strictly maternal.

1.2.2. *Maternal-zygotic genes*

Maternal-zygotic genes are expressed during both oogenesis and embryogenesis, and both maternal and zygotic products provide significant function. The hallmark of this genetic phenomenon is that the phenotype when both maternal and zygotic contributions are mutant (the maternal-zygotic (MZ) phenotype) is stronger than the phenotypes produced by mutation in either the maternal contribution or the zygotic contribution alone. Such a genetic interaction can also show different effects with respect to the *separate* effects of the maternal and zygotic genetic contributions (Table 1, Table 3):

i) Mutations in the zygotic contribution alone, but not the maternal contribution alone, result in a detectable phenotype. This category

Table 1 Maternal and maternal-zygotic effects.

Case	m/m ♀ x +/+ ♂	m/m ♀ x +/m ♂	m/m ♀ x m/m ♂	m/+ ♀ x m/+ ♂	Summary (genotype: phenotype)	Phenotypic strengths
Recessive strictly maternal	100% mut	100% mut	100% mut	100% wt	$M^{+}Z^{-}$:wt $M^{-}Z^{+}$:mut $M^{-}Z^{-}$:mut	n.a.
Recessive maternal—zygotic						
i) only zygotic-effects result in a phenotype	100% wt	50% wt 50% MZmut	100% MZmut	75% wt 25% Zmut	$M^{+}Z^{-}$:Zmut $M^{-}Z^{+}$:wt $M^{-}Z^{-}$:MZmut	MZmut > Zmut
ii) both maternal and zygotic effects result in phenotypes	100% Mmut	50% Mmut 50% MZmut	100% MZmut	75% wt 25% Zmut	$M^{+}Z^{-}$:Zmut $M^{-}Z^{+}$:Mmut $M^{-}Z^{-}$:MZmut	MZmut > Mmut, Zmut
iii) only maternal-effects result in a phenotype	100% Mmut	50% Mmut 50% MZmut	100% MZmut	100% wt	$M^{+}Z^{-}$:wt $M^{-}Z^{+}$:Mmut $M^{-}Z^{-}$:MZmut	MZmut > Mmut
iv) neither maternal nor zygotic effects result in a phenotype	100% wt	50% wt 50% MZmut	100% MZmut	100% wt	$M^{+}Z^{-}$:wt $M^{-}Z^{+}$:wt $M^{-}Z^{-}$:MZmut	n.a.
Dominant maternal — zygotic				25% Zmut 50% MZ^{D}mut 25% wt	$M^{+/-}Z^{-/-}$: Zmut $M^{+/-}Z^{+/-}$: MZ^{D}mut $M^{+/-}Z^{+/+}$:wt	MZ^{D}mut < Zmut[a]

Abbreviations. mut: mutant phenotype; Mmut: maternal-effect mutant phenotype, MZmut: maternal-zygotic mutant phenotype; Zmut: zygotic mutant phenotype; n.a.: not applicable (since there is only one relevant phenotype); M^{+} or M^{-}: genotypically wild-type or homozygous mutant, respectively; in the maternal germ line; Z^{+} or Z^{-}: genotypicaly wild-type or homozygous mutant, respectively, in the zygote; $M^{+}/^{-}$: genotypically heterozygous in the maternal germ line; $Z^{+}/^{-}$: genotypically heterozygous in the zygote.

[a] Maternal-zygotic dominant phenotypes have been observed to be weaker than a full loss of zygotic function. This is consistent with the dominant maternal-zygotic interaction being caused by an overall partial reduction of maternal and zygotic function. In principle, however, other strength

Table 2 Genes with a strict maternal-effect.

Gene name	Process affected	References
futile cycle	Pronuclear fusion	78
ichabod	Induction of the dorsal axis (accumulation of nuclear ßcat protein in dorsal nuclei)	21
janus	Cell adhesion, results in split blastoderms which produce duplicated axes	20
nebel	Organization of the furrow microtubule array, which affects cell adhesion and the segregation of the germ plasm	60
yobo[a]	Dorsal axis convergence and extension	79

[a]Mutations in *yobo* also have a zygotic effect, where homozygotes exhibit a reduction in xanthophores during the larval stages. However, the convergence extension embryonic phenotype appears unrelated and is strictly maternal.

contains recessive mutations in genes such as *lazarus*,[5] *lost-a-fin*,[6,7] *ogon*,[8–10] *one-eyed pinhead*,[11] and *schmalspur*[12,13] (Fig. 2),

ii) Mutations in both maternal and zygotic contribution separately result in a detectable phenotype. Recessive mutations in genes such as *pipetail*[14] and *somitabun*[15] fall in this category.

In principle, there are two other possibilities:

iii) Mutations in the maternal contribution alone, but not the zygotic contribution alone, result in a detectable phenotype. This may be the case for the gene *radar* with respect to its role in dorsoventral patterning,[16] as suggested by the analysis of functional knockdowns (see "Morpholino-mediated functional knockdown" and "Bmp signaling: promotion of ventral cell fates"). However, this finding should be confirmed through genetic analysis of loss of function mutations in the gene.

iv) Mutations in both the maternal and the zygotic contribution, result in a detectable phenotype, but when occurring separately show no phenotype. This situation can occur if either the maternal or the zygotic contribution alone provides sufficient function for normal development. Such redundancy would make these mutations difficult

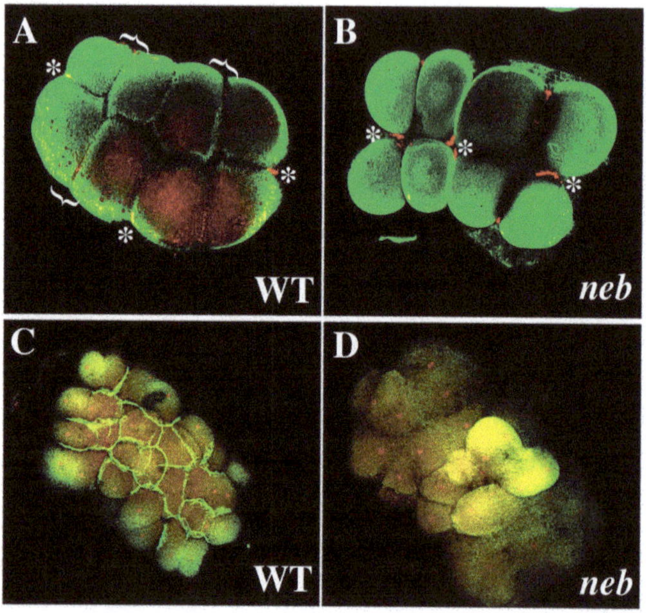

Fig. 1 A recessive, strictly maternal-effect mutation in the gene *nebel* affects germ plasm segregation and cell adhesion. (**A, C**) wild-type embryos; (**B, D**) *nebel* mutant embryos. (**A, B**) confocal images of wild-type and *nebel* mutant embryos labeled using an antibody against α-tubulin (green), to highlight microtubular structures, and in situ hybridization to detect the *vasa* mRNA (red), a component of the germ plasm. Arrays of microtubules at the forming furrows in wild-type embryos (brackets where visible) are reduced or absent in *nebel* mutants. Note that *vasa* mRNA containing aggregates (asterisks), which in wild-type embryos localize at this stage to the peripheral end of the furrow, are mislocalized towards the middle of the furrow in *nebel* mutants. (**C, D**) confocal images of wild-type and *nebel* mutant embryos labeled using an antibody to ß-catenin (green), which highlights secreted membrane at the furrow, and the DNA stain propidium iodide (red). *nebel* mutant embryos lack accumulation of adhesive membrane at the furrow in spite of a normal pattern of nuclear divisions and furrow initiation. Note that some cells in the *nebel* mutant embryo protrude due to defective cell adhesion. See text for details.

to identify, and so far no mutation with this characteristic has been reported.

As opposed to the case of strictly maternal-effect mutations, in all cases of maternal-zygotic effects there is some degree of rescue from the zygotic copy. For recessive maternal-effect mutations, this is noticeable

Table 3 Genes with maternal-zygotic effects.

Gene name	Process affected/molecular identity	References
alk8/lost-a-fin	Ventral cell determination[a]/Type I TGF-ß receptor	6, 7
foxH1/fast1/schmalspur	Mesendoderm induction[a]/Forkhead domain transcription factor	12, 13
half-baked	Epibolic movements of the inner cellular layer[b]/molecular identity unknown	18
one-eyed pinhead	Mesendoderm induction[a]/EGF-CFC family co-receptor for nodal signals	11
pbx/lazarus	Hindbrain segmentation and rhombomere identity [a]/homeodomain transcription factor	5
radar	Ventral cell fate determination[c]/TGF-ß factor of the Bmp family	16
sizzled/ogon	Control of ventral cell determination[a]/secreted Frizzled-related factor involved in feedback inhibition of Bmp signaling	9, 10, 103, 104
smad5/somitabun	Ventral cell determination[b,d]/intracellular factor in TGF-ß signaling	17, 15
tcf-3/headless	Determination of anterior brain structures[a]/HMG box transcription factor	85
wnt5/pipetail	Regulation of dorsal organizer[d]/secreted ligand that activates Wnt/calcium signaling	14

[a]Mutating the zygotic contribution alone, but not in the maternal contribution alone, causes defects in the described process (class (i) in Table 1).
[b]Dominant effect caused by heterozygosity for both maternal and zygotic genetic contributions.
[c]MO and gtMO analysis suggests that mutation of the maternal contribution alone, but not in the zygotic contribution, results in defects in axis formation (class (iii) in Table 1). Genetic experiments are needed to confirm this expectation.
[d]Mutations in both the maternal contribution alone and the zygotic contribution alone, result in defects in the described process (class (ii) in Table 1).

when the wild-type allele is provided by the father, and is referred to as paternal rescue. Such paternal rescue can be complete (classes (i) and (iv) above), when the zygotic copy alone is sufficient for the full function of the gene, or partial (classes (ii) and (iii)), when maternal product is essential and cannot be fully substituted by zygotic product.

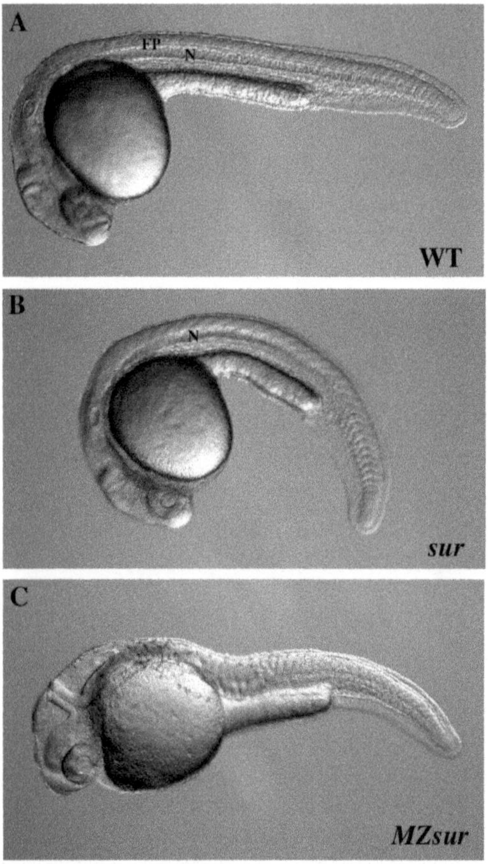

Fig. 2 The gene *schmalspur* has both maternal and zygotic contributions. Wild-type embryo (**A**) and embryos with mutant zygotic (**B,** ***sur***) and mutant maternal and zygotic (**C,** ***MZsur***) contributions. Embryos lacking zygotic *sur* function have an abnormal body curvature and lack a floor plate (FP, visible in the wild-type as a line above the notochord (N)). Embryos lacking both maternal and zygotic function show a more severe phenotype including an anterior truncation of the brain and a shortened axis. Photos courtesy of Dirk Meyer.

In addition, for some mutations in the genes *somitabun*[15,17] and *half-baked*,[18] a dominant maternal-zygotic interaction has been observed where heterozygosity for both maternal and zygotic contributions results in an embryonic phenotype (Table 1).

All maternal-zygotic genetic interactions identified so far correspond to genes where mutation of the zygotic contribution alone results in a detectable phenotype (classes (i) and (ii) above). It is possible that there is an underlying biological basis for this bias, indicating the possibility that for maternal-zygotic genes the zygotic contribution tends to be more important than the maternal contribution. It is also possible, however, that this phenomenon reflects the fact that, to date, most mutations with maternal-zygotic interactions have been identified on the basis of their zygotic phenotype. A more thorough analysis of other mutations, including those identified by virtue of their maternal-effect phenotype, will provide a better estimate of the relative importance of maternal and zygotic contributions of maternal-zygotic genes.

1.3. Underlying Basis for Strictly Maternal and Maternal-Zygotic Genetic Effects

All processes that occur prior to the activation of the zygotic genome at MBT are expected to rely solely on maternal products and, therefore, to be dependent on strictly maternal-effect genes. It is also possible, however, that some functions that occur after zygotic gene activation at MBT are dependent on perduring products derived from strictly maternal effect genes (see "Bmp signaling: promotion of ventral cell fates", "Induction of cell fates along the anteroposterior axis" and "Induction of the mesendodermal layer").

In the case of maternal-zygotic genes, on the other hand, genetic analysis reveals that both maternal and zygotic products can substitute for each other, i.e. both maternal and zygotic contributions can provide significant functions. However, this genetic interaction does not indicate which product is primarily used during normal development. For example, a maternal product may normally form part of a preformed functional complex that is already saturated by the time zygotic product appears in the embryo. When a female is genotypically mutant, such a functional complex would not form, and a zygotic product could in principle still form part of the complex and contribute, partially or completely, to the function of the gene. However, in this case, in wild-type embryos the maternal product would be performing most of the gene function. Conversely, newly made zygotic products may be preferentially used over maternal product under some circumstances. The function of a gene could for example depend on its interaction with newly made products (for example, in a cellular compartment such as the endoplasmic reticulum), so that in wild-type embryos most of the normal function of the gene could be carried out by zygotic product. These situations are, of course, extremes in a more likely continuum where both maternal and zygotic products are utilized to varying degrees. In the simplest scenario, non-complexed protein present in the cell is utilized according to the ratio in which maternal and zygotic products are present in the cytoplasm. Earlier stages may rely primarily on maternally-derived products, while later stages may rely increasingly on zygotic products. The dosage sensitivity of dominant maternal-zygotic interactions, such as for the genes *half-baked* and *somitabun,* is consistent with a scenario of additive maternal and zygotic contributions.

2. Tools to Identify and Study Maternal-Effect Genes

2.1. Forward Screens

With rare exceptions, such as in the case of strongly antimorphic mutations, for example *somitabun,*[15,17] and, potentially, strong sensitivity to gene dosage, the majority of mutations that result in high-penetrance phenotypes act in a recessive manner. In order to create homozygous individuals for newly induced mutations, the identification of new recessive mutations requires several generations of inbreeding. Although continuous genetic inbreeding is not a common practice in the propagation of zebrafish due to weakening of the stocks and resulting abnormal sex ratios,[19] zebrafish can tolerate inbreeding to a limited number of generations, therefore allowing the identification of recessive maternal effect mutations. Several such mutations have been fortuitously identified during routine maintenance of laboratory stocks (*janus,*[20] *ichabod*[21]). A more directed approach, however, purposely aims at producing and screening individuals that are homozygous for newly induced mutations. This can be carried out through either a two-generation parthenogenesis-based strategy or a three-generation inbreeding strategy, and is described in detail elsewhere.[22] Such approaches have allowed the isolation of mutations in strictly maternal genes required for early development (Table 2).

2.2. Testing for the Maternal Contribution of Known Zygotic Genes

Currently, a large number of mutations in zygotic genes have been identified in large-scale screens.[23,24] These mutations pinpoint a pool of factors essential for developmental processes, which in a number of cases, may be provided both zygotically and maternally (Table 3). It is also possible that some of these factors are used maternally and zygotically in a number of apparently unrelated developmental processes (but which may nevertheless depend on the same biological function or pathway), as in the case of the mutation *yobo*. Regardless of the specific details relating to different genes, it is clear that zygotic genes already identified by mutational analysis, especially if they are expressed in the ovary, represent an important pool of essential factors that may have important maternal genetic contributions.

Testing for the maternal contribution of a gene involves creating a female germ line that is genotypically mutant, and determining whether there are particular phenotypes in the offspring derived from this germ line. Fertilization of the eggs with sperm from heterozygous males allows observing potential maternal effects and maternal-zygotic effects (in progeny that carry the paternally derived wild-type and mutant alleles, respectively). Depending on the nature of the zygotic mutation, such mutant germ lines can be created in different ways:

2.2.1. *Viable or semi-viable mutations*

Some zygotic mutations result in relatively weak defects that permit, for at least a fraction of genotypically mutant individuals, development into fertile adult females. Mutant individuals can be identified either due to their characteristic phenotype, which may be compensated during development, or by genotyping, and then tested as adults for maternal effects. This procedure has been used to obtain adult females homozygous for mutations in the zygotic genes *ogon* and *schmalspur* (Fig. 2), for which mutants from only the zygotic contribution result in weak ventralization and cyclopic phenotypes respectively.[9,10,12,13] The severity of these zygotic phenotypes is increased when zygotically mutant embryos are additionally derived from a mutant germ line, indicating a significant maternal contribution.

2.2.2. *Non-viable mutations with a phenotype that can be rescued experimentally*

Some phenotypes caused by zygotic mutations can be rescued by the injection of maternal mRNA coding for products that provide, mimic or bypass the normal function of the mutated gene. This can be typically done by the injection of mRNAs coding for the same gene or a factor within the same pathway or an interacting pathway. For example, the dorsoventral patterning defect caused by homozygosity for mutations in *bmp2b/swirl* can be rescued by injection at the one-cell stage of mRNA coding for wild-type *bmp2b/swirl* or other Bmp genes.[25] In this manner, homozygous mutant individuals (identified by either genotyping or the presence of other aspects of the phenotype) can be grown to adulthood and tested for maternal effects.

2.2.3. *Non-viable mutations using germ line chimeras*

An alternative to growing mutant individuals to adulthood is to create chimeric individuals whose somatic tissues are wild-type, and therefore allow normal development, but whose germ line is mutant for a particular mutation. Such chimeras can be generated by cell transplantation of primordial germ cells (PGCs), which is usually carried out during the late blastula and early gastrula stages when cells can be easily manipulated.[26] However, PGCs are not morphologically distinguishable from somatic cells at these early stages, and current GFP-based methodologies do not allow easily distinguishing PGCs until later stages, so that typically a mixture of somatic and germ line cells is transplanted into host embryos. Once such chimeric individuals become adults, embryos derived from transplanted, mutant germ line cells can be distinguished from sibling embryos derived from host wild-type cells through either the use of visible genetic markers or genotyping.[5,27,28]

Several methods have significantly improved the efficiency of the PGC transplantation technique. Because PGCs can be identified in live embryos at later stages (e.g. 24 hours) by virtue of their location and morphology, it is possible to select for embryos that contain donor-derived PGCs. Such donor-derived PGCs can be identified by a live fluorescent cell marker present in the donor embryo, such as fluorescently labeled dextran or a GFP product that is specifically expressed in the PGCs.[28] In addition, host embryos can be treated with agents that inhibit proper development of their own PGCs, for example, by the functional knockdown (see "Morpholino-mediated functional knockdown") of the gene *nanos*[29] or *dead end*,[30] both of which are required for proper PGC development. This facilitates the expansion of PGCs derived from the donor embryo and allows obtaining chimeric adults where most or all of the germ line has been replaced with mutant germ line cells.[5,28]

A lack of a maternal effect in embryos derived from chimeric females with a mutant germ line and a wild-type soma is suggestive, but does not conclusively rule out the possibility that the gene does not have a maternal effect. This is because, even though most maternal products appear to be produced by the oocyte itself and would therefore be expected to act cell autonomously, maternal products can also be produced by somatic tissues and acquired by the oocyte during oogenesis.[31] Thus, in such chimeras it remains a formal possibility that the genetic function is rescued non-cell autonomously by wild-type somatic cells. Such a possibility can be further addressed by detailed analysis of the expression pattern of the gene in adult females.

2.3. Reverse Genetic Approaches

A variety of approaches have been developed in recent years that allow testing the function of a gene of known molecular identity by reverse genetics.

2.3.1. *Morpholino-mediated functional knockdown*

Morpholino (MO)-conjugated oligonucleotides complementary to regions at or immediately upstream of the translational start site of the mRNAs have been shown to inhibit the translation of specific mRNAs in early embryos.[32] However, because current methodologies involved the injection of MOs at very early stages of development (typically at the one-cell stage), and since MOs are stable and perdure in embryos for several days, MOs affect all transcripts present in the embryo, regardless of maternal or zygotic origin. Therefore, if a MO knockdown indicates a role for a particular gene in development, additional experimental evidence is needed to determine the extent of its maternal contribution. A potentially useful variation of MOs uses an oligonucleotide complementary to the splice donor site of the premature mRNA (gtMO) and therefore interferes with splicing of newly transcribed (zygotic) mRNA, but not with translation of the maternal mRNA.[16,32] A phenotype by the MO that is stronger than that caused by the gtMO can be interpreted as the result of interference with the function of maternally provided mRNA. However, this interpretation is based on the idea that both MOs are equally effective at affecting their respective targets, and therefore careful experimentation is needed to substantiate this assumption. In addition, MOs injected into early embryos do not affect protein products already present in the fertilized egg, and therefore their use cannot rule out the presence of maternal functions provided by such protein factors.

2.3.2. *RNA interference*

Exposure to double-stranded (ds) RNA complementary to the gene product has been widely used in other systems to knock down genes by a dsRNA-triggered degradation pathway.[33] Zebrafish embryos injected with dsRNA at early stages exhibit a non-specific degradation response that affects many transcripts,[34,35] which may be analogous to other non-specific effects known to be caused by dsRNA in vertebrate cells.[36,37] However, in mammalian cells smaller dsRNA regions of less than 30 nucleotides in length have been found to promote specific RNA degradation without triggering non-specific responses.[38,39] It is possible that small dsRNAs will also be found to be effective at triggering the specific degradation of transcripts in zebrafish embryos, which would provide an alternative to the use of MO oligonucleotides. As in the case of MOs, however, RNAi induced by injection of dsRNA into early zebrafish embryos would be ineffective against gene functions that rely on protein products already present in the egg. This limitation could potentially be overcome by using specific hair pin transgenes expressed in the developing oocyte, as has been successfully used in mouse oocytes.[40,41]

2.3.3. *Target selected inactivation*

Recently, a target-selected approach has been described that relies on the direct identification of mutations in specific genes by high-throughput screening at the DNA level of genomes carrying newly induced mutations.[42] This approach allows obtaining mutations in any gene candidate in order to test for potential maternal-effects and could in principle be used to systematically screen for maternal-effects of ovary-specific transcripts.

3. Maternal Products during Egg Activation and Early Embryogenesis

With few exceptions, most maternal products are produced by the oocyte itself during oogenesis. Many of these products, whether mRNA or protein, are localized during oogenesis to different regions of the oocyte. Previous articles have described the production and localization of products in the oocyte.[31,43,44] Here, I will review the redistribution of maternal products during egg activation and early embryogenesis. Activation of the zebrafish oocyte results in the segregation of the ooplasm away from non-membrane bound yolk and towards the animal pole of the oocyte, resulting in the formation of the blastodisc.[45,46] This redistribution appears to involve the propagation of a slow calcium wave, which triggers both the separation of ooplasm and yolk and the movement of ooplasm along yolk-free paths (called streamers) leading towards the blastodisc.[47] Both of these processes are also dependent on an intact actin cytoskeleton and are independent of microtubules.[48–50] However, the redistribution of at least one mRNA product towards the animal pole has been shown to be dependent on microtubules but independent of microfilaments,[51] and the redistribution of the putative dorsal determinant is also thought to occur along cortical microtubules[52] (see "Redistribution of dorsal determinant signal during the early cleavage stages"). This indicates that both cytoskeletal networks have a role in the transport of maternal products toward the blastodisc. However, actin-based mobility may be more important for bulk movements of the ooplasm, while microtubules appear to be involved in the transport of specific maternal products.

3.1. Redistribution of Maternal mRNAs during Egg Activation

Maternal transcripts present in the oocyte have been shown to localize in four different patterns, each of which may involve different mechanisms for redistribution.[53,44]

3.1.1. *mRNAs evenly distributed in the mature oocyte*

Upon egg activation, these mRNAs are transported to the forming blastodisc along the axial streamers (Fig. 3, yellow). The cytoskeletal requirements for this movement have not been systematically studied, and it may depend on the bulk flow of the cytoplasm. However, in one reported case for which these requirements have been determined, *squint* mRNA transport is dependent on intact microtubules but

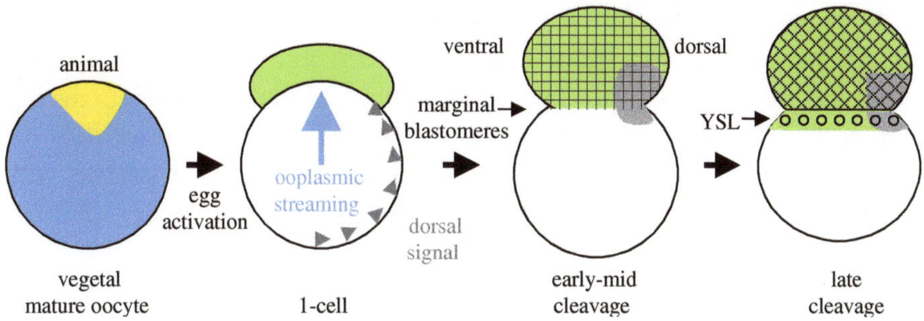

Fig. 3 Redistribution of maternal factors present in the mature oocyte during egg activation and early development. Upon egg activation, transcripts and other ooplasmic components ubiquitously distributed in the mature oocyte (blue) become redistributed to the forming blastodisc along axial streamers and accumulate in the blastodisc (green). Transcripts and other factors already localized at the animal pole of the mature oocyte (yellow) remain in the animal pole region and also become distributed at the forming blastodisc (green). Other transcripts localized during oogenesis, such as the vegetally localized *daz* and *bruno-like* mRNAs and the cortically localized *vasa* mRNA also redistribute towards the animal pole (not shown, see text for details). During egg activation, the putative dorsal signal moves towards the forming blastodisc along the cortex at one side of the embryo (grey arrowheads), and becomes present in dorsal cells and the dorsal YSL, thus specifying the dorsal side of the embryo. Starting at the 16-cell stage, the inner blastomeres begin to be completely enclosed by a membrane. However, the marginal blastomeres maintain their connection to the yolk cell. At the 1000-cell stage, membranes of the marginal blastomeres regress and the nuclei in the resulting layer continues to divide to form the yolk syncytial layer (YSL). The YSL can inherit maternal factors derived from either the yolk cell itself or the marginal blastomeres, and in turn influence cell fate in the overlying cells of the blastoderm.

surprisingly independent of the microfilament network.[51] It remains to be tested whether this is a generalized phenomenon for most mRNAs in this class.

3.1.2. *mRNAs localized to the animal pole of the oocyte during oogenesis*

Even before egg activation and ooplasmic streaming, the animal pole region of the oocyte is relatively devoid of yolk and is enriched in a subclass of mRNAs (Fig. 3, blue).

During oogenesis, tight anchoring of the *zorba* mRNA to the animal cortex appears dependent on an intact actin network.[54] It is unclear whether there are requirements for anchoring of these transcripts during egg activation. However, it is likely that the directed movement of cytoplasm towards the animal pole and the compaction of the yolk would impose additional restrains on the potential diffusion of these products toward more vegetal regions.

3.1.3. *mRNAs localized to the vegetal pole of the oocyte*

Several mRNAs, such as *deleted in azoospermia* (*daz*) and *bruno-like*, have been found to be localized to the vegetal pole of the oocyte.[55,56] Upon egg activation, these mRNAs move towards the animal pole. These mRNAs have been observed to be present in the axial streamers in the activated egg, although it is has not been reported whether this transport is dependent on the microfilament or the microtubule network.

3.1.4. *The mRNA for the gene vasa*

The *vasa* mRNA, a component of the zebrafish germ plasm, is localized during oogenesis to the cortical region of the oocyte.[53,57,58] Upon transport, the *vasa* mRNA becomes localized transiently to the base of the blastodisc, also called the cytokinetic ring.[57] As opposed to the case of *daz* and *bruno-like*, *vasa* mRNA is not observed in axial streamers in the activated egg. It is therefore possible that *vasa* mRNA is transported to the cytokinetic ring along the plane of the cortex, although this remains to be demonstrated. This segregation pattern has only been reported for the *vasa* mRNA, and it is unclear whether it is used by other mRNAs or whether it is characteristic of other components of the germ plasm. During the first cleavage divisions the *vasa* mRNA, together with other germ plasm components, will become localized to the forming furrows (see "Redistribution of components of the germ plasm during the early cleavage stages").

3.2. Redistribution of Components of the Germ Plasm during the Early Cleavage Stages

The maternal mRNAs for three genes have been shown to segregate together with the zebrafish germ plasm: *vasa*, *nanos* and *dead end*, which encode, respectively, a DEAD box protein with homology to RNA helicases, a RNA binding zinc finger protein and a novel putative RNA binding protein.[29,30,59]

The pattern of localization of *vasa* has been the most closely studied[57–60] (see also a prior review[44]). During the first and second cleavage divisions, the *vasa* mRNA moves from its transient location at the cytokinetic ring (see "The mRNA for the gene vasa") towards the forming cleavage furrows. This appears to involve two steps. The first occurs during the initiation of cleavage and consists of the recruitment of *vasa* mRNA as an elongated, rod-like structure along the furrow at a position immediately underneath the plasma membrane. The second step occurs during furrow maturation and consists of the aggregation of this rod-like structure into a compact mass at the peripheral ends of the furrow. During this aggregation process, the *vasa*-containing aggregate is in close association with the distal ends of the tubules of the furrow microtubule array (FMA).[58,60] This structure consists of an array of microtubules parallel to each other and perpendicular to the furrow that forms during furrow maturation and, like the *vasa* mRNA, accumulates at the peripheral ends of the furrow[60,61] (Fig. 1A). A mutation in the gene *nebel*, which specifically affects FMA formation, as well as exposure to microtubule inhibiting drugs during furrow maturation, results in defects in the directed movement of this aggregate towards the periphery[60] (Fig. 1B), confirming that the peripheral movement is dependent on microtubule function.

Ultrastructural analysis has shown that, at least from this stage on, the *vasa* mRNA is part of an electron-dense structure analogous to germ plasm material,[58] which is typically found in association with other subcellular structures such as fibrils and mitochondria and which contain specific mRNA and protein products.[62] The mRNAs for the genes *nanos*[29] and *dead end*[30] also become localized to the cleavage furrows in a pattern very similar to that of the *vasa* mRNA, although it is unclear whether the segregation patterns of *nanos* and *dead end* mRNAs during oogenesis and egg activation are the same as that of the *vasa* mRNA. It seems clear, however, that by the time the early cleavage furrows form the *vasa*, *nanos* and *dead end* mRNAs have become part of a multicomponent germ plasm structure. The analysis of MO-knockdowns for both *nanos* and *dead end* shows that these genes are required for PGC migration during gastrulation, although it is unclear whether this function corresponds to maternal or zygotic transcripts.[29,30] However, there is no experimental evidence suggesting that *vasa*, *nanos* or *dead end* are involved in the initial specification of PGCs.

The four germ plasm aggregates remain in a peripheral position until the 32-cell stage, when they ingress into four cells and remain subcellularly localized.[57–59] This localization appears to occur at or near one of the spindle poles, and during cell division the aggregates segregate asymmetrically so that only one of the daughter cells receives the aggregate.[57,58] This fascinating segregation program ceases at the sphere stage (cell cycles 12–13), when these mRNAs become evenly distributed in the cytoplasm of the PGCs and are evenly distributed during cell division. This transition coincides with the midblastula transition and the initiation of zygotic gene transcription, and in the case of *vasa*, transcriptional activation of the *vasa* gene itself.[58] Surprisingly, this transition appears to be independent of both the nuclear-cytoplasmic ratio and transcriptional initiation,[58] suggesting that it relies on other, insofar unknown, cellular counting mechanisms that are likely driven by maternal products.

Vasa protein is localized in perinuclear patches during oogenesis.[58,63] However, even though the Vasa protein originally present during oogenesis remains ubiquitously distributed in the mature egg, Vasa protein does not localize to the aggregating germ plasm at the furrows of the 2- and 4-cell embryo.[58] Thus, it is unclear what function, if any, this maternally derived product has in germ plasm formation. At the midblastula transition, coincident with the loss of asymmetric *vasa* mRNA segregation, Vasa protein begins to accumulate in PGCs, again in perinuclear patches. Colabeling studies have shown that at these late blastula stages these perinuclear structures also contain Nanos and Dead end proteins,[29,30] suggesting that these proteins form part of a subcellular structure important for PGC development. Observations in embryos where DNA replication has been inhibited suggest that although the majority of Vasa protein observed in PGCs is derived from zygotic transcripts, a fraction of the protein is likely produced by translation of the maternal *vasa* mRNA remaining in the PGCs.[58] It will be interesting to determine the precise function of protein derived from maternally derived, germ plasm-specific mRNAs.

The *vasa*, *nanos* and *dead end* mRNAs and their corresponding products are regulated at multiple levels other than mRNA localization, including mRNA stability,[29,30,63] mRNA translatability[29,30] and protein stability.[63] These precise, multiple regulatory programs suggest an important role for maternally-derived germ plasm products for the early specification of the germ line.

3.3. Redistribution of Dorsal Determinant Signal during the Early Cleavage Stages

Removal by ligation of the vegetal most region of the yolk cell during the early cell cycles results in defects in axis formation,[64,65] suggesting the presence of a putative dorsal signal at this location in the mature egg. If, however, the procedure is carried out at later cell cycles (e.g. at the 8-cell stage in zebrafish embryos), axis formation is not affected, an observation that is consistent with the migration of the putative dorsal signal towards the animal pole during the early cell cycles. Moreover, transplantation of the yolk cell at the midblastula stage onto a host blastula is able to induce the localized ectopic expression of dorsal specific genes.[66] This ability is lost if the embryos from which the yolk cells are derived had undergone early removal of the vegetal yolk.[64] Together, this data indicate that a dorsal-inducing signal is originally present at the vegetal pole of the mature oocyte and, upon egg activation, migrates toward the animal pole in a localized region under the blastoderm which will become the dorsal axis (Fig. 3, grey arrowheads).

The inferred movement of the putative dorsal signal appears to be mimicked by the movement of fluorescent beads injected at the vegetal pole of activated eggs.[52] Such beads can be observed to move towards the animal pole along a localized path on the cortex, presumably the future dorsal side of the embryo. This movement is dependent on microtubule function, and indeed a cortical array of microtubules becomes aligned in the direction of the movement at one side of the early embryo. Moreover, induction of the dorsal axis is sensitive to defects in microtubule function during the early cell cycles. Thus, the movement of the putative dorsal signal from the vegetal pole towards the forming blastoderm in the animal region likely occurs along microtubules. Recently, functional manipulation of the actin filament severing factor Gelsolin, which is provided maternally in the embryo, has suggested a role for the actin network in dorsal axis induction.[67] It will be interesting to test whether this network is needed during these early stages, for example to anchor the putative signal at the cortex, or for steps further downstream in the signaling pathways involved in dorsal induction.

3.4. The Yolk Cell and Maternal Determinants

The anatomy of the zebrafish egg, in particular with respect to accessibility to the yolk cell, has important implications for the availability of maternal determinants to the cells of the developing blastoderm (Fig. 3; see Kimmel *et al.*[68] for a detailed description of the anatomical features of the early embryo). Blastoderm cells are initially connected to the yolk at their base, which lacks a membrane boundary. Such a direct connection begins to be lost at the 16-cell stage, when the innermost four cells of the 4 × 4 cell arrangement become completely surrounded by membrane. However, cells at the margin continue to be connected to the yolk cell through cytoplasmic bridges. At the tenth mitosis, the membranes that partially surround the marginal cells begin to regress, and nuclear division proceeds in the absence of cytokinesis. This results in the formation of an acellular nuclear layer, or yolk syncytial layer (YSL) directly underlying the blastoderm. Because of its origin, the YSL can inherit maternal determinants either from the marginal blastomeres, after membrane regression, or directly from the underlying yolk, and these determinants can in principle in turn influence the fates of cells in the overlying blastoderm. Indeed, the YSL has been shown to have a number of inducing activities, such as dorsal axis induction,[65,69] mesendoderm induction[65,66,70,71] and possibly the induction of an anterior neural pattern organizer.[72] Consistent with these activities, expression of zygotic genes in the YSL appears to be patterned into sub-domains. For example, some genes such as *bozozok*[73–75] and *squint*[76] (see "Integration of pathways regulating dorsoventral patterning") are expressed in the dorsal YSL margin. On the other hand, the gene *hex*, which is potentially involved in anterior development, is expressed in the YSL in a dorsal region extending to the animal pole.[77]

4. Maternal Genes with a Function in Early Development

Previous and ongoing[22] (Pelegri, F, Dekens, M, Schulte–Merker, S, Nüsslein–Volhard, C, unpublished; M. Mullins, personal communication) screens for maternal effects, as well as molecular genetic analysis coupled to reverse genetic methods, are identifying a large number of maternal genes with a role in early zebrafish development. Here, I will summarize the reported analysis of a subset of these factors.

4.1. General Functions

A number of mutations have been identified that affect early cellular functions. A recessive maternal-effect mutation in the gene *futile cycle* (*fue*) causes defects in nuclear fusion during fertilization, so that the two pronuclei remain unfused in the blastodisc.[78] Although DNA replication continues within the pronuclei, the normal cellular division cycles do not occur, resulting in embryos where most cells are devoid of DNA. Surprisingly, however, such embryos can undergo a relatively normal cellular cleavage pattern. Thus, aside from an intrinsic role of *fue* in nuclear fusion, this mutation reveals the presence of a precise cellular division program that functions independently of DNA segregation.

A recessive mutation in the gene *janus* results in a strictly maternal effect where blastoderm cells separate during the first three cellular divisions.[20] A recessive mutation in the gene *nebel* also results in defects in cells with defects in cell adhesion.[60] However, the function of *janus* may be more specific to the early cleavage divisions than that of *nebel*, since *janus* mutant embryos can continue to develop into relatively well-formed embryos, albeit with duplicated axes caused by the initial presence of split blastoderms. On the other hand, *nebel* mutant embryos result in either large syncytial masses or, in cases of reduced expressivity when the embryos survive the early cleavage stages, gastrulae undergoing cell death but without axis duplication. In the case of *nebel*, the cell adhesion defect is caused by the defective formation of the microtubule-based FMA at the forming furrow, which appears to be required for localized exocytosis at the furrow of vesicles containing adhesive membrane. As mentioned in "Redistribution of components of the germ plasm during the early cleavage stages", the FMA is also required for the peripheral movement of the aggregating germ plasm along the furrow, which results in defects in germ plasm segregation in *nebel* mutant embryos. Consistent with a role for *nebel* in microtubule reorganization, the *nebel* phenotype is cold-sensitive. However, the fact that even *nebel* mutant embryos raised at permissive temperatures and with apparently normal cell divisions show extensive cell death after gastrulation suggests that maternally derived *nebel* products are also essential at later stages of development.

Other maternal mutations affect general functions important for morphogenesis and gastrulation. The gene *yobo*, which was originally identified by a recessive viable zygotic phenotype consisting in reduced xanthophore pigmentation, also shows a recessive strictly maternal effect.[79] Embryos from *yobo* homozygous females have shortened and broadened dorsal axes, a phenotype which may be the result of defects in convergence extension movements that form the dorsal axis. Another gene important for gastrulating movements is *half-baked* (*hab*).[18] This gene was identified primarily by virtue of its recessive zygotic phenotype where the internal (deep) cells of the blastoderm, but not the nuclear layer of the YSL or the outermost (enveloping) cellular layer, fail to undergo epibolic movements towards the vegetal pole. Heterozygous embryos from females heterozygous for *hab* mutations exhibit a dominant maternal-zygotic interaction that also results in the slowing down of the epibolic movements, indicating that the maternal Hab product is also involved in these movements.

4.2. Induction of Cell Fates along the Dorsoventral Axis

The establishment of dorsoventral patterning in the zebrafish embryo depends on three distinct but interacting pathways that are initiated by maternal factors: Wnt/ß-catenin, Wnt/calcium and BMP signaling. A more detailed description of these pathways has been provided elsewhere.[44] Here, I will emphasize the components of these pathways that have been shown to have a maternal contribution.

4.2.1. *Local activation of Wnt/ß-catenin signaling: induction of the dorsal organizer*

Translocation of the putative dorsal signal is thought to lead to the activation of the Wnt/ß-catenin signaling pathway in the blastomeres and YSL at the dorsal side of the embryo. However, the molecular details of this activation remain to be determined. Canonical Wnt signaling in turn leads to the stabilization and nuclear accumulation of ß-catenin (ßcat) protein in cells in this dorsal region.[80] A recessive maternal-effect mutation in the gene *ichabod* results in axis deficiencies similar to those caused by the early removal of the putative dorsal signal.[21] Embryos from *ichabod* mutant females lack ßcat accumulation in dorsal nuclei, further implicating ßcat nuclear accumulation in axis induction. Interestingly, the mutant *ichabod* phenotype cannot be rescued by the injection of mRNAs coding for products expected to activate Wnt/ßcat signaling upstream of ßcat, including products that should stabilize ßcat protein. These results suggest that *ichabod* functions in an unknown step independent of ßcat protein stabilization, possibly in promoting its nuclear localization.

ßcat protein is thought to interact with transcription factors of the Tcf/Lef family and result in the activation of dorsal-specific target genes. In *Xenopus*, ßcat appears to act by releasing the repression of Tcf-3 on dorsal genes, thus allowing their expression in dorsal cells.[81,82] It is unclear whether the same mechanism occurs in zebrafish, although a dominant negative Tcf construct inhibits dorsal gene expression.[83] There are at least three maternally-expressed factors of this family: *tcf-3* (also called *headless*;[83–85] *tcf3-b*,[84] and *lef-1*[84]). However, functional analyses with MO-mediated knockdown of these genes[86,87] and, in the case of *tcf-3/hdl*, genetic loss of function,[85] do not result in major defects in axis formation. Further research will be needed to determine the precise roles of Tcf/Lef factors and ßcat in dorsal induction.

4.2.2. *Wnt/calcium signaling: downregulation of the dorsal organizer*

During the blastula stages, precisely when the dorsal axis is being specified, aperiodic calcium fluxes occur distributed in an apparently random manner in the blastula.[88] These calcium fluxes appear to be part of a Wnt signaling pathway, different from the Wnt/ßcat pathway. This non-canonical Wnt signaling pathway relies on the activation of G-proteins and phospholipase C and the production of the second messenger inositol 1,4,5-trisphosphate (IP_3), which in turn binds to receptors in the endoplasmic reticulum and promotes the release of calcium to the cytoplasm.[89] Wnt/calcium signaling appears to be important to regulate the activity of the dorsal activity conferred by the Wnt/ßcat pathway. This is suggested by a number of observations. Expression of Wnt ligands of a different subclass, such as Wnt5, both increases the frequency of calcium fluxes and inhibits the axis-inducing effect of other Wnt ligands.[14,90,91] This suppressive effect on Wnt/ßcat signaling can be mimicked by artificially increasing intracellular calcium levels.[91] Moreover, both the removal of Wnt5 maternal and zygotic function, and the inhibition of Wnt/calcium signaling, result in the ectopic accumulation of nuclear ßcat protein and ectopic expression of dorsal gene expression.[14,92] Thus, calcium released from the endoplasmic reticulum appears to antagonize Wnt/ßcat signaling at a level upstream of ßcat nuclear accumulation.

4.2.3. *Bmp signaling: promotion of ventral cell fates*

A ventral to dorsal gradient of activity of zygotic Bmp genes, themselves members of the TGF-ß family of extracellular factors, has been shown to be instrumental for the determination of the embryonic dorsal axis in vertebrates,[93] including the zebrafish.[94] For example, *bmp2b/swirl* and *bmp7/snailhouse* (*snh*) are expressed in the ventral region of the gastrula and are required in a dosage sensitive manner for the determination of ventral cell fates.[25,95–97] Recent studies have shown that the expression of these zygotic Bmp genes depends on specific maternal factors. The expression of *bmp2b/swirl*, but not that of *bmp7/snh*, is dependent on the presence of the maternal product of the gene *radar*, which encodes another member of the Bmp subfamily and whose mRNA is uniformly distributed in the early embryo.[16,98] Reception of the Radar ligand may be carried out by the Alk8/Lost-a-fin (Laf) TGF-ß type I receptor, which has both maternal and zygotic genetic contributions[6,7] and appears to interact functionally with Radar.[16] While for *bmp2b/swirl*, but not *bmp7/snh*, expression is dependent on radar and Alk8/Laf signaling; genetic experiments suggest that the expression of *bmp7/snh*, but not that of *bmp2b/swirl*, depends on the maternal function of the intracellular TGF-ß signaling factor Smad5/Somitabun.[15] Thus, zygotic Bmp genes may be activated by independent, possibly converging TGF-ß activated pathways. Interestingly, genetic experiments suggest that maternal Alk8/Laf and Smad5/Sbn products perdure in the embryo and may also be required at a later stage for the reception and transduction of the zygotically produced Bmp2b/Swirl and Bmp7/Snh ligands.[6,7,99,100]

The product of the gene *ogon* (also known as *mercedes* and *short tail*), a zygotic gene required for the development of dorsal cell fates,[8,101,102] also has a maternal functional contribution.[9,10] Recently, *ogon* has been shown to encode Sizzled, a secreted factor with homology to the Wnt receptor.[103,104] In spite of its molecular similarity to a regulator of Wnt signaling, zebrafish Sizzled/Ogon does not appear to regulate this signaling pathway, but is instead involved in a negative regulatory feedback loop that attenuates zygotic Bmp signaling in ventral regions.[10,103,104]

4.2.4. *Integration of pathways regulating dorsoventral patterning*

The Wnt/ßcat, Wnt/calcium and Bmp signaling pathways interact to produce the final dorsoventral patterning. During the early cellular cleavages, the putative dorsal determinant redistributes towards the dorsal side of the blastodisc and will eventually induce localized Wnt/ßcat signaling in the dorsal side of the embryo. At the same time, other maternal factors, including products from the *wnt5*, *radar*, *alk8/laf*, *smad5/sbn* and *sizzled/ogon* genes, redistribute so that they are evenly distributed in the blastodisc and its descendant blastodermal cells. Activation of Wnt/ßcat signaling results in the dorsal expression of zygotic genes such as the transcription factor *bozozok* (*boz*), the Bmp antagonist *chordin*, the Wnt antagonist *dkk1*, and the TGF-ß factor *squint* (reviewed by Schier[105]). Concurrently, calcium fluxes triggered by Wnt/calcium signaling may provide a ubiquitous negative regulatory input that could serve to sharpen the boundary of Wnt/ßcat signaling activity (see Meinhardt and Gierer[106] for a theoretical description of such a regulatory interaction). Although ubiquitous Radar protein would, in principle, promote the activation of zygotic Bmp genes throughout the embryo, expression of dorsal-specific genes acts to exclude zygotic Bmp activity from dorsal regions. Specifically, transcription of the *bmp2b* gene has been shown to be directly downregulated in the dorsal region by the Boz protein.[107] In addition, downregulation at the posttranscriptional level by Bmp antagonists such as Ogon and Chordin, continue to stabilize and refine the dorsoventral pattern initiated by maternal factors (see Schier[105] for a review on zygotic interactions that pattern the dorsoventral axis).

4.3. Induction of Cell Fates along the Anteroposterior Axis

It is not yet clear how important maternal factors are for the induction of embryonic centers involved in the organization of the anteroposterior axis of the embryo. An anteriorizing activity has been shown to emanate from the first row of cells of the neural plate,[72] and possibly this activity could in turn be induced by a maternal signal present in the underlying yolk cell. Similarly, a posteriorizing activity has been shown to be present in either the marginal cells or the yolk cell near the margin of the epiblast.[108,109] As described in "The yolk cell and maternal determinants", the yolk cell and the YSL can in principle inherit maternal factors and present them to the overlying blastoderm in order to produce localized gene expression patterns. The identity of such maternal factors, if any, awaits further research.

Genetic analysis has begun to show, however, that some zygotically-expressed genes involved in anteroposterior patterning also have a significant maternal contribution. Thus maternal Hdl/Tcf3 product is essential for the development of anterior structures, apparently by acting in anterior brain regions as a repressor of genes that promote the formation of posterior brain regions.[85] In addition, maternal Lazarus/Pbx4 product is important in hindbrain segmentation and rhombomere identity.[5] Because the function of these gene products occurs during gastrulation, well after MBT, they provide clear examples of developmental control by perduring maternal products even after zygotic gene activation.

4.4. Induction of the Mesendodermal Layer

Genetic analysis has shown that two zygotic genes with a role in mesendoderm induction also have a significant maternal contribution. One of these genes, *one-eyed pinhead* (*oep*), codes for an EGF-CFC family co-receptor for nodal-related extracellular signals. Embryos mutant for both maternal and zygotic *oep* have severe defects in mesendoderm induction.[11] Another gene with a maternal contribution important for mesendoderm induction is *schmalspur* (*sur*), which encodes the forkhead domain transcription factor FoxH1/Fast1, a downstream component of nodal signaling[12,13] (Fig. 2). The mesendodermal defects in *foxH1/fast1/sur* maternal-zygotic mutant embryos, however, are significantly less severe than those observed in *oep* maternal-zygotic mutants, which may indicate that *foxH1/fast1/sur* is only one of several intracellular targets of nodal signaling.

Both *oep* and *sur* have been shown to mediate the response to the nodal-related extracellular factors, Squint and Cyclops, which act redundantly to induce the mesendoderm. The *squint* and *cyclops* genes, however, are strictly zygotic, which provides another example of maternal products acting after MBT and downstream of zygotic gene products.

It is currently unknown what signals trigger *squint* and *cyclops* zygotic expression. In *Xenopus*, *nodal*-related genes involved in mesendoderm induction are themselves activated by the maternal factor VegT, whose mRNA is localized to the vegetal egg cortex (reviewed by Whitman[110]) The zebrafish *vegT* homologue, however, is not expressed before the activation of *squint* and *cyclops* transcription,[111] indicating that the maternal control of mesendoderm induction is different in these two organisms. Further research will be required to understand the initiation of mesendoderm induction in the zebrafish embryo.

5. Determination of the Maternal-Zygotic Transition

The transition from maternal to zygotic developmental control that occurs at MBT is of basic importance to the developing organism. Embryological procedures such as manipulation of ploidy and partial enucleation show that, as in other organisms, changes associated with MBT in zebrafish, such as the onset of cell cycle lengthening, are dependent on the nucleocytoplasmic ratio and not on the absolute cell division number.[3] Injection of foreign DNA into early embryos also results in a premature increase in the cell cycle length,[4] again suggesting a role for the nucleocytoplasmic ratio on MBT. Premature cell cycle lengthening in DNA-injected embryos is abolished by coinjection of the transcriptional inhibitory drug actinomycin D, which suggests that these effects are mediated by newly transcribed zygotic products. Thus, although this has not been directly proven, it seems likely that the nucleoplasmic ratio acts by controlling zygotic transcription, and that newly transcribed gene products in turn result in cell cycle lengthening and other changes associated with MBT.

Lengthening of the cell cycle at MBT reflects an increase in the duration of the interphase period, while the duration of mitosis itself remains constant.[3] The increase in interphase length is itself caused by the appearance of G1 phase and the lengthening of S-phase.[4] Interestingly, the induction of G1 depends on the nucleocytoplasmic ratio and zygotic transcription, but the lengthening of S-phase does not appear to depend on these variables. The latter effect is reminiscent of the cessation of asymmetric germ plasm segregation that roughly coincides with MBT, but which is also independent of the nucleocytoplasmic ratio and new transcription[58] (see "Redistribution of components of the germ plasm during the early cleavage stages"). Thus, the embryo appears to possess several types of cell-counting mechanisms that result in the various changes associated with MBT.

The cellular changes that begin at cell cycle 10 appear to occur gradually. For example, newly synthesized transcript production at cycle 10 is less than 10% of the level of transcription at cycle 13.[3] Similarly, at cycle 10 the increase in the cell cycle length is only 10% and gradually becomes more pronounced so that by cycle 13 it has increased more than three-fold. Moreover, only a fraction of the cells appear to add G1 to their cell cycle in cell cycle 10, while the remaining cells will add it in subsequent cell cycles.[4] The end of the maternal to zygotic transition at MBT appears to occur in late cycle 13. This is suggested by the observation that in embryos that have been treated with a transcriptional inhibitor drug, and thus contain only maternal control factors, the first appearance of cell cycle abnormalities occurs in cycle 14.[112] Thus, the changes associated with MBT appear to occur gradually throughout a window of about two hours, starting at cell cycle 10 and ending at late cycle 13.

It is also interesting to note that, at least in *Xenopus*, there are instances of zygotic genes that are expressed prior to the bulk activation of gene transcription at MBT. For example, ßcat/Tcf-dependent transcription of the *Xenopus* nodal genes *Xnr5* and *Xnr6* occurs as early as the 256-cell stage, four cell cycles before the activation of transcription of the bulk of the genome.[113] It will be interesting to test whether pre-MBT gene activation also occurs in the zebrafish embryo, especially for genes involved in early patterning decisions such the induction of the dorsal axis.

6. Conclusions

Maternal factors are essential to carry out developmental processes required for early development prior to the activation of zygotic transcription at the midblastula transition. Even after this transition occurs, maternal factors can act, sometimes in concert with or even downstream of zygotic factors, to carry out essential developmental functions. Much progress has occurred in the identification of zebrafish maternal factors and the initial characterization of their role in early development. However, our understanding of the processes involved remains at the present patchy and sporadic. Even in the development of the dorsoventral axis, which is arguably the best understood early patterning process in the zebrafish embryo, the nature of the signal that activates the Wnt/ßcat pathway is poorly understood, as are the precise interactions between pathways involved in dorsoventral patterning. Similar gaps in knowledge exist in other patterning processes addressed in this review, such as the determination of the anteroposterior axis and the induction of the mesendoderm. In the case of germ cell determination, although several maternal RNAs, from the genes *vasa*, *nanos* and *dead end*, are known to localize to the zebrafish germ plasm, it is still unclear what role the maternal contribution of these or other genes have in primordial germ cell specification. Thus, much remains to be discovered in this important field of research. The ongoing identification of maternal-effect mutations as well as the implementation of reverse genetic methods to analyze the function of maternal products promises to rapidly expand our knowledge of the role of maternal factors in early zebrafish development.

Acknowledgements

I am grateful to Dirk Meyer for kindly providing photographs for Fig. 2 and to Xiaoyan Ge and Jamie Lyman-Gingerich for critical comments on the manuscript. Work in our laboratory is supported by a March of Dimes Birth Defects Foundation grant #5-FY00-597 and a NIH grant RO1 GM65303.

References

1. Newport J and Kirschner M (1982). *Cell* **30**: 675–686.
2. Newport J and Kirschner M (1982). *Cell* **30**: 687–696.
3. Kane DA and Kimmel CB (1993). *Development* **119**: 447–456.
4. Zamir E, Kam Z and Yarden A (1997). *Mol. Cell. Biol.* **17**: 529–536.
5. Waskiewicz AJ, Rikhof HA and Moens CB (2002). *Dev. Cell* **3**: 723–733.
6. Mintzer KA, Lee MA, Runke G, Trout J, Whitman M and Mullins MC (2001). *Development* **128**: 859–869.
7. Bauer H, Lele Z, Rauch G-J, Geisler R and Hammerschmidt M (2001). *Development* **128**: 849–858.
8. Hammerschmidt M, Pelegri F, Mullins M, Kane DA, van Eeden FJM, Granato M, Brand M, Furutani–Seiki M, Haffter P, Heisenberg C-P, Jiang YJ, Kelsh RN, Odenthal J, Warga RM, Nüsslein–Volhard C (1996). *Development* **123**: 95–102.
9. Miller-Bertoglio V, Carmany-Rampey A, Fürthauer M, Gonzalez EM, Thisse C, Thisse B, Halpern ME and Solnica-Krezel L (1999). *Dev. Biol.* **214**: 72–86.
10. Wagner D and Mullins MC (2002). *Dev. Biol.* **245**: 109–123.
11. Gritsman K, Zhang J, Cheng S, Heckscher E, Talbot WS and Schier AF (1999). *Cell* **97**: 121–132.
12. Pogoda H-M, Solnica-Krezel L, Driever W and Meyer D (2000). *Curr. Biol.* **10**: 1041–1049.
13. Sirotkin HI, Gates MA, Kelly PD, Schier AF and Talbot WS (2000). *Curr. Biol.* **10**: 1051–1054.
14. Westfall TA, Brimeyer R, Twedt J, Gladon J, Olberding A, Furutani-Seiki M and Slusarski D (2003). *J. Cell Biol.* **162**: 889–898.

15. Kramer C, Mayr T, Nowak M, Schumacher J, Runke G, Bauer H, Wagner DS, Schmid B, Imai Y, Talbot WS, Mullins MC and Hammerschmidt M (2002). *Dev. Biol.* **250**: 263–279.
16. Sidi S, Goutel C, Peyrieras N and Rosa FM (2003). *Proc. Natl. Acad. Sci. USA* **100**: 3315–3320.
17. Mullins MC, Hammerschmidt M, Kane DA, Odenthal J, Brand M, van Eeden FJM, Furutani–Seiki M, Granato M, Haffter P, Heisenberg C-P, Jiang Y-J, Kelsh RN and Nüsslein-Volhard C (1996). *Development* **123**: 81–93.
18. Kane DA, Hammerschmidt M, Mullins MC, Maischein H-M, Brand M, van Eeden FJM, Furutani-Seiki M, Granato M, Haffter P, Heisenberg C-P, Jiang Y-J, Kelsh RN, Odenthal J, Warga RM and Nüsslein-Volhard C (1996). *Development* **123**: 47–55.
19. Chan STH and Yeung WSB (1983). Sex control and sex reversal in fish under natural conditions. In Hoar WS, Randall DJ and Donaldson EM (eds.) *Fish Physiology*. Academic Press, New York, pp. 171–222.
20. Abdelilah S, Solnica–Krezel L, Stainier DY and Driever W (1994). *Nature* **370**: 468–471.
21. Kelly C, Chin AJ, Leatherman JL, Kozlowski DJ and Weinberg ES (2000). *Development* **127**: 3899–3911.
22. Pelegri F and Schulte–Merker S (1999). A gynogenesis-based screen for maternal-effect genes in the zebrafish, *Danio rerio*. In: Detrich W, Zon LI and Westerfield M (eds.) *The Zebrafish: Genetics and Genomics:* Vol. 60, Academic Press, San Diego, pp. 1–20.
23. Driever W, Solnica–Krezel L, Schier AF, Neuhauss SCF, Malicki J, Stemple DL, Stainier DYR, Zwartkruis F, Abdelilah S, Rangini Z, Belak J and Boggs C (1996). *Development* **123**: 37–46.
24. Haffter P, Granato M, Brand M, Mullins MC, Hamerschmidt M, Kane DA, Odenthal J, van Eeden FJM, Jiang Y-J, Heisenberg C-P, Kelsh RN, Furutani–Seiki M, Vogelsang E, Beuchle D, Schach U, Fabian C and Nüsslein–Volhard C (1996). *Development* **123**: 1–36.
25. Kishimoto Y, Lee K-H, Zon L, Hammerschmidt M and Schulte–Merker S (1997). *Development* **124**: 4457–4466.
26. Kane DA and Kishimoto Y (2002). Cell labelling and transplantation techniques. In Nüsslein-Volhard C and Dahm R (eds.) *Zebrafish*, Vol. 261, Oxford University Press, Oxford, pp. 95–119.

27. Lin S, Long W, Chen J and Hopkins N (1992). *Proc. Natl. Acad. Sci. USA* **89**: 4519–4523.
28. Ciruna B, Weidinger G, Knaut H, Thisse B, Thisse C and Raz E (2002). *Proc. Natl. Acad. Sci. USA* **99**: 14919–14924.
29. Köprunner M, Thisse C, Thisse B and Raz E (2001). *Genes Dev.* **15**: 2877–2885.
30. Weidinger G, Stebler J, Slanchev K, Dumstrei K, Wise C, Lovell-Badge R, Thisse C, Thisse B and Raz E (2003). *Curr. Biol.* **13**: 1429–1434.
31. Guraya SS (1969). *The cell and molecular biology of fish oogenesis:* Vol. 18 Karger, New York.
32. Nasevisius A and Ekker SC (2000). *Nature Genet.* **26**: 216–220.
33. Hutvágner G and Zamore PD (2002). *Curr. Opin. Genet. Dev.* **12**: 225–232.
34. Oates AC, Bruce AEE and Ho RK (2000). *Dev. Biol.* **224**: 20–28.
35. Zhao Z, Cao Y, Li M and Meng A (2001). *Dev. Biol.* **229**: 215–223.
36. Clemens MJ and Elia A (1997). *J. Interferon Cytokine Res.* **17**: 503–524.
37. Player MR and Torrence PF (1998). *Pharmacol. Ther.* **78**: 55–113.
38. Caplen N, Parrishe S, Imani F, Fire A and Morgan RA (2001). *Proc. Natl. Acad. Sci. USA* **98**: 9742–9747.
39. Elbashir SM, Harborth J, Lendeckel W, Yalcin A, Weber K and Tuschl T (2001). *Nature,* **411**: 494–498.
40. Svoboda P, Stein P, Hayashi H and Schultz RM (2000). *Development* **127**: 4147–4156.
41. Stein P, Svoboda P and Schultz RM (2003). *Dev. Biol.* **256:** 188–193.
42. Wienholds E, Schulte-Merker S, Walderich B and Plasterk RHA (2002). *Science* **297**: 99–102.
43. Selman K, Wallace RA, Sarka A and Qi X (1993). *J. Morphol.* **218**: 203–224.
44. Pelegri F (2003). *Dev. Dyn.* **228**: 535–554.
45. Roosen–Runge EC (1938). *Biol. Bull.* **75**: 119–133.
46. Beams HW, Kessel RG, Shih CY and Tung HN (1985). *J. Morphol.* **184**: 41–49.
47. Leung CF, Webb SE and Miller AL (1998). *Dev. Growth Differ.* **40**: 313–326.
48. Katow H (1983). *Develop. Growth Differ.* **25**: 477–484.
49. Hart NH and Fluck RA (1995). *Curr. Topics Dev. Biol.* **31**: 343–381.
50. Leung CF, Webb SE and Miller AL (2000). *Dev. Growth Differ.* **42**: 29–40.

51. Gore AV and Sampath K (2002). *Mech. Dev.* **112**: 153–156.
52. Jesuthasan S and Strähle U (1997). *Curr. Biol.* 7: 31–42.
53. Howley C and Ho RK (2000). *Mech. Dev.* **92**: 305–309.
54. Bally–Cuif L, Schatz WJ and Ho RK (1998). *Mech. Dev.* 77: 31–47.
55. Maegawa S, Yasuda K and Inoue K (1999). *Mech. Dev.* **81**: 223–226.
56. Suzuki H, Maegawa S, Nishibu T, Sugiyama T, Yasuda K and Inoue K (2000). *Mech. Dev.* **93**: 205–209.
57. Braat AK, Zandbergen T, van de Water S, Goos HJT and Zivkovic D (1999). *Dev. Dyn.* **216**: 153–167.
58. Knaut H, Pelegri F, Bohmann K, Schwarz H and Nüsslein–Volhard C Zebrafish (2000). *J. Cell Biol.* **149**: 875–888.
59. Yoon C, Kawakami K and Hopkins N (1997). *Development* **124**: 3157–3165.
60. Pelegri F, Knaut H, Maischein H-M, Schulte–Merker S and Nüsslein–Volhard C (1999). *Curr. Biol.* **9**: 1431–1440.
61. Jesuthasan S (1998). *J. Cell Sci.* **111**: 3695–3703.
62. Wylie C (2000). *Curr. Opin. Genet. Dev.* **10**: 410–413.
63. Wolke U, Widinger G, Köprunner M and Raz E (2002). *Curr. Biol.* **12**: 289–294.
64. Mizuno T, Yamaha E, Kuroiwa A and Takeda H (1999). *Mech. Dev.* **81**: 51–63.
65. Ober EA and Schulte–Merker S (1999). *Dev. Biol.* **215**: 167–181.
66. Mizuno T, Yamaha E, Wakahara M, Kuroiwa A and Takeda H (1996). *Nature* **383**: 131–132.
67. Kanungo J, Kozmik Z, Swamynathan SK and Piatigorsky J (2003). *Proc. Natl. Acad. Sci. USA* **100**: 3287–3292.
68. Kimmel C, Ballard WW, Kimmel SR, Ullmann B and Schilling TF (1995). *Dev. Dyn.* **203**: 253–310.
69. Mizuno T, Yamaha E and Yamazaki F (1997). *Dev. Genes Evol.* **206**: 389–396.
70. Rodaway A, Takeda H, Koshida S, Broadbent J, Price B, Smith JC, Patient R and Holder N (1999). *Development* **126**: 3067–3078.
71. Chen S-R and Kimelman D (2000). *Development* **127**: 4681–4689.
72. Houart C, Westerfield M and Wilson SW (1998). *Nature* **391**: 788–792.
73. Koos KS and Ho RK (1998). *Curr. Biol.* **8**: 1199–1206.
74. Yamanaka Y, Mizuna T, Sasai Y, Khishi M, Takeda H, Kim C-H, Hibi M and Hirano T (1998). *Genes Dev.* **12**: 2345–2353.

75. Fekany K, Yamanaka Y, Leung T, Sirotkin HI, Topczewski J, Gates MA, Hibi M, Renucci A, Stemple D, Radbill A, Schier AF, Driever W, Hirano T, Talbot WS and Solnica–Krezel L (1999). *Development* **126**: 1427–1438.
76. Schier AF and Talbot WS (2001). *Int. J. Dev. Biol.* **45**: 289–297.
77. Ho C-Y, Houart C, Wilson SW and Stainier DYR (1999). *Curr. Biol.* **9**: 1131–1134.
78. Dekens MPS, Pelegri FJ, Maischein H-M and C. Nüsslein–Volhard (2003). *Development* **130**: 3907–3916.
79. Odenthal J, Rossnagel K, Haffter P, Kelsh RN, Vogelsang E, Brand M, van Eeden FJM, Furutani–Seiki M, Granato M, Hammerschmidt M, Heisenberg C-P, Jiang Y-J, Kane DA, Mullins AC and Nüsslein–Volhard C (1996). *Development* **123**: 391–398.
80. Schneider S, Steinbeisser H, Warga RM and Hausen P (1996). *Mech. Dev.* **57**: 191–198.
81. Brannon M, Gomperts M, Sumoy L, Moon RT and Kimelman D (1997). *Genes Dev.* **11**: 2359–2370.
82. Houston DW, Kofron M, Resnik E, Langland R, Destree O, Wylie C and Heassman J (2002). *Development* **129**: 4015–4025.
83. Pelegri F and Maischein H-M (1998). *Mech. Dev.* 77: 63–74.
84. Dorsky RI, Snyder A, Cretekos CJ, Grunwald DJ, Geisler R, Haffter P, Moon RT and Raible DW (1999). *Mech. Dev.* **86**: 147–150.
85. Kim C-H, Oda T, Itoh M, Jiang D, Artinger KB, Chandrasekharappa SC, Driever W and Chitnis AB (2000). *Nature* **407**: 913–916.
86. Dorsky RI, Sheldahl LC and Moon RT (2002). *Dev. Biol.* **241**: 229–237.
87. Dorsky RI, Itoh M, Moon RT and Chitnis A (2003). *Development* **130**: 1937–1947.
88. Reinhard E, Yokoe H, Niebling KR, Allbritton NL, Kuhn MA and Meyer T (1995). *Dev. Biol.* **170**: 50–61.
89. Kühl M, Sheldahl LC, Park M, Miller JR and Moon RT (2000). *Trends Genet.* **16**: 279–283.
90. Slusarski DC, Corces VG and Moon RT (1997). *Nature* **390**: 410–413.
91. Slusarski DC, Yang-Snyder J, Busa WB and Moon RT (1997). *Dev. Biol.* **182**: 114–120.
92. Westfall TA, Hjertos B and Slusarski DC (2003). *Dev. Biol.* **259**: 380–391.
93. Hogan BL (1996). *Genes Dev.* **10**: 1580–1594.
94. Mullins MC (1998). *Trends Genet.* **14**: 127–129.

95. Nguyen VH, Schmid B, Trout J, Connors SA, Ekker M and Mullins MC (1998). *Dev. Biol.* **199**: 93–110.
96. Dick A, Hild M, H, Bauer, Imaj Y, Maifeld H, Schier AF, Talbot WS, Bouwmeester T and Hammerschmidt M (2000). *Development* **127**: 343–354.
97. Schmid B, Fürthauer M, Connors SA, Trout J, Thisse B, Thisse C and Mullins MC (2000). *Development* **127**: 957–967.
98. Goutel C, Kishimoto Y, Schulte–Merker S and Rosa F (2000). *Mech. Dev.* **99**: 15–27.
99. Dick A, Meier A and Hammerschmidt M (1999). *Dev. Dyn.* **216**: 285–298.
100. Hild M, Dick A, Rauch GJ, Meier A, Bouwmeester T, Haffter P and Hammerschmidt M (1999). *Development* **126**: 2149–2159.
101. Solnica–Krezel L, Stemple DL, Mountcastle–Shah W, Rangini Z, Neuhauss SCF, Malicki J, Schier AF, Stainier DYR, Zwartkruis F, Abdelilah S and Driever W (1996). *Development* **123**: 67–80.
102. Walker C (1999). Haploid screens and gamma-ray mutagenesis. In: Detrich WH, Westerfield M and Zon LI (eds.) *The zebrafish: Genetics and genomics*, Vol. 60, Academic Press, San Diego, pp. 43–70.
103. Yabe T, Shimizu T, Muraoka O, Bae Y-K, Hirata T, Nojima H, Kawakami A, Hirano T and Hibi M (2003). *Development* **130**: 2705–2716.
104. Martyn U and Schulte–Merker S (2003). *Dev. Biol.* **260**: 58–67.
105. Schier AF (2001). *Curr. Opin. Genet. Dev.* **11**: 393–404.
106. Meinhardt H and Gierer A (2000). *Bioessays* **22**: 753–760.
107. Leung T, Bischof J, Söll I, Niessing D, Ahang D, Ma J, Jäckle H and Driever W (2003). *Development* **130**: 3639–3649.
108. Koshida S, Shinya M, Mizuno T, Kuroiwa A and Takeda H (1998). *Development* **125**: 1957–1966.
109. Nikaido M, Tada M, Takeda H, Kuroiwa A and Ueno N (1999). *Development* **126**: 181–190.
110. Whitman M (2001). *Dev. Cell* **1**: 605–617.
111. Ruvinsky I, Silver LM and Ho RK (1998). *Dev. Genes Evol.* **208**: 94–99.
112. Kane DA, Warga RM and Kimmel CB (1992). *Nature* **360**: 735–737.
113. Yang J, Tan C, Darken RS, Wilson PA and Klein PS (2002). *Development* **129**: 5743–5752.

Chapter 2

Gastrulation in Zebrafish

Florian Ulrich and Carl-Philipp Heisenberg
Max-Planck-Institute of Molecular Cell Biology and Genetics
Dresden, Germany

Introduction

Gastrulation is a central process in early development in which a seemingly unstructured blastula rearranges to form an embryo with a distinct head-to-tail, left-to-right and bottom-up morphology. During this process, the three germ layers — ectoderm, mesoderm and endoderm — are formed and progenitor cells are brought into the positions from where they later will form more complex tissues and organs[1] (Fig. 1).

The underlying principles of gastrulation movements are well-conserved among vertebrates and have been most extensively studied in *Xenopus laevis* and zebrafish. Here, the initially spherical embryo achieves its structure with distinct anterior-posterior and dorsal-ventral polarities by the complex interplay of four main types of cell movements and tissue rearrangements: epiboly, by which a multicellular blastoderm cap thins and spreads out to cover the entire yolk cell (Fig. 1A, D); continuous internalization of mesendodermal cell progenitors at the margin of the spreading blastoderm (Fig. 1B, E); and convergence of embryonic cells towards the emerging dorsal axis and extension, by which this axis lengthens along its anterior-posterior extent (Fig. 1C, F). These types of movements are closely linked to one another and cooperatively they shape the early embryo.[2–4]

The zebrafish has become increasingly popular as a model organism to study the cellular and molecular mechanisms that underlie vertebrate gastrulation movements. Zebrafish embryos are fertilized *ex utero* and undergo rapid embryonic development. Their accessibility from the earliest stages of development together with their optical clarity makes

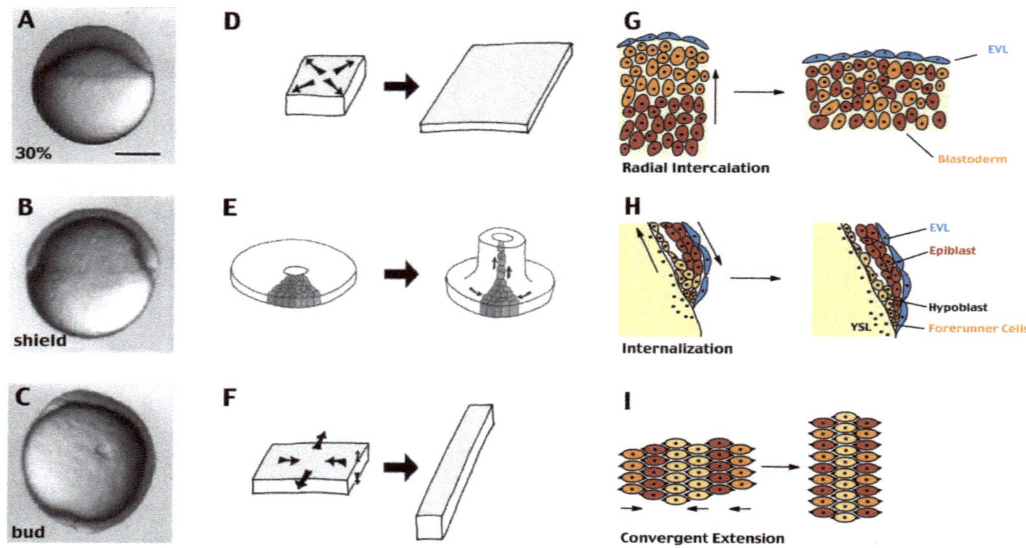

Fig. 1 Cell rearrangements during zebrafish gastrulation. (A–C) DIC images of wild-type embryos at 30% epiboly (A), 60% epiboly or shield stage (B) and bud stage (C). Lateral views with animal to the top and dorsal (B, C) to the right. Scale bar = 250 μm. (D–F) Drawings illustrating the principles of the tissue rearrangements at the stages depicted in (A–C). The orientations in (A–C) and (D–F) are the same. (D) Epiboly. The tissue flattens and spreads outwards, away from the center. (E) Mesendodermal progenitor cell internalization. The tissue leaves its original plane and folds into a direction perpendicular to the original plane. (F) Convergent extension. The tissue narrows mediolaterally (convergence) and lengthens into the perpendicular direction (extension). (G–I) Schematic views of the main cellular rearrangements at the stages depicted in (A–C). (G) Radial intercalations flatten the blastodermal cells flatten the tissue during epiboly and push cells towards the side. (H) Internalization. Hypoblast cells move towards the animal pole in a direction opposite to that of the overlying epiblast, EVL and forerunner cells. (I) Convergent extension. Mediolateral intercalations of cells lead to the extension of the tissue in anterior-posterior direction. The orientations in (A–B) and (G–H) are the same; (I) shows a dorsal view on the cells instead of a lateral view in (C). In (G–I), black dots indicate the cell nuclei. YSL = yolk syncytial layer. Reprinted from Kimmel *et al.*[1] and Keller *et al.*[45]

them ideally suited for an *in vivo* analysis of morphogenetic processes during early development. Moreover, their short generation time, high offspring number and ease of handling have allowed the accomplishment of several large-scale forward genetics screens that uncovered a large number of genes involved in various developmental processes.[5–9] Combined with more recent advances towards "reverse" genetics techniques,[10,12] all of these approaches have led to the identification of many signaling pathways essential for vertebrate embryonic development.

In this review, we will first concentrate on the current status of knowledge about the cellular mechanisms that drive gastrulation movements in zebrafish. We will then focus on the molecular mechanisms that are thought to underlie these processes. Finally, we will try to outline the development of new experimental tools and future approaches that aim to uncover the interplay between the genetic, cellular and molecular mechanisms regulating gastrulation movements.

I. Cellular Mechanisms

Epiboly

Shortly after fertilization of the oocyte, cytoplasm separates from the yolk by an outward streaming, the so-called "lifting", thereby generating the first blastodermal cell. This is followed by rapid and highly synchronous cell divisions with an average cell cycle length of 15 minutes, eventually leading to a multicellular blastoderm cap on top of a large yolk cell (meroblastic or discoidal cleavage; for reference see Kimmel *et al.*[1]). At this stage, the embryo can be subdivided into three distinct domains. One domain is the yolk cell, a syncytium of multiple nuclei positioned on the interface between yolk sac and blastoderm (yolk syncytial layer, YSL). The nuclei of the YSL arise during early cleavage stages by the fusion of marginal blastomeres with the underlying yolk cell, and divide rapidly thereafter. The second domain consists of the rounded and loosely associated deep layer blastomeres (DEL), which eventually will form the embryo proper. The third domain is a thin layer of extraembryonic cells, the enveloping layer (EVL), which covers the DEL and will later form the embryonic periderm.

Epiboly (Greek: epiballein, to throw onto something) describes a process by which a tissue flattens medially and spreads out in lateral directions (Fig. 1D). During gastrulation, this is recognizable by a thinning of the blastoderm cap, accompanied by extensive movements of the EVL, the DEL blastomeres and the nuclei of the YSL from animal or equatorial regions towards the vegetal pole. At the end of gastrulation, the yolk cell is entirely covered by the blastoderm and EVL[2] (Fig. 1C).

Epiboly is driven by distinct cellular rearrangements such as cell intercalations and active cell movements. The first sign of epiboly is that the DEL cells become more loosely associated with each other, round up, exert bleb-like protrusions and exhibit an increased motility. Subsequently, cells deep in the blastoderm intercalate radially into more superficial layers, a movement that pushes epiblast cells towards the vegetal pole. It is thought that these cell rearrangements provide the driving force for the epibolic movements of DEL blastomeres (Fig. 1G). Radial intercalations are most pronounced at the onset of gastrulation when the blastoderm cap thins out and attains a cup-shaped appearance. The thinning of the blastoderm continues relatively uniformly during early epiboly until it reaches a thickness of approximately two to three cells prior to involution[2,12,13] (Fig. 1B). Parallel to the DEL blastomeres, the EVL also thins out, eventually covering the embryo as a tight and squamous epithelium. Interestingly, while labeling of single blastomeres with vital dyes revealed extensive cell intercalations within the DEL, these cells never mix with cells of the EVL and vice versa, suggesting that radial intercalation movements are restricted to the cells within one tissue.[2] In addition to the radial cell intercalations, DEL cells at the blastoderm margin extend pseudopodial and filopodial processes towards the yolk cell, with cells at the very margin of the germ ring showing characteristic bleb-like protrusions towards the yolk sac surface. These observations suggest that epibolic movements of DEL blastomeres are not just the result of radial intercalations at the animal pole that push cells passively towards the vegetal pole, but that marginal blastodermal cells also actively pull the blastoderm over the yolk cell during epiboly (Ulrich F and Heisenberg C-P, unpublished).

There is increasing evidence that the yolk cell itself plays a pivotal role in directing epibolic movements. Most prominently, nuclei of the YSL exhibit epibolic movements similar to the movements of overlying blastodermal cells, suggesting that the movements of these two domains are regulated in a similar way.[14] A tight coordination of the movements between the blastodermal cells and the YSL nuclei appears to be of particular importance, considering that the YSL serves as a source for mesoderm-inducing signals.[15–17] The mechanisms by which the YSL nuclei move during gastrulation are not well-understood. YSL nuclei appear to be connected to a complex microtubule network inside the yolk cell. Furthermore, it has been suggested that the nuclei are pulled over the yolk cell towards the vegetal pole by a microtubule-dependent mechanism.[18,19] It has also been proposed that the epibolic movements of the YSL reflect mitotic movements of the YSL nuclei, which divide several times during the course of gastrulation.[13]

In addition to the movement of the YSL, a contractile force around the margin of the germ ring has been suggested to drive epiboly at later stages of gastrulation. This contractile force could be either mediated by an actin-rich cytoskeletal network forming around a subset of YSL nuclei at the blastoderm margin or, alternatively, by EVL cells close to the blastoderm margin forming an actin-rich 'purse string' at their vegetally oriented sides around the circumference of the germ ring.[13,14,20] However, the functional relevance of these marginal actin accumulations for the progress of epiboly has not yet been tested.

The mechanisms by which EVL cells move during gastrulation and their potential contribution to the general progress of epiboly are still only poorly understood. At the blastoderm margin, these cells are tightly attached to the plasma membrane of the yolk cell. From observations in the related teleost *Fundulus heteroclitus*, it has been suggested that the EVL cells are passively pulled over the embryo by interacting with the YSL nuclei.[1,13,21] However, the observation that EVL cells actively extend filopodia and lamellipodia on the YSL surface suggests that the movement of EVL cells towards the vegetal pole is also driven by active migration of those cells[22] (Ulrich F and Heisenberg C-P, unpublished). Finally, these cells also exhibit extensive cellular rearrangements, indicating that cell intercalations might also contribute to the epibolic movements of the EVL.[23,24]

What is the relationship of the epibolic movements of the YSL nuclei and the cells of the DEL and EVL during gastrulation? In mutants exhibiting defective epibolic movements and in embryos in which the microtubule cytoskeleton is disrupted, the epibolic movements of the DEL, YSL and EVL cells can be uncoupled from each other.[25,26] Thus, these movements appear to be autonomously regulated. This assumption is also supported by earlier observations in *Fundulus*, which show that the YSL nuclei can cell-autonomously accomplish epiboly, independently of the overlying blastoderm.[27] However, although formally independent of each other, these movements might still be highly coordinated with each other. Indeed, DEL cells extend protrusions both along the YSL surface and towards the EVL layer,[13,25,28] suggesting that there is extensive intercellular communication between these cell layers during epiboly.

Finally, endocytosis of yolk sac plasma membrane by cells at the blastoderm margin has been proposed to be involved in the regulation of epibolic movements, both in zebrafish and *Fundulus*.[21,29] Within the marginal cells, such endocytic activity can be best seen in the so-called forerunner cells, a distinct group of 5–10 specialized enveloping and deep layer cells (Fig. 1H). These cells become located at the dorsal blastoderm margin shortly after the germ ring has formed (see below). They subsequently move in front of the margin towards the vegetal pole, where they will form part of Kupffer's vesicle.[14,30,31] However, whether endocytosis of the yolk sac plasma membrane is essential for the general progress of epiboly has not been experimentally tested. It has been speculated that it could serve to release the strain, which is imposed onto the YSL membrane vegetal to the germ ring margin by the epibolic movements of the DEL and EVL cells. Alternatively, endocytosis might not just serve to facilitate epibolic movements, but could also provide part of the driving force for epiboly, since the removal of membrane from the vegetal half of the yolk cell surface could also result in a pulling force, which draws the blastoderm margin along the cortex of the yolk cell towards the vegetal pole.[13]

Internalization

The internalization of mesendodermal progenitor cells constitutes the central element in gastrulation. During this process, cells separate from ectodermal precursors, eventually leading to the establishment of the three germ layers — ectoderm, mesoderm and endoderm. Additionally, precursor cells are also brought into their eventual positions from where they can further develop into more specialized tissues[1,2,13] (Fig. 1B, C). This is reflected by the emergence of a rather defined lineage fate map at the onset of internalization.[32–34]

In zebrafish, internalization of mesendodermal progenitors can be first observed at 50% epiboly, when the blastoderm has covered half of the yolk sac. It becomes apparent by a local thickening at the blastoderm margin, which from thereon is defined as the 'germ ring' (Fig. 1B, H). During formation of the germ ring, epibolic movements slow down but continue towards the vegetal pole after the first mesendodermal progenitors have internalized.[1,13] In early stages of internalization, DEL blastomeres within the germ ring start moving towards the yolk and, after reaching the yolk cell surface, turn around and move towards the animal pole in a direction opposite to their overlying (not internalized) cells. Prospective mesendodermal cells that internalize at later stages of gastrulation do not turn towards the animal pole but instead move towards the vegetal pole.[2,3] The internalization of mesendodermal progenitors forms an internal cell layer, the hypoblast, which is located underneath an outer layer of non-internalizing cells, the epiblast. Whereas the epiblast gives rise to ectodermal tissues, early internalizing hypoblast cells will predominantly form endodermal tissues, while later internalizing cells become mesodermal (Fig. 1H).

Interestingly, accompanying the animal-wards movements of early internalizing hypoblast cells, a subset of the YSL nuclei also reverses its formerly vegetal-wards directed movement (epibolic movement) and moves in the opposite direction towards the animal pole. This anterior movement of the YSL nuclei seems to be at least partially independent from the internalizing hypoblast cells, as the movement of those nuclei is unchanged in embryos maternally and zygotically mutant for the nodal-related gene *one-eyed pinhead*, which lacks mesendodermal cell internalization.[13,14] In contrast to DEL cells, EVL cells are never internalized.[2,32,34]

While cells can internalize at the entire margin of the blastoderm, only the two most marginal rows of cells seem to directly take part in this process.[2,13] Two main types of cellular movements have been proposed to drive mesendodermal cell internalization: involution of a coherent sheet of blastoderm cells directly at the margin of the germ ring and ingression of single blastodermal cells close to the germ ring margin. Studies in *Fundulus* have identified both involution- and ingression-like cell behaviors in the process of mesendodermal cell internalization.[35] Recent studies suggest that in zebrafish, blastodermal cells at the margin of the germ ring first involute synchronously as a coherent sheet of cells and then, as soon as they have reached the yolk cell surface, undergo an epithelial to mesenchymal transition and start to move as individuals, reminiscent of single cell ingression.[13,36] The observation that internalizing cells extend many protrusions towards the yolk cell surface, together with the finding that single blastoderm cells can cell-autonomously move from more superficial into deeper layers of the blastoderm, even if they are not in close proximity to the blastoderm margin, supports the view that cell ingression is the main type of movement involved in zebrafish mesendodermal cell internalization[13,37] (Joubin K, Ulrich F and Heisenberg C-P, unpublished).

Internalizing cells that reach the yolk cell surface turn around and migrate towards the animal pole (Fig. 1H). Cells originating from more lateral regions of the germ ring margin migrate as loosely associated mesenchymal cells, while cells ingressing in the region of the embryonic organizer are tightly clustered together and move as an epithelial-like sheet of cells. These cells constitute the future prechordal plate (Concha M and Heisenberg C-P, unpublished). It is conceivable that the movement of internalized mesendodermal cells towards the animal pole pulls adhering cells in more superficial blastodermal layers passively towards the yolk cell surface, leading to the internalization of those cells. Furthermore, YSL nuclei that move together with the first internalizing cells towards the animal pole might dynamically modify the plasma membrane of the yolk cell. This would generate a force which is able to pull more superficially located mesendodermal progenitors that adhere to the plasma membrane of the yolk cell towards the yolk cell surface.[13]

Recent studies have shown that the orientation and formation of polarized cellular processes in mesendodermal progenitor cells is important to facilitate and stabilize the movement of these cells towards the animal pole. Interestingly, hypoblast cell polarization appears to be dispensable for hypoblast cell ingression and movement *per se*, indicating that other cellular processes may be involved (Montero J-A, Ulrich F and Heisenberg C-P, unpublished). It is likely that the epithelial to mesenchymal transition of the prospective mesendodermal cells within the germ ring provides the driving force that leads to the internalization of these cells and their subsequent movement towards the animal pole. This transition appears to be regulated in a cell-autonomous manner, as individual cells can internalize independent of their neighbors, and single wild-type cells transplanted into *Mzoep* mutants that lack any recognizable hypoblast cell internalization can still undergo normal cell internalization.[38]

Convergence and Extension

Convergent extension (CE) movements, in conjunction with epiboly and the continuous internalization of mesendodermal precursor cells, lead to the formation of an embryonic body axis with distinct anterior-posterior and dorsal-ventral polarity out of an initially spherical gastrula. While convergence narrows the embryo mediolaterally, extension lengthens the tissue into the anterior-posterior direction (Fig. 1F, J). The mechanism of forming an embryonic body axis by CE movements is a central and conserved principle in gastrulation of various species, including sea urchins, *Drosophila*, *C. elegans*, ascidians, mouse, chick, *Xenopus* and zebrafish.[39–46] The extent to which convergence and extension movements are coordinated with each other varies between different species and within specific domains within the gastrula.[45]

In zebrafish, both epiblast and hypoblast cells converge towards the dorsal side of the gastrula, which narrows the germlayers mediolaterally. Convergence first becomes visible shortly after involution starts, when a compaction of blastodermal cells at the dorsal-most region of the embryo leads to the formation of the embryonic organizer or shield[1] (Fig. 1B). Cells converge through all stages of gastrulation until early somitogenesis.[1] This leads to a nearly complete accumulation of all embryonic cells at the dorsal side, where the axis forms, and a concomitant depletion of cells at the ventral side of the gastrula (Fig. 1C). Hypoblast (mesendodermal) cells that exhibit CE movements initially migrate as loosely associated mesenchymal cells on the yolk cell surface, while epiblast (ectodermal) cells move within a sheet of epithelial cells towards the dorsal side[1,37] (Concha M and Heisenberg C-P, unpublished).

Depending on the position within the gastrula, cells seem to exhibit different degrees of CE movements. Cells on the prospective ventral side of the embryo do not significantly contribute to convergence or extension of the body axis. Instead, they migrate over the vegetal half of the yolk towards the vegetal pole of the embryo, where they later become part of the tail. In contrast, cells laterally positioned within the gastrula show increasing convergence and extension movements the closer they are to the dorsal side, while cells very close to the dorsal side show little convergence and high extension movements.[1,32,47,48] On the cellular level, the extent to which cells undergo CE movements is reflected by the degree of stable and persistent movement trajectories these cells show. They elongate along their mediolateral axes, and this elongation appears to be required for the velocity and persistence of CE movements.[49,50] The same cells also extend numerous pseudopodial and filopodial extensions towards the yolk and onto neighboring cells. Whether this protrusive activity is required to stabilize cell movements or to directly mediate them is not yet fully understood[51] (Ulrich F and Heisenberg C-P, unpublished).

In *Xenopus*, movements of convergence and extension in both mesodermal and neurectodermal tissues are thought to be mediated and coupled to each other by cellular rearrangements commonly termed "mediolateral intercalation behavior" (MIB). MIB describes a process by which mediolaterally aligned cells intercalate between each other, thus simultaneously accumulating cells towards the dorsal midline and displacing them apart from each other in an anterior-posterior direction (Fig. 1I). Oriented bipolar protrusive activities in mesodermal cells and directed monopolar protrusive activities in both neurectodermal and mesodermal cells are thought to mediate MIB. These protrusions allow the cells to adhere to and exert traction on their immediate neighbors, which enables them to insert themselves between each other.[45]

In zebrafish, as in other related teleosts like the rosy barb and *Fundulus*, the link between convergence and extension seems to be more complex, and MIB might not be the only means by which the embryonic axis extends. The reasons for this assumption mainly originate from studies in zebrafish, which analyzed the cellular rearrangements underlying CE movements in posterior axial and paraxial mesendodermal tissues, such as the notochord and presomitic mesoderm, respectively.[52] Although MIB can be observed in both of these tissues, the rate of convergence is not always linked to the rate of extension (as it would be the case with MIB), indicating that MIB cannot be the sole cellular mechanism driving CE movements. This is particularly obvious in *no tail* (*ntl*) mutants, where the embryonic axis can still extend even in the near complete absence of convergence movements within the paraxial mesendoderm. What other cellular mechanisms that might mediate CE in zebrafish have not yet been established, but it has been speculated that radial intercalations of mesendodermal cells — thought to drive epibolic movements at earlier stages of gastrulation — might also contribute to the anterior-posterior extension of the emerging embryonic axis. This notion is supported by the phenotypic analysis of the *volcano* mutant, which shows that defects in epibolic cell movements at the onset of gastrulation are followed by reduced CE movements. This suggests that these different movements might be regulated by a common cellular mechanism.[26]

The role of MIB in regulating CE movements in more anterior regions of the embryo is less well-understood. Here, extension of axial mesendodermal tissues, such as the prechordal plate, is mediated primarily by the directed migration of single cells or groups of cells towards the animal pole, while MIB appears to be restricted to paraxial mesendodermal tissues (Ulrich F and Heisenberg C-P, unpublished). The observation that different signaling molecules are required to mediate CE movements in anterior and posterior portions of the embryo also supports the notion that multiple cellular mechanisms might operate to control gastrulation movements along the anterior-posterior axis of the embryo.[51]

Similar to mesendodermal cells, neurectodermal cells also display CE movements. This is first visible at shield stage, when non-internalizing epiblast cells at the blastoderm margin, in addition to their epibolic movements towards the vegetal pole, converge towards the dorsal midline in a highly persistent fashion. With gastrulation proceeding, cells located more animally to the blastoderm margin also start to converge towards the dorsal midline. Simultaneously, epiblast cells become more compact, so that by mid-gastrulation the whole epiblast moves as a coherent sheet of cells. Cells become elongated and extend protrusions into the mediolateral directions, possibly mediating mediolateral intercalations at the dorsal midline. However, it is not yet clear to what extent convergence and extension movements in the epiblast are coupled by MIB.[37]

Parallel to the movements of neurectodermal and mesendodermal tissues, the nuclei of the YSL also undergo CE movements. This can first be seen at shield stage, when a small fraction of these nuclei converges towards the dorsal side. By the time blastoderm cells begin to internalize, the YSL nuclei move together with the first internalizing mesendodermal cells towards the animal pole. This anterior movement is accompanied by CE movements of nuclei in paraxial regions of the gastrula. Interestingly, these nuclei show increasingly faster CE movements as closer they are positioned to the dorsal side, whereas nuclei in axial regions move anteriorly towards the animal pole. Similar to the situation in mesendodermal and neurectodermal germ layers, CE movements of the YSL nuclei appear to be mediated by MIB, suggesting that the movements of the YSL nuclei and of cells of the overlying germ layers are coordinated.[14]

Little is known about the extent to which cells in the EVL exhibit CE movements during gastrulation. In principle, cells of the EVL are capable of extending cellular protrusions and are motile within the plane of the EVL, but whether they actively converge and extend has not yet been experimentally addressed.[22–24] Since the EVL covers the blastoderm to an equal extent along its dorso-ventral axis throughout gastrulation, convergent movements of cells within the EVL might be of less importance.[1]

Future Directions

Many of the studies analyzing the cellular mechanisms underlying gastrulation movements have focused on the cellular rearrangements within specific germ layers and tissues. However, the interaction between these germ layers and tissues is likely to be of equal importance. Evidence for cell interactions between the mesodermal and neurectodermal germ layers can be found in recent studies, which show that during *Xenopus* gastrulation, the mesendoderm is required for normal polarization and protrusive activity of cells within the overlying neurectoderm.[53] Other mechanisms, such as cell adhesion and cell proliferation, are also likely to contribute to the regulation of gastrulation movements, but have not yet been analyzed in detail. Studies on the function of paraxial protocadherin, both in *Xenopus* and zebrafish, have provided evidence that cell adhesion within the posterior paraxial mesoderm is required for normal CE movements within this tissue.[54,55] Finally, a role of cell proliferation in regulating cell and tissue morphogenesis has been proposed by studies that have analyzed cell divisions in the zebrafish neurectoderm. Here, the orientation of the cleavage planes of neurectodermal cells appear to be highly ordered, indicative of a role for cell division in the positioning of cells within the neurectoderm during gastrulation.[37]

II. Molecular Mechanisms

Wnt/PCP Pathway

Wnt proteins constitute a family of secreted glycoproteins that can be found in species throughout the animal kingdom, ranging from *Hydra* to humans. They are involved in a variety of cellular functions during development, such as axis induction and patterning, tissue and organ morphogenesis, asymmetric cell division and, as more recent studies confirm, also in axon guidance.[56,57] Several Wnt signal transduction pathways have been identified. The best characterized of them is the canonical Wnt signaling pathway, originally found to specify segment polarity in *Drosophila* and mediate axis induction in *Xenopus*. This pathway transduces the Wnt signal via binding to its transmembrane receptor Frizzled (Fz), followed by activation of the cytoplasmic protein Dishevelled (Dsh). This leads to the inhibition of a destruction complex containing APC, Axin and GSK3, which in turn allows the stabilization of β-Catenin and its subsequent translocation to the nucleus, where it is involved in the transcriptional regulation of certain target genes[58,59] (Fig. 2A).

During vertebrate gastrulation, both the functional characterization of Dsh in *Xenopus* and the analysis of the *silberblick* (*slb/wnt11*) mutant in zebrafish provided initial evidence that a Wnt signaling pathway different from the canonical pathway is involved in regulating gastrulation movements.[60–62] This pathway shares significant similarities with the planar cell polarity pathway (PCP) in *Drosophila*, a signaling pathway downstream of the putative Wnt receptor *frizzled*, that has been shown to mediate the correct orientation of ommatidia and the polarized outgrowth of sensory bristles and wing hairs in the eye, thorax and wing.[63,64]

Besides *slb/wnt11*, several other zebrafish mutants that exhibit defective gastrulation movements have been shown to encode different components of the Wnt/PCP pathway.[4,65] These mutants include *pipetail* (*ppt*), which encodes Wnt5, *knypek* (*kny*), mutated in the Wnt co-receptor Glypican4/6, and *trilobite* (*tri*), encoding the transmembrane protein Strabismus/VanGogh (Stbm/Vang).[49,62,66,67] The molecular characterization of these mutants and the identification of further downstream components mediating the function of these genes during gastrulation have revealed a signaling pathway with striking homology to the Fz/PCP pathway in *Drosophila*. Shared components include the Wnt receptors Fz2 and Fz7, the intracellular signaling mediator Dsh, the formin homology domain containing cytoplasmic protein Daam1,

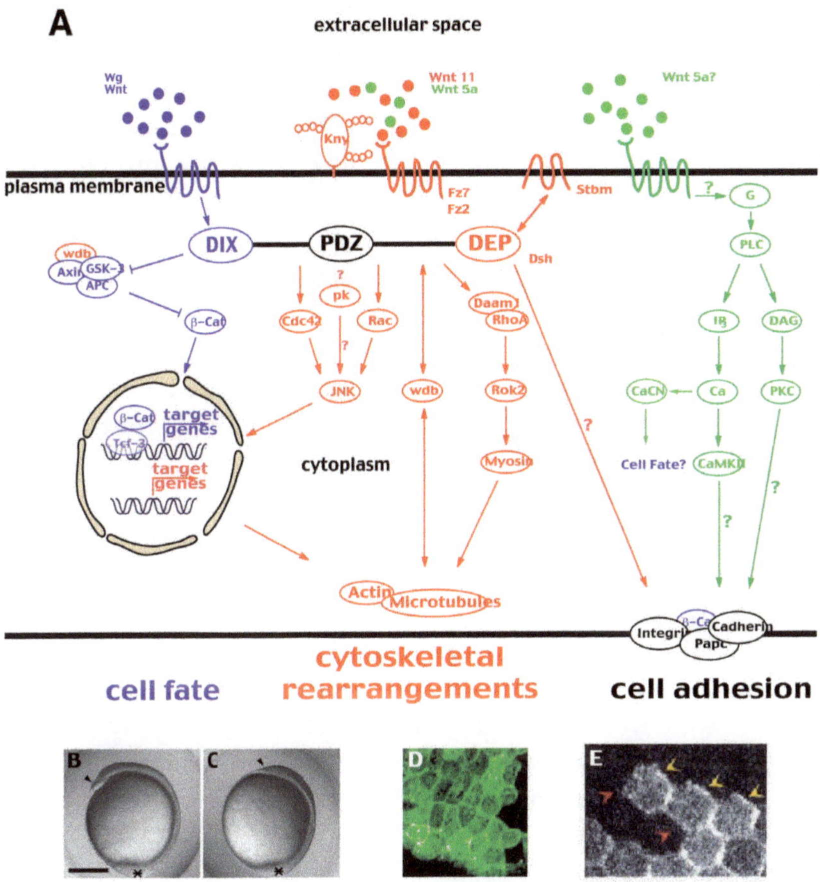

Fig. 2 Wnt signaling during zebrafish gastrulation. (A) Overview of the canonical, the Wnt/PCP and the Wnt/Ca^{2+} pathway. Members of these pathways are shown in blue, red and green, respectively. See text for more details. (B, C) Wild-type (B) and *slb* mutant embryo (C) at bud stage. In *slb* mutants, the prechordal plate is flattened and shifted posteriorly (arrowhead), and the anterior-posterior body axis is shortened (arrowhead and asterisk). Lateral views with anterior to the left. Scale bar = 250 μm. (D) Overexpression of a Dsh-GFP construct in *Xenopus* mesodermal cells during convergent extension. The protein is distributed ubiquitously around the cell membrane. (E) Expression of Dsh-GFP in the wing epithelium of *Drosophila* during prehair formation. In contrast to *Xenopus*, the protein is asymmetrically localized to the distal edge of the cells (yellow arrowheads). Red arrowheads mark the proximal side of the cells. Reprinted from Wallingford *et al.*[61] and Axelrod.[159]

the small GTPases RhoA, Rac and Cdc42 and the Rho effector kinase Rok2.[50,60–62,68–70] Other components, which are related to but not directly part of this signaling pathway, are the JNK module, the ankyrin repeat protein Diego, the transmembrane protein Strabismus/VanGogh, the protein phosphatase PP2A and the cytoplasmic protein Prickle (Pk)[67,71–79] (Fig. 2A).

Although the molecular conservation of the Wnt/PCP pathway between vertebrates and *Drosophila* is evident, the cellular and molecular mechanisms by which this pathway functions in vertebrates are largely unknown. In the *Drosophila* wing epithelium, planar cell polarity is manifested by the outgrowth of a single actin- and tubulin-filled, distally pointing hair at the distal vertex of a wing cell.[64,80] Based on the phenotypic analysis of mutants in genes of the planar cell polarity pathway, these components can be subdivided into two main classes. Upstream members of the pathway, like *fz*, *dsh*, *pk* and *stbm*, lead to wing hairs pointing into the wrong directions when mutated or mis-expressed, indicating that they are required for the polarization of wing cells. In contrast, abolishing or increasing the function of genes, which are more downstream in the pathway, like *rhoA* and the *Drosophila rho-associated kinase* (*Drok*), interferes with the number but not the polarity of forming wing hairs, suggesting that they are needed for the formation but not polarization of wing hairs.[64,81] In *Drosophila*, the polarization and formation of such a hair requires the localized accumulation of PCP proteins, such as Fz and Dsh, at the distal edge of the corresponding wing cell.[82] Subsequently, components of the actin and microtubule cytoskeleton become localized towards this site, eventually leading to the outgrowth of a single hair.[80] It has been hypothesized that upstream components of the PCP pathway specify cell polarity by establishing the site where the wing hair will emerge, whereas downstream components of the pathway regulate or associate with the actin and microtubule cytoskeleton, thereby regulating wing hair formation itself. Further evidence for an interaction of these downstream components with the cytoskeleton results from studies in which embryos were treated with drugs that antagonize the cytoskeleton, resulting in phenotypes looking strikingly similar to mutant phenotypes of those components.[64]

In zebrafish, mutants of the Wnt/PCP pathway have been identified primarily on the basis of a broadened and shortened body axis at the end of gastrulation, indicative of defects in CE movements during gastrulation (Fig. 2B, C). However, in most of the cases the precise cellular basis that gives rise to these phenotypes is not yet understood. It has been speculated that the defects in CE movements in these mutants are due to a failure of epiblast and hypoblast cells to elongate along their mediolateral axes, which is thought to constitute a prerequisite for mediolateral cell intercalation behavior driving CE movements. However, it is not at all clear if the lack of mediolateral cell elongation in these embryos is a mere secondary consequence of these cells not being able to move dorsally and intercalate mediolaterally, or if the Wnt/PCP pathway directly affects the elongation of these cells. Furthermore, it is not clear to which extent the mediolateral elongation of cells during vertebrate gastrulation can be compared to the polarization of *Drosophila* wing epithelial cells. As mentioned above, the formation of a single wing hair at the distal vertex of such a cell is preceded by the asymmetric localization of planar cell polarity pathway components along the cell membrane (Fig. 2E). In contrast, no asymmetric membrane localization of homologous proteins in epiblast and hypoblast cells undergoing CE movements has yet been observed during vertebrate gastrulation (Fig. 2D). One possible reason for this could be that the organization of the tissues in which the Wnt–Fz/PCP pathways function is rather different between *Drosophila* and vertebrates. While wing cells are organized in a tight, stationary epithelium with little space between single cells, gastrulating cells in *Xenopus* and zebrafish appear more loosely associated and show extensive movements relative to each other and on the substrate to which they attach.[45,52] Although several components of the PCP pathways are conserved between *Drosophila* and vertebrates, it could be that the functions of these pathways in *Drosophila* planar cell polarization and vertebrate gastrulation are rather different.

Interestingly, recent studies analyzing the cellular mechanisms by which *slb/wnt11* functions during zebrafish gastrulation have shown that this gene is required in single epiblast and hypoblast cells for the polarized outgrowth of cellular processes into their individual movement directions and that the defect in cell polarization in *slb* mutants is accompanied by slower and less persistent movements of these cells at the onset of gastrulation.[28] This suggests that *slb/wnt11* allows cells to correctly polarize into their individual movement directions, suggesting that cell polarization is required to facilitate and stabilize cell movements. Future studies will need to address how *slb/wnt11* regulates the polarization of epiblast and hypoblast cells and how this polarization can facilitate and stabilize their movements.

A more detailed analysis of the pathways involved in gastrulation with respect to cell morphologies and their relation to the actual movements of these cells should help to further elucidate the complex cellular functions of these pathways during gastrulation. Also, analysis of cells from tissues with both epithelial and mesenchymal characteristics, such as cells mediating dorsal closure in *Drosophila*, should provide further insights into the molecular and cellular conservation of the Wnt–Fz/PCP pathways between vertebrates and flies.[83,84]

Cell Adhesion Molecules

Cell adhesion is thought to play an essential role in the regulation of tissue integrity, cellular morphology and cell movements.[85] Several studies on epithelial tissues and migrating cells have revealed a complex interplay between the actin and microtubule cytoskeleton and sites of cell-cell or cell-substrate adhesion. It is, therefore, very likely that important downstream effectors of the pathways that regulate gastrulation movements in zebrafish encode adhesion molecules.[86–88]

In zebrafish and *Xenopus*, several adhesion molecules have been shown to be involved in the regulation of cell movements and tissue morphogenesis during gastrulation. The best-studied adhesion molecules in this process include members of the Protocadherin family of adhesion molecules, such as Paraxial protocadherin (Papc) and Axial protocadherin (Axpc). Papc is expressed in paraxial mesendodermal tissues and is required for CE movements of cells within this tissue. Moreover, Papc is able to induce cell shape changes when overexpressed in *Xenopus* animal cap explants, suggesting that Papc regulates CE movements by determining cell polarization or elongation.[54,55] Another protocadherin, Axpc, is required to facilitate the homophilic sorting of notochordal precursor cells into forming notochord in *Xenopus*, pointing at an important role of adhesion molecules in tissue rearrangements during gastrulation.[89] Although these studies have revealed an essential function of protocadherins in regulating gastrulation movements, not much is known about the signaling mechanisms controlling their transcription or activation. In mouse and *Xenopus*, Papc has been shown to function downstream of the transcriptional activator Lim1, which is expressed in dorsal regions of the gastrula.[90,91] Studies in zebrafish have found that Papc expression is regulated by *spadetail*, a transcription factor required for morphogenetic movements during gastrulation.[54,92]

In addition to protocadherins, cadherins, such as E-, C- and N-cadherin, have been implicated in the regulation of cell adhesion during gastrulation. Studies in *Xenopus* and zebrafish have shown that E-cadherin is expressed during gastrulation in both anterior mesendodermal cells and cells of the dorsal midline. It has been further postulated that E-cadherin mediates cell adhesion during gastrulation, although the precise molecular and cellular nature of such a function is at present unclear.[93,94] Other cadherins, such as C-cadherin and N-cadherin, are also expressed during zebrafish and *Xenopus* gastrulation, where they have been reported to mediate CE movements in mesodermal and ectodermal tissues.[95–97] In zebrafish, N-cadherin encodes the *parachute* (*pac*) mutant locus and is needed for proper convergence movements of the neural plate and keel.[98,99] However, similar to the situation with Protocadherins, the signaling mechanisms that control the expression and localization of cadherins remain to be elucidated.

In *Xenopus*, integrins and fibronectin, important regulators of cell-substrate interactions, are required for cell polarization and radial cell intercalation movements during early gastrulation. Interestingly, the binding of cells to fibronectin via integrins has been shown to translocate Dsh to the membrane of those cells, pointing at an interaction of cell adhesion and the Wnt-signaling pathway during gastrulation.[100,101] Although the integrin-mediated binding to fibronectin at focal adhesion sites is essential for the establishment of cell-substrate contacts in zebrafish,[22] not much is known about the role of integrins in regulating zebrafish gastrulation movements.

In *Drosophila*, focal adhesion kinase (FAK) is required for cell movements during oogenesis. Interestingly, the activity of FAK in this process is regulated by Wnt4, which signals through a pathway involving both components of the canonical Wnt and the Wnt/PCP pathway. In zebrafish, *wnt4* is weakly expressed at the onset of gastrulation, and misexpression of this gene results in severe morphogenetic defects at later developmental stages.[102] However, neither the functional requirements of Wnt4 nor the signaling pathway and targets through which Wnt4 is exerting its morphogenetic function have been identified yet. Recent studies have reported that zebrafish FAK is expressed during early gastrulation, with more abundant levels towards late gastrulation, indicating that it might also serve as a target molecule for Wnt signaling during zebrafish gastrulation.[103,104]

More evidence for a role of the Wnt-signaling pathway in regulating cell adhesion comes from studies in *Xenopus* that have analyzed the function of the putative Wnt receptor Fz7 in the separation of the germ layers at the onset of gastrulation. Interfering with the function of Fz7 led to the failure of mesendodermal and ectodermal tissues to separate from each other, indicating that Fz7 controls the differential adhesiveness of these tissues. This function is neither mediated by downstream members of the canonical nor the PCP pathway, but rather by a Dsh-independent signaling cascade, which includes Protein kinase C (PKC) and calcium.[105] The identities of further mediators of Fz7 function in this process are not determined, but independent studies suggest that cadherins can serve as targets of the Wnt/PKC pathway[106] (Fig. 2A).

Nodal/TGFβ Signaling

Nodal, a gene first discovered in mice, where it is required for the formation of mesodermal and endodermal tissues, belongs to the Transforming Growth Factor β (TGFβ) superfamily of secreted glycoproteins. Members of the Nodal-signaling pathway have been shown to play important roles for anterior-posterior patterning, positioning of the embryonic axis and the specification of mesodermal and endodermal cell fates during early development in different vertebrate species[107–109] (Fig. 3).

In zebrafish, several mutants encode components of the Nodal-signaling pathway. The Nodal-related ligands Cyclops (Cyc) and Squint (Sqt) bind to their putative receptors and coreceptors Taram-A (Tar) and One-eyed pinhead (Oep), respectively. While Tar encodes a transmembrane receptor with serine-threonine kinase activity, the *oep* gene product belongs to the EGF–CFC family of membrane-attached extracellular glycoproteins that are thought to act as co-receptors for receiving the Nodal signal. Loss of *cyc, sqt* or *oep* function leads to a failure of mesendodermal tissues to develop. Furthermore, constitutive activation of the Taram-A receptor can induce ectopic mesendodermal cell fates in a cell-autonomous manner, indicating that Nodal signaling can instructively induce mesendodermal cell fates in zebrafish.[110–112] The Nodal signal is transduced via Smad proteins as intracellular signal-mediators. These proteins are thought to heteromerize upon phosphorylation by the Nodal receptors and translocate to the nucleus. There, they interact with transcription factors such as Schmalspur (Sur), a member of the Fast/FoxH1 family, and control the transcription of genes involved in mesendoderm induction, such as *no tail* (*ntl*), *goosecoid* (*gsc*) and *floating head* (*flh*).[17] Other transcriptional targets are the Nodal ligands themselves, like Cyc, and the secreted Nodal antagonists Lefty1 and Lefty2, suggesting that the Nodal pathway is regulated by a complex molecular feedback loop[113–116] (Fig. 3A).

In addition to determining the fate of mesendodermal cells, there is increasing evidence that the Nodal pathway also regulates the ingressive behavior of mesendodermal progenitors within the germ ring margin. Wild-type cells transplanted into the germ ring of maternal-zygotic *oep* (*MZoep*) mutant host embryos, which are not able to receive the Nodal signal and therefore lack any mesendodermal progenitors themselves, can internalize cell-autonomously. Conversely, *MZoep* mutant cells grafted into the germ ring of a wild-type host do not internalize.[38] Finally, cells overexpressing a constitutive active form of the receptor

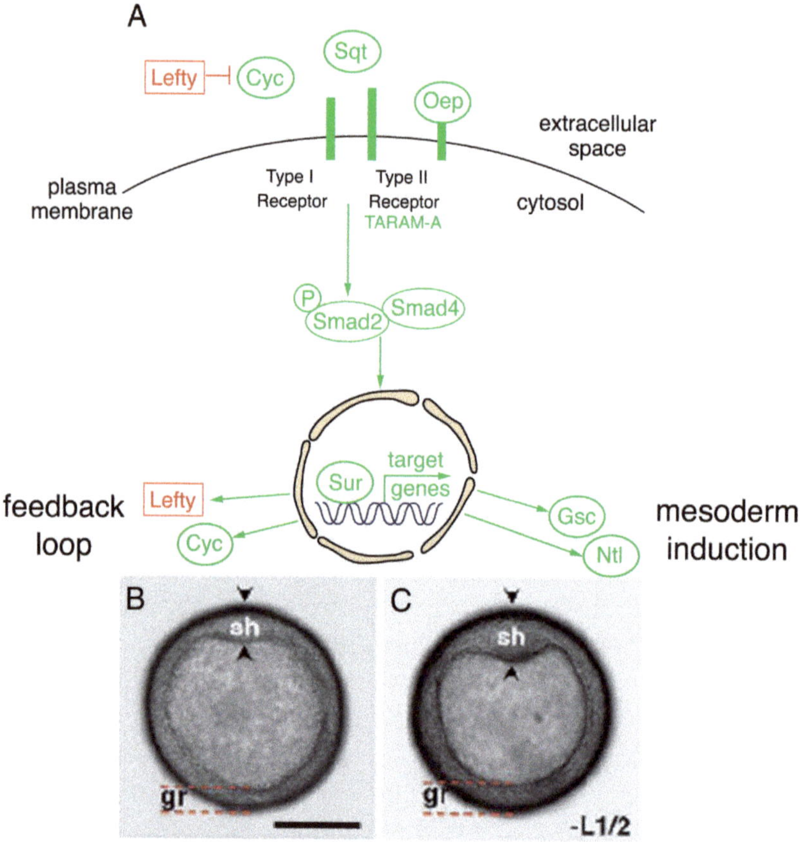

Fig. 3 Nodal signaling during zebrafish gastrulation. (A) The Nodal-signaling pathway with members identified in zebrafish (see text for details). The secreted ligands Cyclops (Cyc) and Squint (Sqt) bind to Type I and Type II receptors with One-eyed pinhead (Oep) as an essential co-receptor. This leads to the activation of Smad proteins, which heteromerize and translocate to the nucleus, where they associate with transcriptional regulators such as Schmalspur (Sur). This induces the expression of mesoderm-inducing signals like Goosecoid (Gsc) and No Tail (Ntl) and a molecular feedback loop involving the nodal ligand Cyc and the Lefty proteins as inhibitory factors. (B) Wild-type zebrafish embryo at the onset of gastrulation. The germ ring (gr) forms by involution and is visible around the equator of the embryo. Arrowheads mark the region of the forming organizer, the shield (sh). (C) Germ ring formation in an embryo, where the nodal signal is upregulated by morpholino antisense oligonucleotide injection directed against the *lefty 1* and *2* transcripts (-L1/2). Animal views with dorsal to the top. Scale bar = 250 μm. Reprinted from Kimelman and Schier[109] and Feldman.[115]

Taram-A that are transplanted into the ectoderm of a wild-type embryo are able to directly move towards the yolk, thereby mimicking internalization behavior normally observed only within the germ ring.[111] These results indicate that Nodal signaling is required and sufficient for mesendodermal cell internalization and that this internalization is a cell-autonomous process. They further suggest that internalization is mediated by the ingression of single cells that undergo an epithelial to mesenchymal transition. This conclusion is also supported by recent studies showing that embryos where the Nodal inhibitors Lefty1 and Lefty2 are inactivated exhibit more pronounced single cell ingression behaviors within the germ ring, leading to an increased thickening of the shield and the germ ring tissue[115] (Fig. 3B, C).

PDGF/PI3K Pathway

Phosphoinositide-3 kinase (PI3K) is required in various types of single cells — such as leukocytes or cells of the slime mold *Dictyostelium* — to polarize and move along a chemotactic gradient.[117,118] During vertebrate gastrulation, PI3K is involved in regulating mesodermal cell movement and morphology.[119]

In response to an extracellular chemoattractant gradient, PI3K becomes activated and localized to the leading edge of single migrating cells. Here, it converts phosphoinositide-4,5-diphosphate ($PI(4,5)P_2$) to phosphoinositide-(3,4,5)-triphosphate (PIP_3), which can bind PH-domain containing proteins like Akt/protein kinase B (Akt/PKB). These proteins in turn can direct the assembly of actin filaments to the leading edge of such cells, which is then used for the localized production of cellular processes. Thus, the function of PI3K is to enable a cell to respond to an extracellular gradient of a chemoattractant by polarizing the outgrowth of cellular processes along this gradient, thereby facilitating its polarized movement[120–122] (Fig. 4).

PI3K plays an important role for cell polarization and migration during vertebrate gastrulation, where it acts in a pathway downstream of Platelet Derived Growth Factor (PDGF). In *Xenopus*, PDGF-A is expressed in the ectoderm above the involuting mesodermal

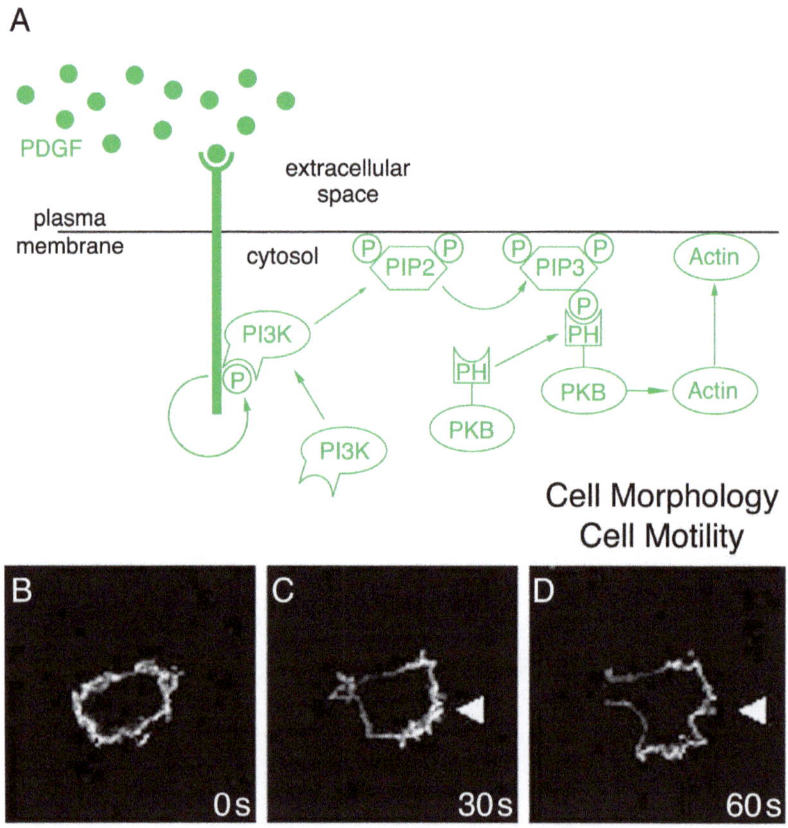

Fig. 4 PDGF/PI3K signaling during zebrafish gastrulation. (A) Schematic view of the pathway. Upon binding of PDGF, the receptor becomes phosphorylated and binds phosphoinositol-3-kinase (PI3K), which then converts Phosphoinositide biphosphate (PIP2) to Phosphoinositide triphosphate (PIP3) at the plasma membrane. Protein kinase B (PKB) binds to PIP3 via its Pleckstrin-homology (PH) domain and mediates cell polarization and motility through the localized polymerization of Actin (see text for more details). (B–D) A construct between GFP and the PH domain of PKB becomes asymmetrically localized to the leading edge of a migrating leukocyte (arrowheads) that respond to a chemoattractant gradient. Reprinted from Weiner *et al.*[118]

progenitors, which themselves express the corresponding PDGF receptor α.[123] It has been speculated that ectodermal cells secrete PDGF to facilitate involution of the underlying mesodermal cells. Supporting this hypothesis are findings that show that PDGF can facilitate the adhesion of mesodermal cells to the overlying ectodermal tissue.[124] Furthermore, when PDGF is applied to *Xenopus* mesendodermal cells *in vitro*, it promotes the outgrowth of lamellipodia and filopodia and the subsequent spreading of these cells on their substrate.[119] This function of PDGF appears to be mediated by PI3K, suggesting that PI3K controls the outgrowth and polarization of mesodermal cell processes during vertebrate gastrulation. PDGF-A and its receptors in mice show a temporal and spatial expression profile during gastrulation, which strikingly resembles the situation in *Xenopus*, indicating that PDGF signaling has conserved functions during vertebrate gastrulation.[125,126]

Recent studies have shown that in zebrafish, both PDGF and PI3Ks are required for cell polarization and process formation of mesendodermal cells *in vivo* at the onset of gastrulation.[127] Similar to the results obtained in mammalian neutrophils and *Dictyostelium* cells, PI3Ks appear to control mesendodermal cell polarization and process formation by asymmetrically localizing Protein Kinase B (PKB) to the plasma membrane at the leading edge of those cells (Fig. 4B, C). This is accompanied by an accumulation of actin at the front of the cells. Interestingly, although cell polarization and process formation is lost in mesendodermal cells in the absence of PI3K, these cells still retain the ability to move in a relatively coordinated and directed way. This suggests that PI3K-dependent cell polarization and process formation are at least to some degree dispensable for directed cell movements at the onset of gastrulation. Furthermore, these observations indicate that the PDGF/PI3K pathway, similar to the Wnt/PCP pathway, is predominantly required to facilitate and stabilize mesendodermal cell movements but not to determine their general movement direction. These conclusions are also corroborated by findings in *Xenopus* showing that PDGF/PI3K signaling in mesodermal cells is required not for mediating CE movements during gastrulation, but rather for the stabilization of cell-substrate adhesion during involution at the onset of gastrulation.[119,124] Future studies will have to address if and how the PDGF/PI3K and Wnt/PCP pathways might interact in the regulation of mesendodermal cell polarization and movement.

Other Pathways

Wnt signals can influence gastrulation movements not just through the Wnt/PCP pathway mentioned above, but also by signaling through a pathway regulating intracellular calcium levels. Co-expression of *Xenopus wnt5a* with the rat *fz2* receptor in zebrafish embryos can raise the levels of intracellular Ca^{2+} via the activation of heteromeric G-proteins and inositoltriphosphate (IP_3).[128] More recent studies have shown that Wnt/Ca^{2+} signaling is involved in the regulation of tissue separation during *Xenopus* gastrulation, suggesting that the Wnt/Ca^{2+} pathway in vertebrates is needed for the regulation of differential cell adhesiveness (see above and Winklbauer[129]). Possible downstream targets mediating this and other effects of the Wnt/Ca^{2+} signaling pathway in *Xenopus* include calcineurin (CaCN) and calmodulin-dependent kinase II (CamKII). However, the precise functions of these proteins in respect to cell rearrangements during gastrulation are not yet fully understood[105,130,131] (Fig. 2A). In addition to its role for the intracellular Wnt signal transduction, Ca^{2+} release into the extracellular space might play an important role in the cell-cell communication underlying the coordination of cell movements during gastrulation.[128,132]

The JAnus Kinase family of Signal Transducers and Activators of Transcription (JAK/STAT) is thought to regulate a variety of different processes during invertebrate and vertebrate development, such as hematopoiesis, cell proliferation, cell fate specification, planar cell polarity and cell migration.[133] The binding of several cytokines and growth factors to their respective receptors induces a conformational change that activates JAK proteins, which are constitutively associated to the cytoplasmic side of the receptor. Upon activation, JAK proteins autophosphorylate themselves and specific tyrosine residues in the cytoplasmic part of their receptors. These residues serve as docking sites for the SH2 domains of STAT proteins, which themselves become tyrosine-phosphorylated and activated. They subsequently dimerize and translocate to the nucleus, where they regulate the expression of certain target genes[133] (Fig. 5). STAT activity is involved in the regulation of cell migration in a variety of species. In mice, *stat3* loss-of-function leads to impaired movements of mesodermal tissues during gastrulation and disrupts keratocyte

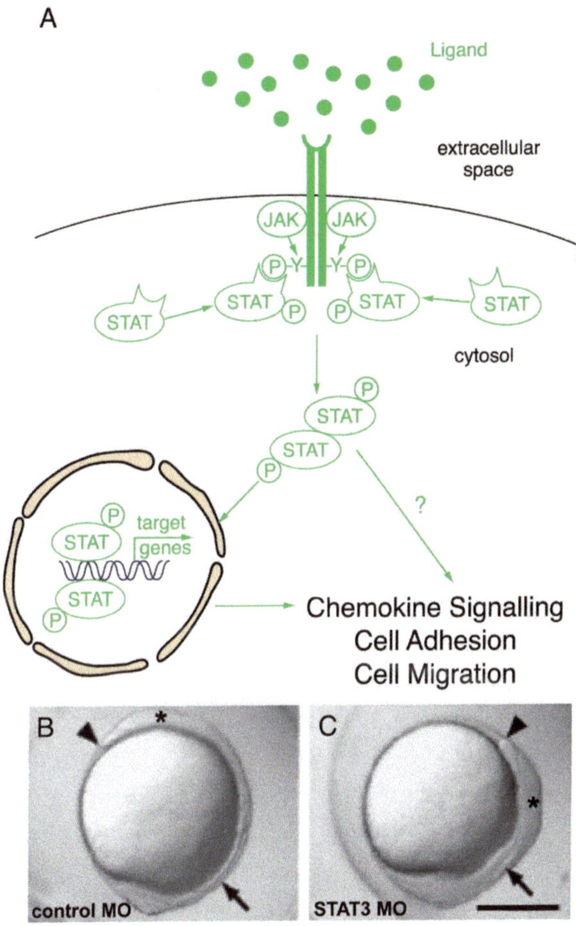

Fig. 5 JAK/STAT signaling during zebrafish gastrulation. (A) Upon ligand binding, JAK proteins phosphorylate tyrosine residues at the receptor that serve as docking sites for STAT proteins, which themselves become phosphorylated. This enables their dimerization and translocation to the nucleus, where they regulate the transcription of certain target genes (see text for more details). (B) Zebrafish embryo, which has been injected with a control morpholino antisense oligonucleotide, at the end of gastrulation. (C) Zebrafish embryo depleted of Stat3 protein by *morpholino* injection at the end of gastrulation. Prechordal plate and neural plate (arrowhead and asterisk, respectively) are shifted posteriorly, and the chordal mesoderm appears broader (arrow). Lateral views with anterior to the left. Scale bar = 250 μm. Reprinted from Hou *et al.*[133] and Yamashita *et al.*[139]

movements during wound repair. In *Drosophila*, interfering with JAK/STAT signaling impairs border cell migration during egg chamber development, while in *Dictyostelium*, cells that are mutant for *STATa* exhibit a reduced ability to form a multicellular stalk, probably due to defects in their chemotactic cell movements.[135–138]

In zebrafish, the JAK/STAT pathway functions as a regulator of cell movements during gastrulation. Misexpression of a dominant-negative form of *jak1* interferes with normal radial cell intercalation movements during epiboly. Furthermore, *stat3* is required cell-autonomously for the anterior movement of hypoblast cells and non-autonomously for the dorsal convergence of paraxial mesodermal cells[137–139] (Fig. 5B, C). The observations that the expression of both FAK and E-Cadherin is dependent on STAT protein function in *Drosophila* border cells and that STAT3 can directly bind to Rac1 suggests that JAK/STAT functions during gastrulation by controlling both cell adhesion and cytoskeletal rearrangements.[140,141]

Upstream regulators of JAK/STAT signaling during zebrafish gastrulation have not yet been identified. The activation of Stat3 on the dorsal side of zebrafish embryos depends on the activity of the maternal Wnt/β-catenin pathway.[133] However, no specific ligand/receptor pair has been found to function directly upstream of Stat3. The recent findings that the chemokine receptor CXCR4 and its proposed ligand SDF-1 are needed for the migration of primordial germ cells during gastrulation hint that these or similar factors may also be involved in the regulation of Stat3-dependent hypoblast cell movements.[144,145]

Signaling through ephrins and their corresponding receptors has been shown to mediate various morphogenetic processes in development, such as axonal guidance, cell migration, boundary formation and angiogenesis.[143,144]

Ephrins are extracellular signaling molecules attached to the cell membrane either via a GPI anchor (class A) or through a transmembrane domain (class B) and are specifically recognized by class A or class B receptors, respectively. The interaction of ephrins with a globular domain at the N-terminus of their corresponding receptors induces the phosphorylation of intracellular tyrosine residues of both the receptor and the ligand, followed by the subsequent assembly of adaptor proteins

(Fig. 5A). More importantly, ephrin signaling occurs via the interaction between different cells, not through the action of a secreted molecule on a specific target tissue, and is usually bidirectional, involving both the source and target cells.

Several ephrin receptors and ephrin ligands are expressed during zebrafish gastrulation.[145,146] They interfere with cell movements during gastrulation without altering cell fates[146,147] (Fig. 5B, C). However, the precise function and molecular targets of ephrin signaling during gastrulation still remain to be elucidated. The results from other developmental systems suggest a role of ephrin signaling in controlling cell adhesion or modulating the cytoskeleton.[144,146,148] Interestingly, Ephrin-B ligands are enriched in lipid-raft microdomains of cultured hippocampal neurons, suggesting that plasma membrane compartmentalization and cell polarization might be involved in the regulation of ephrin signaling.[149]

Zebrafish homologues of the *Drosophila slit* and *robo* genes are involved in the regulation of convergence and extension movements during gastrulation[150] (Fig. 6). In *Drosophila*, Slit is secreted by midline glia cells within the nervous system, where it inhibits the crossing of commissural axons expressing the Slit receptor Robo.[151–154] During zebrafish gastrulation, Slit homologues are expressed in axial mesendodermal tissues such as the notochord and prechordal plate,[150] while Robo homologues are expressed throughout the whole gastrula.[155,156] Misexpression of the zebrafish Slit-2 homologue leads to defects in convergence and extension movements during gastrulation [150] (Fig. 6B, C) suggesting that Slit, secreted by axial mesendodermal cells, can regulate the movement of axial and paraxial mesendodermal cells expressing Robo. Which cellular interactions precisely depend on *slit/robo* function and what molecular target mechanisms mediate the Slit/Robo signal remain to be elucidated.

III. Future Prospects

'New' Methods

Although many genes with a function in regulating gastrulation movements have been identified, the interaction of these genes with

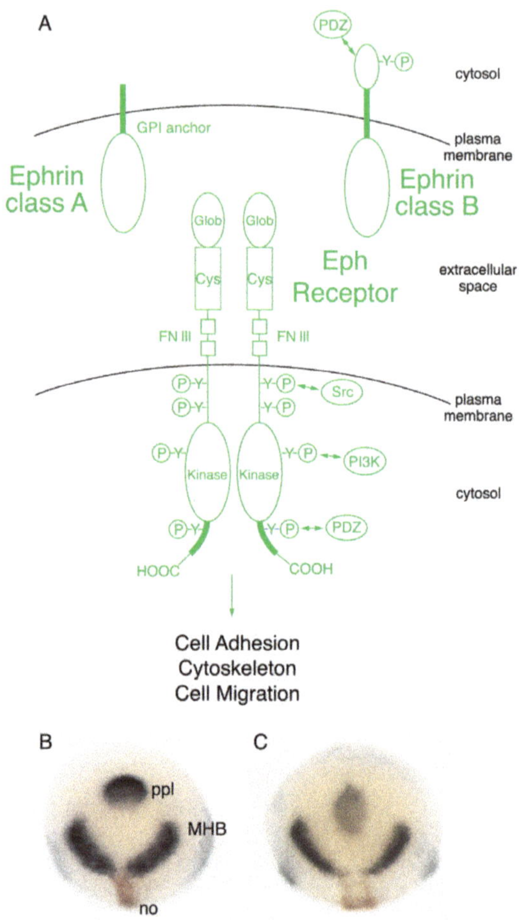

Fig. 6 Ephrin-signaling during zebrafish gastrulation. (A) Class A Ephrins are bound to the plasma membrane via a GPI anchor, whereas class B Ephrins are transmembrane proteins with a cytosolic kinase domain. Upon activation of the receptor/ligand, various proteins, such as PI3K and Src homology- or PDZ domain-containing proteins, can bind to the cytosolic part of ephrins or ephrin receptors (see text for more details). (B) Wild-type zebrafish embryo at the end of gastrulation, stained with markers outlining the prechordal plate (ppl), the mid-hindbrain boundary (MHB) and the notochord (no). (C) Embryo injected with a dominant negative ephrin ligand at the end of gastrulation. The prechordal plate is displaced posteriorly and the notochord appears broadened. Animal views with anterior to the top. Scale bar = 250 μm. Cys = cysteine rich domains; FN III = fibronectin type III like repeats. Reprinted from Holder and Klein[143] and Chan *et al.*[146]

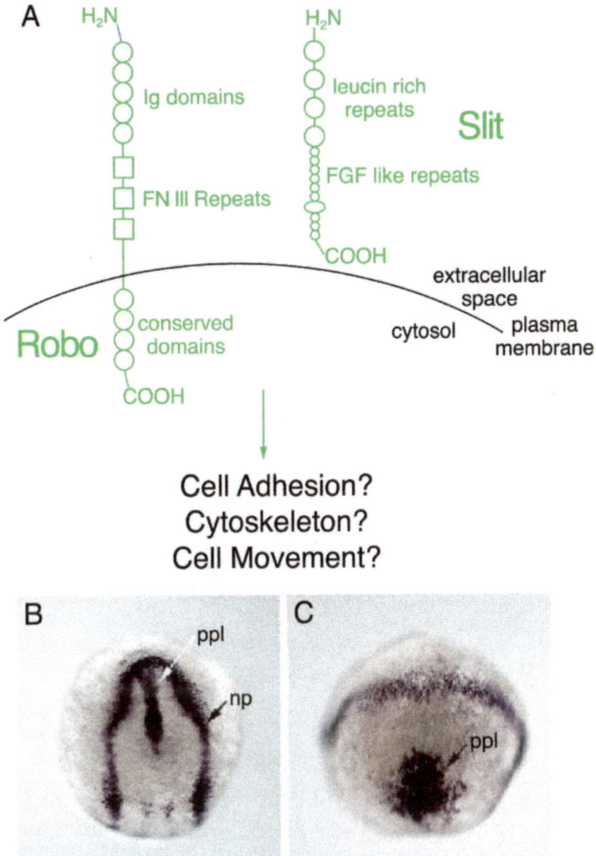

Fig. 7 Slit/Robo signaling during zebrafish gastrulation. (A) The secreted molecule Slit binds to the Robo receptor, thereby possibly triggering changes in cell adhesion or the cytoskeleton. (B) Wild type embryo at bud stage, stained with markers outlining the prechordal plate (ppl) and the neural plate (np). (C) Bud stage embryo ubiquitously overexpressing *slit2*. The neural plate is broadened and the prechordal plate is shifted posteriorly. Animal views with anterior to the top. Scale bar = 250 μm. Ig = Immunoglobulin, FGF = Fibroblast Growth Factor, FN III = Fibronectin Type III repeats. Reprinted from Holder and Klein[143] and Yeo.[150]

each other is only poorly understood. Furthermore, effector genes mediating the function of the various signaling pathways implicated are still not known. To obtain insight into the molecular and cellular function of these signaling pathways, strategies must be developed that allow a better characterization of already existing genes and the identification of downstream effectors of those signaling pathways.

One way of identifying potential effectors is to compare the expression profile between cells or tissues in which a specific pathway is activated versus those cells or tissues in which this pathway is inactive. Alternatively, cells of different origin or fate can be sorted by labeling subsets of cells, e.g. through the expression of GFP under the control of tissue specific promoters, followed by FACS-sorting of those cells. Using cDNA arrays or cDNA subtraction protocols, genes that are differentially expressed between different tissues can then be identified. The functional and molecular characterization of transcriptional targets will add more to our understanding of how different morphogenetic processes during gastrulation are regulated on the molecular and cellular level.

Morphogenetically active signaling pathways are likely to exert their function not only by regulating the transcription of specific target genes but also by directly modifying intracellular proteins and lipids. For instance, the Fz/PCP pathway in *Drosophila* triggers the phosphorylation of the Myosin Light Chain Regulatory subunit (MLRC), thereby modulating the cytoskeletal architecture[81] (Fig. 2A). Identifying protein phosphorylation targets of specific morphogenetically active signaling pathways during gastrulation requires the adaptation of a whole series of protein-biochemical techniques in zebrafish. Methods such as heat-shock-promoter driven transgenic lines that allow the conditional activation of a specific signaling pathway at a certain time during gastrulation must be established.[157,158] In addition, methods to detect phosphorylation of specific target proteins, e.g. through the introduction of radioactively labeled phosphate, need to be developed. The separation of subcellular fractions of the proteome by two-dimensional gel electrophoresis, followed by mass-spectrometric analysis, will then help in identifying differentially phosphorylated proteins. Once specifically-phosphorylated proteins are identified, a functional analysis of their role during gastrulation will yield a deeper understanding of how cell movements are regulated during this process.

A main challenge beyond the identification of target genes or proteins is the development of assays that can be used for analyzing cell morphology and movement during gastrulation. A frequent difficulty is that phenotypes are subtle, therefore quantitative and sufficiently sensitive assays need to be developed. As quantitative approaches have not yet been a stronghold in developmental biology, the establishment of such assays will significantly rely on finding new ways of incorporating biophysical methods in those assays. High resolution three-dimensional timelapse imaging, which allows the quantification of cell morphological phenotypes, might be one way of providing the starting material.[28] The determination of various morphometric parameters by such imaging approaches, combined with the use of sophisticated computer programs, will be an important first step in elucidating the functional role of cell morphology and movement during gastrulation.

Remaining and Arising Questions

Several genes have been identified that affect cellular movements and morphology during gastrulation.[3] In most cases, however, it is still unclear how these genes function during gastrulation. They might either instructively determine the pattern of cellular rearrangements within the embryo or they might function permissively by allowing cells to interpret any patterning cues. Although these modes are not always easy to distinguish — as the cellular output might look very much alike — they fulfill fundamentally different functions. In the case of a gene involved in instructively patterning morphogenetic movements, the gene product itself must confer the patterning information that regulates cell movement and morphology during gastrulation. In contrast, genes that function permissively only have to possess the ability to interpret a patterning signal but are not needed to confer any patterning information themselves. More specific experiments, such as selective misexpression of candidate genes, will help to distinguish between these different functional possibilities.

Not only the mode of action but also the place within the gastrula and the time during gastrulation where they are required has not yet been determined for most of the genes implicated in regulating gastrulation movements.[3] Although the expression pattern of most of these genes has been established during gastrulation, this has provided only a rather global picture about the distribution of RNAs coding for those genes. In contrast, the endogenous distribution of protein products for nearly all of those genes is yet unknown. Studies in *Drosophila* have clearly shown that the intracellular localization of planar cell polarity proteins, for example, is the essential step in polarizing cells within the wing epithelium.[82] The generation of functional antibodies directed against genes with a morphogenetic function during vertebrate gastrulation will be required to visualize the intra- and intercellular distribution of those proteins, which in turn might help to obtain insight into the mechanisms by which these genes function during gastrulation.

One of the biggest challenges ahead is to identify the complete spectrum of direct and indirect effects a gene exerts on the regulation of gastrulation movements. These cell-autonomous and non-autonomous gene functions might depend on the interaction of cells within a certain germ layer and also between populations of cells in different tissues or germ layers. Studies in *Xenopus* have shown that mesodermal cells are required for the proper polarization and movement of ectodermal cells during gastrulation, indicating that the cellular interactions between the germ layers are important.[53] Similarly, in the case of zebrafish, the yolk cell appears to provide a substrate for the movement of hypoblast (mesendodermal) and epiblast (ectodermal) cells during gastrulation.[14,28] It was also shown that the movement of yolk syncytial cell nuclei during gastrulation resembles the convergence and extension movements seen in the overlying hypoblast and epiblast cell layers during gastrulation, suggesting that cell movements are coordinated between yolk cell and germ layers.[14] Tissue- or cell-specific manipulation of cellular movement and morphology and the subsequent analysis of its effects on other tissues during gastrulation will be required to obtain insight into the interaction between different cellular populations during gastrulation.

How quickly all those questions can be answered will depend on the development and adaptation of new experimental techniques. As many of these techniques are already available (although not necessarily in the gastrulation field), one can expect significant progress in understanding the molecular and cellular mechanisms that regulate zebrafish gastrulation movements in the near future.

Acknowledgments

We thank Sarah Bowman and Mathias Köppen for careful reading of the manuscript and Franziska Friedrich and Christine Panagiotidis for help with the figure layout. F. U. and C.-P. H. are supported by an Emmy-Noether fellowship from the DFG.

References

1. Kimmel CB, Ballard WW, Kimmel SR, Ullmann B and Schilling TF (1995). Stages of embryonic development of the zebrafish. *Dev. Dyn.* **203**: 253–310.
2. Warga RM and Kimmel CB (1990). Cell movements during epiboly and gastrulation in zebrafish. *Development* **108**: 569–580.
3. Heisenberg CP and Tada M (2002). Zebrafish gastrulation movements: bridging cell and developmental biology. *Semin. Cell Dev. Biol.* **13**: 471–479.
4. Wallingford JB, Fraser SE and Harland RM (2002). Convergent Extension. The Molecular Control of Polarized Cell Movement during Embryonic Development. *Dev. Cell* **2**: 695–706.
5. Haffter P, Granato M, Brand M, Mullins MC, Hammerschmidt M, Kane DA, Odenthal J, van Eeden FJ, Jiang YJ, Heisenberg CP, Kelsh RN, Furutani–Seiki M, Vogelsang E, Beuchle D, Schach U, Fabian C and Nusslein–Volhard C (1996). The identification of genes with unique and essential functions in the development of the zebrafish, *Danio rerio. Development* **123**: 1–36.
6. Driever W, Solnica–Krezel L, Schier AF, Neuhauss SC, Malicki J, Stemple DL, Stainier DY, Zwartkruis F, Abdelilah S, Rangini Z, Belak J

and Boggs C (1996). A genetic screen for mutations affecting embryogenesis in zebrafish. *Development* **123**: 37–46.
7. Amsterdam A, Burgess S, Golling G, Chen W, Sun Z, Townsend K, Farrington S, Haldi M and Hopkins N (1999). A large-scale insertional mutagenesis screen in zebrafish. *Genes. Dev.* **13**: 2713–2724.
8. Farber SA, Pack M, Ho SY, Johnson ID, Wagner DS, Dosch R, Mullins MC, Hendrickson HS, Hendrickson EK and Halpern ME (2001). Genetic analysis of digestive physiology using fluorescent phospholipid reporters. *Science* **292**: 1385–1388.
9. Patton EE and Zon LI (2001). The art and design of genetic screens: zebrafish. *Nat. Rev. Genet.* **2**: 956–966.
10. Nasevicius A and Ekker SC (2000). Effective targeted gene 'knockdown' in zebrafish. *Nat. Genet.* **26**: 216–220.
11. Lele Z, Bakkers J and Hammerschmidt M (2001). Morpholino phenocopies of the swirl, snailhouse, somitabun, minifin, silberblick, and pipetail mutations. *Genesis* **30**: 190–194.
12. Wilson ET, Cretekos CJ and Helde KA (1995). Cell mixing during early epiboly in the zebrafish embryo. *Dev. Genet.* **17**: 6–15.
13. Kane D and Adams R (2002). Life at the edge: epiboly and involution in the zebrafish. *Results Probl. Cell Differ.* **40**: 117–135.
14. D'Amico LA and Cooper MS (2001). Morphogenetic domains in the yolk syncytial layer of axiating zebrafish embryos. *Dev. Dyn.* **222**: 611–624.
15. Chen S and Kimelman D (2000). The role of the yolk syncytial layer in germ layer patterning in zebrafish. *Development* **127**: 4681–4689.
16. Kimelman D and Griffin KJ (2000). Vertebrate mesendoderm induction and patterning. *Curr. Opin. Genet Dev.* **10**: 350–356.
17. Schier AF (2001). Axis formation and patterning in zebrafish. *Curr. Opin. Genet. Dev.* **11**: 393–404.
18. Strahle U and Jesuthasan S (1993). Ultraviolet irradiation impairs epiboly in zebrafish embryos: evidence for a microtubule-dependent mechanism of epiboly. *Development* **119**: 909–919.
19. Solnica–Krezel L and Driever W (1994). Microtubule arrays of the zebrafish yolk cell: organization and function during epiboly. *Development* **120**: 2443–2455.
20. Jesuthasan S and Stahle U (1997). Dynamic microtubules and specification of the zebrafish embryonic axis. *Curr. Biol.* 7: 31–42.

21. Betchaku T and Trinkaus JP (1978). Contact relations, surface activity, and cortical microfilaments of marginal cells of the enveloping layer and of the yolk syncytial and yolk cytoplasmic layers of fundulus before and during epiboly. *J. Exp. Zool.* **206**: 381–426.
22. Zalik SE, Lewandowski E, Kam Z and Geiger B (1999). Cell adhesion and the actin cytoskeleton of the enveloping layer in the zebrafish embryo during epiboly. *Biochem. Cell Biol.* 77: 527–542.
23. Keller RE and Trinkaus JP (1987). Rearrangement of enveloping layer cells without disruption of the epithelial permeability barrier as a factor in Fundulus epiboly. *Dev. Biol.* **120**: 12–24.
24. Fink RD and Cooper MS (1996). Apical membrane turnover is accelerated near cell-cell contacts in an embryonic epithelium. *Dev. Biol.* **174**: 180–189.
25. Kane DA, Hammerschmidt M, Mullins MC, Maischein HM, Brand M, van Eeden FJ, Furutani–Seiki M, Granato M, Haffter P, Heisenberg CP, Jiang YJ, Kelsh RN, Odenthal J, Warga RM and Nusslein–Volhard C (1996). The zebrafish epiboly mutants. *Development* **123**: 47–55.
26. Solnica–Krezel L, Stemple DL, Mountcastle–Shah E, Rangini Z, Neuhauss SC, Malicki J, Schier AF, Stainier DY, Zwartkruis F, Abdelilah S and Driever W (1996). Mutations affecting cell fates and cellular rearrangements during gastrulation in zebrafish. *Development* **123**: 67–80.
27. Trinkaus JP (1951). A study of the mechanism of epiboly in the egg of *Fundulus heteroclitus. Journal of Exp. Zool.* **118**: 260–320.
28. Ulrich F, Concha ML, Heid PJ, Voss E, Witzel S, Roehl H, Tada M, Wilson SW, Adams RJ, Soll DR and Heisenberg CP. Slb/Wnt11 regulates cell polarity and directed cell movement during zebrafish gastrulation (submitted).
29. Solnica–Krezel L, Stemple DL and Driever W (1995). Transparent things: cell fates and cell movements during early embryogenesis of zebrafish. *Bioessays* **17**: 931–939.
30. Cooper MS and D'Amico LA (1996). A cluster of noninvoluting endocytic cells at the margin of the zebrafish blastoderm marks the site of embryonic shield formation. *Dev. Biol.* **180**: 184–198.
31. Melby AE, Warga RM and Kimmel CB (1996). Specification of cell fates at the dorsal margin of the zebrafish gastrula. *Development* **122**: 2225–2237.
32. Kimmel CB, Warga RM and Schilling TF (1990). Origin and organization of the zebrafish fate map. *Development* **108**: 581–594.

33. Woo K and Fraser SE (1997). Specification of the zebrafish nervous system by nonaxial signals. *Science* **277**: 254–257.
34. Warga RM and Nusslein–Volhard C (1999). Origin and development of the zebrafish endoderm. *Development* **126**: 827–838.
35. Trinkaus JP (1996). Ingression during early gastrulation of fundulus. *Dev. Biol.* **177**: 356–370.
36. D'Amico LA and Cooper MS (1997). Spatially distinct domains of cell behavior in the zebrafish organizer region. *Biochem. Cell Biol.* **75**: 563–577.
37. Concha ML and Adams RJ (1998). Oriented cell divisions and cellular morphogenesis in the zebrafish gastrula and neurula: a time-lapse analysis. *Development* **125**: 983–994.
38. Carmany–Rampey A and Schier AF (2001). Single-cell internalization during zebrafish gastrulation. *Curr. Biol.* **11**: 1261–1265.
39. Hardin J (1996). The cellular basis of sea urchin gastrulation. *Curr. Top. Dev. Biol.* **33**: 159–262.
40. Irvine KD and Wieschaus E (1994). Cell intercalation during Drosophila germband extension and its regulation by pair-rule segmentation genes. *Development* **120**: 827–841.
41. Williams–Masson EM, Malik AN and Hardin J (1997). An actin-mediated two-step mechanism is required for ventral enclosure of the *C. elegans* hypodermis. *Development* **124**: 2889–2901.
42. Miyamoto DM and Crowther RJ (1985). Formation of the notochord in living ascidian embryos. *J. Embryol. Exp. Morphol.* **86**: 1–17.
43. Sausedo RA and Schoenwolf GC (1994). Quantitative analyses of cell behaviors underlying notochord formation and extension in mouse embryos. *Anat. Rec.* **239**: 103–112.
44. Sausedo RA and Schoenwolf GC (1993). Cell behaviors underlying notochord formation and extension in avian embryos: quantitative and immunocytochemical studies. *Anat. Rec.* **237**: 58–70.
45. Keller R, Davidson L, Edlund A, Elul T, Ezin M, Shook D and Skoglund P (2000). Mechanisms of convergence and extension by cell intercalation. *Philos. Trans. R. Soc. Lond. B. Biol. Sci.* **355**: 897–922.
46. Myers D, Sepich D and Solnica–Krezel L (2002). Convergence and extension in vertebrate gastrulae: cell movements according to or in search of identity? *Trends Genet.* **18**: 447.

47. Sepich DS, Myers DC, Short R, Topczewski J, Marlow F and Solnica–Krezel L (2000). Role of the zebrafish trilobite locus in gastrulation movements of convergence and extension. *Genesis* **27**: 159–173.
48. Myers DC, Sepich DS and Solnica–Krezel L (2002). Bmp activity gradient regulates convergent extension during zebrafish gastrulation. *Dev. Biol.* **243**: 81–98.
49. Topczewski J, Sepich DS, Myers DC, Walker C, Amores A, Lele Z, Hammerschmidt M, Postlethwait J and Solnica–Krezel L (2001). The zebrafish glypican knypek controls cell polarity during gastrulation movements of convergent extension. *Dev. Cell* **1**: 251–264.
50. Marlow F, Topczewski J, Sepich D and Solnica–Krezel L (2002). Zebrafish rho kinase 2 acts downstream of wnt11 to mediate cell polarity and effective convergence and extension movements. *Curr. Biol.* **12**: 876–884.
51. Kilian B, Mansukoski H, Barbosa FC, Ulrich F, Tada M and Heisenberg CP (2003). The role of Ppt/Wnt5 in regulating cell shape and movement during zebrafish gastrulation. *Mech. Dev.* **120**: 467–476.
52. Glickman NS, Kimmel CB, Jones MA and Adams RJ (2003). Shaping the zebrafish notochord. *Development* **130**: 873–887.
53. Elul T and Keller R (2000). Monopolar protrusive activity: a new morphogenic cell behavior in the neural plate dependent on vertical interactions with the mesoderm in Xenopus. *Dev. Biol.* **224**: 3–19.
54. Yamamoto A, Amacher SL, Kim SH, Geissert D, Kimmel CB and De Robertis EM (1998). Zebrafish paraxial protocadherin is a downstream target of spadetail involved in morphogenesis of gastrula mesoderm. *Development* **125**: 3389–3397.
55. Kim SH, Yamamoto A, Bouwmeester T, Agius E and Robertis EM (1998). The role of paraxial protocadherin in selective adhesion and cell movements of the mesoderm during Xenopus gastrulation. *Development* **125**: 4681–4690.
56. Wodarz A and Nusse R (1998). Mechanisms of Wnt signaling in development. *Annu. Rev. Cell Dev. Biol.* **14**: 59–88.
57. Yoshikawa S, McKinnon RD, Kokel M and Thomas JB (2003). Wnt-mediated axon guidance via the Drosophila derailed receptor. *Nature* **422**: 583–588.
58. Cadigan KM and Nusse R (1997). Wnt Signaling — A Common Theme in Animal Development [Review]. *Genes & Development* **11**: 3286–3305.

59. Huelsken J and Birchmeier W (2001). New aspects of Wnt signaling pathways in higher vertebrates. *Curr. Opin. Genet. Dev.* **11**: 547–553.
60. Tada M and Smith JC (2000). Xwnt11 is a target of Xenopus Brachyury: regulation of gastrulation movements via Dishevelled, but not through the canonical Wnt pathway. *Development* **127**: 2227–2238.
61. Wallingford JB, Rowning BA, Vogeli KM, Rothbacher U, Fraser SE and Harland RM (2000). Dishevelled controls cell polarity during Xenopus gastrulation. *Nature* **405**: 81–85.
62. Heisenberg CP, Tada M, Rauch GJ, Saude L, Concha ML, Geisler R, Stemple DL, Smith JC and Wilson SW (2000). Silberblick/Wnt11 mediates convergent extension movements during zebrafish gastrulation. *Nature* **405**: 76–81.
63. Adler PN and Lee H (2001). Frizzled signaling and cell-cell interactions in planar polarity. *Curr. Opin. Cell Biol.* **13**: 635–640.
64. Adler PN (2002). Planar signaling and morphogenesis in Drosophila. *Dev. Cell* **2**: 525–535.
65. Tada M, Concha ML and Heisenberg CP (2002). Non-canonical Wnt signaling and regulation of gastrulation movements. *Semin. Cell Dev. Biol.* **13**: 251–260.
66. Rauch GJ, Hammerschmidt M, Blader P, Schauerte HE, Strahle U, Ingham PW, McMahon AP and Haffter P (1997). Wnt5 is required for tail formation in the zebrafish embryo. *Cold Spring Harbor Symposia on Quantitative Biology* **62**: 227–234.
67. Jessen JR, Topczewski J, Bingham S, Sepich DS, Marlow F, Chandrasekhar A and Solnica-Krezel L (2002). Zebrafish *trilobite* identifies new roles for Strabismus in gastrulation and neuronal movements. *Nat. Cell Biol.* **4**: 610–615.
68. Djiane A, Riou J, Umbhauer M, Boucaut J and Shi D (2000). Role of frizzled 7 in the regulation of convergent extension movements during gastrulation in *Xenopus laevis. Development* **127**: 3091–3100.
69. Habas R, Kato Y and He X (2001). Wnt/Frizzled activation of rho regulates vertebrate gastrulation and requires a novel formin homology protein daam1. *Cell* **107**: 843–854.
70. Habas R, Dawid IB and He X (2003). Coactivation of Rac and Rho by Wnt/Frizzled signaling is required for vertebrate gastrulation. *Genes Dev.* **17**: 295–309.

71. Yamanaka H, Moriguchi T, Masuyama N, Kusakabe M, Hanafusa H, Takada R, Takada S and Nishida E (2002). JNK functions in the non-canonical Wnt pathway to regulate convergent extension movements in vertebrates. *EMBO Rep* **3**: 69–75.
72. Schwarz–Romond T, Asbrand C, Bakkers J, Kuhl M, Schaeffer HJ, Huelsken J, Behrens J, Hammerschmidt M and Birchmeier W (2002). The ankyrin repeat protein Diversin recruits Casein kinase Iepsilon to the beta-catenin degradation complex and acts in both canonical Wnt and Wnt/JNK signaling. *Genes Dev.* **16**: 2073–2084.
73. Park M and Moon RT (2002). The planar cell-polarity gene stbm regulates cell behaviour and cell fate in vertebrate embryos. *Nat. Cell Biol.* **4**: 20–25.
74. Goto T and Keller R (2002). The planar cell polarity gene strabismus regulates convergence and extension and neural fold closure in Xenopus. *Dev. Biol.* **247**: 165–181.
75. Darken RS, Scola AM, Rakeman AS, Das G, Mlodzik M and Wilson PA (2002). The planar polarity gene strabismus regulates convergent extension movements in Xenopus. *EMBO J.* **21**: 976–985.
76. Hannus M, Feiguin F, Heisenberg CP and Eaton S (2002). Planar cell polarization requires Widerborst, a B′ regulatory subunit of protein phosphatase 2A. *Development* **129**: 3493–3503.
77. Veeman MT, Slusarski DC, Kaykas A, Louie SH and Moon RT (2003). Zebrafish prickle, a modulator of noncanonical wnt/fz signaling, regulates gastrulation movements. *Curr. Biol.* **13**: 680–685.
78. Takeuchi M, Nakabayashi J, Sakaguchi T, Yamamoto TS, Takahashi H, Takeda H and Ueno N (2003). The prickle-related gene in vertebrates is essential for gastrulation cell movements. *Curr. Biol.* **13**: 674–679.
79. Carreira–Barbosa F, Concha M, Takeuchi M, Ueno N, Wilson SW and Tada M (2003). Prickle 1 regulates cell movements during gastrulation and neuronal migration in zebrafish. *Development* (in press).
80. Eaton S (1997). Planar polarization of Drosophila and vertebrate epithelia. *Curr. Opin. Cell Biol.* **9**: 860–866.
81. Winter CG, Wang B, Ballew A, Royou A, Karess R, Axelrod JD and Luo L (2001). Drosophila Rho-associated kinase (Drok) links Frizzled-mediated planar cell polarity signaling to the actin cytoskeleton. *Cell* **105**: 81–91.
82. Strutt DI (2002). The asymmetric subcellular localisation of components of the planar polarity pathway. *Semin. Cell Dev. Biol.* **13**: 225–231.

83. Jacinto A, Wood W, Balayo T, Turmaine M, Martinez–Arias A and Martin P (2000). Dynamic actin-based epithelial adhesion and cell matching during Drosophila dorsal closure. *Curr. Biol.* **10**: 1420–1426.
84. Kaltschmidt JA, Lawrence N, Morel V, Balayo T, Fernandez BG, Pelissier A, Jacinto A and Martinez Arias A (2002). Planar polarity and actin dynamics in the epidermis of Drosophila. *Nat. Cell Biol.* **4**: 937–944.
85. Gumbiner BM (1996). Cell adhesion: the molecular basis of tissue architecture and morphogenesis. *Cell* **84**: 345–357.
86. Wedlich D (2002). The polarising role of cell adhesion molecules in early development. *Curr. Opin. Cell Biol.* **14**: 563–568.
87. Small JV and Kaverina I (2003). Microtubules meet substrate adhesions to arrange cell polarity. *Curr. Opin. Cell Biol.* **15**: 40–47.
88. Perez–Moreno M, Jamora C and Fuchs E (2003). Sticky business: orchestrating cellular signals at adherens junctions. *Cell* **112**: 535–548.
89. Kuroda H, Inui M, Sugimoto K, Hayata T and Asashima M (2002). Axial protocadherin is a mediator of prenotochord cell sorting in Xenopus. *Dev. Biol.* **244**: 267–277.
90. Taira M, Jamrich M, Good PJ and Dawid IB (1992). The LIM domain-containing homeo box gene Xlim-1 is expressed specifically in the organizer region of Xenopus gastrula embryos. *Genes Dev.* **6**: 356–366.
91. Hukriede NA, Tsang TE, Habas R, Khoo PL, Steiner K, Weeks DL, Tam PP and Dawid IB (2003). Conserved requirement of Lim1 function for cell movements during gastrulation. *Dev. Cell* **4**: 83–94.
92. Ho RK and Kane DA (1990). Cell-autonomous action of zebrafish spt-1 mutation in specific mesodermal precursors. *Nature* **348**: 728–730.
93. Levine E, Lee CH, Kintner C and Gumbiner BM (1994). Selective disruption of E-cadherin function in early Xenopus embryos by a dominant negative mutant. *Development* **120**: 901–909.
94. Babb SG, Barnett J, Doedens AL, Cobb N, Liu Q, Sorkin BC, Yelick PC, Raymond PA and Marrs JA (2001). Zebrafish E-cadherin: expression during early embryogenesis and regulation during brain development. *Dev. Dyn.* **221**: 231–237.
95. Lee CH and Gumbiner BM (1995). Disruption of gastrulation movements in Xenopus by a dominant-negative mutant for C-cadherin. *Dev. Biol.* **171**: 363–373.
96. Zhong C, Kinch MS and Burridge K (1997). Rho-stimulated contractility contributes to the fibroblastic phenotype of Ras-transformed epithelial cells. *Mol. Biol. Cell* **8**: 2329–2344.

97. Tepass U, Truong K, Godt D, Ikura M and Peifer M (2000). Cadherins in embryonic and neural morphogenesis. *Nat. Rev. Mol. Cell Biol.* **1**: 91–100.
98. Lele Z, Folchert A, Concha M, Rauch GJ, Geisler R, Rosa F, Wilson SW, Hammerschmidt M and Bally–Cuif L (2002). Parachute/n-cadherin is required for morphogenesis and maintained integrity of the zebrafish neural tube. *Development* **129**: 3281–3294.
99. Bitzur S, Kam Z and Geiger B (1994). Structure and distribution of N-cadherin in developing zebrafish embryos: morphogenetic effects of ectopic over-expression. *Dev. Dyn.* **201**: 121–136.
100. Marsden M and DeSimone DW (2001). Regulation of cell polarity, radial intercalation and epiboly in Xenopus: novel roles for integrin and fibronectin. *Development* **128**: 3635–3647.
101. Davidson LA, Hoffstrom BG, Keller R and DeSimone DW (2002). Mesendoderm extension and mantle closure in Xenopus laevis gastrulation: combined roles for integrin alpha(5)beta(1), fibronectin, and tissue geometry. *Dev. Biol.* **242**: 109–129.
102. Ungar AR, Kelly GM and Moon RT (1995). Wnt4 affects morphogenesis when mis-expressed in the zebrafish embryo. *Mech Dev* **52**: 153–164.
103. Henry CA, Crawford BD, Yan YL, Postlethwait J, Cooper MS and Hille MB (2001). Roles for zebrafish focal adhesion kinase in notochord and somite morphogenesis. *Dev. Biol.* **240**: 474–487.
104. Cohen ED, Mariol MC, Wallace RM, Weyers J, Kamberov YG, Pradel J and Wilder EL (2002). DWnt4 regulates cell movement and focal adhesion kinase during Drosophila ovarian morphogenesis. *Dev. Cell* **2**: 437–448.
105. Kuhl M, Sheldahl LC, Park M, Miller JR and Moon RT (2000). The Wnt/Ca2+ pathway: a new vertebrate Wnt signaling pathway takes shape. *Trends Genet.* **16**: 279–283.
106. Wacker S, Grimm K, Joos T and Winklbauer R (2000). Development and control of tissue separation at gastrulation in Xenopus. *Dev. Biol.* **224**: 428–439.
107. Shen MM and Schier AF (2000). The EGF-CFC gene family in vertebrate development. *Trends Genet.* **16**: 303–309.
108. Whitman M (2001). Nodal signaling in early vertebrate embryos: themes and variations. *Dev Cell* **1**: 605–617.
109. Kimelman D and Schier AF (2002). Mesoderm induction and patterning. *Results Probl. Cell Differ.* **40**: 15–27.

110. Gritsman K, Zhang J, Cheng S, Heckscher E, Talbot WS and Schier AF (1999). The EGF-CFC protein one-eyed pinhead is essential for nodal signaling. *Cell* **97**: 121–132.
111. David NB and Rosa FM (2001). Cell autonomous commitment to an endodermal fate and behaviour by activation of Nodal signaling. *Development* **128**: 3937–3947.
112. Yeo C and Whitman M (2001). Nodal signals to Smads through Cripto-dependent and Cripto-independent mechanisms. *Mol. Cell* **7**:949–957.
113. Bisgrove BW, Essner JJ and Yost HJ (1999). Regulation of midline development by antagonism of lefty and nodal signaling. *Development* **126**: 3253–3262.
114. Meno C, Gritsman K, Ohishi S, Ohfuji Y, Heckscher E, Mochida K, Shimono A, Kondoh H, Talbot WS, Robertson EJ, Schier AF and Hamada H (1999). Mouse Lefty2 and zebrafish antivin are feedback inhibitors of nodal signaling during vertebrate gastrulation. *Mol. Cell* **4**: 287–298.
115. Feldman B, Concha ML, Saude L, Parsons MJ, Adams RJ, Wilson SW and Stemple DL (2002). Lefty antagonism of squint is essential for normal gastrulation. *Curr. Biol.* **12**: 2129–2135.
116. Chen Y and Schier AF (2002). Lefty proteins are long-range inhibitors of squint-mediated nodal signaling. *Curr. Biol.* **12**: 2124–2128.
117. Wang F, Herzmark P, Weiner OD, Srinivasan S, Servant G and Bourne HR (2002). Lipid products of PI(3)Ks maintain persistent cell polarity and directed motility in neutrophils. *Nat. Cell. Biol.* **4**: 513–518.
118. Weiner OD, Neilsen PO, Prestwich GD, Kirschner MW, Cantley LC and Bourne HR (2002). A PtdInsP(3)- and Rho GTPase-mediated positive feedback loop regulates neutrophil polarity. *Nat. Cell Biol.* **4**: 509–513.
119. Symes K and Mercola M (1996). Embryonic mesoderm cells spread in response to platelet-derived growth factor and signaling by phosphatidylinositol 3-kinase. *Proc. Natl. Acad. Sci. USA* **93**: 9641–9644.
120. Weiner OD (2002). Regulation of cell polarity during eukaryotic chemotaxis: the chemotactic compass. *Curr Opin Cell Biol* **14**: 196–202.
121. Comer FI and Parent CA (2002). PI 3-kinases and PTEN: how opposites chemoattract. *Cell* **109**: 541–544.
122. Iijima M, Huang YE and Devreotes P (2002). Temporal and spatial regulation of chemotaxis. *Dev. Cell* **3**: 469–478.

123. Ho L, Symes K, Yordan C, Gudas LJ and Mercola M (1994). Localization of PDGF A and PDGFR alpha mRNA in Xenopus embryos suggests signaling from neural ectoderm and pharyngeal endoderm to neural crest cells. *Mech. Dev.* **48**: 165–174.
124. Ataliotis P, Symes K, Chou MM, Ho L and Mercola M (1995). PDGF signaling is required for gastrulation of Xenopus laevis. *Development* **121**: 3099–3110.
125. Orr–Urtreger A and Lonai P (1992). Platelet-derived growth factor-A and its receptor are expressed in separate, but adjacent cell layers of the mouse embryo. *Development* **115**: 1045–1058.
126. Orr–Urtreger A, Bedford MT, Do MS, Eisenbach L and Lonai P (1992). Developmental expression of the alpha receptor for platelet-derived growth factor, which is deleted in the embryonic lethal Patch mutation. *Development* **115**: 289–303.
127. Montero JA, Kilian B, Chan J, Bayliss PE, Heisenberg CP. Phosphoinositide 3-kinase is required for process outgrowth and cell polarisation of gastrulating mesendodermal cells (submitted).
128. Slusarski DC, Yang–Snyder J, Busa WB and Moon RT (1997). Modulation of embryonic intracellular Ca2+ signaling by Wnt-5A. *Dev. Biol.* **182**: 114–120.
129. Winklbauer R, Medina A, Swain RK and Steinbeisser H (2001). Frizzled-7 signaling controls tissue separation during Xenopus gastrulation. *Nature* **413**: 856–860.
130. Miller JR, Hocking AM, Brown JD and Moon RT (1999). Mechanism and function of signal transduction by the Wnt/beta-catenin and Wnt/Ca2+ pathways. *Oncogene* **18**: 7860–7872.
131. Pandur P, Lasche M, Eisenberg LM and Kuhl M (2002). Wnt-11 activation of a non-canonical Wnt signaling pathway is required for cardiogenesis. *Nature* **418**: 636–641.
132. Wallingford JB, Ewald AJ, Harland RM and Fraser SE (2001). Calcium signaling during convergent extension in Xenopus. *Curr. Biol.* **11**: 652–661.
133. Hou SX, Zheng Z, Chen X and Perrimon N (2002). The Jak/STAT pathway in model organisms: emerging roles in cell movement. *Dev. Cell* **3**: 765–778.
134. Takeda K, Noguchi K, Shi W, Tanaka T, Matsumoto M, Yoshida N, Kishimoto T and Akira S (1997). Targeted disruption of the mouse

Stat3 gene leads to early embryonic lethality. *Proc. Natl. Acad. Sci. USA* **94**: 3801–3804.

135. Mohanty S, Jermyn KA, Early A, Kawata T, Aubry L, Ceccarelli A, Schaap P, Williams JG and Firtel RA (1999). Evidence that the Dictyostelium Dd-STATa protein is a repressor that regulates commitment to stalk cell differentiation and is also required for efficient chemotaxis. *Development* **126**: 3391–3405.
136. Sano S, Itami S, Takeda K, Tarutani M, Yamaguchi Y, Miura H, Yoshikawa K, Akira S and Takeda J (1999). Keratinocyte-specific ablation of Stat3 exhibits impaired skin remodeling, but does not affect skin morphogenesis. *EMBO J.* **18**: 4657–4668.
137. Conway G, Margoliath A, Wong–Madden S, Roberts RJ and Gilbert W (1997). Jak1 kinase is required for cell migrations and anterior specification in zebrafish embryos. *Proc. Natl. Acad. Sci. USA* **94**: 3082–3087.
138. Oates AC, Wollberg P, Pratt SJ, Paw BH, Johnson SL, Ho RK, Postlethwait JH, Zon LI and Wilks AF (1999). Zebrafish stat3 is expressed in restricted tissues during embryogenesis and stat1 rescues cytokine signaling in a STAT1-deficient human cell line. *Dev. Dyn.* **215**: 352–370.
139. Yamashita S, Miyagi C, Carmany–Rampey A, Shimizu T, Fujii R, Schier AF and Hirano T (2002). Stat3 controls cell movements during Zebrafish gastrulation. *Dev. Cell* **2**: 363–375.
140. Simon AR, Vikis HG, Stewart S, Fanburg BL, Cochran BH and Guan KL (2000). Regulation of STAT3 by direct binding to the Rac1 GTPase. *Science* **290**: 144–147.
141. Silver DL and Montell DJ (2001). Paracrine signaling through the JAK/STAT pathway activates invasive behavior of ovarian epithelial cells in Drosophila. *Cell* **107**: 831–841.
142. Doitsidou M, Reichman–Fried M, Stebler J, Koprunner M, Dorries J, Meyer D, Esguerra CV, Leung T and Raz E (2002). Guidance of primordial germ cell migration by the chemokine SDF-1. *Cell* **111**: 647–659.
143. Holder N and Klein R (1999). Eph receptors and ephrins: effectors of morphogenesis. *Development* **126**: 2033–2044.
144. Klein R (2001). Excitatory Eph receptors and adhesive ephrin ligands. *Curr. Opin. Cell Biol.* **13**: 196–203.

145. Xu Q, Holder N, Patient R and Wilson SW (1994). Spatially regulated expression of three receptor tyrosine kinase genes during gastrulation in the zebrafish. *Development* **120**: 287–299.
146. Chan J, Mably JD, Serluca FC, Chen JN, Goldstein NB, Thomas MC, Cleary JA, Brennan C, Fishman MC and Roberts TM (2001). Morphogenesis of prechordal plate and notochord requires intact Eph/ephrin B signaling. *Dev. Biol.* **234**: 470–482.
147. Oates AC, Lackmann M, Power MA, Brennan C, Down LM, Do C, Evans B, Holder N and Boyd AW (1999). An early developmental role for eph-ephrin interaction during vertebrate gastrulation. *Mech. Dev.* **83**: 77–94.
148. Kalo MS and Pasquale EB (1999). Signal transfer by Eph receptors. *Cell Tissue Res.* **298**: 1–9.
149. Bruckner K, Pablo Labrador J, Scheiffele P, Herb A, Seeburg PH and Klein R (1999). EphrinB ligands recruit GRIP family PDZ adaptor proteins into raft membrane microdomains. *Neuron* **22**: 511–524.
150. Yeo SY, Little MH, Yamada T, Miyashita T, Halloran MC, Kuwada JY, Huh TL and Okamoto H (2001). Overexpression of a slit homologue impairs convergent extension of the mesoderm and causes cyclopia in embryonic zebrafish. *Dev. Biol.* **230**: 1–17.
151. Simpson JH, Bland KS, Fetter RD and Goodman CS (2000). Short-range and long-range guidance by Slit and its Robo receptors: a combinatorial code of Robo receptors controls lateral position. *Cell* **103**: 1019–1032.
152. Rajagopalan S, Vivancos V, Nicolas E and Dickson BJ (2000). Selecting a longitudinal pathway: Robo receptors specify the lateral position of axons in the Drosophila CNS. *Cell* **103**: 1033–1045.
153. Simpson JH, Kidd T, Bland KS and Goodman CS (2000). Short-range and long-range guidance by slit and its Robo receptors. Robo and Robo2 play distinct roles in midline guidance. *Neuron* **28**: 753–766.
154. Rajagopalan S, Nicolas E, Vivancos V, Berger J and Dickson BJ (2000). Crossing the midline: roles and regulation of Robo receptors. *Neuron* **28**: 767–777.
155. Lee JS, Ray R and Chien CB (2001). Cloning and expression of three zebrafish roundabout homologs suggest roles in axon guidance and cell migration. *Dev. Dyn.* **221**: 216–230.

156. Challa AK, Beattie CE and Seeger MA (2001). Identification and characterization of roundabout orthologs in zebrafish. *Mech. Dev.* **101**: 249–253.
157. Halloran MC, Sato-Maeda M, Warren JT, Su F, Lele Z, Krone PH, Kuwada JY and Shoji W (2000). Laser-induced gene expression in specific cells of transgenic zebrafish. *Development* **127**: 1953–1960.
158. Adam A, Bartfai R, Lele Z, Krone PH and Orban L (2000). Heat-inducible expression of a reporter gene detected by transient assay in zebrafish. *Exp. Cell Res.* **256**: 282–290.
159. Axelrod JD (2001). Unipolar membrane association of Dishevelled mediates Frizzled planar cell polarity signaling. *Genes Dev.* **15**: 1182–1187.

Chapter 3

Development of the Zebrafish Organizer and Notochord

Kevin A. Thomas* and Derek L. Stemple*
Division of Developmental Biology
National Institute for Medical Research
The Ridgeway, Mill Hill
London, NW7 1AA
United Kingdom

The major axes of a zebrafish embryo are established early in its development. One key structure, centrally involved in the specification of these axes, is the dorsal Organizer. The Organizer becomes apparent at the beginning of gastrulation and generates signals that pattern the mesoderm to generate dorsal structures and specifies the neurectoderm. In addition to providing inductive signals, the Organizer itself will eventually differentiate to form the axial mesendoderm tissues prechordal plate and notochord. These tissues provide signals to pattern surrounding tissues and, in the case of the notochord, mechanical support required for locomotion of zebrafish larvae. Thus, how the Organizer forms and functions has been a source of great interest in developmental biology. Modern molecular and genetic studies are now providing a detailed picture of the events controlling both formation of the Organizer and its activity. Focusing on these events during zebrafish development, but drawing on results from a variety of other experimental systems, we review dorsal Organizer establishment and function, differentiation of Organizer tissue to chordamesoderm and notochord and finally, the patterning and mechanical functions of the notochord.

1. Organizer

1.1. Introduction

Since its publication in 1924 the dorsal Organizer experiment of Spemann and Mangold has stimulated embryologists and molecular Biologists to

*Current Address: Vertebrate Development and Genetics (Team 31), Wellcome Trust Sanger Institute, Wellcome Trust Genome Campus, Hinxton, Cambridge, CB10 1SA, United Kingdom, Email: ds4@sanger.ac.uk

provide a detailed explanation of its activity. With the genetics and molecular Biology available to zebrafish and *Xenopus* researchers, many details concerning the genes and molecules that establish, maintain and prosecute Organizer activity have been put forth. Nevertheless, there are important aspects of Organizer activity that remain unexplained. What then is known about the Organizer and what questions remain?

The dorsal Organizer of zebrafish first becomes evident during gastrulation. Its formation, however, is dependent on a cascade of events that begin before fertilization. Initially, a profoundly polarized egg containing all of the activities to produce a zebrafish embryo is produced. After fertilization, zebrafish development can be divided into distinct phases. The cytoplasm and yolk of the activated egg becomes separated into animal and vegetal hemispheres respectively. Early meroblastic cleavages of the cytoplasm produce blastomeres, whose connections to the yolk are maintained by large fenestrae. By the end of the cleavage stage, blastomeres exist in three distinct groups. The centrally located deep-cells will form the embryo proper. The superficial enveloping layer (EVL) cells will form the outer covering of the blastoderm. Finally, the marginal cells, which maintain their connections to the large yolk cell, formally become part of the yolk, contributing their nuclei and thus forming the yolk syncytial layer (YSL).[1–4]

In amphibia, the Spemann Organizer was identified by virtue of its ability to induce a secondary axis when transplanted to the ventral side of a host embryo. The defining activities of the Organizer are intercellular signals that perform several distinguishable functions: the Organizer provides dorsal patterning to the mesoderm; the Organizer induces convergent-extension movements of ectoderm and mesoderm; and the Organizer induces neurectoderm and provides signals to pattern the neurectoderm along the anterior-posterior axis.[5] Transplantation studies have shown that structures equivalent to the amphibian Spemann Organizer are present in the embryos representing the major vertebrate phyla. In teleost fish, such as zebrafish, the dorsal Organizer is known as the embryonic shield.[6–8] In avians, the dorsal Organizer is known as Hensen's node, and in mammals, the node.[9,10] The Organizer is not only a source of its defining signaling activities, but it also normally gives rise to the axial tissues, prechordal plate and notochord, which are essential for specification of midline structures, such as the ventral part of the brain and spinal cord. How then is the Organizer specified?

1.2. Specification of the Organizer

To understand how the Organizer is specified, it is convenient to split the problem into two component processes: dorsal determination and mesoderm induction. Both embryological experiments and molecular analyses support the notion that the processes can be separated and that they work together to generate the specialized mesoderm that constitutes the dorsal Organizer.

Dorsal specification

By the first cleavage of an amphibian zygote, dorsal determinants are segregated by a process known as cortical rotation.[11] This event establishes a group of vegetal cells shown by Nieuwkoop to induce a full secondary axis without contributing to axial tissues.[12] This group of vegetal cells constitutes a signaling center, termed the Nieuwkoop center, which induces the formation of Organizer tissue. Neither the dorsal determinants nor the Nieuwkoop center signals are known in precise molecular detail. A strong clue as to their identity was provided by the observation that overexpression of the secreted signaling molecule, Wnt1, could induce a secondary axis in *Xenopus laevis.*[13] At about the same time the molecular details of Wingless/Wnt signaling were being reported in *Drosophila melanogaster* molecular genetic studies. In particular, the protein armadillo was shown to play a key role in Wingless signal transduction.[14–16] The vertebrate homolog of armadillo is a protein called β-catenin, which was known to be associated with the cell adhesion complexes of the Cadherin class. Antibodies directed against β-catenin were found to induce axis duplication in *Xenopus.*[17] Indeed overexpression of β-catenin itself in either *Xenopus laevis* or zebrafish was found to induce formation of a full secondary axis.[18,19] In concert with transcription factors of the TCF/LEF family, β-catenin induces the expression of genes such as *siamois* and *twin* in *Xenopus laevis*, which are thought to participate in Organizer specification.[20–23]

The cortical rotation in amphibia, which is known to be microtubule-dependent, leads to the activation of β-catenin and the subsequent formation of a Nieuwkoop center. The equivalent process in teleost fish is not clear, but does apparently culminate in the localization of β-catenin at the dorsal side.[24] Direct manipulations of developing zebrafish embryos have been used to define the zebrafish equivalent of a Nieuwkoop center. For example, in studies where the vegetal third of the yolk cell is removed within 20 minutes post-fertilization, the embryo becomes completely ventralized.[25] Such embryos lack all dorsal mesoderm, neurectoderm and the most anterior 14–15 somites, indicating that a vegetal determinant localized within the yolk cell acts to specify the Organizer. In other studies, disruption of microtubules in the early embryo has shown that an activity located in the vegetal hemisphere, dependent on microtubule transport, is necessary for shield formation and correct axis specification.[26] Thus, although no obvious cortical rotation takes place in activated zebrafish eggs, a microtubule dependent process is apparently required for the proper activation of β-catenin in the correct region.

Additional clues as to the nature of β-catenin localization and activation are given by analysis of the maternal mutation *ichabod*. Mutant embryos are severely ventralized and are phenotypically similar to ventralized embryos generated by ablation of the vegetal yolk cell region. Mutant embryos from a homozygous *ichabod* mutant female can be rescued by injection of β-catenin.[27] Thus, activation of β-catenin on the dorsal side by some unknown factor, possibly involving *ichabod*, produces Organizer-inducing activity that may reside in the YSL, marginal blastomeres or both.[24]

Cellular and molecular evidence suggests that the zebrafish functional equivalent of the Nieuwkoop center may be distributed between the YSL and dorsal marginal blastomeres. By injection of RNAse, Chen and Kimelman have shown that the RNAs in the YSL are required for its ventrolateral and mesodermal inductive capabilities, as well as for the induction of Nodal-related gene expression in the ventrolateral marginal blastomeres.[28] This study further demonstrated, however, that the YSL-derived mRNA is not essential for the induction of the dorsal mesoderm, suggesting that dorsal specification is due to the stabilization of β-catenin in dorsal marginal blastomeres. So what then are the targets of activated β-catenin?

In *Xenopus laevis*, dorsal activation of β-catenin is known to induce Organizer specific homeodomain transcription factors such as *siamois* and *twin*.[20–23] In zebrafish, severely affected *boz* mutant embryos show complete loss of the axial mesendoderm tissues prechordal plate and notochord.[29,30] The *bozozok/dharma/nieuwkoid* (*boz*) gene, encoding a homeodomain containing protein, is also regulated by activated β-catenin.[31–33] Although *siamois* and *twin* diverge from the zebrafish *boz* in primary sequence, they appear to play a similar role to *boz* in Organizer specification and there are several lines of evidence that place *boz* downstream of β-catenin in this pathway.[34] Overexpression of cRNA encoding a constitutively activated β-catenin will induce *boz* expression in wild-type embryos, as well as induce axis duplication in *boz* mutant embryos, but does not rescue axial mesendoderm. In contrast, overexpression of constitutively activated Taram-A, a type I activin receptor, in *boz* mutant embryos is sufficient to induce both axis duplication as well as axial mesendoderm.[29,35] Finally, injection of *boz* cRNA is sufficient to rescue ventralized *ichabod* mutants.[27]

Although *boz* is clearly involved in dorsal specification, there are key Organizer activities that *boz* does not control. By morphological criteria, severely affected *boz* mutants have an incomplete Organizer. For example, *boz* mutants fail to express dorsal determinants such as *chordin* and *dkk1*, lack axial mesoderm and are defective in anterior neural specification. This range of defects is similar to those produced by complete surgical ablation of the shield region, which results in a loss of normal shield-derived tissues and central nervous system (CNS) patterning defects.[7] Despite the CNS defects seen in either *boz* mutant or shield-ablated embryos, both the anterior-posterior (AP) and dorsal-ventral (DV) axes of the embryo are specified. This suggests that the primary action of *boz* is the specification of axial mesendoderm, and that other factors specify the Organizer's neural inductive and neurectodermal patterning activities.

Mesendoderm induction

The Organizer is made of cells that will give rise to mesoderm of the midline. To understand Organizer formation, it is therefore essential to understand how mesoderm is induced. Among his other achievements, Nieuwkoop demonstrated that a signal, emanating from the vegetal region of the embryo, is responsible for the induction of mesoderm in the overlying cells at the embryonic equator. This observation has been exploited to identify secreted molecules that control mesoderm formation. Screens for secreted mesoderm-inducing factors identified members of the fibroblast growth factor (FGF) family, and transforming growth factor β (TGFβ) superfamily. Among these factors Activin was shown to possess morphogen activity, since it was able to induce different mesoderm types at varying concentrations.[36,37] In addition, Activin was shown to be sufficient to induce the formation of dorsal mesoderm, i.e. Organizer.[37] More recent studies implicate Nodal-related proteins as essential inducers, rather than Activin, where the Nodals and Activin operate though a common signal transduction mechanism reviewed by Schier and Shen.[37] To understand how Nodal gene expression is controlled then becomes an important issue.

Loss-of-function studies in *Xenopus* implicate a member of the T-box transcription factor family known as VegT to be in control of the initial expression of Nodal-related genes.[38–41] VegT is localized to the vegetal region and at mid-blastula transition activates zygotic signals essential for correct patterning of the developing embryo. Indeed in the absence of VegT activity, Nodal-related growth factors fail to be expressed.[42,43] Although there is a zebrafish homolog of VegT, encoded by the *spadetail* locus, it is not expressed maternally and loss of *spadetail* function does not produce the same range and severity as loss of VegT function in *Xenopus laevis*.[44] Hence a T-box protein functionally analogous to VegT has yet to be identified in zebrafish.

Genetic studies in mouse and zebrafish demonstrate the essential nature of Nodals in mesoderm induction.[45–49] Two zebrafish Nodal-related proteins, Squint and Cyclops, play essential though redundant roles in the specification of zebrafish mesendoderm. The combined loss of Squint and Cyclops leads to complete loss of endoderm and a loss of all mesoderm except for a few somites in the tail.[46] This phenotype is copied by the maternal and zygotic loss of the Nodal co-receptor One-eyed pinhead or the overexpression of Nodal antagonists, such as Antivin/Lefty-1.[50,51] Despite the lack of mesoderm in Nodal mutants these embryos possess a neuraxis with distinct anterior and posterior identities.[46,52] Hence, at least two key elements of Organizer activity, neural induction and neural AP patterning, are manifested in the absence of the Nodal-derived Organizer tissue. In contrast, ventralized embryos generated by removal of the vegetal yolk cell region lack not only those tissues absent in the Nodal mutants, but also neurectoderm, suggesting that other signals, such as an FGF or another unidentified signal, induce and pattern the neurectoderm.[53,54]

The differentiation of mesodermal tissue in response to Nodal signaling is complicated by the presence of mesoderm inducers of the bone morphogenic protein (BMPs) class. A variety of BMPs have been shown to induce mesoderm of a ventral/posterior character. In addition, when overexpressed, BMPs suppress the formation of dorsal mesoderm. Given that several BMPs are expressed in the lateral and ventral margin it is reasonable to conclude that the BMPs may normally play an antagonistic role favoring formation of ventral or lateral mesodermal fates over the most dorsal. Indeed secreted inhibitors of BMPs are among the earliest dorsal-specific genes expressed. It is not clear, however, whether BMPs and their antagonists play a definitive role in the establishment of Organizer tissue; this role may be more pertinent to Organizer function, with other types of antagonist modulating dorsal specification.

While activation of the canonical Wnt-signaling pathway will specify dorsal identity during cleavage stages, zygotic activation of the pathway can suppress Organizer formation. Again, as with the BMPs, it is not clear whether Wnts and their antagonists have a definitive role in the establishment of Organizer tissue, though they clearly are important in Organizer function.[5]

In summary, the earliest post-fertilization events establish a gradient of activated, nuclear-localized β-catenin whose peak activity predicts the future dorsal side of the embryo. Independently, vegetal signals specify marginal, mesendodermal fates. The coincidence of high levels of activated β-catenin with vegetal-derived signals serve to specify the Organizer, as distinct from ventrolateral mesendoderm. At the dorsal side high levels of Nodal activity are sufficient to specify dorsal Organizer fate. Lateral and ventral Nodals are critical for mesoderm formation. Parallels are readily drawn between zebrafish and amphibian dorsal specification and mesoderm induction (Fig. 1).

1.3. Structure and Patterning of the Organizer

Soon after the work of Spemann and Mangold, Oppenheimer demonstrated that the teleost embryonic shield is the equivalent of the amphibian Organizer. This was confirmed in more recent zebrafish studies.[7,8] Importantly, the early studies were extended by micro-dissection of Organizer tissue, which demonstrated that the Organizer has separable head and trunk/tail Organizer activities.[7,55] The zebrafish embryonic shield consists of a superficial epiblast layer and a deeper hypoblast layer sitting on the yolk cell, both covered with the tight-epithelial EVL. Donor tissue dissected to enrich for deeper layer cells was often able to induce second axes possessing anterior structures but completely lacking posterior structures, while superficial layer donor tissue was often found to induce axes consisting only of posterior structures. When the two layers are transplanted together a complete second axis is induced in the majority of experiments.[7]

Expression patterns of dorsal-specific genes complement the experimental embryology. By the time the morphological shield is apparent, expression of the homeobox genes *goosecoid* (*gsc*) and *floating head* (*flh*) is confined to the deep and superficial layers respectively. The regions fated to become prechordal plate and notochord are distinguishable through the expression of *gsc* and *flh* in the respectively

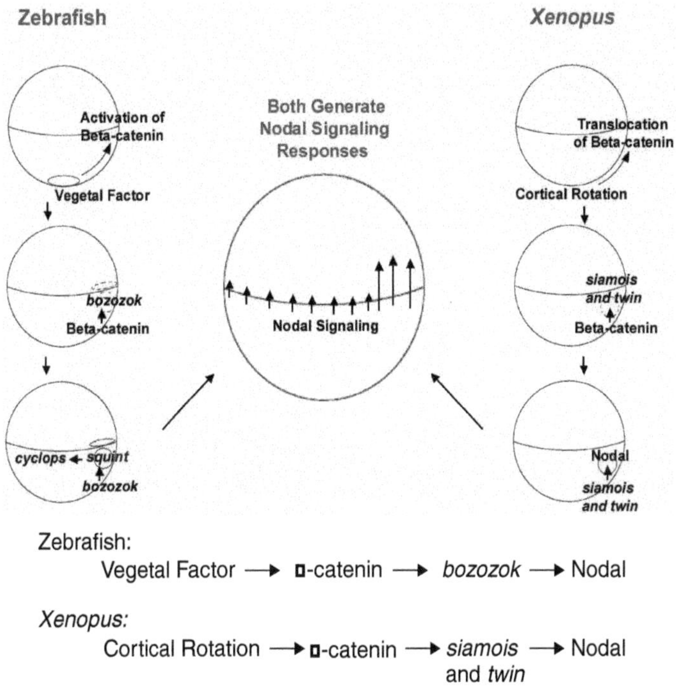

Fig. 1 Early development in fish and frogs. A comparison of events underlying early establishment of the dorsal-ventral axis in *Xenopus* and zebrafish is shown. Vegetal factors in zebrafish or cortical rotation in *Xenopus* result in the activation of β-catenin at the dorsal side, which will constitute the YSL/Nieuwkoop center. Nieuwkoop center activity leads to expression of *boz* in the zebrafish and *siamois* and *twin* in frog, which are thought to act to amplify the maternal signal and results in the induction of Nodal expression. Nodal signaling then acts to pattern the developing mesoderm, which include the developing Organizer.

fated regions.[56–58] Prior to the formation of the embryonic shield the dorsal region fated to form prechordal plate resides close to the blastoderm margin and expresses *gsc*, whereas the notochord progenitors are situated further from the margin and express *flh*.[56,59] Studies on the induction of both *gsc* and *flh* in the Organizer have shown that the differential activity of Nodals is necessary for the correct patterning of the Organizer AP axis.[56] Overexpression of *sqt* and *cyc* at different doses induces *flh* at low doses and both *flh* and *gsc* at higher doses, showing that Nodal signaling is vital for the patterning of the Organizer

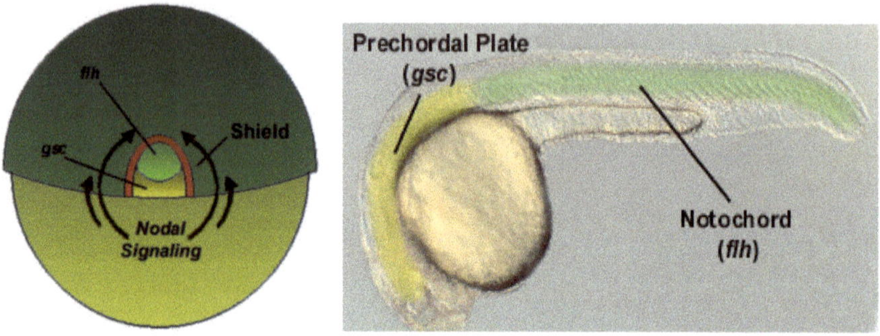

Fig. 2 Organizer structure and fate. Left Panel: Nodal signals pattern the Organizer (shield) at shield stage to form two distinct types of tissue. The highest levels of Nodal signaling give rise to the deep prechordal plate, *gsc* expressing, domain; while lower levels of Nodal give rise to the superficial notochord domain expressing *flh*. Right Panel: In a 24-hour embryo, the prechordal plate and notochord are highlighted to show the fates of the deep and superficial shield regions. The deep, *gsc*, cells in yellow give rise to the prechordal plate; and the superficial, *flh*, cells in green give rise to the notochord.

before gastrulation (Fig. 2). This is consistent with the fact that both *sqt* and *cyc* are expressed in the most marginal dorsal blastomeres where they would expose the marginal, *gsc* region of the shield, to higher levels of Nodal signaling.

1.4. Organizer Activities

The defining inductive properties of the Organizer are understood primarily in the context of grafting experiments (reviewed in Harland and Gerhart.[5]) In such assays, Organizer tissue induces the formation of neural tissue from tissue that would otherwise form non-neural ectoderm and patterns adjacent mesoderm imparting a more dorsal character. One successful approach has been to screen cDNA libraries to identify proteins able to induce dorsal structures in *Xenopus laevis* overexpression assays. Several groups have taken such an approach using cDNAs from *bona fide* Organizer tissues or from embryos substantially dorsalized by treatment with Li^+, a treatment that leads to activation of β-catenin. Many genes thus identified are specifically expressed within Organizer and have demonstrated roles in the patterning.[21,37,60,61] Among the most abundant type of molecules identified in such screens have been secreted antagonists of canonical signaling molecules. Noggin, Chordin and Follistatin, for example, antagonize BMP activity thus preventing ventralization and promoting the expression of more dorsal mesoderm and neural fates.[62–64] Similarly, several antagonists of Wnt signaling have been implicated in the control of DV patterning of mesoderm, AP patterning of the ectoderm or both. This growing list of molecules includes Dickkopf (Dkk1) and secreted forms of Fizzled receptors.

Complementing the overexpression approaches, genetic screens in zebrafish have yielded several genes underlying the Organizer's inductive activities. The mutants *swirl*/BMP2b, *snailhouse*/BMP7 and *somitabun*/Smad5 all encode components of the BMP-signaling pathway and result in substantially dorsalized embryos.[65–67] Recently, the zebrafish locus *ogon* was found to encode Sizzled. Similar to activities reported for the *Xenopus* Sizzled, zebrafish Sizzled was found not to inhibit Wnt8 activity but instead found to modulate BMP signaling, in a chordin-dependent fashion.[68–70] Thus the emerging model of Organizer activity is one in which secreted factors that antagonize BMP and Wnt signaling establish a DV gradient within the mesoderm specifying different fates at different threshold activity levels.[5,71] While such a simple model is attractive, it doesn't fit several observations concerning the specification of fates. For example, specification of what is considered to be the most dorsal mesoderm fate, trunk chordamesoderm, is relatively unaffected by increased or decreased levels of BMP signaling seen in the host of zebrafish mutants affecting BMP signaling. Thus is seems that BMPs and zygotic Wnts act in a complicated and not yet fully understood mechanism, to pattern the established mesendoderm (Fig. 3).

Other constraints on the timing and nature of Organizer activity arise from experimental embryology. Direct ablation of Organizer tissue has been achieved both genetically as seen in *boz* mutant embryos as well as surgically.[7,29,72] In either case, despite the lack of Organizer-derived tissue, embryos develop with an essentially complete AP axis,

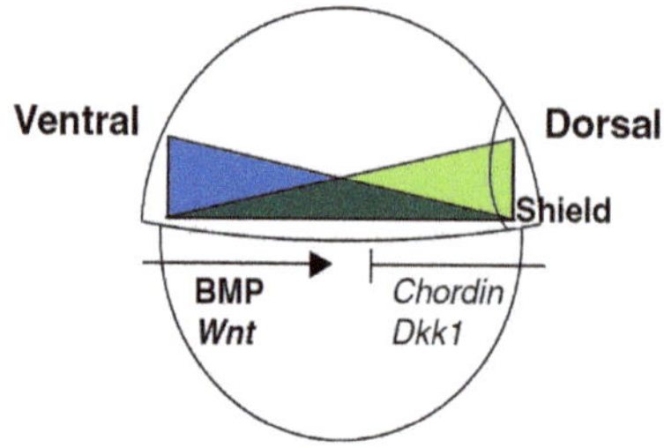

Fig. 3 Antagonistic dorsal and ventral activities. BMP and Wnt signals from the ventral side of the developing embryo are antagonised by factors expressed within the shield region. Thus a gradient of signals is created, allowing varying fates to be specified in an activity-level dependent fashion. This model, though appealing in its simplicity, is unlikely to be complete and further work will undoubtedly reveal more complex roles for BMP, Wnts and their antagonists in DV patterning.

i.e. there is a head with an eye, a spinal cord and trunk and tail somites. Although some embryos lack anterior-most neural tissue, neural induction and patterning clearly occurs and somites are formed indicating that the mesoderm has been patterned. Yet the removed Organizer tissue is fully capable of patterning a secondary axis in hosts. Thus, either the Organizer, as defined by transplantation assays, is only transiently required to induce surrounding tissues, or alternatively, the zebrafish Organizer is a dynamic, possibly regenerative entity as Hensen's node seems to be in chick.[73–75]

After acting to establish the initial body pattern, the Organizer differentiates and develops to form the axial mesoderm. In the anterior, Organizer tissue forms the prechordal plate and hatching gland, while in the posterior, Organizer tissue produces chordamesoderm, which differentiates to become notochord. Identification of the mutant *flh* provided the first real insights into chordamesoderm specification. This mutation was isolated from the background of pet store zebrafish stocks and was found to encode the zebrafish homolog of the *Xenopus Xnot* gene.[58] In these mutants notochord does not form, but other mesoderm derivatives, such as prechordal plate and somites are still produced. In *flh* mutants, tissue that would develop to chordamesoderm is mis-specified to become somite and tissues dependent on notochord signaling, such as hypochord and floorplate, largely fail to form.[76] In the control of chordamesoderm specification *flh* was found to interact genetically with another locus *spt*, which we have discussed previously in another context. In the absence of *spt* gene product, somitic tissue normally destined for the trunk is mis-localised to the tail,[77] a function somewhat distinct from the known function of the closest *Xenopus* homolog.[42–44] The interaction between *flh* and *spt* was identified in double mutant embryos. While *flh* mutants lack notochord, *flh/spt* double mutants possess trunk notochord. Thus the *spt* mutation is able to suppress the *flh* mutation, suggesting that *flh* acts in midline development to promote chordamesoderm and notochord fate by suppressing the induction of somatic fates in this region by *spt*.[78]

2. Notochord

2.1. Introduction

The major tissue derivative of the Organizer is the notochord, which is the defining structure of the phylum chordata. The notochord serves two main roles in vertebrate development. First, as a mechanical structure the notochord is the main embryonic skeletal element of lower vertebrates, important for locomotion. Second, the notochord is essential for normal development of all vertebrates, providing signals that pattern adjacent tissues such as the gut, somites and spinal cord. Notochord development in zebrafish is relatively simple, as the notochord comprises a single cell type that undergoes a characteristic series of differentiation events, marked by dramatic morphological changes. Our understanding of notochord differentiation has been significantly informed by studies of mutant zebrafish. Phenotypically, the notochord differentiation process can be broken into two discrete transitions. In the first step, as we have discussed, chordamesoderm is specified as a specialized midline mesoderm, involving *boz* and *flh* among other loci. The second step is the transition from chordamesoderm to notochord, which we term "notochord differentiation".

2.2. Differentiation of the Notochord

There are two morphological features that mark the differentiation of the notochord. First, the cells of the chordamesoderm develop a thick basement membrane, which forms a sheath surrounding the notochord. Second, coupled to basement membrane formation, each cell acquires a large vacuole that exerts turgor pressure against the sheath. Failure to inflate the vacuoles properly leads to a substantially shortened embryo that is easily scored. For this reason many mutations affecting notochord differentiation were identified in the 1996 zebrafish mutagenesis screens.[79,80] Efforts to identify the mutated genes have lead to some understanding of the role the basement membrane in notochord differentiation. For example the *sleepy* (*sly*) and *grumpy* (*gup*) loci were found to encode the Laminin $\gamma 1$ and $\beta 1$ chains respectively.[81] Laminins are well-known essential components of basement membranes, where a given laminin isoform is a heterotrimer of an α, β and γ chain. In humans there are five α chains, four β chains and three γ chains (reviewed in Colognato and Yurchenco[82]). In zebrafish notochord, the absence of either the Laminin $\gamma 1$ or $\beta 1$ chain leads to a complete failure to form a basement membrane surrounding the notochord. Consequently, mutant notochords fail to differentiate as marked by both a failure of vacuole inflation and the persistent expression of early marker genes such as *echidna hedgehog*, *sonic hedgehog*, *α1-collagen Type II* and *no tail.*[80] Transplantation studies showed that the missing Laminin chain could be supplied either by the notochord or by non-notochordal sources to rescue notochord differentiation. Furthermore, these studies suggest that either Laminin or another basement membrane-dependent signal is the notochord differentiation signal. While pursuit of Laminin receptors has not yielded the notochord differentiation signal, it has led to the development of a zebrafish model of muscular dystrophy.[83]

Three other notochord differentiation mutants, *dopey* (*dop*), *happy* (*hap*) and *sneezy* (*sny*) were grouped because, in addition to the notochord defects, much later in development embryos undergo widespread apoptosis. As with the Laminin mutants, *dop, hap* and *sny* notochords fail to form normal vacuoles, persistently express early markers, such as *sonic hedgehog* and have a disrupted sheath. Recent work indicates that these loci encode essential components of intracellular vesicular transport machinery, important for the secretion of basement membrane components and vacuole formation.[84] Thus, as with the Laminin mutants, a link is drawn between basement membrane formation and notochord differentiation.

In the last class of notochord differentiation mutations are *doc* and *no tail* (*ntl*), which both lead to the persistent expression of some early markers and failure of vacuole formation, but possess normal basement membranes. Transplantation experiments show that the notochord differentiation defect is cell-autonomous for both *ntl* and *doc*.[79,85] Of these two loci, *doc* has the most notochord-specific defects. While *ntl* mutants fail to generate tails, in the trunk region they are phenotypically very similar to *doc* mutants, where the only observed defects are in notochord differentiation coupled to a failure to signal surrounding tissues. A detailed understanding of the upstream factors controlling *doc* and *ntl* should elucidate the nature of the notochord differentiation signal. An understanding of their downstream effectors will tell us how differentiation is manifested.

In *ntl* mutants the chordamesoderm develops normally but arrests development prior to notochord differentiation. In contrast to *flh* mutant, in which chordamesoderm is converted to somitic mesoderm, the fate of *ntl* mutant chordamesoderm is not clear. Some cells may die by apoptosis but other cells end up in the spinal cord and have been interpreted to form the medial floorplate, although some of these cells inappropriately express *ntl* mRNA at stages when ntl expression is normally extinguished.[80] What is clear, however, is that *ntl* encodes a zebrafish homolog of the mouse T brachyury, a T-box transcription factor.[86,87] Moreover, there is good evidence that *ntl* expression, like its counterpart in *Xenopus*, Xbra, is substantially controlled by FGF signaling.[44,88] During normal development, *ntl* is first expressed by marginal cells in late blastulae and early gastrulae, and then in internalized deep cells and is maintained only in chordamesoderm. Double mutant studies of *ntl*, *flh* and *cyc* have helped establish the relationship between these genes in control of midline identities. Despite the dramatic loss of floorplate cells in *cyc* mutant embryos, double mutant *ntl/cyc* embryos display an apparent rescue of floorplate. Similarly, the majority of *ntl/flh* double mutants were found to be most similar to *ntl* single mutants with a rescue of midline tissue not found in *flh* single mutants.[89] In the case of *ntl/flh* double mutants, since no ambiguous marker of floorplate was used in the analysis, it is formally possible that undifferentiated chordamesoderm, persistently expressing early marker genes, has infiltrated the ventral neural tube. It is clear, however, that midline tissue not present in *flh* mutant embryos is rescued in the *ntl/flh* double mutants. While *ntl* single mutants suggest a role for *ntl* in notochord differentiation the double mutant results show that *ntl* also has a role in chordamesoderm specification. Considering that rescue of midline mesoderm also occurs in *spt/flh* double mutants and that *ntl/spt* double mutants have no trunk mesoderm, it appears as though ntl has some function partially overlapping with other T-box genes.[78,90] One hypothesis is that *ntl*, *spt* and *flh* are controlling the choice between medial floorplate and chordamesoderm fate, as seen with the *ntl/flh* double mutants, and between medial and lateral fate, as seen with the *spt/flh* double mutants, and the three competing activities are balanced through feedback loops, possibly involving Nodal or FGF signaling, to ensure the appropriate amount of each tissue is specified.[91–93]

2.3. Patterning of Surrounding Tissues by the Notochord

The most studied signaling role of the notochord is in the patterning of the neural tube. The neural tube develops distinct cell types at specific locations along its dorsal-ventral axis. The notochord, situated just ventral to the neural tube, was thus considered a strong candidate for a source of patterning signals. Embryological work performed with chick demonstrated that the notochord is able to coordinate correct neural tube formation, and that the absence of notochord results in abnormal formation of the neural tube.[94,95] Ablation of the notochord and the floorplate, which is itself dependent on notochord-derived signals, prevents the differentiation of motor neurons and other ventral neuronal cell types in chicken as well as zebrafish.[7,96,97] Moreover, grafting either the notochord or the floorplate to the dorsal midline of the neural tube suppresses dorsal neural tube fates and promotes the ectopic formation of ventral neuronal cell types.[97,98] Similar studies demonstrated that a diffusible signal, derived first from the notochord and later from the floorplate, patterns the neural tube.[99]

The diffusible signal involved in neural tube patterning was identified as Sonic hedgehog (Shh).[100,101] In zebrafish there are three hedgehogs expressed in the midline: *echidna hedgehog* in the chordamesoderm, *tiggywinkle hedgehog* in the floorplate and *sonic hedgehog* in both.[102–104] Shh is essential for correct patterning of the neural tube and also formation of the floorplate, since blocking Shh function with antibodies prevents floorplate formation and causes incorrect patterning of the neural tube.[105] Genetically null mice lacking Shh also fail to form floorplate and proper neural tube.[106] It was observed, however, that ectopic Shh alone cannot induce formation of the floorplate.[107] Explants of chick neural plate treated *in vitro* with a combination of Shh and Chordin, a BMP antagonist normally expressed by the notochord, developed floorplates suggesting a mechanism of floorplate induction whereby the notochord produces Chordin to inhibit the dorsally-derived BMPs, generating a permissive environment in which Shh can induce floorplate. The prevailing view holds that the combination of Shh produced ventrally and BMPs produced dorsally establish opposing gradients that provide DV position information in neural tube. Shh is initially expressed by the notochord and then by the floorplate, with its expression becoming confined to the floorplate later in development. This establishes a gradient of Shh that promotes specification of ventral cell types while repressing dorsal identities in the ventral neural tube (Fig. 4).

Both muscle fiber type and the characteristic chevron shape of zebrafish somites is controlled in part by notochord-derived hedgehog signals. Normally adaxial cells, which form immediately adjacent to the chordamesoderm and express *myoD*, will migrate to the outer surface of the developing muscle and differentiate to form slow-twitch muscle fibers.[108] A few adaxial cells eventually express Engrailed and become

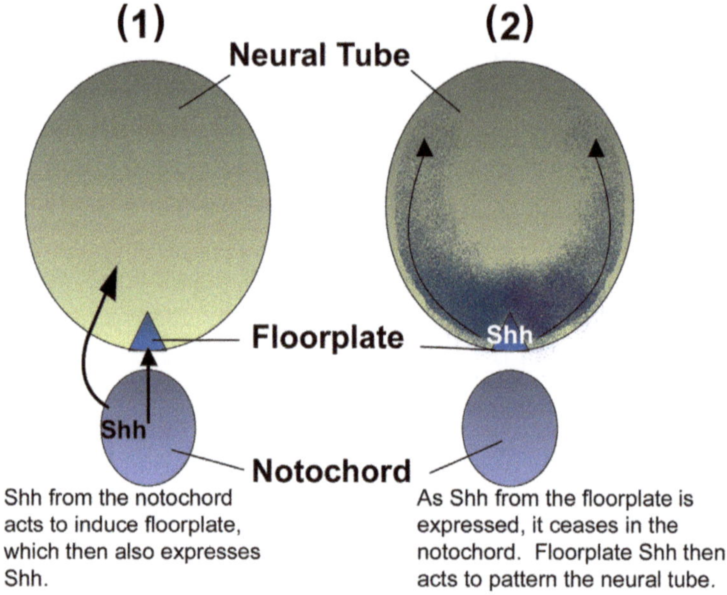

Fig. 4 Patterning of the ventral neural tube. (1) Hedgehog proteins, such as Shh, produced by the notochord act early to pattern to induce floorplate. (2) Once the floor plate is formed, Shh, as well as Echidna hedgehog expression is extinguished in the notochord but Shh and Tiggywinkle hedgehog continue to be expressed in the floorplate. This generates a gradient of Hedgehog activity in the neural tube, which specifies ventral fates while suppressing dorsal fates.

the muscle pioneer cells that define the horizontal myoseptum and the chevron shape of the somite. When ligand-activated hedgehog signaling is abolished as in *slow-muscles-omitted* (*smu*) mutants, which lack the hedgehog signal transduction component Smoothened, slow-twitch muscle fibers as well as the Engrailed-positive muscle pioneers fail to form.[109] Similarly, mutants lacking Shh (*sonic you*) or Gli2 (*you-too*), a transcription factor that mediates hedgehog signaling, fail to form muscle pioneers and slow-twitch muscle fibers.[110–113] In notochord differentiation mutants, the somites take on an abnormal 'U' shape due to their failure to form horizontal myosepta and show compromised Engrailed expression despite the persistent expression of the midline hedgehogs in undifferentiated notochord.[79,80] This most likely results from a diminished capacity to transmit the signal from the notochord to the forming somites.[81]

A role has also been demonstrated for the notochord in the development of the heart and vasculature. Removal of the anterior region of the notochord has been shown to cause an increase in the size of the expression domain of *Nkx2.5*, a marker for the region fated to become the heart, indicating a role for notochord in suppressing heart formation thereby defining the posterior limit of the heart field.[114] Several lines of evidence demonstrate the role for the notochord in formation of the major blood vessels of the trunk. In both *ntl* and *flh* mutants the dorsal aorta (DA) fails to form.[115,116] The DA and posterior cardinal vein (PCV) form in a highly conserved fashion in vertebrates, with the DA forming just ventral to the notochord and the PCV forming dorsal to the trunk endoderm. Vascular endothelial growth factor (VegF) is vital for the correct formation of these vessels and is thought to be sufficient for arterial specification. Overexpression of VegF mRNA in zebrafish embryos leads to ectopic expression of *ephrin-b2a*, an arterial marker, in tissue that would otherwise be venous.[117] Recent work also indicates a role for Shh in blood vessel formation. Mutants deficient in Shh were found to lack *ephrin-b2a* in the vasculature. Interestingly VegF overexpression was sufficient to rescue arterial differentiation in the absence of Shh. In contrast, VegF was unable to rescue arterial defects in notch-signaling mutants. Taken together these data suggest a model of blood vessel formation whereby Shh emitted from the notochord induces the expression of VegF in the somites, with VegF then acting in the DA in a Notch-signaling dependent fashion to induce proper arterial development.

The notochord has instructive roles in development of both the pancreas and the hypochord. By mechanically separating notochord from endoderm, expression of markers normally associated with pancreatic development are extinguished.[118] In culture, presumptive pancreatic endoderm, unable to express pancreatic markers on its own, will express them in the presence of notochord. When cultured with other endoderm, however, pancreatic markers are not induced, suggesting that the notochord is only able to induce pancreatic development permissively in preconditioned endoderm.

The hypochord is a transient rod-like structure situated immediately ventral to the notochord. The hypochord also expresses high levels *VegF* and so may well be an important source of signals in the development of the vasculature.[119] It has been noted that removal of the notochord during early neurulation stages results in a failure of hypochord formation, whereas removal of the notochord later in development does not.[120] Thus notochord-dependent hypochord induction is complete by late neurula stages. Chick transplantation studies in which notochord is grafted adjacent the endoderm, have demonstrated that the ability of endoderm to form hypochord is restricted to the dorsal-most region of endoderm. Moreover Notch signaling is essential for hypochord development.[121] Although specific roles have not been assigned, candidate notochord-derived signals controlling hypochord induction include Shh, Activin-βB and FGF2.[122]

To summarize, the signaling activities of the notochord include: patterning of ectoderm; specification of DV pattern in the neural tube;

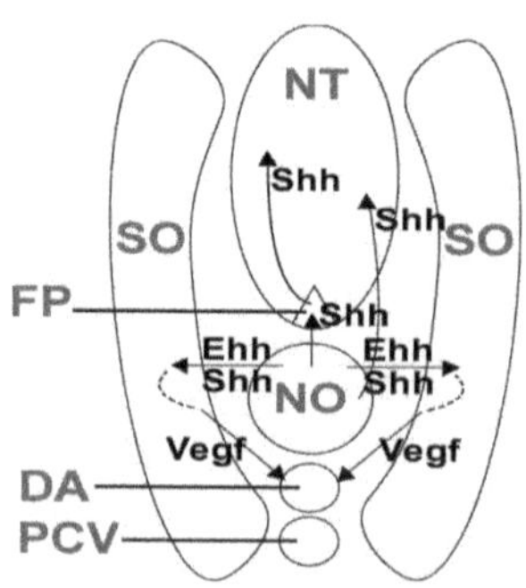

Fig. 5 Overview of the patterning activities of the notochord. Hedgehogs from the notochord induce floorplate and act in early patterning of the neural tube. Once floorplate is induced, *hedgehog* expression is extinguished in the notochord and Hedgehogs produced by the floorplate continue to pattern the neural tube. Sonic hedgehog and Ehh (Echidna hedgehog) are also involved in patterning the somites and signaling to the somites is able to induce VegF in the somites which then acts to pattern the dorsal aorta (NT, neural tube; SO, somites; FP, floorplate; NO, notochord; DA, dorsal aorta; PCV, pericardinal vein).

induction of somite, vascular and cardiac mesodermal tissues; and patterning of the pancreas and hypochord endodermal tissues (Fig. 5).

2.4. Mechanics of Notochord Structure and Development

The notochord plays an important mechanical role in the development of early embryos, especially in lower vertebrates in which it acts as the major skeletal element important for locomotion. The notochord consists of a stack of single cells, each of which acquires a large vacuole, surrounded by a thick sheath of basement membrane. This sheath serves as a physical boundary to limit and control the length and shape of

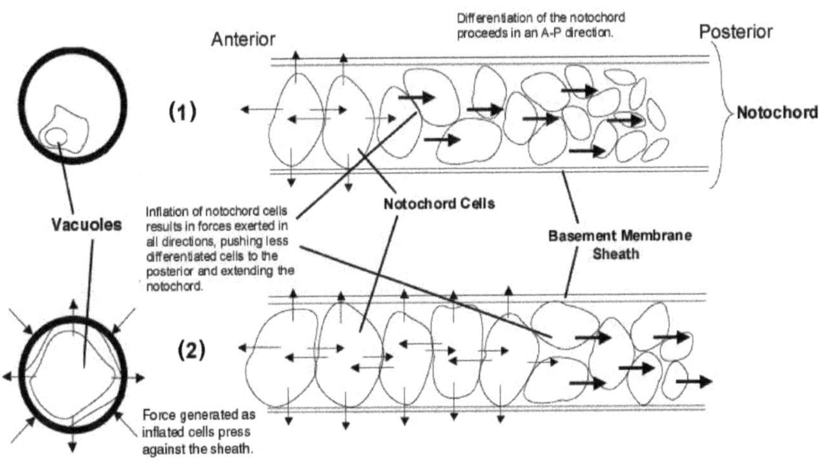

Fig. 6 Mechanical aspects of notochord development. (1) Early in notochord development, the vacuoles of the cells have not inflated and no pressure is exerted on the sheath. As such, the notochord lacks rigidity and provides no support. Cells at the anterior of the notochord inflate and differentiate first, pushing the cells further posterior. (2) As differentiation proceeds, the vacuoles inflate and cells push against the sheath generating hydrostatic force, and the notochord is able to provide support to the developing embryo. As more cells inflate there is a continued movement of cells to the posterior. This elongation, combined with the generated stiffness resulting from inflation causes elongation of the embryo. Once cells have fully inflated, the forces between the cells and sheath are equal and inflation ceases.

the notochord. Turgor pressure, generated by the vacuoles, is constrained by the fibrous sheath, acting to strengthen and stiffen the notochord (Fig. 6). *In vitro* experiments with *Xenopus* notochord show that notochord vacuoles will respond to environmental osmolarity, causing the notochord to lengthen and stiffen under physiological osmolarities, and to become flaccid under conditions of higher osmolarity.[123] The lengthening and stiffening of notochord was not observed at stages prior to sheath formation.

Notochord cells differentiate in an anterior to posterior wave. Consequently, the large change in cell volume of anterior cells pushes more posterior cells toward the tail, thus extending the notochord (Fig. 6). This extension is driven by inflation of the vacuoles constrained by the sheath. This stiffens the notochord, preventing buckling caused by movements of the embryo. Notochord cells are effectively pushed to the posterior, within the tube formed by the sheath, since strong mechanical connections, in the form of hemidesnosomes, between notochord cells and the sheath are not formed until notochord cells are mature.[124]

Fibers of the sheath are arranged carefully and deliberately. Electron micrographs of transverse sections through the notochord show that the fibers are arranged to run both parallel and perpendicular to the notochord (Fig. 7).[81] Studies of the fiber angle in the notochords of *Xenopus* embryos demonstrated that the average fiber angle in the sheath is 54°. This angle means that the sheath is able to resist longitudinal and circumferential stress equally, so that if the shape of the notochord were determined solely by the inflation of the notochord cells, then the length/diameter ratio would always remain constant.[123]

The notochord is constrained in another way that limits the type of tail movements an early embryo can make. If the structure of the notochord consisted only of a thick sheath filled with vacuolated cells it would be able to bend in any direction. The notochord, however, is mechanically coupled to two other structures that serve as restraints. Dorsal to the notochord is the floorplate, which expresses many of the same extracellular proteins as the notochord, such as $\alpha 1$ Collagen Type II.[125] Ventrally, the hypochord expresses similar proteins. These

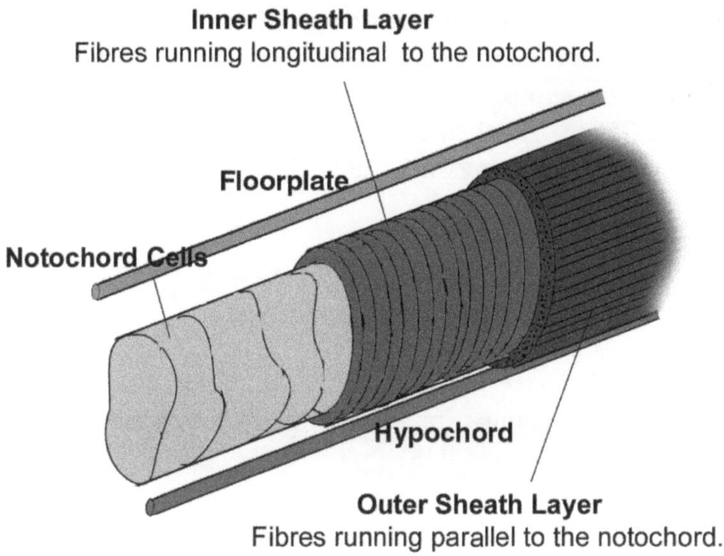

Fig. 7 Notochord structure. The basement membrane sheath that surrounds the notochord is vital for the generation of hydrostatic force, which enables the notochord to act in support and elongation. To maximize the strength of the sheath it is formed of two layers of longitudinal and parallel to the AP axis of the notochord.

Situated dorsally and ventrally to the notochord respectively, the floorplate and hypochord are able to resist compression and tension and therefore provide support on both the dorsal and ventral side. This then limits the movement of the notochord to only the lateral plane and as such, focuses any movement from the somites into this plane, resulting in the swimming movements of the embryo.

two structures serve as cables running along the top and bottom of the notochord limiting the notochord to movement in the horizontal plane (Fig. 7). Thus any force exerted on the notochord by surrounding muscle will only result in a left-right movement of the tail, consistent with the requirements for forward locomotion.

In summary, cells of the notochord act, by inflation of their large characteristic vacuoles, to generate a force to support the embryo. The cells enlarge and exert pressure on the thick sheath of basement membrane that surrounds the notochord generating a hydrostatic force. This inflation also acts to elongate the embryo, since an absence of this inflation leads to drastically shortened embryos. The inflation, which begins at the anterior end of the notochord and proceeds towards the posterior, effectively pushes the posterior cells towards the posterior as they expand. These cells then expand and exert the same force on their neighboring cells, resulting in a general extension of the embryonic AP axis.

Acknowledgments

Stuart Smith, Branko Latinkic and Mike Sargent made helpful suggestions for the manuscript. KAT and DLS were supported by the Medical Research Council, UK.

References

1. Kimmel CB, Ballard WW, Kimmel SR, Ullmann B, and Schilling TF (1995). Stages of embryonic development of the zebrafish. *Dev. Dyn.* **203**: 253–310.
2. Kimmel CB and Law RD (1985a). Cell lineage of zebrafish blastomeres. I. Cleavage pattern and cytoplasmic bridges between cells. *Dev. Biol.* **108**: 78–85.
3. Kimmel CB and Law RD (1985b). Cell lineage of zebrafish blastomeres. II. Formation of the yolk syncytial layer. *Dev. Biol.* **108**: 86–93.
4. Kimmel CB and Law RD (1985c). Cell lineage of zebrafish blastomeres. III. Clonal analyses of the blastula and gastrula stages. *Dev. Biol.* **108**: 94–101.
5. Harland R and Gerhart J (1997). Formation and function of Spemann's organizer. *Annu Rev Cell Dev. Biol.* **13**: 611–667.
6. Oppenheimer JM (1936). Transplantation experiments on developing telelosts (*Fundulus* and *Perca*). *J Exp Zool* **72**: 409–437.
7. Saude L, Woolley K, Martin P, Driever W and Stemple DL (2000). Axis-inducing activities and cell fates of the zebrafish organizer. *Development* **127**: 3407–3417.
8. Shih J and Fraser SE (1996a). Characterizing the zebrafish organizer: microsurgical analysis at the early-shield stage. *Development* **122**: 1313–1322.

9. Beddington RS (1994). Induction of a second neural axis by the mouse node. *Development* **120**: 613–620.
10. Waddington CH (1932). Experiments on the development of the chick and the duck embryo cultivated *in vitro. Proc. Trans. R. Soc. Lond. (B)* **211**: 179–230.
11. Gerhart J, Danilchik M, Doniach T, Roberts S, Rowning B and Stewart R (1989). Cortical rotation of the Xenopus egg: consequences for the anteroposterior pattern of embryonic dorsal development. *Development* **107** (*Suppl*): 37–51.
12. Nieuwkoop PD (1973). The organization center of the amphibian embryo: its origin, spatial organization, and morphogenetic action. *Adv. Morphog.* **10**: 1–39.
13. McMahon AP and Moon RT (1989). Ectopic expression of the proto-oncogene int-1 in Xenopus embryos leads to duplication of the embryonic axis. *Cell* **58**: 1075–1084.
14. Peifer M, Rauskolb C, Williams M, Riggleman B and Wieschaus E (1991). The segment polarity gene armadillo interacts with the wingless signaling pathway in both embryonic and adult pattern formation. *Development* **111**: 1029–1043.
15. Peifer M and Wieschaus E (1990). The segment polarity gene armadillo encodes a functionally modular protein that is the *Drosophila* homolog of human plakoglobin. *Cell* **63**: 1167–1176.
16. Riggleman B, Wieschaus E and Schedl P (1989). Molecular analysis of the armadillo locus: uniformly distributed transcripts and a protein with novel internal repeats are associated with a *Drosophila* segment polarity gene. *Genes. Dev.* **3**: 96–113.
17. McCrea PD, Brieher WM and Gumbiner BM (1993). Induction of a secondary body axis in Xenopus by antibodies to beta-catenin. *J. Cell Biol.* **123**: 477–484.
18. Funayama N, Fagotto F, McCrea P and Gumbiner BM (1995). Embryonic axis induction by the armadillo repeat domain of beta-catenin: evidence for intracellular signaling. *J. Cell Biol.* **128**: 959–968.
19. Kelly GM, Erezyilmaz DF and Moon RT (1995). Induction of a secondary embryonic axis in zebrafish occurs following the overexpression of beta-catenin. *Mech. Dev.* **53**: 261–273.
20. Laurent MN, Blitz IL, Hashimoto C, Rothbacher U and Cho KW (1997). The Xenopus homeobox gene twin mediates Wnt induction of

goosecoid in establishment of Spemann's organizer. *Development* **124**: 4905–4916.

21. Lemaire P, Garrett N and Gurdon JB (1995). Expression cloning of Siamois, a Xenopus homeobox gene expressed in dorsal-vegetal cells of blastulae and able to induce a complete secondary axis. *Cell* **81**: 85–94.
22. Moon RT and Kimelman D (1998). From cortical rotation to organizer gene expression: toward a molecular explanation of axis specification in Xenopus. *Bioessays* **20**: 536–545.
23. Nelson RW and Gumbiner BM (1998). Beta-catenin directly induces expression of the Siamois gene, and can initiate signaling indirectly via a membrane-tethered form. *Ann. N. Y. Acad. Sci.* **857**: 86–98.
24. Schneider S, Steinbeisser H, Warga RM and Hausen P (1996). Beta-catenin translocation into nuclei demarcates the dorsalizing centers in frog and fish embryos. *Mech. Dev.* **57**: 191–198.
25. Ober EA and Schulte–Merker S (1999). Signals from the yolk cell induce mesoderm, neuroectoderm, the trunk organizer, and the notochord in zebrafish. *Dev. Biol.* **215**: 167–181.
26. Jesuthasan S and Stahle U (1997). Dynamic microtubules and specification of the zebrafish embryonic axis. *Curr. Biol.* **7**: 31–42.
27. Kelly C, Chin AJ, Leatherman JL, Kozlowski DJ and Weinberg ES (2000). Maternally controlled (beta)-catenin-mediated signaling is required for organizer formation in the zebrafish. *Development* **127**: 3899–3911.
28. Chen S and Kimelman D (2000). The role of the yolk syncytial layer in germ layer patterning in zebrafish. *Development* **127**: 4681–4689.
29. Fekany K, Yamanaka Y, Leung T, Sirotkin HI, Topczewski J, Gates MA, Hibi M, Renucci A, Stemple D, Radbill A *et al.* (1999). The zebrafish bozozok locus encodes Dharma, a homeodomain protein essential for induction of gastrula organizer and dorsoanterior embryonic structures. *Development* **126**: 1427–1438.
30. Koos DS and Ho RK (1999). The nieuwkoid/dharma homeobox gene is essential for bmp2b repression in the zebrafish pregastrula. *Dev. Biol.* **215**: 190–207.
31. Koos DS and Ho RK (1998). The nieuwkoid gene characterizes and mediates a Nieuwkoop-center-like activity in the zebrafish. *Curr. Biol.* **8**: 1199–1206.
32. Shimizu T, Yamanaka Y, Ryu SL, Hashimoto H, Yabe T, Hirata T, Bae YK, Hibi M and Hirano T (2000). Cooperative roles of Bozozok/

Dharma and Nodal-related proteins in the formation of the dorsal organizer in zebrafish. *Mech. Dev.* **91**: 293–303.

33. Yamanaka Y, Mizuno T, Sasai Y, Kishi M, Takeda H, Kim CH, Hibi M, and Hirano T (1998). A novel homeobox gene, dharma, can induce the organizer in a non-cell-autonomous manner. *Genes. Dev.* **12**: 2345–2353.
34. Ryu SL, Fujii R, Yamanaka Y, Shimizu T, Yabe T, Hirata T, Hibi M and Hirano T (2001). Regulation of dharma/bozozok by the Wnt pathway. *Dev. Biol.* **231**: 397–409.
35. Renucci A, Lemarchandel V and Rosa F (1996). An activated form of type I serine/threonine kinase receptor TARAM-A reveals a specific signalling pathway involved in fish head organiser formation. *Development* **122**: 3735–3743.
36. Green JB and Smith JC (1990). Graded changes in dose of a Xenopus activin A homologue elicit stepwise transitions in embryonic cell fate. *Nature* **347**: 391–394.
37. Schier AF and Shen MM (2000). Nodal signalling in vertebrate development. *Nature* **403**: 385–389.
37. Smith JC, Price BM, Van Nimmen K and Huylebroeck D (1990). Identification of a potent Xenopus mesoderm-inducing factor as a homologue of activin A. *Nature* **345**: 729–731.
38. Horb ME and Thomsen GH (1997). A vegetally localized T-box transcription factor in Xenopus eggs specifies mesoderm and endoderm and is essential for embryonic mesoderm formation. *Development* **124**: 1689–1698.
39. Lustig KD, Kroll KL, Sun EE and Kirschner MW (1996). Expression cloning of a Xenopus T-related gene (Xombi) involved in mesodermal patterning and blastopore lip formation. *Development* **122**: 4001–4012.
40. Stennard F, Carnac G and Gurdon JB (1996). The Xenopus T-box gene, Antipodean, encodes a vegetally localised maternal mRNA and can trigger mesoderm formation. *Development* **122**: 4179–4188.
41. Zhang J and King ML (1996). Xenopus VegT RNA is localized to the vegetal cortex during oogenesis and encodes a novel T-box transcription factor involved in mesodermal patterning. *Development* **122**: 4119–4129.
42. Kofron M, Demel T, Xanthos J, Lohr J, Sun B, Sive H, Osada S, Wright C, Wylie C and Heasman J (1999). Mesoderm induction in Xenopus is a

zygotic event regulated by maternal VegT via TGFbeta growth factors. *Development* **126**: 5759–5770.

43. Xanthos JB, Kofron M, Wylie C and Heasman J (2001). Maternal VegT is the initiator of a molecular network specifying endoderm in *Xenopus laevis. Development* **128**: 167–180.
44. Griffin KJ, Amacher SL, Kimmel CB and Kimelman D (1998). Molecular identification of spadetail: regulation of zebrafish trunk and tail mesoderm formation by T-box genes. *Development* **125**: 3379–3388.
45. Conlon FL, Lyons KM, Takaesu N, Barth KS, Kispert A, Herrmann B, and Robertson EJ (1994). A primary requirement for nodal in the formation and maintenance of the primitive streak in the mouse. *Development* **120**: 1919–1928.
46. Feldman B, Gates MA, Egan ES, Dougan ST, Rennebeck G, Sirotkin HI, Schier AF and Talbot WS (1998). Zebrafish organizer development and germ-layer formation require nodal-related signals. *Nature* **395**: 181–185.
47. Rebagliati MR, Toyama R, Haffter P and Dawid IB (1998). Cyclops encodes a nodal-related factor involved in midline signaling. *Proc. Natl. Acad. Sci. USA* **95**: 9932–9937.
48. Sampath K, Rubinstein AL, Cheng AM, Liang JO, Fekany K, Solnica-Krezel L, Korzh V, Halpern ME and Wright CV (1998). Induction of the zebrafish ventral brain and floorplate requires cyclops/nodal signalling. *Nature* **395**: 185–189.
49. Zhou X, Sasaki H, Lowe L, Hogan BL and Kuehn MR (1993). Nodal is a novel TGF-beta-like gene expressed in the mouse node during gastrulation. *Nature* **361**: 543–547.
50. Gritsman K, Zhang J, Cheng S, Heckscher E, Talbot WS and Schier AF (1999). The EGF-CFC protein one-eyed pinhead is essential for nodal signaling. *Cell* **97**: 121–132.
51. Thisse C and Thisse B (1999). Antivin, a novel and divergent member of the TGFbeta superfamily, negatively regulates mesoderm induction. *Development* **126**: 229–240.
52. Feldman B, Dougan ST, Schier AF and Talbot WS (2000). Nodal-related signals establish mesendodermal fate and trunk neural identity in zebrafish. *Curr. Biol.* **10**: 531–534.
53. Reim G and Brand M (2002). Spiel-ohne-grenzen/pou2 mediates regional competence to respond to Fgf8 during zebrafish early neural development. *Development* **129**: 917–933.

54. Streit A, Berliner AJ, Papanayotou C, Sirulnik A and Stern CD (2000). Initiation of neural induction by FGF signalling before gastrulation. *Nature* **406**: 74–78.
55. Zoltewicz JS and Gerhart JC (1997). The Spemann organizer of *Xenopus* is patterned along its anteroposterior axis at the earliest gastrula stage. *Dev. Biol.* **192**: 482–491.
56. Gritsman K, Talbot WS and Schier AF (2000). Nodal signaling patterns the organizer. *Development* **127**: 921–932.
57. Stachel SE, Grunwald DJ and Myers PZ (1993). Lithium perturbation and goosecoid expression identify a dorsal specification pathway in the pregastrula zebrafish. *Development* **117**: 1261–1274.
58. Talbot WS, Trevarrow B, Halpern ME, Melby AE, Farr G, Postlethwait JH, Jowett T, Kimmel CB and Kimelman D (1995). A homeobox gene essential for zebrafish notochord development. *Nature* **378**: 150–157.
59. Melby AE, Warga RM and Kimmel CB (1996). Specification of cell fates at the dorsal margin of the zebrafish gastrula. *Development* **122**: 2225–2237.
60. Blumberg B, Wright CV, De Robertis EM and Cho KW (1991). Organizer-specific homeobox genes in *Xenopus laevis* embryos. *Science* **253**: 194–196.
61. Smith WC and Harland RM (1991). Injected Xwnt-8 RNA acts early in *Xenopus* embryos to promote formation of a vegetal dorsalizing center. *Cell* **67**: 753–765.
62. Hemmati–Brivanlou A, Kelly OG and Melton DA (1994). Follistatin, antagonist of activin, expressed in the Spemann organizer and displays direct neuralizing activity. *Cell* **77**: 283–295.
63. Piccolo S, Sasai Y, Lu B and De Robertis EM (1996). Dorsoventral patterning in *Xenopus*: inhibition of ventral signals by direct binding of chordin to BMP-4. *Cell* **86**: 589–598.
64. Zimmerman LB, De Jesus–Escobar JM and Harland RM (1996). The Spemann organizer signal noggin binds and inactivates bone morphogenetic protein 4. *Cell* **86**: 599–606.
65. Hild M, Dick A, Rauch GJ, Meier A, Bouwmeester T, Haffter P and Hammerschmidt M (1999). The smad5 mutation somitabun blocks Bmp2b signaling during early dorsoventral patterning of the zebrafish embryo. *Development* **126**: 2149–2159.
66. Kishimoto Y, Lee KH, Zon L, Hammerschmidt M and Schulte–Merker S (1997). The molecular nature of zebrafish swirl: BMP2 function is essential during early dorsoventral patterning. *Development* **124**: 4457–4466.

67. Schmid B, Furthauer M, Connors SA, Trout J, Thisse B, Thisse C and Mullins MC (2000). Equivalent genetic roles for bmp7/snailhouse and bmp2b/swirl in dorsoventral pattern formation. *Development* **127**: 957–967.
68. Collavin L and Kirschner MW (2003). The secreted Frizzled-related protein Sizzled functions as a negative feedback regulator of extreme ventral mesoderm. *Development* **130**: 805–816.
69. Salic AN, Kroll KL, Evans LM and Kirschner MW (1997). Sizzled: a secreted Xwnt8 antagonist expressed in the ventral marginal zone of *Xenopus* embryos. *Development* **124**: 4739–4748.
70. Yabe T, Shimizu T, Muraoka O, Bae YK, Hirata T, Nojima H, Kawakami A, Hirano T and Hibi M (2003). Ogon/Secreted Frizzled functions as a negative feedback regulator of Bmp signaling. *Development* **130**: 2705–2716.
71. De Robertis EM, Wessely O, Oelgeschlager M, Brizuela B, Pera E, Larrain J, Abreu J and Bachiller D (2001). Molecular mechanisms of cell-cell signaling by the Spemann–Mangold organizer. *Int. J. Dev. Biol.* **45**: 189–197.
72. Shih J and Fraser SE (1996b). Characterizing the zebrafish organizer: microsurgical analysis at the early-shield stage. *Development* **122**: 1313–1322.
73. Joubin K and Stern CD (1999). Molecular interactions continuously define the organizer during the cell movements of gastrulation. *Cell* **98**: 559–571.
74. Yuan S, Darnell DK and Schoenwolf GC (1995). Identification of inducing, responding, and suppressing regions in an experimental model of notochord formation in avian embryos. *Dev. Biol.* **172**: 567–584.
75. Yuan S and Schoenwolf GC (1999). Reconstitution of the organizer is both sufficient and required to re-establish a fully patterned body plan in avian embryos. *Development* **126**: 2461–2473.
76. Halpern ME, Thisse C, Ho RK, Thisse B, Riggleman B, Trevarrow B, Weinberg ES, Postlethwait JH and Kimmel CB (1995). Cell-autonomous shift from axial to paraxial mesodermal development in zebrafish floating head mutants. *Development* **121**: 4257–4264.
77. Ho RK and Kane DA (1990). Cell-autonomous action of zebrafish spt-1 mutation in specific mesodermal precursors. *Nature* **348**: 728–730.

78. Amacher SL and Kimmel CB (1998). Promoting notochord fate and repressing muscle development in zebrafish axial mesoderm. *Development* **125**: 1397–1406.
79. Odenthal J, Haffter P, Vogelsang E, Brand M, van Eeden FJ, Furutani-Seiki M, Granato M, Hammerschmidt M, Heisenberg CP, Jiang YJ *et al.* (1996). Mutations affecting the formation of the notochord in the zebrafish, Danio rerio. *Development* **123**: 103–115.
80. Stemple DL, Solnica–Krezel L, Zwartkruis F, Neuhauss SC, Schier AF, Malicki J, Stainier DY, Abdelilah S, Rangini Z, Mountcastle–Shah E and Driever W (1996). Mutations affecting development of the notochord in zebrafish. *Development* **123**: 117–128.
81. Parsons MJ, Pollard SM, Saude L, Feldman B, Coutinho P, Hirst EM and Stemple DL (2002a). Zebrafish mutants identify an essential role for laminins in notochord formation. *Development* **129**: 3137–3146.
82. Colognato H and Yurchenco PD (2000). Form and function: the laminin family of heterotrimers. *Dev. Dyn.* **218**: 213–234.
83. Parsons MJ, Campos I, Hirst EM and Stemple DL (2002b). Removal of dystroglycan causes severe muscular dystrophy in zebrafish embryos. *Development* **129**: 3505–3512.
84. Coutinho P. *et al.* (submitted).
85. Halpern ME, Ho RK, Walker C and Kimmel CB (1993). Induction of muscle pioneers and floor plate is distinguished by the zebrafish no tail mutation. *Cell* **75**: 99–111.
86. Schulte–Merker S, Ho RK, Herrmann BG and Nusslein–Volhard C (1992). The protein product of the zebrafish homologue of the mouse T gene is expressed in nuclei of the germ ring and the notochord of the early embryo. *Development* **116**: 1021–1032.
87. Schulte–Merker S, van Eeden FJ, Halpern ME, Kimmel CB and Nusslein–Volhard C (1994). No tail (ntl) is the zebrafish homologue of the mouse T (Brachyury) gene. *Development* **120**: 1009–1015.
88. Schulte–Merker S and Smith JC (1995). Mesoderm formation in response to Brachyury requires FGF signalling. *Curr. Biol.* **5**: 62–67.
89. Halpern ME, Hatta K, Amacher SL, Talbot WS, Yan YL, Thisse B, Thisse C, Postlethwait JH and Kimmel CB (1997). Genetic interactions in zebrafish midline development. *Dev. Biol.* **187**: 154–170.
90. Amacher SL, Draper BW, Summers BR, and Kimmel CB (2002). The zebrafish T-box genes no tail and spadetail are required for development

of trunk and tail mesoderm and medial floor plate. *Development* **129**: 3311–3323.

91. Griffin K, Patient R and Holder N (1995). Analysis of FGF function in normal and no tail zebrafish embryos reveals separate mechanisms for formation of the trunk and the tail. *Development* **121**: 2983–2994.
92. Griffin KJ and Kimelman D (2002). One-Eyed Pinhead and Spadetail are essential for heart and somite formation. *Nat. Cell. Biol.* **4**: 821–825.
93. Schier AF, Neuhauss SC, Helde KA, Talbot WS and Driever W (1997). The one-eyed pinhead gene functions in mesoderm and endoderm formation in zebrafish and interacts with no tail. *Development* **124**: 327–342.
94. Smith JL and Schoenwolf GC (1989). Notochordal induction of cell wedging in the chick neural plate and its role in neural tube formation. *J. Exp. Zool.* **250**: 49–62.
95. van Straaten HW, Hekking JW, Wiertz–Hoessels EJ, Thors F and Drukker J (1988). Effect of the notochord on the differentiation of a floor plate area in the neural tube of the chick embryo. *Anat. Embryol.* (*Berl.*) **177**: 317–324.
96. van Straaten HW and Hekking JW (1991). Development of floor plate, neurons and axonal outgrowth pattern in the early spinal cord of the notochord-deficient chick embryo. *Anat. Embryol.* (*Berl.*) **184**: 55–63.
97. Yamada T, Placzek M, Tanaka H, Dodd J and Jessell TM (1991). Control of cell pattern in the developing nervous system: polarizing activity of the floor plate and notochord. *Cell* **64**: 635–647.
98. Monsoro–Burq AH, Bontoux M, Vincent C and Le Douarin NM (1995). The developmental relationships of the neural tube and the notochord: short and long term effects of the notochord on the dorsal spinal cord. *Mech. Dev.* **53**: 157–170.
99. Yamada T, Pfaff SL, Edlund T and Jessell TM (1993). Control of cell pattern in the neural tube: motor neuron induction by diffusible factors from notochord and floor plate. *Cell* **73**: 673–686.
100. Echelard Y, Epstein DJ, St–Jacques B, Shen L, Mohler J, McMahon JA and McMahon AP (1993). Sonic hedgehog, a member of a family of putative signaling molecules iiplicated in the regulation of CNS polarity. *Cell* **75**: 1417–1430.
101. Roelink H, Augsburger A, Heemskerk J, Korzh V, Norlin S, Ruiz i Altaba A, Tanabe Y, Placzek M, Edlund T, Jessell TM *et al.* (1994).

Floor plate and motor neuron induction by vhh-1, a vertebrate homolog of hedgehog expressed by the notochord. *Cell* **76**: 761–775.
102. Currie PD and Ingham PW (1996). Induction of a specific muscle cell type by a hedgehog-like protein in zebrafish. *Nature* **382**: 452–455.
103. Ekker SC, Ungar AR, Greenstein P, von Kessler DP, Porter JA, Moon RT and Beachy PA (1995). Patterning activities of vertebrate hedgehog proteins in the developing eye and brain. *Curr. Biol.* **5**: 944–955.
104. Schauerte HE, van Eeden FJ, Fricke C, Odenthal J, Strahle U and Haffter P (1998). Sonic hedgehog is not required for the induction of medial floor plate cells in the zebrafish. *Development* **125**: 2983–2993.
105. Ericson J, Morton S, Kawakami A, Roelink H and Jessell TM (1996). Two critical periods of Sonic Hedgehog signaling required for the specification of motor neuron identity. *Cell* **87**: 661–673.
106. Matise MP, Epstein DJ, Park HL, Platt KA and Joyner AL (1998). Gli2 is required for induction of floor plate and adjacent cells, but not most ventral neurons in the mouse central nervous system. *Development* **125**: 2759–2770.
107. Patten I and Placzek M (2002). Opponent activities of Shh and BMP signaling during floor plate induction *in vivo*. *Curr. Biol.* **12**: 47–52.
108. Devoto SH, Melancon E, Eisen JS and Westerfield M (1996). Identification of separate slow and fast muscle precursor cells *in vivo*, prior to somite formation. *Development* **122**: 3371–3380.
109. Barresi MJ, Stickney HL and Devoto SH (2000). The zebrafish slow-muscle-omitted gene product is required for Hedgehog signal transduction and the development of slow muscle identity. *Development* **127**: 2189–2199.
110. Blagden CS, Currie PD, Ingham PW and Hughes SM (1997). Notochord induction of zebrafish slow muscle mediated by Sonic hedgehog. *Genes. Dev.* **11**: 2163–2175.
111. Karlstrom RO, Talbot WS and Schier AF (1999). Comparative synteny cloning of zebrafish you-too: mutations in the Hedgehog target gli2 affect ventral forebrain patterning. *Genes. Dev.* **13**: 388–393.
112. Pownall ME, Strunk KE and Emerson CP, Jr. (1996). Notochord signals control the transcriptional cascade of myogenic bHLH genes in somites of quail embryos. *Development* **122**: 1475–1488.

113. Xue XJ and Xue ZG (1996). Spatial and temporal effects of axial structures on myogenesis of developing somites. *Mech. Dev.* **60**: 73–82.
114. Goldstein AM and Fishman MC (1998). Notochord regulates cardiac lineage in zebrafish embryos. *Dev. Biol.* **201**: 247–252.
115. Fouquet B, Weinstein BM, Serluca FC and Fishman MC (1997). Vessel patterning in the embryo of the zebrafish: guidance by notochord. *Dev. Biol.* **183**: 37–48.
116. Sumoy L, Keasey JB, Dittman TD and Kimelman D (1997). A role for notochord in axial vascular development revealed by analysis of phenotype and the expression of VEGR-2 in zebrafish flh and ntl mutant embryos. *Mech. Dev.* **63**: 15–27.
117. Lawson ND, Vogel AM and Weinstein BM (2002). Sonic hedgehog and vascular endothelial growth factor act upstream of the Notch pathway during arterial endothelial differentiation. *Dev. Cell* **3**: 127–136.
118. Kim SK, Hebrok M and Melton DA (1997). Notochord to endoderm signaling is required for pancreas development. *Development* **124**: 4243–4252.
119. Cleaver O and Krieg PA (1998). VEGF mediates angioblast migration during development of the dorsal aorta in *Xenopus*. *Development* **125**: 3905–3914.
120. Cleaver O, Seufert DW and Krieg PA (2000). Endoderm patterning by the notochord: development of the hypochord in *Xenopus*. *Development* **127**: 869–879.
121. Latimer AJ, Dong X, Markov Y and Appel B (2002). Delta-Notch signaling induces hypochord development in zebrafish. *Development* **129**: 2555–2563.
122. Hebrok M, Kim SK and Melton DA (1998). Notochord repression of endodermal Sonic hedgehog permits pancreas development. *Genes. Dev.* **12**: 1705–1713.
123. Adams DS, Keller R and Koehl MA (1990). The mechanics of notochord elongation, straightening and stiffening in the embryo of *Xenopus laevis*. *Development* **110**: 115–130.
124. Coutinho, Parsons, Hirst and Stemple, unpublished.
125. Yan YL, Hatta K, Riggleman B and Postlethwait JH (1995). Expression of a type II collagen gene in the zebrafish embryonic axis. *Dev. Dyn* **203**: 363–376.

Chapter 4

Formation and Functions of the Floor Plate

Jing Tian and Karuna Sampath
Temasek Life Sciences Laboratory
1 Research Link, NUS, Singapore 117604

The most ventral cells of the vertebrate neural tube, the *floor plate*, comprise a specialized group of cells distinct in form and function from the rest of the neural tube. These cells have been ascribed many functions, ranging from the differentiation of motor neurons that innervate specific muscle cell types in the body, to providing cues for the correct path-finding of various axons. Here, we summarize the process by which the floor plate develops in a few model vertebrate organisms, the various functions of the floor plate, and the molecular nature of signals emanating from the floor plate. Finally, we assess the prevailing models of how the cells of the floor plate are thought to arise.

1. Introduction

Understanding the development of an organism that can think, remember, and coordinate conscious and unconscious bodily processes has been one of the most challenging problems in biological research. Over the past several decades, a lot of effort has been directed towards understanding the process of neurulation, one of the most important steps during vertebrate embryonic development. Neurulation includes the induction of the neural plate and the formation of the neural tube, the rudiment of the central nervous system (CNS). The anterior portion of the neural tube gives rise to the brain, and the more posterior parts give rise to the spinal cord — the simplest and most conserved region in the vertebrate CNS.

At early stages in the development of the spinal cord, three major classes of cells are generated in the ventral neural tube: floor plate cells at the ventral midline, motor neurons at ventrolateral positions, and interneurons at more dorsal locations. On the other hand, the dorsal neural tube initially gives rise to neural crest cells, and subsequently, to roof plate cells, commissural neurons, and several classes of dorsal sensory relay interneurons. Despite the complexity of the vertebrate CNS, detailed studies of the individual cell types that comprise the neural tube have made advances towards the understanding of the diversity and pattern in the developing CNS. In this chapter, we focus on a transient group of cells in the CNS, the cells of the floor plate, a structure that occupies the ventral midline of the developing spinal cord, hindbrain, midbrain, and caudal forebrain.

Located along the ventral midline of the neural tube, the floor plate acts as a transient embryonic organizing center and as a source of signals that patterns adjacent cells. These activities of the floor plate are critical for the proper development of the CNS. For instance, the organization of axons in the CNS, the regulation of cell differentiation along the dorsoventral axis of the neural tube, and the induction and differentiatin of motor neurons, are all functions that require a functional floor plate.

The identification of important functions of this ventral neural cell type has also led to a lot of interest in the origin of the floor plate cells. Many studies have provided evidence that the differentiation of the floor plate requires inductive signals provided by notochord, a rod-like group of cells running from the level of the hindbrain in embryonic head to the tail in vertebrates. *In vivo* and *in vitro* experiments performed in several different model systems have provided evidence to support the concept that the notochord is a key cellular source of inductive signals for floor plate differentiation. However, more recent data challenges this textbook version of the organizational chart of notochord for the origin of the floor plate. Experimental evidence from the chick and zebrafish suggest that there may be more to floor plate differentiation than a single inductive signal provided by the notochord. In this chapter, we focus on the roles of floor plate in CNS and discuss the recent advances in the understanding of the molecular steps in floor plate development.

2. Overview of the Neural Tube

2.1. Formation of the Neural Tube: Neurulation

In vertebrates, gastrulation results in an embryo with an internal endodermal layer, an intermediate mesodermal layer, and an external ectodermal layer of cells. The epidermis, the central and peripheral nervous systems, and some non-neuronal cells of the head and heart are derived from ectoderm. During the third week of gestation in chick, a portion of the dorsal ectoderm is specified to become neural ectoderm — this region of the embryo is called the neural plate. The process by which the neural plate forms the neural tube is called neurulation.

In vertebrates, the process of neurulation occurs differently in different regions of the body. The head and trunk regions both undergo variants of primary neurulation, and this process occurs in four steps[1–3]:

(I) *Formation of the neural plate.* The process that transforms the general embryonic ectoderm into a thickened neural plate is typically described as neural induction. The first morphological response of the embryonic ectoderm to neural specification that can be detected is an increase in the height of cells destined to become components of the nervous system. These transformed cells, known as neuroepithelial cells or neuroectoderm, are evident as a thickened neural plate that is visible on the medial dorsal surface of the early embryo (Fig. 1A and 1a).

(II) *Shaping of the neural plate.* This step involves the concomitant narrowing and lengthening of the neural plate. At the time of its formation, the neural plate is shaped like a spade, being relatively wide mediolaterally and short rostrocaudally.

(III) *Bending of the neural tube.* The term "neurulation" specifically refers to this stage. The major portion of the neural tube is thus formed through the process of primary neurulation. In primary neurulation, lateral folding or bending of the neural plate results in elevation of two walls, the neural folds, flanking a midline ventral depression, the neural groove (Fig.1 B, C and 1b, 2, 3). The original ectoderm is divided into three sets of cells: (i) the

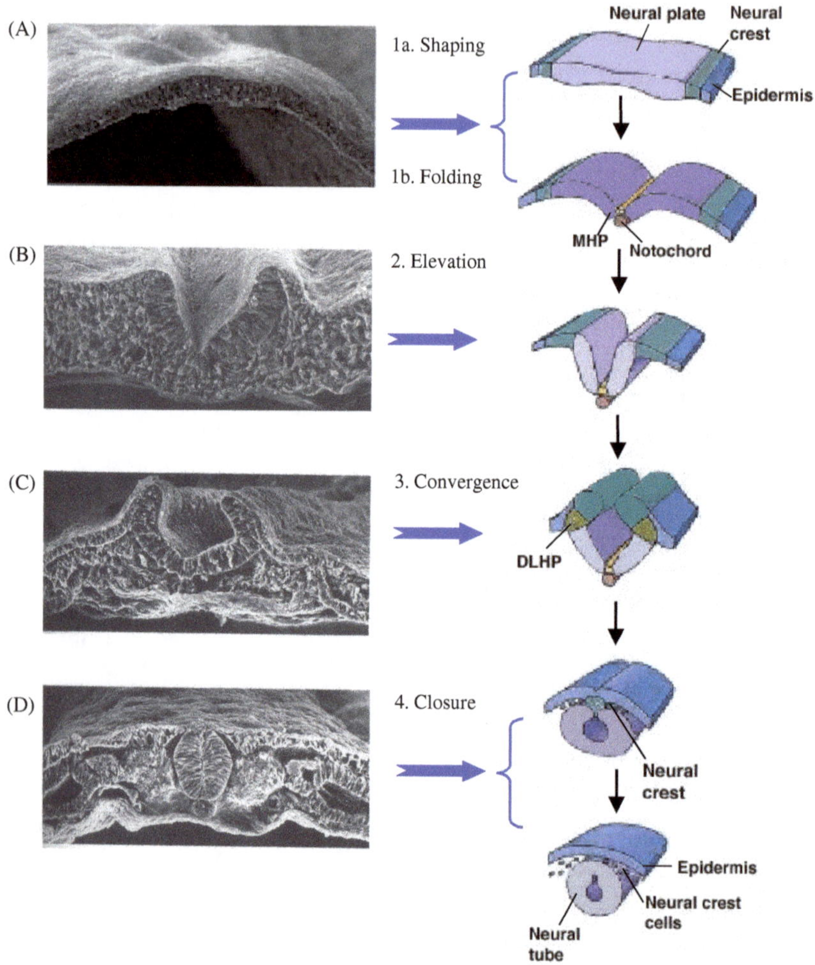

Fig. 1 Primary neurulation. (A–D): Scanning electron micrographs (A–D) and diagrammatic representations (1–4) of neural tube formation in the chick embryo. (A, 1) Cells of the neural plate can be distinguished as elongated cells in the dorsal region of the ectoderm. Folding begins as the medial neural hinge point (MHP) cells anchor to notochord. (B, 2) The neural folds are elevated as presumptive epidermis moves toward the dorsal midline. (C, 3) Convergence of the neural folds occurs as the dorsolateral hinge point (DLHP) cells become wedge-shaped and epidermal cells push toward the center. (D, 4) The neural folds fuse, and the neural crest cells disperse, leaving the neural tube separate from the epidermis. (Plates 1–4, Kathryn Tosney; reproduced with permission from *Developmental Biology*, 5th ed., Gilbert S.)

internally positioned neural plate; (ii) the externally positioned epidermis of the skin; and (iii) the neural crest cells that connect the neural plate and epidermis.

(IV) *Closure of the neural groove and formation of the neural tube* (Fig. 1 C, D and 3, 4). This consists of apposition of the two dorsolateral apical surfaces of the neural folds, their fusion at the dorsal midline, and separation of the completed segment of the neural tube from the overlying ectoderm.

Caudal to the posterior neuropore, the neural tube is formed by secondary neurulation. Secondary neurulation usually occurs during the production of the lumbar and tail vertebrate. It can been seen as a continuation of gastrulation. In the secondary neurulation, a rod-like condensation of mesenchymal cells forms beneath the dorsal ectoderm of the tail bud. Within the mesenchymal rod, a central canal forms by cavitation. This central canal becomes continuous with the one formed during primary neurulation and closure of the posterior neuropore.[4,5] In avian systems, the anterior portions of the neural tube are constructed by primary neurulation, while the neural tube caudal to the 27-somite pair is made by secondary neurulation.[2,6] In mice, similar to the chick, secondary neurulation begins at or around the level of 35-somite stage.[7,8] Neurulation in fish is thought to be exclusively secondary. In amphibians, most of the tadpole neural tube is made by primary neurulation, but the tail neural tube is derived from secondary neurulation.[9]

2.2. The Floor Plate: A Transient Structure in the CNS Which Forms during Neurulation

At early stages of ventral neural tube development, cells of the floor plate, an epithelial structure located ventrally in the neural tube, are among the first to differentiate at the ventral midline soon after neural plate formation (Fig. 2). Morphologically the floor plate is made up of columnar ependymal cells, recognizable by their characteristic wedge-shaped appearance that span the width of the neural tube at its ventral midline.[10] In higher vertebrates, this group of cells is about

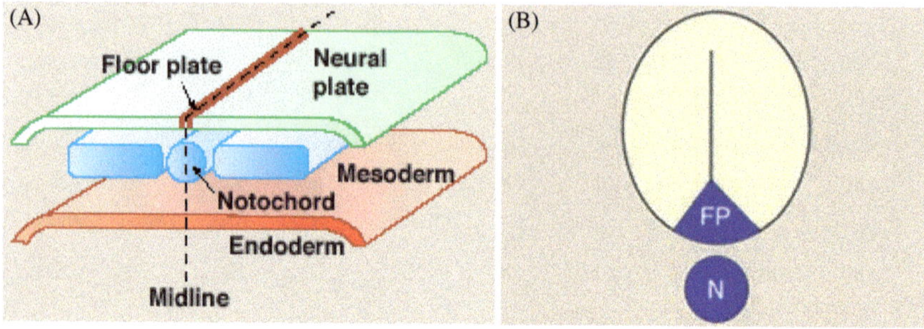

Fig. 2 Schematic views of the position of the floor plate in the neural plate and neural tube. (A) Structures formed during embryogenesis. Early embryogenesis is characterized by a series of morphogenetic movements during gastrulation, which establish the three primary germ layers — the ectoderm, mesoderm and endoderm. At the end of this process, the mesoderm (the future muscle and bone) comes to rest sandwiched between the ectoderm (the future nervous system) on the outside of the embryo, and the endoderm (the future gut) on the inside. (B) Transverse section through the neural tube showing the position of notochord (N) and floor plate (FP) at late stages of neural tube development. (A, modified from Blader and Strahle, 1998).

15–20 cells wide in day 11–12 rat embryos.[11,12] The analysis of the expression of various immunocytochemical and molecular markers of the ventral neural tube in several vertebrate species has shown spatially restricted expression patterns. The analyses of gene expression patterns within the ventral neural tube indicates that this group of cells is heterogeneous and includes two distinct cell populations, the medial floor plate (MFP) and the lateral floor plate (LFP).[13] Along the embryonic body axis, the floor plate extends through the midbrain into the caudal forebrain and ends near the mammillary region.[12,14–18] In zebrafish, the floor plate extends from the spinal cord through the hindbrain and midbrain but not apparently into the caudal forebrain.[19] In transverse sections of the spinal cord, the floor plate comprises a single large cell at the ventral midline called MFP which is distinct morphologically and antigenically from other spinal cord cells.[19] In zebrafish, the floor plate also includes the cells immediately flanking the MFP cell, called LFP cells, and in transverse sections, the entire floor plate appears to be 3–4 cells wide.[20,21]

2.3. Signals Involved in Polarization within the Neural Tube

2.3.1. *Shh, a major signaling molecule in ventral neural tube patterning*

At early stages of ventral neural tube development, three main classes of cells are generated: floor plate cells, motor neurons and ventral interneurons[22] (Fig. 3C). The molecular basis of neurulation has begun to be elucidated with the identification of several candidate genes, which when mutated, perturb neurulation. Sonic hedgehog (Shh) is an important signaling molecule required for proper patterning of the ventral neutral tube.[23,24] Shh is a member of the Hedgehog family of signaling molecules identified by homology to the Drosophila Hedgehog (HH).[25] Shh is proteolytically cleaved to produce two secreted proteins[26,27]: a 19KDa N-terminal peptide (Shh-N) that mediates all signaling activities in several vertebrate and invertebrate species,[28] and a 25 KDa C-terminal protein (Shh-C) that possesses protease activity.[27,29] The patterning of ventral cell fates is controlled by Shh-N that represses the expression of several transcription factors in cells at medial positions within the neural plate as described below. When the caudal neural plate is formed, cells at mediolateral positions express the homeodomain-containing transcription factors, Pax3, Pax7, Msx1, and Msx2.[30–34] The expression of these genes is rapidly repressed in medial neural plate cells by a Shh-mediated signal from the floor plate and notochord (Fig. 3B).[30,31,34] After neural tube closure, the expression of these transcription factors is restricted to proliferating cells in the dorsal neural tube (Fig. 3C).[35]

2.3.2. *BMPs: key signaling factors in patterning the dorsal neural tube*

Dorsal cell types in the neural tube, including the roof plate at the dorsal midline, and several classes of dorsal sensory relay interneurons, are generated in response to signals derived from the epidermal ectoderm that flanks the lateral margins of the neural plate (Fig. 3C).[22] Bone Morphogenetic Proteins (BMPs), members of the Transforming Growth

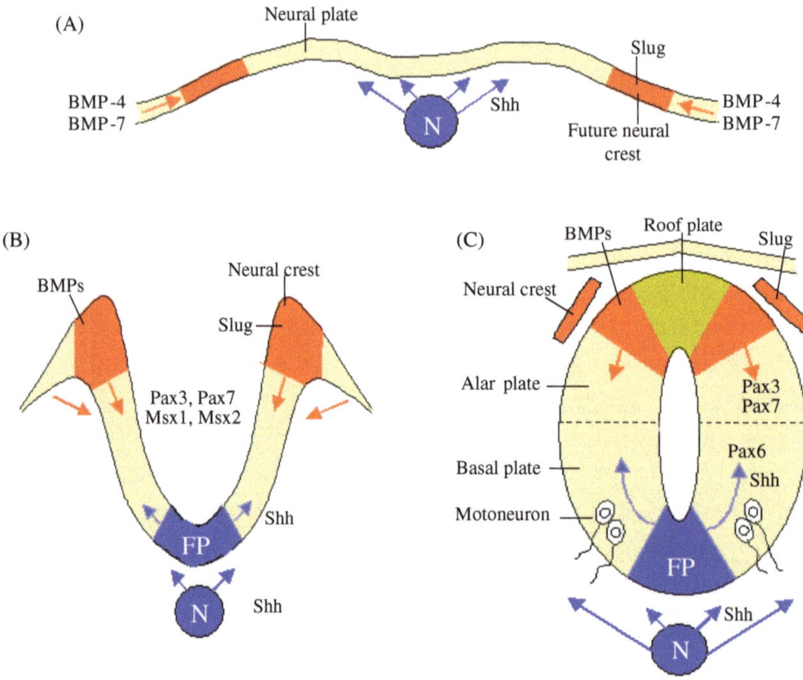

Fig. 3 Dorsal and ventral signaling in the early central nervous system. (A) One model proposes that signals such as Sonic hedgehog (Shh) (blue arrows) from the notochord (N) induce the floor plate. (B) In the dorsal part of the future neural tube, BMP-4 and BMP-7 (orange arrows) from the ectoderm adjacent to the neural tube are thought to induce *slug* expression in the future neural crest and maintain Pax3 and Pax7 expression dorsally. Ventrally, Sonic hedgehog, now produced by the floor plate, induces motoneurons. (C) Sonic hedgehog, produced by the floor plate, suppressed the expression of dorsal *Pax* genes (*Pax3 and Pax7*) in the ventral half of the neural tube. (Adapted from *Human Embryology and Developmental Biology*, Carlson BM)

Factor-β (TGF-β) family of secreted proteins, are key signals to mediate this ectodermal signal.[34] In the dorsal portions of the future neural tube, BMP4 and BMP7 secreted by the epidermal ectoderm adjacent to the neural tube induce Slug, a zinc finger transcription factor,[36] in the future neural crest cells, and maintain *pax3* and *pax7* expression (Fig. 3). These genes are required for the appropriate differentiation of neural crest cells and are involved in dorsal cell differentiation.[32] Shh signaling from the floor plate suppresses the expression of dorsal *pax* genes in the ventral half of the neural tube where motor neurons develop.

3. The Roles of the Floor Plate in the Central Nervous System

3.1. Axon Guidance

3.1.1. *Overview of axon projections which are influenced by the floor plate*

Neural connections form during embryonic development when each differentiating neuron sends out an axon, extending long distances along stereotypical paths to reach their final target. To generate this precise pattern of connections, each axon must integrate many guidance cues in its environment and react in a specific fashion to recognize its particular path. The floor plate is a key landmark to act as an intermediate cellular target for extending axons due to its midline location and its early differentiation within the neural tube.

In vertebrates, there are two types of growth patterns in the early neuronal populations of the neural tube — a circumferential pattern and a longitudinal pattern. Circumferential neurons are found in the more dorsal aspects of the spinal cord and comprise two classes of cells, commissural and association. These two groups of axons initially extend in the same direction ventrally along the lateral edge of the spinal cord but then diverge upon reaching the lateral or ventrolateral marginal zone. Axons from association neurons turn at right angles to join the ipsilateral longitudinal pathway, whereas axons from the commissural neurons cross the basal portion of the floor plate and turn longitudinally. After executing the turn, the axons fasciculate among themselves and with other longitudinally-oriented axons running in the ventral marginal zone (Fig. 4).[37]

There are different classes of neurons interacting with the floor plate in different species (Fig. 4). In the chick, axons of the circumferential neurons grow ventrally along the lateral margin of the spinal cord (but not in direct contact with the external limiting membrane) in the transverse plane.[38–42] Upon reaching the ventrolateral spinal cord, axons of the ipsilateral projecting

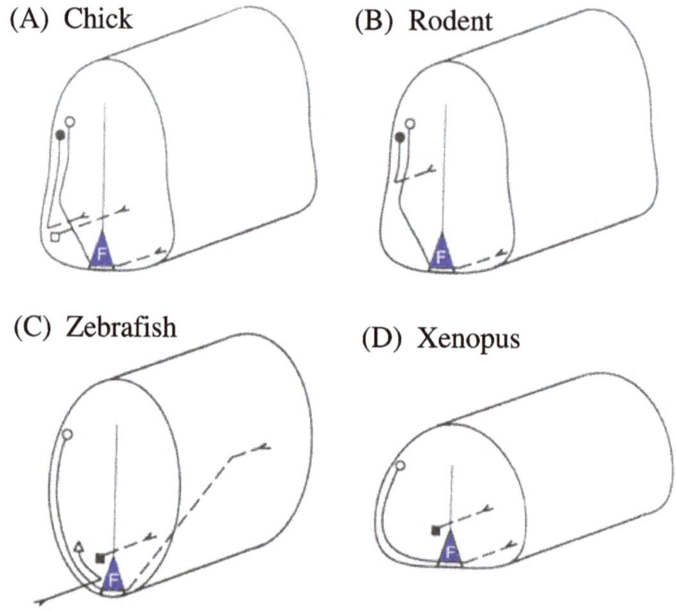

Fig. 4 Schematic diagram summarizing early axonal populations in the developing spinal cord whose growth is influenced by the floor plate. There are two patterns of axonal projections — circumferential and longitudinal — characteristic of the earliest neurons. The blue area represents the floor plate (F). Dorsal is up; caudal is to the left. Dashed lines indicate rostrally directed projections. □ — association neurons; □ — commissural neurons; □ — primitive logitudinal (PL) cells; □ — Kolmer-Agduhr (KA) neurons; Δ — VeLD neurons. The developmental stage listed for each species is that at which the first axons begin to extend. (A) Chick, stage 15. (B) Rodent, rat E11; mouse E9. (C) Zebrafish, 14-somite (16h). (D) *Xenopus*, stage 25.[37]

association neurons turn at right angles and project longitudinally. Axons of the contralaterally projecting commissural neurons grow towards the floor plate and cross the ventral midline before turning to project longitudinally. Commissural neurons apparently take two different trajectories to reach the floor plate. The earliest-born commissural neurons follow the lateral edge of the neural tube until they arrive at the ventral midline. The later-born commissural axons break away from the edge in the ventral spinal cord and grow ventro–medially to the floor plate. Primitive longitudinal (PL) cells have longitudinally directed axons that extend either rostrally or caudally in the ventrolateral spinal cord (Fig. 4A).[38,41,43,44] In rodents, commissural neurons are similar to the later-born commissural neurons in the chick, except that in rodents, they only project rostrally, whereas in the chick, these neurons have been found to turn both rostrally and caudally. Association neurons are also similar, except that the earliest neurons in rodents project longitudinally in the lateral rather than ventrolateral marginal zone (Fig, 4B).[38,43,44] In zebrafish, commissural (CoPA and CoSA) neurons project circumferentially along the edge of the spinal cord to the ventral midline, just like early-born commissural neurons in the chick.[20,45,46] After crossing the floor plate they turn rostrally and ascend obliquely for approximately one segment to join the dorsal longitudinal pathway. Axons of the VeLD neurons initially extend circumferentially to contact the floor plate but do not cross the midline and instead turn to project caudally. Kolmer–Agduhr (KA) neurons have longitudinally directed axons that extend rostrally in the ventral spinal cord (Fig. 4C). Similarly, in *Xenopus*, commissural neurons project circumferentially along the edge of the spinal cord to the ventral midline. After they cross the floor plate, they project longitudinally either rostrally, caudally, or with branches in both directions (Fig. 4D).[47–50]

Axons that grow to the ventral midline, cross the midline, and then turn to project longitudinally, are also found at higher axial levels and have been particularly well characterized in zebrafish.[51,52] Notably, the projection patterns of ventrally decussating commissural axons are highly conserved between the spinal cord and the caudal hindbrain in both chick and zebrafish embryos,[53–56] between the spinal cord and midbrain of chick embryos,[57,58] and between the metencephalic axons and the spinal cord commissural axons in the rat.[59]

3.1.2. *Chemoattraction by the floor plate*

Evidence accumulated in the last decade indicates that the floor plate has a potent chemotropic effect on commissural axons. Spinal commissural axons show reoriented growth towards floor plate explants[60] when co-cultured with rat floor plate tissue in collagen gel which establishes a gradient of diffusible substances.[61–63] Moreover, the floor plate's ability to induce turning of commissural axons is not mimicked by explants of any other portion of the neural tube.[60,64] Consistent with these *in vitro* findings are the *in vivo* observations in chick embryos. Rotation of a segment of the spinal cord results in an ectopic floor plate, which becomes apposed to the dorsal half of the remainder of the neural tube and redirects commissural axons towards to the ectopic floor plate.[65] A floor plate grafted alongside the spinal cord of a developing chick embryo *in ovo* also can induce the abnormal growth of commissural axons, out of the spinal cord, towards to the graft.[66] Similarly, these axons show abnormal trajectories in the mouse mutant, *Danforth short-tail*, which lacks the floor plate at the most caudal levels,[67] and in the zebrafish mutant, *cyclops*, which lacks the midline floor plate cells at all axial levels.[45,46]

The floor plate in the brain also provides a chemoattractive guidance cue for crossed axons. Early evidence had shown that cerebellofugal axons were attracted by floor plate explants from the metencephalon, where the axons crossed the midline.[59] Recent data also indicates that the floor plate in the rhombencephalon chemoattracts rhombencephalic alar plate axons just as the floor plate in the spinal cord attracts spinal commissural axons. These data suggest that the floor plate attracts various kinds of crossed axons in both the metencephalon and myelencephalon, probably contributing to the formation of crossed projections in these regions of the brain (Fig. 5).[68]

3.1.3. *Chemorepulsion by the floor plate*

In the zebrafish mutant, *cyclops*, which lacks MFP cells, normally uncrossed axons in the spinal cord[45] and the hindbrain[69] often cross the ventral midline. Fifteen percent of VeLD neurons, which extend to the midline and turn posterior without crossing the midline, aberrantly cross the midline in *cyclops* mutant embryos, suggesting that the floor plate releases diffusible chemorepulsant(s) that prevents axons from crossing the midline. Studies *in vitro* have demonstrated that the floor plate has the ability to repel subsets of axons from a distance.[37,70–72] These include axons of the visceral motor and branchiomotor neuronal subclasses, and posterior commissural axons. Notably, posterior commissural axons and axons of the midbrain alar and basal plate are repelled by the floor

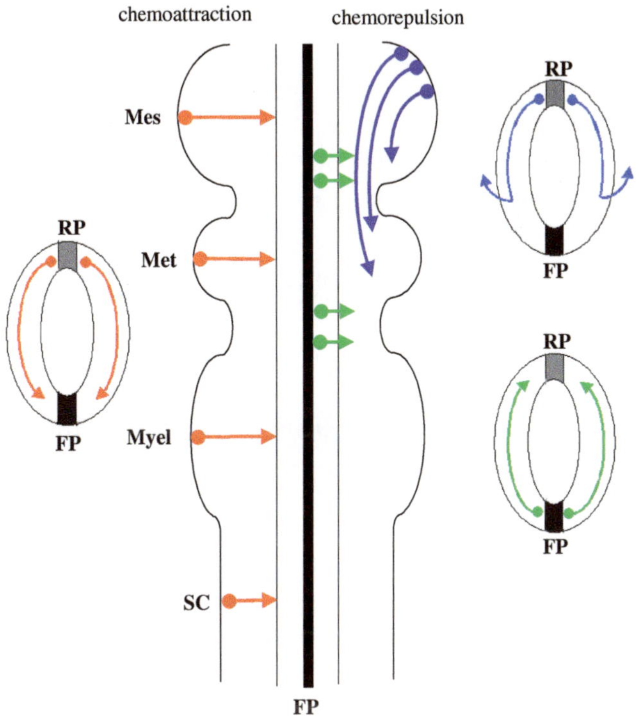

Fig. 5 Schematic diagram showing the global role of the floor plate in the axonal guidance in the vertebrate CNS by chemoattraction and chemorepulsion. The diagram represents axonal trajectories in flat whole-mount preparations. Insets (left and right) show axon trajectories in a transverse plane. The FP, shown at the center, possesses both chemoattractive and chemorepulsive activities along its entire length. Regions containing chemoattracted axons (red) are shown on the left, while those containing chemorepulsed axons (blue and green) are illustrated on the right. FP, floor plate; RP, roof plate; Mes, Mesencephalon; Met, metencephalon; Myel, myelencephalon; SC, spinal cord.

plate at all axial levels,[71,72] suggesting that the floor plate provides a global guidance cue for a variety of axons by repelling them (Fig. 5).

3.1.4. *Guidance at the midline by the floor plate*

Guidance events in the midline are more complex. Commissural axons, once having arrived at the midline, have to decide to cross the midline or not, in order not to stall at the floor plate. In the mouse mutant, *Danforth short-tail*, in which the floor plate and the underlying notochord are missing, the commissural axons can reach the midline, but they make incorrect pathway choices, often failing to turn longitudinally, and projecting out of the spinal cord (Fig. 6A). Consistent with this observation, the errors in the migration of the commissural primary ascending (CoPA) neurons and the association-like ventral longitudinal descending (VeLD) neurons occur in the zebrafish *cyclops* (ajc) mutant as well. These migration errors also can be observed in embryos in which the floor plate has been ablated by laser.[45,73] CoPA neurons normally cross the midline floor plate cells and turn anteriorly, while VeLD neurons extend to the midline and

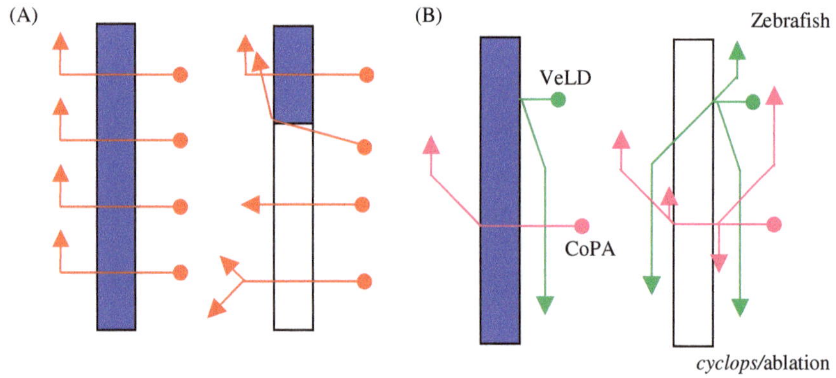

Fig. 6 Schematic diagrams illustrating aberrarct axon behavior caused by the loss of midline floor plate cells in mouse Danforth short-tail and zebrafish *cyclops* mutant embryos. (A) In the mouse, commissural axons cross the midline and extend rostrally. In the Danforth short-tail mutant, much of the floor plate and notochord fails to develop. Many commissural axons can reach the midline, often taking an aberrant pathway along the circumference of the neural tube, but then fail to make the correct turning decisions. Where floor plate tissue remains, commissural axons will turn to project directly to these cells. (B) In the zebrafish, both commissural neurons (CoPA) and ipsilaterally projecting neurons (VeLD) are affected by the removal of the midline by laser ablation or in *cyclops* mutants. CoPA neurons reach the midline but can then make incorrect turning decisions, extending either longitudinally on their own side or close to the midline on the contralateral side. The VeLD neurons can cross the midline when midline cells are missing and usually make their correct turn on the contralateral side. The VeLD neurons also make incorrect decisions on their own side.

turn posteriorly without crossing the midline. In the embryos with deficiencies in the floor plate, approximately 25% of the CoPA axons fail to cross the midline and 15% of the VeLD axons incorrectly cross the midline (Fig. 6B).[74]

3.1.5. *Molecular mechanisms of axon guidance by the floor plate*

Recently, Netrins (Netrin-1 and Netrin-2), which are diffusible chemotropic factors, have been identified during the purification of a chemotropic activity from extracts of embryonic chick brain. Netrins define a family of vertebrate homologs of the *C. elegans* unc-6 gene, mutations in which disrupt the circumferential dorsal and ventral growth of axons in the body wall of the worm.[75,76] Vertebrate Netrins can attract spinal commissural axons towards the floor plate *in vitro.* It has been shown that similar to the floor plate, Netrins can promote out-growth of commissural axons from explants of rat dorsal spinal cord into collagen gel.[77] In all vertebrates examined thus far, cells at the ventral midline express at least one Netrin family member. For instance, the floor plate of the spinal cord is a source of the diffusible attractant, Netrin-1. This molecule attracts commissural growth cones, which navigate to the floor plate and then cross to the contralateral side. Loss of *netrin-1* function at the midline results in the misrouting of many axons and their failure to grow to the midline.[75–79] Netrin-1 also acts as a repellent for some axons that grow away from the midline.[80] Thus, in vertebrates, Netrin-1 produced by the floor plate cells appears to be a bi-functional molecule that acts as an attractant for some axons, and as a repellent for others, depending on the Netrin receptor expressed by the cell. Since Netrin-1 is expressed in the floor plate at all axial levels, from the caudal diencephalon to the spinal cord, it is likely to contribute to the global guidance functions of the floor plate in the CNS.

Once at the midline, growth cones have to make a variety of decisions, to stop or continue growing, to cross the midline or not. Studies in axon guidance have suggested that the floor plate clearly provides some information necessary to direct the proper routing of axons at the midline.

So far, many growth-promoting molecules are known to be enriched in the floor plate, and could possibly aid midline crossing by serving as permissive signals. N-Cadherin, Neurofascin, N-CAM and Nr-CAM are examples of adhesion molecules that are concentrated on the commissural segment of the crossing axons and are also expressed by the floor plate.[81–84] For instance, in the chick, commissural axons and growth cone express Axonin-1, whereas the floor plate cells express the receptor Nr-CAM. Function-blocking reagents disturb the heterophilic interaction between Nr-CAM at the midline and Axonin-1 in commissural axons, and the floor plate becomes inhibitory to the commissural growth cone.[75,85]

3.2. Floor Plate in Neuronal Differentiation

3.2.1. *Ventral neural cell type differentiation*

Distinct neural cell types are generated along the dorsal/ventral (D/V) axis of the neural tube. In vertebrates, the specification of neural subtypes in the spinal cord becomes evident during early neurulation. At early stages in the development of the neural tube, three main classes of cells are generated in the ventral region: **the floor plate**, a specialized class of cells, differentiate at the ventral midline soon after the neural plate formation, whereas **motor neurons** and **interneurons** are generated at more dorsal positions (Fig. 7).

3.2.2. *Motor neuron differentiation by the floor plate*

Motor neurons project axons to muscles and autonomic neurons, and mediate the central control of movement, using acetylcholine as a neurotransmitter. This leads to the formation of ordered connections with appropriate targets and their correct topographical organization. On the basis of neuroanatomical studies, individual motor neuron subtypes are defined in two ways: (1) by their axonal path and target choice, and (2) by their cell-body position within the spinal cord.[80,86,87] All motor neurons share some common properties, such as neurotransmitter expression, neurotrophin sensitivity, and neuromuscular

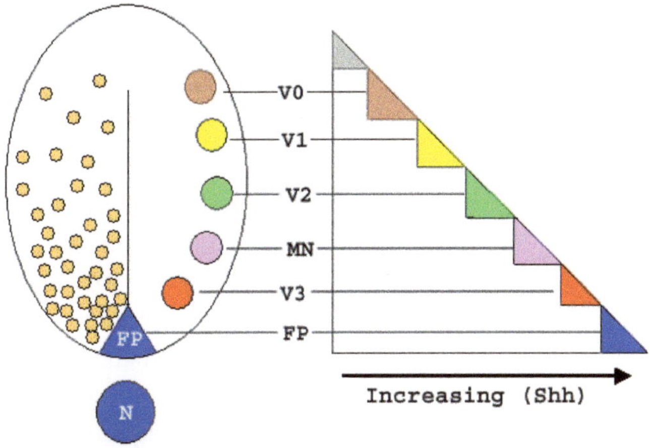

Fig. 7 Gradient model for the induction of ventral neural tube cell types by increasing concentrations of Shh protein. The concentration of Shh required to induce specific ventral cell types *in vitro* correlates directly with their dorso-ventral position *in vivo*. Proposed gradient of Shh signal moving from its sources of expression in the ventral neural tube and notochord shows on the left. FP, floor plate; MN, motor neurons; V0–V3, different classes of ventral interneurons generated at spinal cord levels.

synapse formation. Although different motor neuron subtypes acquire specific properties that mediate the formation of specific synaptic connections, several lines of evidence show that all motor neurons have a common developmental origin. Thus, the establishment of motor neuron properties occurs first in precursor cells before their specification into further cell types. Due to the position of the floor plate and notochord at the midline, it has been suggested that they trigger these differentiation events by sending signals and serve as the source of the diffusible signals that pattern the ventral neural tube.[88]

The main signaling activities of the floor plate and notochord are mediated by a secreted protein Sonic hedgehog (Shh).[89] In vertebrates, Shh is expressed highly in the floor plate and notochord, suggesting therefore that it plays a key role in motor neuron differentiation.[30,90–93] Explants of the floor plate or the notochord have been shown to induce the ectopic expression of motor neuron markers in the chick. This activity is mimicked by recombinant Shh protein and blocked by antibodies against the Shh protein.[30,92,94] In addition, loss-of-function mutations in the mouse *shh* locus result in the complete lack of expression of the motor neuron marker, *Islet 1*.[95] These experiments indicate that in the chick and mouse, Shh from the floor plate and notochord is both sufficient and necessary for motor neuron development (Fig. 7).[89]

In zebrafish, motor neuron differentiation is a little different from that described in amniotes. Early during gastrulation, similar to other vertebrates, zebrafish *shh* is expressed, starting at 60% epiboly in the organizer region, the embryonic shield, and subsequently, in the floor plate and notochord. But unlike amniotes, which possess a single *hedgehog* homolog, *shh*, zebrafish expresses two additional *hh* genes in different subsets of the amniote *shh* expression domain, during the time when motor neurons are likely to be specified: *tiggywinkle hedgehog* (*twhh*), which is expressed exclusively in the floor plate, and *echidna hedgehog* (*ehh*), which is expressed in the notochord domain alone.[96,97] Zebrafish embryos with mutations in *cyc* lack the medial floor plate and show no expression of s*hh*, at the midline of the neural tube[91] and have reduced *twhh* expression.[96] Consequently, the development of branchiomotor motor neurons, which are located in the hindbrain and innervate muscles that differentiate in the pharyngeal arches, is severely affected in *cyclops* mutant embryos.[98] Embryos with mutations in *sonic-you* (*syu*), which delete the gene s*hh*, also have significantly reduced branchiomotor neurons. These neurons are also completely absent in *syu* mutants injected with Morphlinos against the *twhh*.[99] Therefore, *shh* and *twhh* function synergistically during branchiomotor neuron development.[98,99]

However their function does not seem to be essential for spinal motor neuron development, since *cyclops* mutant embryos still possess spinal motor neurons.[19,69] Moreover, *syu* mutant embryos still have normal numbers of both primary and secondary motor neurons,[100] though the axon tracts of many of these neurons are aberrant. Genetic and cell biology studies involving a direct or indirect reduction of all three Hh signals and antisense morpholinos suggest that shh, twhh and ehh can all act redundantly to specify motor neurons, indicating that zebrafish motor neuron differentiation does require Hh signaling.[101]

3.2.3. *Ventral neural tube patterning by Shh secreted by the floor plate and notochord*

In addition to its role in motor neuron specification, studies have also demonstrated that Shh functions in the formation of ventral neural tube cell types at all rostrocaudal levels.[90,102–104] In the prospective spinal cord and hindbrain, in addition to the induction of motor neurons, the action of Shh can also induce the differentiation of a variety of ventral interneurons and oligodendrocytes.[105,106] Together, these groups of neurons will function in the adult in the direct regulation of motor function, and in the integration of sensory information. In the prospective midbrain and hindbrain regions, Shh is involved in the induction of dopaminergic and serotonergic neurons.[107–110] These neurons later have roles in both emotional regulation and higher-level control of movement, and are directly implicated in movement disorders such as Parkinson's disease and a variety of psychiatric conditions.

How can a single factor induce the differentiation of such a range of diverse cell types? Evidence derived from *in vitro* studies suggest that Shh acts as a morphogen, forming a gradient in the ventral neural tube, in response to which cells differentiate in a concentration-dependent fashion (Fig. 7).[89,105] Neural explants that are exposed to two-fold incremental increases in Shh concentration differentiate into specific ventral cell types in a concentration-dependent manner. The highest concentration of Shh can induce ventral midline cells, while lower concentrations induce cell types found *in vivo* to lie further away from the notochord and ventral midline (Fig. 7).

4. The Origin of the Floor Plate

4.1. Model One: Shh-Mediated Induction of Floor Plate by the Notochord

4.1.1. *Vertical induction of floor plate by Shh secreted from notochord*

During the process of neurulation in vertebrate embryos, two successive steps can be distinguished. The first consists of the determination of two domains in the presumptive ectoderm: a ventral domain fated to become the epidermis, and a dorsal domain from which the primordium of the nervous system will develop. The second step involves morphogenetic transformations. The ectoderm, which is committed to a neural fate, forms a neural tube, and subsequently, neuronal and glial differentiation take place. Studies in amphibian embryos have shown that the critical event through which neurogenesis is initiated is the involution of notochordal material through the dorsal blastoporal tip during gastrulation. The notochord, an axial structure of mesodermal origin, has been suggested as a signaling source for several critical events in early development. It is a rod-shaped mass of vacuolated cells, and lies immediately below the neural tube (Fig. 1). Its presence during embryonic development is a definitive feature of the chordate phylum. The notochord is one of the earliest embryonic structures to be formed, and functions as a support structure for the entire organism, either transiently (in higher vertebrates) or persistently (in some lower vertebrates). From early neural stages, the notochord forms by elongation of a distinct population of cells located in the late blastoporal lip, from the rostral to caudal end of the embryo. This group of cells has been designated as the 'chordoneural hinge' (CNH), also known as the Hensen's node in amniotes. The notochord emerges from Hensen's node of the chick and the mouse,[6,111] the blastopore lip of the amphibian embryos[9] and the shield of zebrafish embryos.[112]

In chick embryos, Hensen's node is a structure that represents the organizer region. It undergoes a characteristic rostrocaudal movement known as "regression", which takes place as the embryos elongate along the anteroposterior axis. The notochord is derived from Hensen's node.[113,114] The floor plate is also derived, in part, from cells within Hensen's node, although a major contribution for the floor plate is from cells in a region of the epiblast (termed "region A") immediately anterior to Hensen's node prior to its regression. Unlike the Hensen's node, this group of cells does not contribute to the notochord.[115–118] As Hensen's node regresses posteriorly during gastrulation, cells that are laid down at the midline in its wake form the notochord. Concurrently, region A cells which stream posteriorly and populate the midline of the neural tube over the newly formed notochord will form the floor plate.[115,118] Notochord grafts placed next to the neural tube can induce the morphological,[119,120] antigenic[107] and functional[121] properties of the floor plate in adjacent neural cells. Other ventral neural cell types, including motor neurons, also differentiate ectopically in response to notochord grafts (Fig. 8).[107,122] Conversely, floor plate cells and motor neurons do not develop in the absence of the notochord in amniotes.[67,107,121,123–126]

During vertebrate embryogenesis, one of the major strategies by which generation of diverse cell types is achieved is by inductive interactions, in which signals from one group of cells control the fate of adjacent cells.[127,128] The induction of the floor plate at the ventral midline of the neural tube is one of the earliest events in the establishment of D/V polarity in the vertebrate central nervous system.[11,129] The first neural cells to exhibit overt differentiation are located at the midline of the neural plate[130] and these later give rise

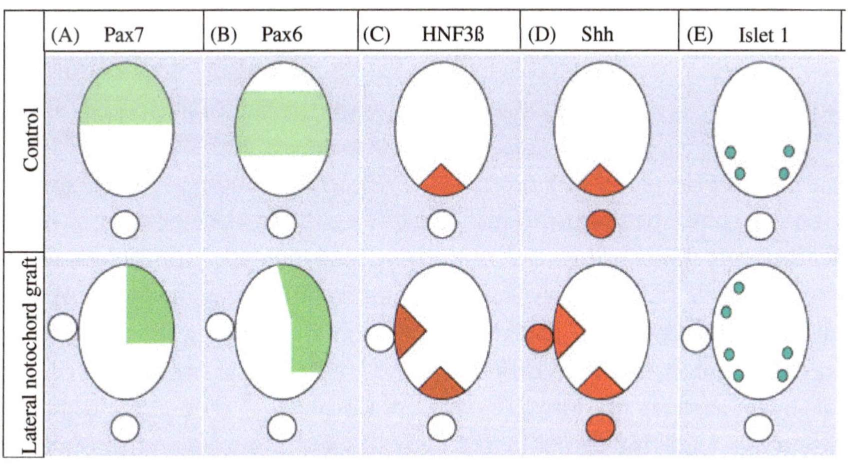

Fig. 8 Notochord grafts ventralize adjacent neuroepithelium. Serial adjacent transverse sections of neural tube, after a lateral notochord graft. (A) and (B) show the repression of Pax7 and Pax6 in the neural tube adjacent to the graft. (C), (D) and (E) show the induction of floor plate (HNF3β and Shh) and motor neurons (Islet 1) (modified from Patten and Placzek, 2000).

to the floor plate at the ventral midline of the neural tube. The cells in region A which are destined to populate the floor plate exhibit convergent-extension movements similar to those displayed by cells of the Hensen's node.[115,116,131] However, they neither express definitive floor plate markers, such as HNF3β, an indicator of early floor plate differentiation,[132] nor do they acquire floor plate properties when grown in isolation.[133] Thus, the differentiation of these anterior cells into floor plate appears to take place outside the node, and at a later developmental stage, following exposure to inductive signals from notochord.[134]

The secreted molecule, Sonic hedgehog, has been shown to be both necessary and sufficient for this important process. Shh is expressed initially by cells in the node, and later by axial mesodermal cells, i.e. the notochord, and finally by floor plate cells themselves. Gain-of-function genetic studies have shown that ectopic Shh signaling can lead to the ectopic differentiation of floor plate and other ventral cells from neural precursors *in vivo* and *in vitro*.[91,92,94,135] Mis-expression of the zinc-finger transcription factor, *Gli1*, and the winged helix-loop-helix transcription factor, HNF3β, both downstream effectors of the Shh-signaling pathway, cause the ectopic differentiation of ventral floor plate cells.[136,137] Such mis-expression studies have been complemented by loss-of-function studies. Inactivation of Shh signaling through the use of antibodies against SHH protein or by the targeted inactivation of the *shh* gene leads to failure of floor plate differentiation.[30,95] In the mouse three kinds of mutations have been shown to block floor plate development without affecting the specification of notochord cells. Firstly, mutations in the *shh* gene itself block floor plate differentiation. The generation of Shh-null mice reveals that in the absence of Shh signaling *in vivo*, floor plate and motor neurons fail to differentiate. However, the notochord of Shh-null mice appears to initially develop normally.[95] The second line of evidence is that mutations in the mouse zinc-finger transcription factor gene *Gli2*, a key intracellular mediator of Shh signaling, also block floor plate formation without affecting the expression of Shh by the notochord.[138,139] The third set of mutations, in the PS1 and PS2 Presenilin genes, also block floor plate differentiation at spinal cord levels while leaving Shh expression in the notochord intact.[140] Together, these experiments suggest that while the molecular pathways of floor plate and notochord differentiation are separable, Shh signaling is able to induce floor plate differentiation, and is necessary and sufficient for this process. This induction is thought to depend on the setting of a concentration gradient of Shh along the D/V axis of the neural tube. Every 2–3-fold increase in Shh concentration leads to the differentiation of a more ventral neural cell type and the most ventral cell type, the floor plate, needs the highest levels of Shh (Fig. 7).

4.1.2. *Homeogenetic induction of the floor plate*

Classical embryological experiments where the notochord was removed at different stages of the development in amphibian and chick embryos showed that late stages of floor plate differentiation can occur independently of the notochord.[141–143] Recent data from grafting studies in the chick also support the ability of the floor plate to induce floor plate differentiation.[107,144] In zebrafish, wild-type cells introduced into the neural plate of *cyclops* mutant embryos are able to confer floor plate fates in adjacent mutant cells.[19] These experiments suggest that floor plate induction may be initiated in the organizer, but appears to be continued through the neural plate by the propagation of a floor plate-derived signal. Homeogenetic induction, the induction of a tissue by the same type of tissue ("like-begets-like"), may underlie the marked increase in the number of the floor plate cells that occurs after neural tube closure and may ultimately re-induce floor plate after notochord removal.[12] The floor plate does not appear to serve as a passive carrier of notochord-derived inductive signals, and indeed, the floor plate appears to synthesize inducing signals for a longer period than the notochord. Thus, a substantial proportion of floor plate cells appears to be specified relatively late, and are thought to be derived from progenitor cells. The division and subsequent rostrocaudal extension of these progenitors is likely to operate together with homeogenetic induction to generate the final dimensions of the floor plate.

One model proposed suggests that the induction of the floor plate includes both vertical and homeogenetic induction. The induction of the floor plate occurs primarily after neural tube closure, and patterning of the ventral neural tube occurs in response to contact-mediated and diffusible signals derived first from the notochord and then from the floor plate itself. The notochord is the source of both contact-dependent and contact-independent signals. Cells immediately above the notochord receive both types of signals and form the floor plate, while those positioned more laterally receive only the diffusible signals and form motor neurons. At later stages, the floor plate moves apart from the notochord and becomes the other signaling center that can pattern the neural tube. Like the notochord, the cells of the floor plate can induce additional floor plate cells through a contact-dependent signal, and motor neurons and other ventral cell types via long-range diffusible factors (Fig. 9).[12,145–148]

4.2. Model Two: Floor Plate Induction Occurs Independent of the Notochord

4.2.1. *Node/organizer: the common source of floor plate and notochord cells*

In avian embryos, neurulation proceeds according to two different morphogenetic mechanisms in the anterior and posterior regions of the body. Anteriorly, in the cephalic, cervical and dorsal regions, the neural epithelium forms a neural plate whose lateral ridges fuse in the dorsal midline, thus generating the basis of the central nervous system: the neural tube and the neural crest, which forms the origin of the peripheral nervous system (PNS). This process, called primary neurulation, takes place from the anterior to the posterior neuropore (also see Fig. 1). Posteriorly, the neural primodium is formed during the elongation of the tail bud, from a cord of epithelial cells in which the lumen of the neural tube appears by cavitation. This type of neurulation, called secondary neurulation, is limited to the lumbo-sacro-caudal part of the body in amniotes.[2,5.]

New and more sensitive technologies have led to advances in our understanding of the mechanisms underlying cell commitment and

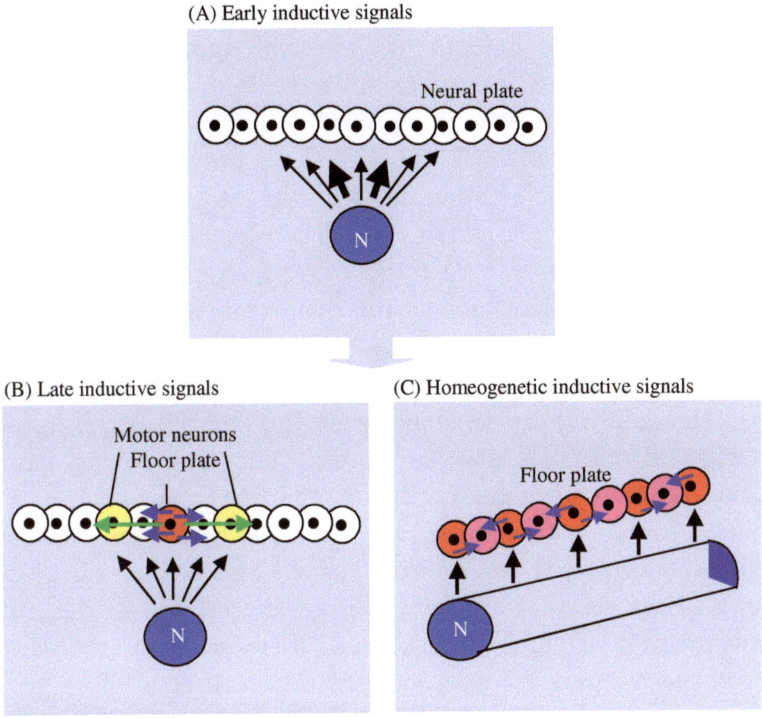

Fig. 9 Vertical induction and homeogenetic induction models for floor plate formation. Contact-dependent and contact-independent signals pattern the neural tube and induce the floor plate. (A) Early inductive signals from notochord, when the notochord is in contact with the neural tube. (B) Late inductive signals from notochord and floor plate. The notochord is separated from the neural plate at this stage. (C) Lateral view of the neural tube shows that floor plate can induce additional floor plate cells through a contact-dependent signal by homeogenetic induction. Contact-dependent signals are shown with blue arrows and contact-independent signals with green arrows.

patterning in early vertebrate embryos. Accurate, detailed, and high-resolution prospective fates maps, which reveal the sites of origins of relevant populations of cells during normal development, have been very important for this. These maps are very useful in tracing patterns of cell displacement that collectively constitute the morphogenetic movements that occur during gastrulation, cardiogenesis, neurulation and somitogenesis. Moreover, the availability of detailed prospective fate maps at multiple critical stages allow comparison between the expression patterns of inductive factors and signaling intermediaries, providing insights into the intercellular interactions occurring during formation of the vertebrate body axis.[149–151]

Among avians, quail-chick chimeras have provided a good model system to generate high-resolution fate maps at critical stages of early development to study the patterns of cell displacement. Quail and chick embryos develop in very similar ways (especially during early development). When portions of the quail embryo are grafted into a similar region of the chick embryo, the cells become integrated into the host embryo and participate in the construction of the appropriate organs. This grafting can be done while the embryo is still inside the egg shell, and the chick that hatches is a 'chimera', having a portion of its body composed of quail cells.[152,153] However, the chick and quail cells differ, in two critical ways. First, the quail heterochromatin in the nucleus is concentrated around the nucleoli. This creates a large deeply staining mass that is easily distinguishable from the diffuse heterochromatin of chick cells. Second, there are some antigens that are quail-specific and are not detected on chick cells. Both of these criteria allow one to readily distinguish individual quail cells, even when the majority of the host cell population is derived from chick.[151–153]

Fate maps of the avian blastoderm using fluorescently-labeled grafts and antibodies specific for grafted cells at intermediate primitive-streak stages (which later becomes Hensen's node) show the patterns of cell displacements that occur during progression of the primitive streak. Fine fate mapping results show that the future floor plate of the neural tube arises exclusively from a midline, circumscribed area just rostral to the primitive streak (Cr).[151] Floor plate cells originate rostral to the primitive streak and become incorporated into its cranial end during primitive streak progression, and the rostral end of primitive streak contributes cells to the notochord. Thus, in embryos at stages 3 and 4, the rostral end of primitive streak (Hensen's node) contains cells of both the prospective notochord and the prospective floor plate of the neural tube, as well as cells of the head mesenchyme and foregut endoderm.[151]

Cell lineage studies of the chick Hensen's node during secondary neurulation have been performed using the chick-quail chimera system in 6-somite stage embryos. At this stage, Hensen's node appears as a median depression situated in the middle of the sinus rhomboidalis. Replacement of the node and the underlying endoderm at this stage in chick by the quail counterpart demonstrates that during its rostrocaudal regression, the node material becomes inserted into the overlying neural plate, forming the future floor plate dorsally, and the notochord ventrally. At this stage, notochord and floor plate rudiments are intimately associated. Afterwards both of them become separated by a basement membrane. The notochord slides caudally in comparison to the floor plate, accounting for the different rostral levels of the graft-derived notochord and floor plate. Thus, Hensen's node is the structure containing the midline precursor cells of both floor plate and notochord (Fig. 10).[2,5,154,155]

In zebrafish, another very good vertebrate model system, cell lineage analyses at the onset of gastrulation suggest that zebrafish also possess a midline precursor cell population in the embryo shield,[156,157] a structure similar to the chick Hensen's node. By this stage, cells of the organizer region have already been specified to develop into particular tissue types and are under the control of zygotically expressed genes. Therefore, this model proposes that floor plate specification commences at gastrulation, and that it occurs independently of notochord development. Evidence from the study of zebrafish mutants provide further support for this model. A number of genes with defects in midline structures or signaling have been identified. Mutations in the *floating head* (*flh*) and *no tail* (*ntl*) genes result in embryos that lack notochord. The dorsal mesoderm cells that are notochord progenitors are mis-specified in these embryos, but patchy floor plate cells are detected in *flh* mutant embryos,[158] and a wider floor plate can be observed in *ntl* mutant embryos.[159] The *ntl/Brachyury* gene encodes a T-box transcription factor that is expressed in the rudiments of both the notochord and the tail, and is essential for proper development of both domains.[160–166] Fate mapping in *ntl* mutant embryos using caged fluorescein shows that the cells in the wider floor plate originate from

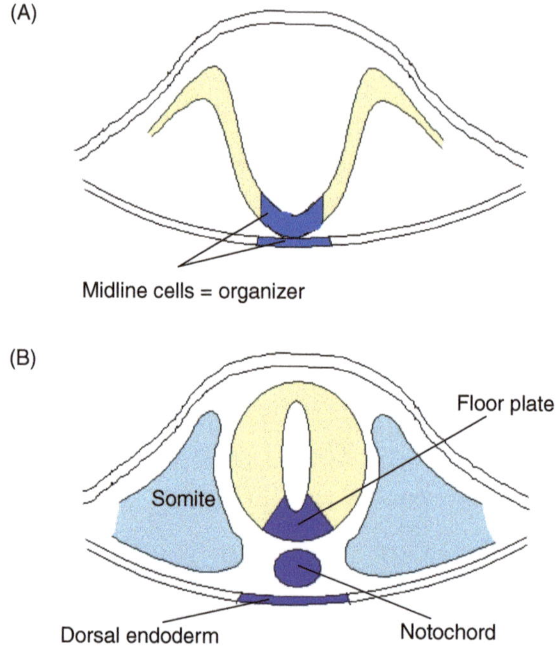

Fig. 10 Schematic representation of the rostrocaudal movement of Hensen's node (HN) in the chick. Hensen's node cells (blue) bisect the superficial layer of the sinus rhomboidalis which will become the alar plates of the spinal cord. Later on, the bulk of Hensen's node becomes segregated into three layers: the floor plate, the notochord and the dorsal endoderm.

a midline precursor population, and that *ntl* function is required during early gastrulation in cells that normally make notochord to repress floor plate and promote notochord fates.[167]

Even more intriguing mutants in zebrafish for the studies of floor plate induction are *cyclops* (*cyc*) which encodes a nodal-related TGF-β signaling factor, and its co-receptor *one-eyed pinhead* (*oep*), which encodes an epidermal growth factor EGF-related protein.[19,168–172] In these two mutants, floor plate cells are missing even though the notochord is present (Fig. 11). These cells can be restored in *cyc* mutants when wild-type *cyc* RNA is overexpressed in early embryos. The *cyclops* mutant phenotype is much like the *shh* mutant in mouse, implicating

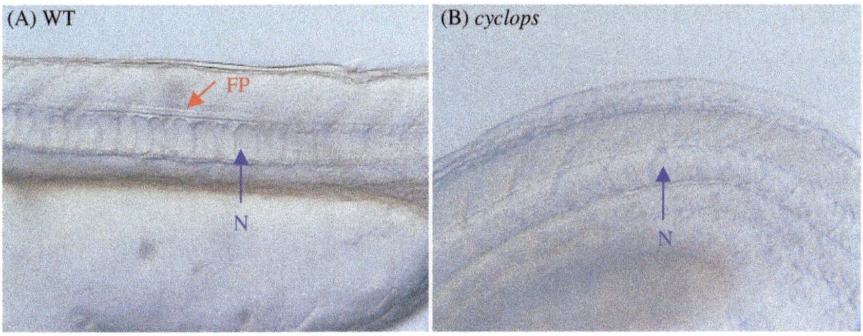

Fig. 11 The floor plate in a zebrafish wild-type embryo and its absence in a *cyclops* mutant embryo. (A) Lateral view of the trunk region of a wild-type zebrafish embryo showing the floor plate lying above the notochord. (B) In *cyclops* mutant embryos the floor plate is absent.

that similar to the function of the *shh* gene in the murine floor plate, Nodal signaling is required for floor plate specification at early gastrulation in zebrafish. Analyses of the shh promoter in zebrafish has also identified elements that are responsive to Cyclops/Nodal signaling.[173]

4.2.2. *When is the floor plate induced?*

Studies of Shh signaling in mouse and chick embryos have not resolved the major questions of the time and place at which floor plate differentiation begins. Fate mapping in chick and zebrafish have shown that the floor plate originates from Hensen's node during secondary neurulation and from the embryonic shield during gastrulation. Recently, a temperature-sensitive allele in *cyclops* (cyc^{sg1}) has been isolated.[174] In contrast to null mutations in cyclops, cyc^{sg1} mutants manifest variable *cyclops* phenotypes at 22°C. At 28°C, cyc^{sg1} homozygous mutant embryos show a complete lack of medial floor plate cells. Temperature shift experiments to abrogate Cyclops function in this mutant allele at various stages of gastrulation and segmentation show that the floor plate in zebrafish is induced during mid-gastrulation stages (between 70–80% epiboly). Furthermore, transient temperature-pulse experiments at the permissive or restrictive temperature using this mutant allele show that while the critical window for floor plate induction in zebrafish is during mid-gastrulation (70–80% epiboly), the formation of a complete floor plate requires continuous signaling by the Cyclops protein throughout gastrulation.[174] Thus, in zebrafish, floor plate induction occurs at early gastrulation. At these stages, Cyclops function is required for induction of the floor plate from its precursors, as well as the complete development of the ventral neural tube throughout the length of the neural axis.[174]

4.2.3. *Where are the floor plate cells induced: functional domains within Hensen's node*

Fine fate mapping studies analysing gene expression in different domains, the distinct cell morphologies and arrangement of cells in the node, and excisions and grafts of Hensen's node sub-regions at the 5–6 somite-stage indicate that Hensen's node is made up of three distinct functional domains, along the rostrocaudal axis.[175] A caudal region, designated as **zone a**, comprises cells of the presumptive floor plate, and is composed of epithelial cells closely apposed to more ventral cells that are randomly organized and are in continuity with the already individualized notochord. These two cellular compartments are separated by a discrete basement membrane. **Zone b** lies in the median pit, where the future floor plate and notochord domains are recognizable but not yet delaminated. **Zone c** includes the extreme caudal tip of the node and contains cells that are randomly distributed. Graft experiments using chick-quail chimeras have demonstrated that the floor plate and notochord are derived from a common group of cells present in **zones b and c** of Hensen's node. These cells are responsible for the formation of midline structures along the whole neural axis from the diencephalon down to the tail end. Cells of **zone c** seem to function as stem cells since they form the whole length of this midline structure without addition of cells from more lateral regions of the embryo during the process of node regression (Fig. 12).[175,176]

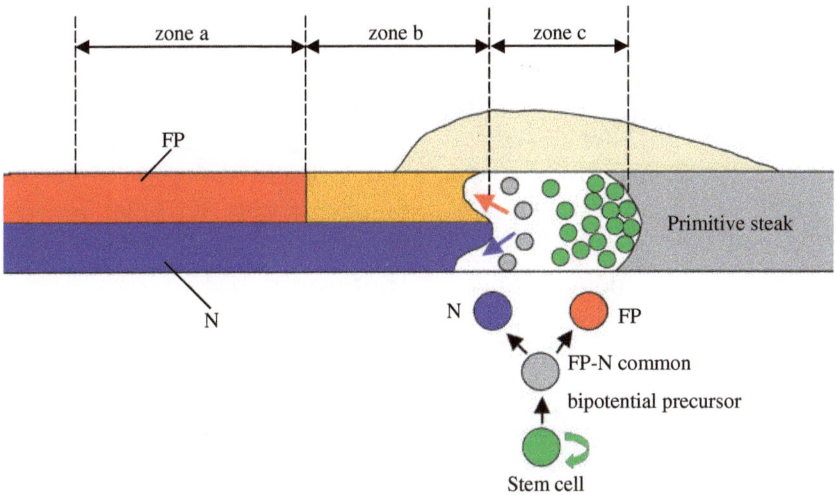

Fig. 12 The bipotential precursor model for midline cell development. The distribution of the three cellular areas forming Hensen's node. In this model, zone c has a pool of stem cells with a bipotential fate which can yield both floor plate (FP) and notochord (N) cells.

4.2.4. *The different origins of medial floor plate and lateral floor plate*

From analyses of immunocytochemical and molecular markers that label different cells in the ventral neural tube of several vertebrate species, two distinctive cell populations in the floor plate have been distinguished. Early studies in the chick ventral neural tube at day 3 show that a medial region where cells expressed both SC1 and FP1 antigens can be distinguished from lateral areas where cells express FP1 but not SC1.[107,144] In rat embryos, all floor plate cells express antigen FP3, whereas only the medial floor plate cells (MFP) express the FP4 antigen.[12,92] In mouse and rat embryos, shh, a very important floor plate marker gene, is expressed in the medial floor plate. In contrast, HNF3β, a gene that functions downstream of Shh signaling, and whose ectopic expression can induce ectopic floor plate marker gene expression, is detected in a larger region of the ventral neural tube.[18,177]

Further gene expression studies using chick-quail grafts suggest that during normal development of the chick embryo, the floor plate is heterogeneous and composed of regions that can be distinguished on the basis of their embryological origin and molecular characteristics.[2] During neurulation, the early neural tube, formed by planar induction in the dorsal ectoderm, lacks a floor plate territory. When Hensen's node regresses, the floor plate becomes intercalated into the neural ectoderm of the neural tube, the notochord becomes part of the mesoderm, and then the definitive neural plate is formed.[2] The gene activities of these two domains of definitive neural plate are different, and genes expressed in the neural ectoderm are not expressed in the node-derived midline structure (floor plate and notochord) during neurogenesis.[178] The node-derived floor plate is made up of polarized cylindrical epithelial cells that express *shh* and *HNF3β* and are fated to become MFP. By contrast, cells of the adjacent neural ectoderm flank the MFP and are designated as lateral floor plate (LFP). LFP is a pseudo-stratified structure of the neuro-epithelium and does not acquire the polarized morphology of the MFP.

In contrast to early data showing induction of floor-plate-like structures in the lateral neural tube of chick embryo by notochord or floor plate grafts, revisited studies show that fragments of notochord or MFP can induce both MFP and LFP in the neural epithelium if applied to embryos at stages ranging from 7–15 somite-stage. Moreover, the MFP is induced only over a short length of the neural tube of the host, located in close vicinity to Hensen's node. So the competence of the neuro-epithelium in responding to notochord or MFP signals is restricted to the posterior-most region of the neural tube in a short time window. More rostrally, the grafts merely induced LFP-type gene activities. Revisited studies that ectopically express Shh also show that MFP and Shh-producing cells only induce LFP-type cells. This suggests that in chick, Shh is not involved in specifying the MFP itself but is essential for inducing the LFP.[13] In grafts from the epiblast anterior to **zone a**, application of Shh protein together with the TGFβ-related Nodal protein resulted in floor plate markers at the same dose that each individual protein was unable to induce.

These experiments have shown that signaling by Nodal factors can function synergistically with Shh in this process. Thus, Shh alone may not be sufficient to confer MFP fates.[179] Furthermore, floor plate in the anterior versus the posterior neural tube may be induced by distinct mechanisms.

In the zebrafish, the MFP consists of a single row of cells flanked on each side by one or two additional rows of LFP cells.[21,180] MFP and LFP can be distinguished by the differential expression patterns of a number of genes. Cells of the MFP express *netrin1*,[181] *shh*,[91] *twhh*,[96] *col2a1*[182] and several forkhead family members: *foxA2* (*axial/HNF3B*), *fkd7* and *fkd4*.[169,180,183] *FoxA2* and *fkh4*[180] are expressed in both MFP and LFP. The *nk2.2* is expressed in the LFP only.[184] Genetic analysis has shown that embryos with mutations in the zebrafish *shh* gene, *sonic-you* (*syu*), unlike mouse *shh* mutants, do form MFP cells and motor neurons, and lack LFP cells. Similarly, the MFP is not abolished by mutations in *you-too* (*yot*), the zebrafish homolog of mouse Gli2,[185] or *slow-muscle-omitted* (*smu*), the zebrafish homolog of *smoothened*, which encodes the transmembrane receptor for Shh.[186,187] In zebrafish, three hedgehog genes are expressed in the midline. *Shh* is expressed in the notochord, the MFP, the ventral midline of the brain and the posterior fin bud of embryo[91]; *twhh* is restricted to the MFP and the ventral midline of the brain during early somitogenesis,[96] and *ehh* is expressed in the notochord exclusively.[97] Inhibition of Hh expression or signaling by pharmacologic interference using the pan-Hh inhibitor, cyclopamine[188] or by antisense morpholino-mediated knockdown approaches[99, 101, 186, 189] does not inhibit differentiation of MFP in 24 hpf embryos. Thus, Hh-signaling pathway seems neither required nor sufficient for the induction of MFP. However, it is necessary for the formation of LFP, and for the recovery of MFP cells in *cyclops* mutant embryos after 48 hpf.[100]

Mutations that affect the Nodal-signaling pathway, *cyc* and *oep*, lack MFP and lack expression of MFP marker genes (*shh*, *twhh*, *netrin1* and *F-spondin*) at 24-pdf. The LFP and notochord are formed and *shh* mRNA is transcribed in both *cyc* and *oep* mutants.[91,168,181] This suggests that MFP differentiation is dependent on Nodal signaling. Recent data from floor plate marker genes analysis in *cyc* mutant embryos at later stages (48-hpf) has shown that MFP markers (*netrin1*, *shh* and *F-spondin*) are expressed in the posterior body axis of mutant embryos, and require intact Shh signaling.[190] This delayed differentiation of floor plate in *cyc* mutant embryos suggests an involvement of Hh signaling during late stages. Early data has shown that inhibition of Hh signaling by either knockdown, inhibitor expression, or genetic lesions can cause defects in MFP-specific gene expression in stages older than 1 day.[186,187] Thus, the late function of Hhs is responsible for the delayed differentiation of MFP cells in *cyc* mutant embryos to maintain the floor plate as a coherent structure. In *cyc* mutants, the notochord expresses *shh* and *ehh* and is close to the neural tube, implicating the notochord as the source of Hh signals. The other signaling source may be the floor plate itself, which has homeogenetic-induction properties. Since cells expressing MFP genes in *cyc* mutant also express the LFP marker nk2.2, and the LFP cells occupy the ventral-most aspect of the neural tube of 24-hpf *cyc* and *oep* mutants, it is likely that LFP is the source of MFP precursor cells. These studies suggest that in zebrafish, Cyclops/Nodal signaling is required for the early MFP differentiation and is also required for the discrimination between MFP and LFP, whereas Hh signaling may be required later.[190]

5. Concluding Remarks

The origin of the floor plate: reconciling planar, vertical and homeogenetic induction

The two models of floor plate differentiation provide very strong evidence to support the two different pathways for floor plate induction. Model one combines the genetic, cellular and embryological studies to draw three main conclusions. First, separate molecular pathways control floor plate and notochord differentiation. Second, Shh signaling is necessary for floor plate induction and differentiation in amniote embryos. Third, signals provided by the notochord, a source of Shh, are required for induction of the floor plate. On the other hand, cell lineage analysis and genetic studies in zebrafish mutants show that the floor plate and notochord come from the same origin, and more importantly, Shh is not required for floor plate induction, at least not for initial induction of the medial floor plate. Thus far, with seemingly contradictory evidence from amniotes versus fish model systems, we still do not have a complete understanding of floor plate differentiation. Many aspects of the early cellular interactions that control the decision of axial midline cells to embark upon distinct pathways of notochord and floor plate differentiation need to be defined more clearly.

However, comparing the data that support the two models, we can find that they are not mutually exclusive. There occurs some overlap in these two models. For instance, the vertical induction of floor plate by Shh secreted from notochord also occurs in model two for lateral floor plate differentiation. Homeogenetic induction plays an important role for floor plate differentiation in late stages in both the models. All of these show that the floor plate differentiation is not a simple process. It needs many interactions that combine the planar, vertical and homeogenetic signals and in addition to HH signaling, more signaling pathways, notably the Nodal-signaling pathway, are involved in the differentiation of a complete and functional floor plate. As more is learnt about the interplay between various signaling pathways, and as more downstream effectors of the pathways that determine floor plate fates are identified, we will be able to understand better how this group of cells achieves its form and functions.

Acknowledgments

We thank Aniket Gore, Vladimir Korzh, Rida Padmashree, and Mahendra Wagle for critical reading, and members of our laboratory for discussions. Work in the laboratory of K.S. is supported by the Temasek Life Sciences Laboratory.

References

1. Schoenwolf GC (1991). *Development* **2** (Suppl.): 157–168.
2. Catala M, Teillet M-A, De Robertis EM and Le Douarin NM (1996). *Development* **122**: 2599–2610.

3. Gilbert SF (1997). *Dev. Biol.* Sinauer Associates, Inc, Sunderland, pp. 253–306.
4. Schoenwolf GC (1991). *Cell Movements in the Epiblast during Gastrulation and Neurulation in Avian Embryos,* Plenum, New York, pp. 1–28.
5. Catala M, Teillet M-A and Le Douarin NM (1995). *Mech. Dev.* **51**: 51–65.
6. Pasteels J, (1937). *Arch. Biol.* **48**: 381–488.
7. Gont LK, Steinbeisser H, Blumberg B and De Robertis EM (1993). *Development* **119**: 991–1004.
8. Schoenwolf GC and Desmond NE (1984). *Anat. Rec.* **209**: 251–263.
9. Nievelstein RAJ, Hartwig NG, Vermeij-keers C and Valk J (1993). *Teratology* **48**: 21–31.
10. His W (1886). *Ges. Wiss.* BD **13**: S. 477.
11. Jessell TM and Dodd J (1992). *Harvey Lect.* **86**: 87–128.
12. Placzek M, Jessell TM and Dodd J (1993). *Development* **117**: 205–218.
13. Charrier JB, Lapointe F, Le douarin NM and Teillet MA (2002). *Development* **129**: 4785–4796.
14. Klar A, Baldassare M and Jessell TM (1992). *cell* **69**: 95–110.
15. Matsui Y, Zsebo KM and Hogan BLM (1990). *Nature* **347**: 667–669.
16. Puelles L, Amat JA and Martinez M (1987). *Neurol. J. Comp.* **266**: 247–268.
17. Ruiz i Altaba A, Prezioso VR, Darnell J and Jessell TM (1993). *Mech. Dev.* **44**: 91–108.
18. Sasaki H and Hogan BLM (1993). *Development* **118**: 47–59.
19. Hatta K, Kimmel CB, Ho RK and Walker C (1991). *Nature* **350**: 339–341.
20. Kuwada JY, Bernhardt RR and Chitnis AB (1990). *J. Neurosci.* **10**: 1299–1308.
21. Odenthal J, van Eeden FJ, Haffter P, Ingham PW and Nussllein–Volhard C (2000). *Dev. Biol.* **219**: 350–363.
22. Tanabe Y and Jessell TM (1996). *Science* **274**: 1115–1123.
23. Briscoe J, Pierani A, Jessell TM and Ericson J (2000). *Cell* **101**: 435–445.
24. Jessell TM (2000). *Nature review* **1**: 20–29.
25. Fietz MJ, Concordet JP, Barbosa R, Johnson R, Krauss S, McMahon AP, Tabin C and Ingham PW (1994). *Dev. Suppl.* pp. 43–51.
26. Bumcrot DA, Takada R and McMahon AP (1995). *Mol. Cell Biol.* **15**: 2294–2303.
27. Porter JA, von Kessler DP, Ekker SC, Young KE, Lee JJ, Moses K and Beachy PA (1995). *Nature* **374**(6520): 363–366.

28. Hammerschmidt M, Brook A and McMahon AP (1997). *Trends Genet.* **13**: 14–21.
29. Porter JA, Ekker SC, Park WJ, von Kessler DP, Young KE, Chen CH, Ma Y, Woods AS, Cotter RJ, Koonin EV, and Beachy PA (1996). *Cell* **86**: 21–34.
30. Ericson J, Morton S, Kawakami A, Roelink H and Jessell TM (1996). *Cell* **87**: 661–673.
31. Goulding MD, Lumsden A and Gruss P (1993). *Development* **117**: 1001–1016.
32. Stuart ET, Kioussi C and Gruss P (1994). *Annu. Rev. Genet.* 219–236.
33. Robert B, Lyons G, Simandl BK, Kuroiwa A and Buckingham M (1991). *Genes Dev.* **5**: 2363–2374.
34. Liem KF Jr, Tremml G, Roelink H and Jessell TM (1995). *Cell* **82**: 969–979.
35. Carlson BM (1999). *Human Embryology and Developmental Biology* Mosby-Year Book, 2nd. ed.
36. Nieto MA, Sargent MG, Wilkinson DG and Cooke J (1994). *Science* **264**: 835–839.
37. Colamarino SA and Tessier–Lavigne M (1995). *Annu. Rev. Neurosci.* **18**: 497–529.
38. Holley JA (1982). *J. Comp. Neurol.* **205**: 371–382.
39. Holley JA and Silver J (1987). *Dev. Biol.* **123**: 375–388.
40. Oppenheim RW, Shneiderman A, Shimizu I and Yaginuma H (1988). *J. Comp. Neurol.* **275**: 159–180.
41. Yaginuma H, Shiga T, Homma S, Ishihara R and Oppenheim RW (1990). *Development* **108**: 705–716.
42. Yaginuma H, Homma S, Kunzi R and Oppenheim RW (1991). *J. Comp. Neurol.* **304**: 78–102.
43. Altman J, Bayer SA (1984). *Adv. Anat. Embryol. Cell Biol.* **85**: 1–164.
44. Wentworth LW (1984). *Neurol. J. Comp.* **222**: 96–115.
45. Bernhardt RR, Nguyen N and Kuwada JY (1992). *Neuron* **8**: 869–882.
46. Bernhardt RR, Patel CK, Wilson SW and Kuwada JY (1992). *J. Comp. Neurol.* **326**: 263–272.
47. Dale N, Roberts A, Ottersen OP and Storm–Mathisen J (1987). *Proc. R. Soc. Lond. B. Biol. Sci.* **232**: 193–203.
48. Jacobson M and Huang S (1985). *Dev. Biol.* **110**: 102–113.
49. Roberts A and Clarke JDW (1982). *Soc. Trans. R. B. Lond. Ser.* **296**: 195–212.
50. Roberts A, Dale N, Ottersen OP and Storm–Mathisen J (1988). *Development* **103**: 447–461.

51. Kimmel CB, Powell SL and Metcalfe WK (1982). *J. Comp. Neurol.* **205**: 112–127.
52. Metcalfe WK, Mendelson B and Kimmel CB (1986). *J. Comp. Neurol.* **251**: 147–159.
53. Clarke JD and Lumsden A (1993). *Development* **118**: 151–162.
54. Glover JC (1993). *Brain Res. Bull.* **30**: 265–271.
55. Kimmel CB (1993). *Annu. Rev. Neurosci.* **16**: 707–732.
56. Trevarrow B, Marks DL and Kimmel CB (1990). *Neuron* **4**: 669–679.
57. Kroger S and Schwarz U (1990). *J. Neurosci.* **10**: 3118–3134.
58. Shepherd IT and Taylor JS (1995). *J. Comp. Neurol.* **354**: 501–510.
59. Shirasaki R, Tamada A, Katsumata R and Murakami F (1995). *Neuron.* **14**: 961–972.
60. Tessier–Lavigne M, Placzek M, Lumsden AG, Dodd J and Jessell TM (1988). *Nature* **336**: 775–778.
61. Ebendal T, Jacobson CO (1977). *Exp. Cell Res.* **105**: 379–387.
62. Lumsden AG and Davies AM (1983). *Nature* **306**: 786–788.
63. Lumsden AG and Davies AM (1986). *Nature* **323**: 538–539.
64. Placzek M, Tessier–Lavigne M, Jessell T and Dodd J (1990). *Development* **110**: 19–30.
65. Yaginuma H and Oppenheim RW (1991). *J. Neurosci.* **11**: 2598–2613.
66. Placzek M, Tessier–Lavigne M, Yamada T, Dodd J and Jessell TM (1990). *Cold Spring Harb. Symp. Quant. Biol.* **55**: 279–289.
67. Bovolenta P and Dodd J (1991). *Development* **113**: 625–639.
68. Murakami F and Shirasaki R (1997). *Cell Tissue Res.* **290**: 323–330.
69. Hatta K (1992). *Neuron* **9**: 629–642.
70. Guthrie S and Pini A (1995). *Neuron* **14**: 1117–1130.
71. Shirasaki R, Mirzayan C, Tessier–Lavigne M and Murakami F (1996). *Neuron* **17**: 1079–1088.
72. Tamada A, Shirasaki R and Murakami F (1995). *Neuron* **14**: 1083–1093.
73. Greenspoon S, Patel CK, Hashmi S, Bernhardt RR and Kuwada JY (1995). *J. Neurosci.* **15**: 5956–5965.
74. Tear G (1999). *Cell Mol. Life Sci.* **55**: 1365–1376.
75. Serafini T, Kennedy TE, Galko MJ, Mirzayan C, Jessell TM and Tessier–Lavigne M (1994). *Cell* **78**: 409–424.
76. Hedgecock EM, Culotti JG and Hall DH (1990). *Neuron* **4**: 61–85.
77. Ishii N, Wadsworth WG, Stern BD, Culotti JG and Hedgecock EM (1992). *Neuron* **9**: 873–881.

78. Mitchell KJ, Doyle JL, Serafini T, Kennedy TE, Tessier–Lavigne M, Goodman CS and Dickson BJ (1996). *Neuron* **17**: 203–215.
79. Serafini T, Colamarino SA, Leonardo ED, Wang H, Beddington R, Skarnes WC and Tessier–Lavigne M (1996). *Cell* **87**: 1001–1014.
80. Colamarino SA and Tessier–Lavigne M (1995). *Cell* **81**: 621–629.
81. Boisseau S, Nedelec J, Poirier V, Rougon G and Simonneau M (1991). *Development* **112**: 69–82.
82. Dodd J, Morton SB, Karagogeos D, Yamamoto M and Jessell TM (1988). *Neuron* **1**: 105–116.
83. Krushel LA, Prieto AL, Cunningham BA and Edelman GM (1993). *Neuroscience* **53**: 797–812.
84. Shiga T and Oppenheim RW (1991). *J Comp Neurol.* **310**: 234–52.
85. Stoeckli ET and Landmesser LT (1995). *Neuron* **14**: 1165–1179.
86. Landmesser LT (1992). *The Nerve Growth Cone* K.S. Letourneau PC, Macagno ER, Eds. Raven Press, New York, pp 373–385.
87. Varela-Echavarria A, Tucker A, Puschel AW and Guthrie S (1997). *Neuron* **18**: 193–207.
88. Pfaff S and Kintner C (1998). *Curr. Opin. Neurobiol.* **8**: 27–36.
89. Patten I and Placzek M (2000). *Cell Mol. Life Sci.* **57**: 1695–1708.
90. Echelard Y, Epstein DJ, St-Jacques B, Shen L, Mohler J, McMahon JA and McMahon AP (1993). *Cell* **75**: 1417–1430.
91. Krauss S, Concordet JP and Ingham PW (1993). *Cell* **75**: 1431–1444.
92. Roelink H, Augsburger A, Heemskerk J, Korzh V, Norlin S, Ruiz i Altaba A, Tanabe Y, Placzek M, Edlund T and Jessell TM (1994). *Cell* **76**: 761–775.
93. Yamada T, Pfaff SL, Edlund T and Jessell TM (1993). *Cell* **73**: 673–686.
94. Marti E, Takada R, Bumcrot DA, Sasaki H and McMahon AP (1995). *Development* **121**: 2537–2547.
95. Chiang C, Litingtung Y, Lee E, Young KE, Corden JL, Westphal H and Beachy PA (1996). *Nature* **383**: 407–413.
96. Ekker SC, Ungar AR, Greenstein P, von Kessler DP, Porter JA, Moon RT and Beachy PA (1995). *Curr Biol.* **5**: 944–955.
97. Currie PD and Ingham PW (1996). *Nature* **382**: 452–455.
98. Chandrasekhar A, Moens CB, Warren JT Jr, Kimmel CB and Kuwada JY (1997). *Development* **124**: 2633–2644.
99. Etheridge LA, Wu T, Liang JO, Ekker SC and Halpern ME (2001). *Genesis* **30**: 164–169.

100. Schauerte HE, van Eeden FJ, Fricke C, Odenthal J, Strahle U and Haffter P (1998). *Development* **125**: 2983–2993.
101. Lewis KE and Eisen JS (2001). *Development* **128**: 3485–3495.
102. Ericson J, Muhr J, Placzek M, Lints T, Jessell TM and Edlund T (1995). *Cell* **81**: 747–756.
103. Pringle NP, Yu WP, Guthrie S, Roelink H, Lumsden A, Peterson AC and Richardson WD (1996). *Dev. Biol.* **177**: 30–42.
104. Shimamura K and Rubenstein JL (1997). *Development* **124**: 2709–2718.
105. Ericson J, Briscoe J, Rashbass P, van Heyningen V and Jessell TM (1997). *Cold Spring Harb. Symp. Quant. Biol.* **62**: 451–466.
106. Orentas DM, Hayes JE, Dyer KL, and Miller RH (1999). *Development* **126**: 2419–2429.
107. Yamada T, Placzek M, Tanaka H, Dodd J and Jessell TM (1991). *Cell* **64**: 635–647.
108. Hynes M, Porter JA, Chiang C, Chang D, Tessier–Lavigne M, Beachy PA and Rosenthal A (1995). *Neuron* **15**: 35–44.
109. Hynes M, Poulsen K, Tessier–Lavigne M and Rosenthal A (1995). *Cell* **80**: 95–101.
110. Ye W, Shimamura K, Rubenstein JL, Hynes MA and Rosenthal A (1998). *Cell* **93**: 755–766.
111. Wilson V and Beddington RS (1996). *Mech. Dev.* **55**: 79–89.
112. Kanki JP, Ho RK (1997). *Development* **124**: 881–893.
113. Spemann H (1938). *Embryonic Development and Induction.* Yale University Press, New Haven.
114. Hara K (1978). In Nakamura STO (ed.), *Organizer — A Milestone of a Half Century from Spemann.* Elsevier, Amsterdam, pp. 221–265.
115. Schoenwolf GC, Bortier H and Vakaet L (1989). *J. Exp. Zool.* **249**: 271–278.
116. Schoenwolf GC, Everaert S, Bortier H and Vakaet L (1989). *Anat. Embryol. (Berl).* **179**: 541–549.
117. Schoenwolf GC, Garcia-Martinez V and Dias MS (1992). *Dev. Dyn.* **193**: 235–248.
118. Selleck MA and Stern CD (1991). *Development* **112**: 615–626.
119. van Straaten HW, Hekking JW, Wiertz–Hoessels EJ, Thors F and Drukker J (1988). *Anat. Embryol. (Berl)* **177**: 317–324.
120. Smith JL and Schoenwolf GC (1989). *J. Exp. Zool.* **250**: 49–62.
121. Placzek M, Tessier–Lavigne M, Yamada T, Jessell T and Dodd J (1990). *Science* **250**: 985–988.

122. Ericson J, Thor S, Edlund T, Jessell TM and Yamada T (1992). *Science* **256**: 1555–1560.
123. Clarke JD, Holder N, Soffe SR and Storm–Mathisen J (1991). *Development* **112**: 499–516.
124. Hirano S, Fuse S and Sohal GS (1991). *Science* **251**: 310–313.
125. van Straaten HW and Hekking JW (1991). *Anat. Embryol. (Berl.)* **184**: 55–63.
126. Ruiz i Altaba A (1992). *Development* **116**: 67–80.
127. Gurdon JB (1992). *Cell* **68**: 185–199.
128. Jessell TM and Melton DA (1992). *Cell* **68**: 257–270.
129. Ruiz i Altaba A and Jessell TM (1993). *Curr. Opin. Genet. Dev.* **3**: 633–640.
130. Schoenwolf GC and Smith JL (1990). *Development* **109**: 243–270.
131. Rosenquist GC (1966). *Carnegie Contrib. Embryol Inst.* **38**: 71–110.
132. Ruiz i Altaba A, Jessell TM and Roelink H (1995). *Mol. Cell Neurosci.* **6**: 106–121.
133. Placzek M (1995). *Curr. Opin. Genet. Dev.* **5**: 499–506.
134. Schoenwolf GC and Sheard P (1990). *J. Exp. Zool.* **255**: 323–339.
135. Rowitch DH, St-Jacques B, Lee SM, Flax JD, Snyder EY and McMahon AP (1999). *J. Neurosci.* **19**: 8954–8965.
136. Hynes M, Stone DM, Dowd M, Pitts–Meek P, Goddard A, Gurney A and Rosenthal A (1997). *Neuron* **19**: 15–26.
137. Sasaki H and Hogan BL (1994). *Cell* **76**: 103–115.
138. Matise MP, Epstein DJ, Park HL, Platt KA and Joyner AL (1998). *Development* **125**: 2759–70.
139. Ding Q, Motoyama J, Gasca S, Mo R, Sasaki H, Rossant J and Hui CC (1998). *Development* **125**: 2533–2543.
140. Donoviel DB, Hadjantonakis AK, Ikeda M, Zheng H, Hyslop PS and Bernstein A (1999). *Genes Dev.* **13**: 2801–2810.
141. Horstadius S (1944). *Zool.* **25**: 257–265.
142. Kitchin IC (1949). *Zool. J. Exp.* **112**: 393–415.
143. Watterson RL, Fowler I and Fowler BJ (1954). *Anat. Am. J.* **95**: 337–400.
144. Placzek M, Yamada T, Tessier–Lavigne M, Jessell T and Dodd J (1991). *Development* **2**(Suppl): 105–122.
145. Smith JC (1993). *Cur. Biol.* **3**: 582–585.
146. Smith JC (1994). *Cell* **76**: 193–196.
147. Dodd J, Jessell TM and Placzek M (1998). *Science* **282**: 1654–1657.

148. Placzek M, Dodd J and Jessell TM (2000). *Curr. Opin. Neurobiol.* **10**: 15–22.
149. Garcia–Martinez V, Alvarez IS and Schoenwolf GC (1993). *J. Exp. Zool.* **267**: 431–446.
150. Garcia–Martinez V and Schoenwolf GC (1993). *Dev. Biol.* **159**: 706–719.
151. Lopez–Sanchez C, Garcia–Martinez V and Schoenwolf GC (2001). *Cells Tissues Organs* **169**: 334–346.
152. Le Douarin NM, (1969). *Fr. Bull. Biol. Belg.* **103**: 435–452.
153. Le Douarin NM and Teillet MA (1973). *J. Embryol. Exp. Morphol.* **30**: 31–48.
154. Teillet MA, Lapointe F and Le Douarin NM (1998). *Proc. Natl. Acad. Sci. USA* **95**: 11733–11738.
155. Le Douarin NM and Halpern ME (2000). *Curr. Opin. Neurobiol.* **10**: 23–30.
156. Melby AE, Warga RM and Kimmel CB (1996). *Development* **122**: 2225–2237.
157. Shih J and Fraser SE (1995). *Development* **121**: 2755–2765.
158. Talbot WS, Trevarrow B, Halpern ME, Melby AE, Farr G, Postlethwait JH, Jowett T, Kimmel CB and Kimelman D (1995). *Nature* **378**: 150–157.
159. Halpern ME, Ho RK, Walker C and Kimmel CB (1993). *Cell* **75**: 99–111.
160. Kispert A and Herrmann BG (1993). *EMBO J.* **12**: 3211–3220.
161. Kispert A, Koschorz B and Herrmann BG (1995). *EMBO J.* **14**: 4763–4772.
162. Kispert A, Ortner H, Cooke J and Herrmann BG (1995). *Dev. Biol.* **168**: 406–415.
163. Conlon FL, Sedgwick SG, Weston KM and Smith JC (1996). *Development* **122**: 2427–2435.
164. Wilkinson DG, Bhatt S and Herrmann BG (1990). *Nature* **343**: 657–659.
165. Smith JC, Price BM, Green JB, Weigel D and Herrmann BG (1991). *Cell* **67**: 79–87.
166. Schulte–Merker S, van Eeden FJ, Halpern ME, Kimmel CB and Nusslein–Volhard C (1994). *Development* **120**: 1009–15.
167. Amacher SL, Draper BW, Summers BR and Kimmel CB (2002). *Development* **129**: 3311–3323.
168. Schier AF, Neuhauss SC, Helde KA, Talbot WS and Driever W (1997). *Development* **124**: 327–342.
169. Strahle U, Blader P and Ingham PW (1996). *Int. J. Dev. Biol.* **40**: 929–940.

170. Sampath K, Rubinstein AL, Cheng AM, Liang JO, Fekany K, Solnica-Krezel L, Korzh V, Halpern ME and Wright CV (1998). *Nature* **395**: 185–189.
171. Rebagliati MR, Toyama R, Haffter P and Dawid IB (1998). *Proc. Natl. Acad. Sci. USA*. **95**: 9932–9937.
172. Zhang J, Talbot WS and Schier AF (1998). *Cell* **92**: 241–251.
173. Muller F, Albert S, Blader P, Fischer N, Hallonet M and Strahle U (2000). *Development* **127**: 3889–3897.
174. Tian J, Yam C, Balasundaram G, Wang H, Gore A and Sampath K (2003). *Development* **130**: 3331–3342.
175. Charrier JB, Teillet MA, Lapointe F and Le Douarin NM (1999). *Development* **126**: 4771–4783.
176. Le Douarin NM (2001). *Int. J. Dev. Biol.* **45**: 373–378.
177. Monaghan AP, Kaestner KH, Grau E and Schutz G (1993). *Development* **119**: 567–578.
178. Duprez D, Leyns L, Bonnin MA, Lapointe F, Etchevers H, De Robertis EM and Le Douarin N (1999). *Mech. Dev.* **89**: 179–183.
179. Patten P, Kulesa MM, Shen S, Fraser and Placzek M (2003). *Development* **130**: 4809–4821.
180. Odenthal J and Nusslein–Volhard C (1998). *Dev. Genes. Evol.* **208**: 245–258.
181. Strahle U, Fischer N and Blader P (1997). *Mech. Dev.* **62**: 147–160.
182. Yan YL, Hatta K, Riggleman B and Postlethwait JH (1995). *Dev. Dyn.* **203**: 363–376.
183. Strahle U, Blader P, Henrique D and Ingham PW (1993). *Genes. Dev.* 7: 1436–1446.
184. Barth KA and Wilson SW (1995). *Development* **121**: 1755–1768.
185. Karlstrom RO, Talbot WS and Schier AF (1999). *Genes. Dev.* **13**: 388–393.
186. Chen W, Burgess S and Hopkins N (2001). *Development* **128**: 2385–2396.
187. Varga ZM, Amores A, Lewis KE, Yan YL, Postlethwait JH, Eisen JS and Westerfield M (2001). *Development* **128**: 3497–3509.
188. Neumann CJ, Grandel H, Gaffield W, Schulte–Merker S and Nusslein–Volhard C (1999). *Development* **126**: 4817–4826.
189. Nasevicius A and Ekker SC (2000). *Nat. Genet.* **26**: 216–220.
190. Albert S, Muller F, Fischer N, Biellmann D, Neumann C, Blader P and Strahle U (2003). *Dev. Dyn.* **226**: 59–66.

Chapter 5

Form and Function in the Zebrafish Nervous System

Michael Hendricks and Suresh Jesuthasan
Developmental Neurobiology Group
Temasek Life Sciences Laboratory
1 Research Link, NUS, Singapore 117604

Introduction

George Streisinger originally conceived of the zebrafish as a vertebrate model organism for the study of the molecular and genetic basis of neuronal architecture.[1] The relative simplicity of the larval nervous system, the optical transparency and rapid external development of the embryo, its accessibility to embryological manipulation, and the potential for forward genetics made it a promising system for such an undertaking. These and other experimental advantages of zebrafish have been described often in recent years. And while there is a growing literature on zebrafish neurobiology, a brief search of the PubMed online database shows that only about 15% of articles on zebrafish are related to the nervous system. Most published papers focus on early patterning at the gastrula and neurula stages. In this chapter, however, we will highlight the zebrafish's unique potential for addressing questions related to the development and function of the nervous system.

A satisfying model of how the nervous system functions is difficult to envision. It would have to comprise elements on multiple levels, ranging from the molecular and genetic to the physiological and behavioral. An immense body of work exists on neurophysiology, anatomy, psychobiology, as well as the genetics and molecular biology of the nervous system. However, integration of this work into models that can account for even simple animal behaviors is still rare, particularly in vertebrates. We are just beginning to get a glimpse of the intense phenotypic diversity of neuronal types and the intricacy of the system's architecture at the molecular level, let alone understand precisely how information is represented and processed and functional behaviors are produced. The ontogenetic processes involved are so complex and subtle that — for an example — one breed of dog can be born predisposed to herding behavior, while another will retrieve.

Scientists have worked on many organisms in their bid to understand how the nervous system forms and functions. Simple model organisms such as the fly and worm have been used to identify genes that regulate neurogenesis,[2,3] axon pathfinding,[4,5] olfaction,[6] mechanotransduction,[7] vision,[8] and learning and memory.[9] Vertebrates such as the mouse and chick have also been invaluable in the isolation of molecules involved in axon guidance, primarily through biochemical and *in vitro* assays. Whole animal studies have been essential in understanding sensory perception and learning. With this backdrop of intensive scientific activity, a useful question to keep in mind is: "Why zebrafish?" What can it contribute to neurobiology that other model systems cannot? The zebrafish clearly has some weaknesses: unlike the mouse there is no established method for targeted recombination, which would make brain-specific knock-outs possible, for example. Biochemical approaches, which led to the identification of ephrins[10] and the recently cloned Repulsive Guidance Molecule (RGM),[11] can more easily be done in model systems with larger brains, not in zebrafish. The MARCM system, which enables homozygous mutant clones to be generated in the fly and thus allows the identification of genes that might have pleiotropic effects,[12] is not available.

One thing the zebrafish does provide, however, is the possibility of physiology in the intact animal, either by high-resolution optical recording or electrophysiology, combined with genetics. Imaging of living neurons, whether of whole growth cones or of molecules within, is also possible. Additionally, with the genome sequence data that is already available, and with sequences from Fugu[13] and other species, it is already possible to carry out large-scale screens using antisense morpholino oligonucleotides[14] against a large number of known and predicted genes. Gene-trap approaches[15] can be used to identify genes expressed in the nervous system, and potentially to analyze their function. In this chapter, we hope to demonstrate the niche in neurobiology that the zebrafish is just beginning to fill.

Development of the Nervous System: Axon Guidance

Axon guidance is an active — and actively reviewed — field. Much of the molecular and cellular machinery involved has come to light in recent years. For depth and breadth, the reader is referred to a number of excellent reviews, both general[16,17] and specific to zebrafish.[18–20] What follows is a qualitative overview of the process, which should serve to place in context some of the research highlighted later.

To make appropriate connections in the nervous system, axons must extend long distances and form functional connections with postsynaptic cells. The paths taken by axons can be tortuous; how billions of ordered, specific connections are formed within the Byzantine network of the nervous system is a daunting question. From decades of biological psychology and anatomy, much more is known about the functional significance of connectivity in the brain than about how this architecture arises.

Cells of the nervous system are divided into two major classes: neurons and glia. Neurons are the functional information-transmitting medium of the nervous system, while glia perform a diverse array of structural and physiological support roles. The paradigmatic neuron consists of a soma and two sets of processes: the dendrites, which receive stimuli from other neurons and sensory organs; and an axon, which can branch and form multiple synapses on target cells. Neurons are morphologically diverse, and the organization and relative sizes and shapes of these components vary a great deal. While much more effort to date has been spent studying axon guidance, dendrites face similar challenges in projecting to appropriate targets, and indeed seem to use similar mechanisms.

Newborn neurons exit the cell cycle and migrate to their final locations in the nervous system. As their differentiation continues they extend neurites, the dendritic and axonal processes. At the distal end of the extending axon is the growth cone; a motile, amoeboid structure that navigates to its target through the complex embryological environment, constantly extending and retracting multiple filopodia and lamellipodia. While dynamics at the leading edge of the growth cone are primarily actin-mediated, an internal microtubule framework stabilizes axons. As the axon elongates, this scaffold is extended within the growth cone. A number of specialized cytoskeletal components are associated with growing axons, and vesicle fusion contributes to plasma membrane extension. Branching can occur at the growth cone or well behind it along the axon; many branches are transient and later retract, while others are selectively stabilized. The cytoskeletal underpinnings of these more complex growth cone behaviors are just beginning to be described and analyzed.

Growth cones compute their behavior based on a variety of internal and external factors. Environmental influences can broadly be described as permissive or instructive. Permissive elements allowing neurite growth include suitable substrate composition and topology and the presence of appropriate growth and survival factors. Instructive guidance signals can be attractive or repulsive, or induce more specialized responses. Signals that alter a growth cone's behavior are presumed to do so through the regulation of cytoskeleton, cell adhesion molecules, and cell surface receptors.

As pathfinding continues, the growth cone integrates multiple guidance cues present in the environment. Its reaction to its surroundings is dependent on the developmental state of the neuron, and appears to involve multiple interacting signaling pathways. While the precise interactions of the signaling networks are not well understood (and vary a great deal according to neuronal class), a wide range of research has provided significant glimpses of some of the key elements. The availability, concentration, and subcellular localization of receptors, downstream signal transduction components, translational factors, and cytoskeletal regulators all seem to be tightly controlled and contribute to the growth cone's interpretation of its surroundings. The readout of the interaction between external and internal elements can be continued growth, stopping, turning, collapse, retraction, branching or synaptogenesis. The internal state of a growth cone is dynamic, and responses to a given cue can change quickly. For example, repulsive responses to a guidance molecule can be suppressed by the internalization or degradation of its receptor. Such changes in responsiveness can be brought about by a particular event, for example, reaching a particular point in the embryo (e.g. guide post cells).

Advances in our understanding of growth cone guidance have come from two major approaches — genetics and *in vitro* assays. In *Drosophila*, genetic screens for mutations affecting the midline crossing of commissural axons led to the isolation of a number of genes involved in axon guidance, and provided one example of how axons can change their responsiveness to guidance cues.[21–23] A key feature of midline crossing, both in vertebrates and invertebrates, is that it normally happens just once. Axons are attracted to the midline, cross it, but then never cross again. In the *roundabout* mutant, axons cross multiple times.[4] As it turns out, Roundabout is a receptor for a repellent at the midline, named Slit.[24] Before crossing the midline, Roundabout is expressed at low levels. Another protein, Commissureless, regulates Roundabout levels via an ubiquitin-mediated process such that the repulsive Slit cue is transduced only after midline crossing.[25] The mechanisms unearthed by this series of studies, from the existence of the Slit-Roundabout system to the importance of ubiquitin-dependent mechanisms in axon guidance, appear to be conserved in evolution.[26]

In vitro assays, using *Xenopus* spinal and retinal neurons, have led to different insights into how growth cones can change their response to guidance cues. In these assays, axons growing on a glass cover slip are exposed to gradients of soluble guidance molecules, such as netrin.[27,28] Spinal neurons are normally attracted to netrin. Remarkably, when the level of intracellular cAMP is changed, the axons are repelled.[29] The response to a number of other cues, such as Sema3A, depends on the level of cGMP.[30] Hence, there is no such thing as an intrinsically attractive or repulsive cue. The response is very much dependent on the axon itself.

Several genetic screens have been carried out in zebrafish to investigate the mechanisms of axon guidance and synapse formation.

Among these are the retinotectal screen[31–33] and the primary motoneuron screen,[34] both of which were initiated before the power of genetics was made obvious with the *Drosophila* midline screen.[22] Both screens have yielded a large number of mutants in which the projection of axons to their target is disrupted. In the retinotectal system of wild-type fish, retinal ganglion cells (RGCs) extend axons from the eye to the opposite brain hemisphere. Most of the axons are targeted to the contralateral optic tectum, a prominent midbrain structure, where they branch in very restricted regions and synapse with tectal neurons. Mutations were found that disrupt a number of steps in axon targeting, ranging from exiting the eye, outgrowth, and midline crossing to topographic mapping in the tectum (the synaptic innervation pattern in the tectum that recapitulates cell location in the retina). One of the few retinotectal mutants that has been positionally cloned is *astray*, which has a striking phenotype whereby retinal axons wander in the brain. As it happens, *astray* encodes a zebrafish ortholog of Roundabout,[35] underscoring the fact that many proteins used in setting up the nervous system are evolutionary conserved.

The primary motoneurons of the zebrafish innervate the axial muscles (which mediate swimming), between 18 and 24 hours post-fertilization.[36] Precisely three neurons innervate each myotome, and each follows a stereotyped pathway and forms synapses with a particular region of the myotome. Homology cloning and *in vivo* manipulation[37,38] have been used to identify mechanisms by which this precision is achieved. Several evolutionarily conserved proteins, such as semaphorins, have been found to be expressed in the myotome in restricted regions. Ectopic expression of one of these cues, using laser activation of a heat-shock promoter driving Sema3A1 expression, caused abnormal pathfinding of motoneurons.

Genetic screens have provided a number of mutants with abnormal innervation of myotomes. Two types of screens — a motility test and antibody labeling of motoneurons — have been used. One mutant with motility defects is *ache*, which encodes the acetylcholinesterase gene.[39] This enzyme is required to remove the neurotransmitter acetylcholine from the synaptic cleft. Mutants, in addition to defects in the neuromuscular junction, have reduced dendritic outgrowth in Rohon–Beard cells, a transient population of sensory neurons in the larva. This mutant provides a good example of how zebrafish has provided a unique insight into vertebrate nervous system development in the vertebrate. A role for acetylcholinesterase in dendrite formation was not suspected before, as the mouse mutant had far more subtle defects, presumably owing to the presence of a compensating enzyme not present in zebrafish.

An exciting possibility for zebrafish would be to combine genetics with *in vitro* assays and *in vivo* imaging. For example, it would be informative to carry out growth cone turning assays with mutants that have cell-autonomous guidance defects. This should be possible with both spinal neurons and retinal neurons, as culture systems have now been described. Additionally, the transparency of zebrafish should be further exploited to image signaling events during axon guidance and synapse formation. A number of GFP-based reporters are now available for calcium[40,41] as well as for cAMP.[42] Expression of these genes in individual neurons, either transiently by lipofection or electroporation, or stably in transgenic lines, combined with confocal microscopy, can be used to monitor signaling events *in vivo*.[43] GFP-synaptobrevin, which has been used successfully in *Xenopus*[44] and *C. elegans*[45] to visualize synapses, will be a useful marker for studies on synapse formation.

Function of the Nervous System: Sensory Perception

Mechanotransduction

The zebrafish develops a functional nervous system within days, enabling larvae to escape predators and detect food. Among the quickly developing sense organs are the inner ear, which is required for balance, and the lateral line organ, which detects movement in the water. Both these sense organs are dependent on mechanotransduction, which is the conversion of mechanical signals into biochemical signals, in sensory hair cells. Each sensory hair cell contains an apical bundle of stereocilia linked to each other by fine tip links. Biophysical experiments have led to two theories of how mechanotransduction could occur. In one model, the "gating-spring" hypothesis, the tip links are thought to be directly connected to transduction channels and act like springs that open the channels when the cilia bend in response to vibration.[46] In the second model, membrane junctions near the cilia tips are thought to be involved in stretching the membrane and opening channels.[47] To unravel the molecular mechanism of mechanotransduction in hair cells, and to distinguish between these two models, genetic screens have been carried out in flies, worms and zebrafish.

A number of zebrafish mutants with defects in hair cell function were found in the original Tübingen screen.[48] These mutants were isolated on the basis of defects in balance, which caused them to swim in circles. 72-hour-old mutants were also subsequently found to be non-responsive to vibrations in the water, which can be administered by simply tapping the dish. Several observations confirmed that defects in hair cells, and not other parts of the nervous system, caused the behaviorial abnormalities. To discount the possibility that the Mauthner neuron or reticulospinal neurons were non-functional, calcium imaging was carried out on live fish. The entire embryo was labeled by injecting embryos at early cleavage stages with calcium-green dextran.[49] The hindbrain was then imaged while the fish was stimulated by touch or vibration; only touch caused neural activity in the hindbrain. The technique of microphonics[50] was then used to prove that these mutants were deficient in mechanotransduction. Microphonics, which measures the electrical field in the water near a hair cell bundle, is especially well suited to the analysis of systems where hair cells are exposed, such as the hair cells of the zebrafish lateral line. In combination with whole-cell patch-clamp, three mutants with morphologically normal hair cells were found to be defective in mechanotransduction.

Analysis of the zebrafish genome has been used to address a major question in the field of mechanotransduction, namely the molecular identity of the transduction channel. Analysis of the human and mouse genomes have failed to reveal any homolog to the best characterized mechanotransduction channel in invertebrates — nOMPC. Remarkably, this gene was found in a bioinformatics analysis of the zebrafish genome.[51] A morpholino knockdown of this gene, combined with physiological analysis of morphants, provided strong evidence that this is indeed the mechanotransduction channel in vertebrates.

Both forward and reverse genetics in zebrafish, in conjunction with physiology, have thus been instrumental in increasing our understanding of mechanotransduction, and hence hearing, in vertebrates. Although hearing mutants have been characterized in mice, it is much more difficult to physiologically analyze hair cells that are deeply embedded within the cochlea of higher mammals. It is the presence of hair cells on the outside of the embryo that makes zebrafish an experimentally powerful system. The zebrafish *mariner* gene encodes a myosin VIIa,[52] which is also mutated in the human deafness syndrome, Usher 1B. Zebrafish circler mutants, such as *mariner*, may thus provide useful models for human deafness.

Odor Perception

Aside from the hair cells of the lateral line organ, there is one other class of neurons in the zebrafish that directly contacts the external environment. These are the sensory neurons of the olfactory system. Olfactory sensory neurons are tightly embedded within the epithelium of the olfactory pits.[53] Their dendrites, which contain odorant receptors, contact the surrounding water, and their axons project to the olfactory bulb. In mammals, each sensory neuron appears to express only one allele of one of a thousand odorant receptors.[54] The zebrafish is less complex, possessing only one hundred or so odorant receptors.[55] Neurons expressing a given receptor are scattered within one of four zones in the olfactory epithelium, but appear to converge to one glomerulus in the olfactory bulb.

A number of advances in our understanding of the development of the olfactory system have come from studies on mammals. For example, the use of targeted recombination technology has enabled receptor swap experiments, which led to the conclusion that the odorant receptors themselves have a key role in targeting olfactory axons to specific glomeruli, and not only in reception.[56] Indeed, the receptors themselves were cloned initially from rats.[57] It is possible that odorant receptors can transduce some sort of directional cue, and have recently been shown to play a role in sperm chemotaxis.[58]

One question in the olfaction field has been how odor stimuli in the external world are converted into neural representations in the brain, enabling conscious perception of the chemical world and an appropriate response. A traditional method to investigate this question has been to use electrophysiology to record responses in the olfactory bulb. An alternative would be to use optical recording, as this allows the analysis of large populations of neurons simultaneously; this method of analysis has been used successfully in zebrafish.

The first step in understanding the response of olfactory neurons came from imaging calcium transients in glomeruli.[59] A calcium dye was loaded into neurons of living adults by removing cilia with Triton X-100. After recovery of the cilia (which is where odorant receptors are located), the olfactory bulb together with the epithelium was removed and the response to odorants imaged. Different odorants were found to activate a different combination of glomeruli. In this way, odor information is converted into spatial information within the brain, confirming what had been predicted by electrophysiological experiments. Remarkably, insects appear to utilize a similar mechanism of encoding odor information.

Electrophysiology can provide temporal information that is very difficult to obtain with optical approaches. However, with the use of a lipophilic voltage-sensitive dye, it has been possible to obtain temporal data on the response of populations of olfactory sensory neurons in the zebrafish.[60] Indeed, imaging neuronal activity at high speed with voltage-sensitive dyes, has reinforced the concept of time and not just of space, in the perception of smell.[61] The concentration of an odor, for example, is converted into frequency.

To further understand the processing of odorants in the brain, it will be necessary to look at other neurons in the circuit, specifically the mitral/tufted cells and their downstream synaptic partners. This can be done in a relatively crude way by injecting calcium-responsive membrane permeable dyes into the forebrain. A more refined way would be to create transgenic lines expressing activity reporters in these specific neurons. Another important step for analysis of the olfactory system would be a large-scale mutant screen. Behavioral screens, or morphological screens using GFP lines with expression in olfactory sensory neurons, should provide a more detailed molecular understanding of how vertebrates perceive the chemical world.

Network Function

An organism's behavioral repertoire, from simple reflexes to higher cognition, derives from the patterns of interconnectedness among its neurons. The zebrafish provides an opportunity for an extensive integrated approach towards understanding behavior, given its tractability at all stages and levels of development, using a very broad range of techniques.[62–67] Here we will highlight work on the sensory-motor feedback systems controlling reflexive eye movements, and the motor control network mediating escape responses.

The Visual System

Visual cues can evoke a number of reflexive responses in zebrafish, such as the optokinetic response, the optomotor response and the escape response.[62] The optokinetic response refers to the involuntary movement of eyes while tracking a moving object. In the optomotor response, zebrafish adjust their swimming to ensure that they do not move relative to their surroundings; forward swimming is thus triggered when backward movement is simulated by moving the surroundings. The visually-evoked escape response occurs when a dark object or shadow suddenly appears, as would occur in the wild when a predator looms overhead.

These robust responses have been used as the basis for a number of genetic screens.[68,69] Such screens are useful in identifying which retinal ganglion cell population, or which target area in the midbrain, might mediate a particular response. They may also be useful in investigating neuronal circuits further downstream. In zebrafish, retinal ganglion cells project their axons to ten targets, the largest of which is the optic tectum, described earlier. Most other targets are in the pre-tectal area, and one is at the optic chiasm. At present, it is not clear what determines the choice of target, nor what the functions of the targets are. The optic tectum itself has been suggested to function in learning and memory,[70–72] while the pre-tectal targets are thought to be involved in reflexive responses. The supra-chiasmatic nucleus, as in higher vertebrates, is probably involved in the circadian rhythm.

Twelve mutants with defective or abnormal OKRs have been isolated; these seem to be due to specific defects in the visual pathway, rather than general brain development abnormalities, motor apparatus, or improper retina formation. The majority of these mutations await molecular identification. The few that have been molecularly identified to date are *grumpy* and *sleepy*,[73] and *lakritz*.[74]

As one might expect, some of the OKR mutants exhibit aberrant retinotopic mapping, implying that the visual input is not properly represented in the tectum, or have altered retinal electrophysiology or anatomy. Interestingly, many other mutants that show abnormal retinotectal projections show normal OKR, indicating that the tectum may not be required for mediating this behavior. This was recently confirmed by laser ablation of the tectal neuropil, the RGC innervation site in the optic tectum. OKR in these treated fish was normal[72] furthering earlier results based on surgical ablation.[75]

A screen for genes that affect the escape response in low light yielded the *night blindness b* mutant.[76] This mutant has reduced dopaminergic interplexiform cells, suggesting that this population of retinal cells is required for dark adaptation. Remarkably, this response requires innervation of the terminal nerve from the olfactory bulb, as ablation of the bulb phenocopies the *nbb b* mutation.

The Touch Response

Zebrafish larvae exhibit both visual and tactile stimulus-induced escape responses very early in development. Later, they swim in short spontaneous bursts. There are over 30 locomotor mutants in about 10 phenotypic classes resulting from the 1996 Tübingen screen.[77] This set expectedly overlaps to some extent with the axon guidance mutants.

The touch-induced escape response consists of a rapid coiling of the tail, and by 27-hpf shows directional sensitivity: a tap on the head results in a full coil and reorientation away from the stimulus, while a tail touch results in a partial coiling response.

The larval skin is extensively innervated by two classes of sensory neurons that transduce the initial input to the escape response circuity. The Rohon–Beard cells are a transient array of neurons positioned along the trunk of the fish, while the cells of the trigeminal ganglion innervate the head. Both of these cell types send outputs to the Mauthner array in the hindbrain.[78] This array consists of 3 pairs of reticulospinal neurons, the two prominent Mauthner cells and two additional pairs of homologous neurons in adjacent segments. From the hindbrain, signals are sent to muscles to effect the escape behavior.

Due to the relative ease of *in vivo* functional imaging, this system has proven itself amenable to detailed analysis. By loading the reticulospinal neurons of live embryos with a Ca^{2+} sensitive dye (by injection into the caudal spinal cord to avoid lesioning the hindbrain), O'Malley and colleagues[79] showed that the differential response to anterior and posterior stimuli corresponded to differential activation of the cells of the Mauthner array. While a tail touch resulted in activation of only the Mauthner neuron (resulting in a small contraction on the contralateral side), a head touch activated all three homolog and produced a large contraction.

This same group has mapped the location of all reticulospinal neurons of the hindbrain, which can be identified from fish to fish by position and morphology. This has allowed a systematic approach for determining function in restrained, behaving fish using *in vivo* imaging. The functional relevance of the observed correlations between neuronal activity and behaviors can be tested by single-cell laser ablations of particular hindbrain neurons.

Transgenesis techniques that target subpopulations of cells promise an extension of this type of study. The availability of genetically encoded fluorescent calcium indicators means that more difficult to access neuronal populations can be targeted for analysis. To this end, Higashijima *et al.*[43] recently used such a construct under the neuronal promoters of *HuC* and *islet-1* to show that imaging could be done at single action potential resolution. Calcium transients were measured from Rohon–Beard neurons during stimulation of the skin, and from primary motoneurons and spinal interneurons during an escape response.

Ultimately, determining the functional organization of neural networks requires approaching a system on multiple levels. In the zebrafish, several labs have been able to address questions in this way, combining genetics, embryology, and molecular biology. The *macho* (*mao*) mutant was described in the 1996 screen as possessing abnormalities in the retinotectal projection, touch response, and locomotion.[35,77] In 1998 it was determined by electrophysiology that *mao* fish were defective in sodium channel function, and were thus unable to propagate action potentials normally.[80] The 1999 OKR screen described earlier showed that this mutant was also defective in visuomotor responses.[69] The utility of this mutant is obvious: it provides researchers with zebrafish in which normal neuronal function has been genetically compromised, essentially mimicking the effects of activity-blocking drugs. In the last few years, *mao* mutants have been employed to demonstrate the activity dependence of the refinement of the visual map in the tectum[81] and in the programmed cell death of the larval-specific Rohan–Beard neurons.[82]

Space cadet embryos also show defects in locomotion and retinal axon pathfinding.[77] The guidance errors of *space cadet* RGCs led Lorent *et al.*[83] to investigate the possibility that similar errors in a different neuronal subpopulation underlie the locomotion defect. This group used high-speed video analysis to demonstrate that the mutants fail to properly initiate escape responses, and discovered axonal defects in the spiral fiber neurons. These cells normally extend commissural axons that wind around and synapse on the Mauthner cell axons and modulate their activity. However, in *space cadet* fish, spiral fiber axons were absent from specific commissures, and failed to synapse with Mauthner cells.

Severing the affected commissures in wild-type larvae phenocopied the aberrant *space cadet* escape response. The study of this mutant demonstrates how network formation and function can be dissected by integrating genetic, embryological, and behavioral approaches. The range of techniques available in zebrafish allows for sophisticated and detailed analyses of such systems.

Concluding Remarks

The coming years should see the development of more sophisticated tools, for example the silencing or activation of genes in different parts of a circuit. This, combined with more ingenious screens for genes involved in neuronal function, will surely lead to a considerable advance in our understanding of the brain.

References

1. Grunwald DJ and Eisen JS (2002). Headwaters of the zebrafish — emergence of a new model vertebrate. *Nat. Rev. Genet.* **3**: 717–724.
2. Cabrera CV (1992). The generation of cell diversity during early neurogenesis in *Drosophila. Development* **115**: 893–901.
3. Campos-Ortega JA (1997). Neurogenesis in *Drosophila*: an historical perspective and some prospects. *Perspect. Dev. Neurobiol.* **4**: 267–271.
4. Kidd T, Brose K, Mitchell KJ, Fetter RD, Tessier-Lavigne M, Goodman CS and Tear G (1998). Roundabout controls axon crossing of the CNS midline and defines a novel subfamily of evolutionarily conserved guidance receptors. *Cell* **92**: 205–215.
5. Tessier-Lavigne M and Goodman CS (1996). The molecular biology of axon guidance. *Science* **274**: 1123–1133.
6. Chou JH, Troemel ER, Sengupta P, Colbert HA, Tong L, Tobin DM, Roayaie K, Crump JG, Dwyer ND and Bargmann CI (1996). Olfactory recognition and discrimination in *Caenorhabditis elegans. Cold Spring Harb Symp. Quant. Biol.* **61**: 157–164.
7. Goodman MB and Schwarz EM (2003). Transducing touch in *Caenorhabditis elegans. Annu. Rev. Physiol.* **65**: 429–452.

8. van Swinderen B and Greenspan RJ (2003). Salience modulates 20–30 Hz brain activity in *Drosophila. Nat. Neurosci.* **6**: 579–586.
9. Dubnau J, Chiang AS, Grady L, Barditch J, Gossweiler S, McNeil J, Smith P, Buldoc F, Scott R, Certa U *et al.* (2003). The staufen/pumilio pathway is involved in *Drosophila* long-term memory. *Curr. Biol.* **13**: 286–296.
10. Drescher U, Kremoser C, Handwerker C, Loschinger J, Noda M and Bonhoeffer F (1995). *In vitro* guidance of retinal ganglion cell axons by RAGS, a 25 kDa tectal protein related to ligands for Eph receptor tyrosine kinases. *Cell* **82**: 359–370.
11. Monnier PP, Sierra A, Macchi P, Deitinghoff L, Andersen JS, Mann M, Flad M, Hornberger MR, Stahl B, Bonhoeffer F *et al.* (2002). RGM is a repulsive guidance molecule for retinal axons. *Nature* **419**: 392–395.
12. Lee T and Luo L (2001). Mosaic analysis with a repressible cell marker (MARCM) for *Drosophila* neural development. *Trends Neurosci.* **24**: 251–254.
13. Aparicio S, Chapman J, Stupka E, Putnam N, Chia JM, Dehal P, Christoffels A, Rash S, Hoon S, Smit A *et al.* (2002). Whole-genome shotgun assembly and analysis of the genome of *Fugu rubripes. Science* **297**: 1301–1310.
14. Heasman J (2002). Morpholino oligos: making sense of antisense? *Dev Biol* **243**: 209–214.
15. Chen W, Burgess S, Golling G, Amsterdam A and Hopkins N (2002). High-throughput selection of retrovirus producer cell lines leads to markedly improved efficiency of germ line-transmissible insertions in zebra fish. *J. Virol.* **76**: 2192–2198.
16. Dickson BJ (2002). Molecular mechanisms of axon guidance. *Science* **298**: 1959–1964.
17. Mueller BK (1999). Growth cone guidance: first steps towards a deeper understanding. *Annu. Rev. Neurosci.* **22**: 351–388.
18. Beattie CE, Granato M and Kuwada JY (2002). Cellular, genetic and molecular mechanisms of axonal guidance in the zebrafish. *Results Probl Cell Differ.* **40**: 252–269.
19. Culverwell J and Karlstrom RO (2002). Making the connection: retinal axon guidance in the zebrafish. *Semin. Cell. Dev. Biol.* **13**: 497–506.
20. Hutson LD and Chien CB (2002). Wiring the zebrafish: axon guidance and synaptogenesis. *Curr. Opin. Neurobiol.* **12**: 87–92.

21. Klambt C (1993). The *Drosophila* gene pointed encodes two ETS-like proteins which are involved in the development of the midline glial cells. *Development* **117**: 163–176.
22. Seeger M, Tear G, Ferres–Marco D and Goodman CS (1993). Mutations affecting growth cone guidance in *Drosophila*: genes necessary for guidance toward or away from the midline. *Neuron* **10**: 409–426.
23. Tear G, Seeger M and Goodman CS (1993). To cross or not to cross: a genetic analysis of guidance at the midline. *Perspect. Dev. Neurobiol.* **1**: 183–194.
24. Kidd T, Bland KS and Goodman CS (1999). Slit is the midline repellent for the robo receptor in *Drosophila. Cell* **96**: 785–794.
25. Myat A, Henry P, McCabe V, Flintoft L, Rotin D and Tear G (2002). *Drosophila* Nedd4, a ubiquitin ligase, is recruited by commissureless to control cell surface levels of the roundabout receptor. *Neuron* **35**: 447–459.
26. Chisholm A and Tessier–Lavigne M (1999). Conservation and divergence of axon guidance mechanisms. *Curr. Opin. Neurobiol.* **9**: 603–615.
27. de la Torre JR, Hopker VH, Ming GL, Poo MM, Tessier–Lavigne M, Hemmati–Brivanlou A and Holt CE (1997). Turning of retinal growth cones in a netrin-1 gradient mediated by the netrin receptor DCC. *Neuron* **19**: 1211–1224.
28. Lohof AM, Quillan M, Dan Y and Poo MM (1992). Asymmetric modulation of cytosolic cAMP activity induces growth cone turning. *J. Neurosci.* **12**: 1253–1261.
29. Ming GL, Song HJ, Berninger B, Holt CE, Tessier–Lavigne M and Poo MM (1997). cAMP-dependent growth cone guidance by netrin-1. *Neuron* **19**: 1225–1235.
30. Song H, Ming G, He Z, Lehmann M, McKerracher L, Tessier–Lavigne M and Poo M (1998). Conversion of neuronal growth cone responses from repulsion to attraction by cyclic nucleotides. *Science* **281**: 1515–1518.
31. Baier H, Klostermann S, Trowe T, Karlstrom RO, Nusslein–Volhard C and Bonhoeffer F (1996). Genetic dissection of the retinotectal projection. *Development* **123**: 415–425.
32. Karlstrom RO, Trowe T, Klostermann S, Baier H, Brand M, Crawford AD, Grunewald B, Haffter P, Hoffmann H, Meyer SU *et al.* (1996). Zebrafish mutations affecting retinotectal axon pathfinding. *Development* **123**: 427–438.

33. Trowe T, Klostermann S, Baier H, Granato M, Crawford AD, Grunewald B, Hoffmann H, Karlstrom RO, Meyer SU, Muller B *et al.* (1996). Mutations disrupting the ordering and topographic mapping of axons in the retinotectal projection of the zebrafish, *Danio rerio*. *Development* **123**: 439–450.
34. Beattie CE, Melancon E and Eisen JS (2000). Mutations in the stumpy gene reveal intermediate targets for zebrafish motor axons. *Development* **127**: 2653–2662.
35. Fricke C, Lee JS, Geiger–Rudolph S, Bonhoeffer F and Chien CB (2001). Astray, a zebrafish roundabout homolog required for retinal axon guidance. *Science* **292**: 507–510.
36. Myers PZ, Eisen JS and Westerfield M (1986). Development and axonal outgrowth of identified motoneurons in the zebrafish. *J. Neurosci.* **6**: 2278–2289.
37. Halloran MC, Sato–Maeda M, Warren JT, Su F, Lele Z, Krone PH, Kuwada JY and Shoji W (2000). Laser-induced gene expression in specific cells of transgenic zebrafish. *Development* **127**: 1953–1960.
38. Roos M, Schachner M and Bernhardt RR (1999). Zebrafish semaphorin Z1b inhibits growing motor axons *in vivo*. *Mech. Dev.* **87**: 103–117.
39. Behra M, Cousin X, Bertrand C, Vonesch JL, Biellmann D, Chatonnet A and Strahle U (2002). Acetylcholinesterase is required for neuronal and muscular development in the zebrafish embryo. *Nat. Neurosci.* **5**: 111–118.
40. Miyawaki A, Griesbeck O, Heim R and Tsien RY (1999). Dynamic and quantitative Ca2+ measurements using improved cameleons. *Proc. Natl. Acad. Sci. USA* **96**: 2135–2140.
41. Nakai J, Ohkura M and Imoto K (2001). A high signal-to-noise Ca(2+) probe composed of a single green fluorescent protein. *Nat. Biotechnol.* **19**: 137–141.
42. Zaccolo M and Pozzan T (2002). Discrete microdomains with high concentration of cAMP in stimulated rat neonatal cardiac myocytes. *Science* **295**: 1711–1715.
43. Higashijima SI, Masino MA, Mandel G and Fetcho JR (2003). Imaging neuronal activity during zebrafish behavior with a genetically encoded calcium indicator. *J. Neurophysiol.*
44. Alsina B, Vu T and Cohen–Cory S (2001). Visualizing synapse formation in arborizing optic axons *in vivo*: dynamics and modulation by BDNF. *Nat. Neurosci.* **4**: 1093–1101.

45. Nonet ML (1999). Visualization of synaptic specializations in live *C. elegans* with synaptic vesicle protein-GFP fusions. *J. Neurosci. Methods.* **89**: 33–40.
46. Corey DP and Hudspeth AJ (1983). Kinetics of the receptor current in bullfrog saccular hair cells. *J. Neurosci.* **3**: 962–976.
47. Hackney CM and Furness DN (1995). Mechanotransduction in vertebrate hair cells: structure and function of the stereociliary bundle. *Am. J. Physiol.* **268**: C1–13.
48. Nicolson T, Rusch A, Friedrich RW, Granato M, Ruppersberg JP and Nusslein–Volhard C (1998). Genetic analysis of vertebrate sensory hair cell mechanosensation: the zebrafish circler mutants. *Neuron* **20**: 271–283.
49. Cox KJ and Fetcho JR (1996). Labeling blastomeres with a calcium indicator: a non-invasive method of visualizing neuronal activity in zebrafish. *J. Neurosci. Methods* **68**: 185–191.
50. Flock A (1965). Transducing mechanisms in the lateral line canal organ receptors. *Cold Spring Harb. Symp. Quant. Biol.* **30**: 133–145.
51. Sidi S, Friedrich RW and Nicolson T (2003). NompC TRP channel required for vertebrate sensory hair cell mechanotransduction. *Science* **301**: 96–99.
52. Ernest S, Rauch GJ, Haffter P, Geisler R, Petit C and Nicolson T (2000). Mariner is defective in myosin VIIA: a zebrafish model for human hereditary deafness. *Hum. Mol. Genet.* **9**: 2189–2196.
53. Hansen A and Zeiske E (1993). Development of the olfactory organ in the zebrafish, *Brachydanio rerio. J. Comp. Neurol.* **333**: 289–300.
54. Chess A, Simon I, Cedar H and Axel R (1994). Allelic inactivation regulates olfactory receptor gene expression. *Cell* **78**: 823–834.
55. Weth F, Nadler W and Korsching S (1996). Nested expression domains for odorant receptors in zebrafish olfactory epithelium. *Proc. Natl. Acad. Sci. USA* **93**: 13321–13326.
56. Wang F, Nemes A, Mendelsohn M and Axel R (1998). Odorant receptors govern the formation of a precise topographic map. *Cell* **93**: 47–60.
57. Buck L and Axel R (1991). A novel multigene family may encode odorant receptors: a molecular basis for odor recognition. *Cell* **65**: 175–187.
58. Spehr M, Gisselmann G, Poplawski A, Riffell JA, Wetzel CH, Zimmer RK and Hatt H (2003). Identification of a testicular odorant receptor mediating human sperm chemotaxis. *Science* **299**: 2054–2058.
59. Friedrich RW and Korsching SI (1997). Combinatorial and chemotopic odorant coding in the zebrafish olfactory bulb visualized by optical imaging. *Neuron* **18**: 737–752.

60. Friedrich RW and Korsching SI (1998). Chemotopic, combinatorial, and noncombinatorial odorant representations in the olfactory bulb revealed using a voltage-sensitive axon tracer. *J. Neurosci.* **18**: 9977–9988.
61. Friedrich RW and Laurent G (2001). Dynamic optimization of odor representations by slow temporal patterning of mitral cell activity. *Science* **291**: 889–894.
62. Baier H (2000). Zebrafish on the move: towards a behavior-genetic analysis of vertebrate vision. *Curr. Opin. Neurobiol.* **10**: 451–45.
63. Baier H and Copenhagen D (2000). Combining physiology and genetics in the zebrafish retina. *J. Physiol.* **524**(1): 1.
64. Drapeau P, Saint–Amant L, Buss RR, Chong M, McDearmid JR and Brustein E (2002). Development of the locomotor network in zebrafish. *Prog. Neurobiol.* **68**: 85–111.
65. Fetcho JR and Liu KS (1998). Zebrafish as a model system for studying neuronal circuits and behavior. *Ann. N Y Acad. Sci.* **860**: 333–345.
66. Neuhauss SC (2003). Behavioral genetic approaches to visual system development and function in zebrafish. *J. Neurobiol.* **54**: 148–160.
67. O'Malley DM, Zhou Q and Gahtan E (2003). Probing neural circuits in the zebrafish: a suite of optical techniques. *Methods* **30**: 49–63.
68. Brockerhoff SE, Hurley JB, Janssen–Bienhold U, Neuhauss SC, Driever W and Dowling JE (1995). A behavioral screen for isolating zebrafish mutants with visual system defects. *Proc. Natl. Acad. Sci. U S A* **92**: 10545–10549.
69. Neuhauss SC, Biehlmaier O, Seeliger MW, Das T, Kohler K, Harris WA and Baier H (1999). Genetic disorders of vision revealed by a behavioral screen of 400 essential loci in zebrafish. *J. Neurosci.* **19**: 8603–8615.
70. Pradel G, Schachner M and Schmidt R (1999). Inhibition of memory consolidation by antibodies against cell adhesion molecules after active avoidance conditioning in zebrafish. *J. Neurobiol.* **39**: 197–206.
71. Pradel G, Schmidt R and Schachner M (2000). Involvement of L1.1 in memory consolidation after active avoidance conditioning in zebrafish. *J. Neurobiol.* **43**: 389–403.
72. Roeser T and Baier H (2003). Visuomotor Behaviors in Larval Zebrafish after GFP-guided laser ablation of the optic tectum. *J. Neurosci.* **23**: 3726–3734.
73. Parsons MJ, Pollard SM, Saude L, Feldman B, Coutinho P, Hirst EM and Stemple DL (2002). Zebrafish mutants identify an essential role for laminins in notochord formation. *Development* **129**: 3137–3146.

74. Kay JN, Finger–Baier KC, Roeser T, Staub W and Baier H (2001). Retinal ganglion cell genesis requires lakritz, a zebrafish atonal Homolog. *Neuron* **30**: 725–736.
75. Springer AD, Easter SS, Jr. and Agranoff BW (1977). The role of the optic tectum in various visually mediated behaviors of goldfish. *Brain. Res.* **128**: 393–404.
76. Li L and Dowling JE (2000). Disruption of the olfactoretinal centrifugal pathway may relate to the visual system defect in night blindness b mutant zebrafish. *J. Neurosci.* **20**: 1883–1892.
77. Granato M, van Eeden FJ, Schach U, Trowe T, Brand M, Furutani-Seiki M, Haffter P, Hammerschmidt M, Heisenberg CP, Jiang YJ *et al.* (1996). Genes controlling and mediating locomotion behavior of the zebrafish embryo and larva. *Development* **123**: 399–413.
78. Fetcho JR (1991). Spinal network of the Mauthner cell. *Brain Behav Evol* **37**: 298–316.
79. O'Malley DM, Kao YH and Fetcho JR (1996). Imaging the functional organization of zebrafish hindbrain segments during escape behaviors. *Neuron* **17**: 1145–1155.
80. Ribera AB and Nusslein–Volhard C (1998). Zebrafish touch-insensitive mutants reveal an essential role for the developmental regulation of sodium current. *J. Neurosci.* **18**: 9181–9191.
81. Gnuegge L, Schmid S and Neuhauss SC (2001). Analysis of the activity-deprived zebrafish mutant *macho* reveals an essential requirement of neuronal activity for the development of a fine-grained visuotopic map. *J. Neurosci.* **21**: 3542–3548.
82. Svoboda KR, Linares AE and Ribera AB (2001). Activity regulates programmed cell death of zebrafish Rohon-Beard neurons. *Development* **128**: 3511–3520.
83. Lorent K, Liu KS, Fetcho JR and Granato M (2001). The zebrafish space cadet gene controls axonal pathfinding of neurons that modulate fast turning movements. *Development* **128**: 2131–2142.

Chapter 6

Development of the Primary Nervous System of the Zebrafish Embryo

Uwe Strähle[a] and Vladimir Korzh[b]

[a]*Institute of Toxicology and Genetics, University of Karlsruhe, Germany*

[b]*Institute of Molecular and Cell Biology, National University of Singapore, Singapore*

The early zebrafish embryo develops a simple primary nervous system that controls motility. Similar to that of the nematode *C. elegans*, this primary nervous system is composed of a relatively low number of neurons and is amenable to observation and experimental manipulation at the single cell level. The primary nervous system is derived from neurogenic regions in the neural plate. Several conserved gene loci are required for the development of the primary neurons, suggesting that the molecular mechanisms underlying neurogenesis are strongly related to that of the fruitfly *Drosophila melanogaster* and higher vertebrates. Since inhibition of gene activity or mutations in the zebrafish emulate many human hereditary disorders, it renders the zebrafish an attractive model for the study of vertebrate nervous system development and human hereditary diseases.

Introduction

Zebrafish embryos begin to move towards the end of the first day of development, showing spontaneous bends of the body axis. By two days of age, embryos have a stereotypic escape behavior, the so-called startle or fast-start response, which entails a rapid bend of the body axis away from the source of the tactile stimulus followed by swimming movements.[1]

As a prerequisite, the nervous system of the zebrafish embryo develops very early. The first neurons are born during gastrulation, soon after the specification of the neurectoderm at the dorsal side of the embryo.[2,3] Many more neurons exit the cell cycle in subsequent neurula stages during which the neural plate is transformed into the neural tube.[4,5] At the end of the first day of development, the early-born neurons have formed a scaffold of axon tracts and neuronal connections that control embryonic motility.[6–9] More neurons are added to the nervous system in subsequent stages.[10] These later-born neurons are frequently referred to as secondary neurons to distinguish them from the early-born primary neurons.[11] Unlike teleost and amphibian embryos, higher vertebrates do not form a primary nervous system. The early-developing, primitive nervous system of the zebrafish may thus be a survival strategy for free-swimming vertebrate larvae. It was thought until recently that the regulatory principles underlying the development of the primary nervous system are distinct from those of secondary neurogenesis. However, recent findings suggest many mechanistic similarities raising the question of evolutionary relation of the two nervous systems.

We will summarize here the current knowledge on the mechanisms underlying the development of the primary nervous system of the zebrafish embryo. We will first provide a sketch of the anatomy of the nervous system and then introduce regulatory genes and their function in neurogenesis with emphasis on the primary nervous system. A detailed description of axonogenesis and the development of the nervous system during late embryonic and larval stages are beyond the scope of this review (see Chapter 5 by Jesuthasan S in this volume).

Different Classes of Primary Neurons Occupy Characteristic Positions in the Embryonic Nervous System

As that of other vertebrates, the neural tube of the zebrafish embryo is highly polarized with distinct neuronal cell types along the dorsoventral and anteroposterior axis. The large cell bodies of the primary sensory neurons are localized dorsally in the spinal cord.[12] In the one-day old embryos, these neurons, which are also called Rohon–Beard sensory (RB) neurons, express the LIM homeobox transcription factors Islet-1 (Fig. 1A), Islet-2 (Fig.1B), the T-box transcription factor Tbx2b (also Tbx-c) and the basic helix-loop-helix (bHLH) transcription factors Neurogenin1 (Ngn1, Fig. 1H) and NeuroD.[3,13–18] RB sensory neurons mediate touch response in the trunk and tail of the embryo and early larvae.[12] They extend long, highly branched dendrites under the surface of the skin and project via long ascending axons into the dorsal longitudinal fascicles, early forming axon tracts at the dorsolateral aspects of the spinal cord.[12] RB sensory neurons are transient structures and are replaced by the peripherally located sensory neurons of the dorsal root ganglia during later stages.[19]

Intermediate positions of the spinal cord are occupied by interneurons (Fig. 1C, D). Interneurons comprise a heterogeneous class of cells (Fig. 1E) that differ in morphology, axonal and dendritic projections and gene expression.[7,20] One can distinguish at least eight different types (Fig. 1E). For example, commissural primary ascending or CoPA neurons have a large T-shaped cell body with dendrites extending into the dorsolateral longitudinal fascicle (DLF) at the dorsolateral aspects of the spinal cord (Fig. 1I). The CoPA neuron projects an axon ventrally that crosses the ventral midline above the medial longitudinal fascicles in the ventrolateral spinal cord. After crossing the ventral midline of the spinal cord, the CoPA axon turns dorsally and rostrally and ascends in the DLF of the contralateral side.[7,20] Other interneurons may have unipolar cell bodies, cross the midline below the Mauthner neuron or have descending processes.[7,20] (see Fig. 1E for an overview). Little is known about the function of the individual types of interneurons of the spinal cord. The smooth wave-like swimming movements of the trunk and tail entail a complex interplay of activating and inhibitory input.[21,22] It is assumed that interneurons play a role in the co-ordination of the startle response and the swimming movements.

Specific gene markers were reported only for a few classes of interneurons and we are far from such a comprehensive description of interneuronal gene expression as that reported for the different classes of interneurons in the mammalian spinal cord.[23] The bHLH factor Zash1b is broadly expressed in many different interneurons.[24] (Fig. 1C). CoPA interneurons are characterized by the antigen

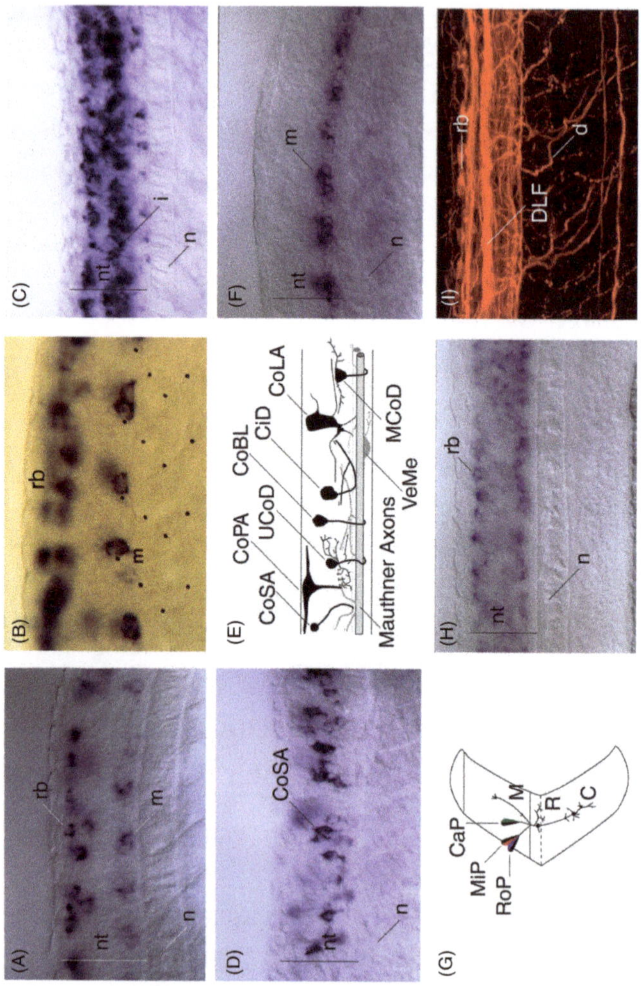

Fig. 1 Primary neurons of the spinal cord of a one-day-old zebrafish embryo.
A: The LIM-homeodomain transcription factor Islet-1 is expressed in the Rohon–Beard sensory neurons (rb). In a subset of primary motor neurons (m) this gene is initially expressed in all three motor neurons and later on its expression disappears in the CaP/VaP after they initiate expression of Islet-2, but maintains in the RoP and MiP motor neurons.[14] B: The LIM-homeodomain transcription factor Islet-2 is expressed in the Rohon–Beard sensory neurons (rb) and in VaP and CaP primary motor neurons (m).[14]
C: The bHLH transcription factor Zash1b is expressed in many if not all interneurons (i) that occupy intermediate positions of the spinal cord.[24].

recognized by the monoclonal antibody CON1.[7] The dorsal longitudinal ascending (DoLA) interneurons are marked by expression of the T-box gene *spadetail*,[25] while commissural secondary ascending (CoSA) interneurons express *pax2.1 (pax-a*, Fig. 1D).[26]

Motor neurons are located at the ventrolateral aspects of the spinal cord slightly above the floor plate (Fig. 1F, G). Three primary motor neurons (named RoP, MiP and CaP) are usually formed per hemisegment. A fourth motor neuron (VaP) appears transiently but is normally eliminated by apoptosis.[14,27,28] Motor neurons can be distinguished by their position within a hemisegment and pattern of innervation of the adjacent somitic muscle (Fig. 1G). While CaP extends

---

Fig. 1 Continued

D: Expression of the paired/homeodomain transcription factor Pax2.1 marks commissural secondary ascending interneurons (CoSA).[26] E: At least eight different types of primary interneurons were identified in the spinal cord.[7,20] These differ in morphology, projections and dorsoventral position within the spinal cord. Scheme was redrawn from Hale *et al.*[20] F: The LIM-homeodomain transcription factor Lim3 is expressed in motor neurons (m). Note that besides of the three primary neurons also the first secondary neurons are positive in respect of this marker. G: Scheme illustrating the axonal projections of the three primary neurons RoP, MiP and CaP. The axon of RoP grows into the medial aspects (R) of the adjacent somitic muscle, while CaP and MiP project to the ventral ("C") and dorsal aspects ("M") of the somite, respectively. H: Expression of the bHLH transcription factor neurogenin1 (ngn1). *ngn1* transcripts are present at low levels in most neurons of the spinal cord. Rohon–Beard sensory neurons (rb) and some ventral cells show higher transcript levels. I: Rohon–Beard sensory neurons (rb) extend dendrites (d) under the epidermis. The dorsal longitudinal fascicle (DLF) runs along the dorsolateral aspect of the spinal cord.[12]

Embryos are shown in lateral views at the level of the hindgut extension. Dorsal is up and anterior is to the left. Abbreviations: CiD, circumferential descending interneuron; CoBL, commissural bifurcating longitudinal interneuron; CoLA, commissural longitudinal ascending interneuron; CoPA, commissural primary ascending interneuron; CoSA, commissural secondary ascending interneuron; d, dendrite; DLF, dorsal longitudinal fascicle; m, motor neuron; McoD, multipolar commissural descending interneuron; n, notochord; nt, neural tube; rb, Rohon–Beard sensory neuron; UcoD, unipolar commissural descending interneuron; VeMe, ventral medial interneuron.

its axon ventrally, the other two neurons, MiP and RoP, innervate the dorsal and intermediate muscle fibers of the somite, respectively (Fig. 1G). The axon tracts of primary motor neurons mark the path along which later-born secondary motor neurons extend their axons.[29,30] Recently, the first mutant affecting axonogenesis of the primary motor neurons was identified.[31] In the *stumpy* mutant, motor axons extend to the horizontal myoseptum similar to the axons in wild-type embryos, but then fail to reach their final targets in time. While the molecular defect in the mutant is still unknown, this analysis supports the idea that successful axonogenesis in vertebrates, similar to that in invertebrates, depends on intermediate targets.[32]

As in other vertebrates, the hindbrain or rhombenencephalon of the zebrafish embryo is a highly segmented structure.[33–36] Early developing reticulospinal interneurons (Fig. 2D) and later differentiating branchiomotor neurons develop in the specific hindbrain segments or rhombomeres.[2,37–40] Gene expression domains have also been proposed to divide the forebrain (comprising telencephalon and diencephalon) and midbrain (mesencephalon) into segment-like compartments. However, they are less obvious on the basis of their morphology and pattern of neurogenesis (Fig. 2A, B) than the hindbrain segments, which are divided by well-characterized intersegmental boundaries and express specific molecular markers.[35,41–43] In contrast to higher vertebrates, primary motor neurons of the zebrafish spinal cord are also arranged in a highly segmented fashion with three primary neurons developing adjacent to each somite block.[27,36] Some, but less strict, segmental organization is obvious in the arrangement of the interneurons,[7,20] while RB sensory neurons appear to develop at random positions with respect to the somite boundaries.[12] However, in contrast to the hindbrain and forebrain, a segmented pattern of gene expression such as that of *krox20*[44] and the homeobox genes[45] found in the hindbrain has so far not been demonstrated in the spinal cord. The segmental arrangement of certain primary neurons in the spinal cord may thus be imposed by the mechanistic requirements or inductive influence of the adjacent somitic tissue and does not reflect an intrinsic segmented organization of the spinal cord.

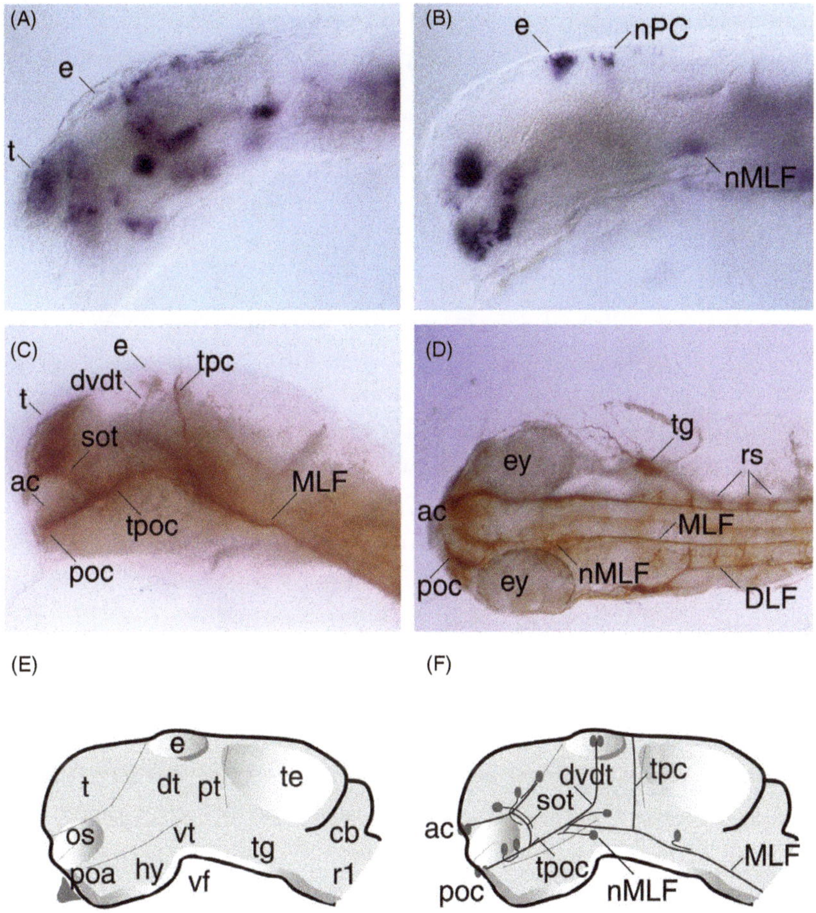

Fig. 2 Neural patterning in the brain of a one-day-old zebrafish embryo.
A: *ngn1* transcripts in the head. B: Expression of *islet-1* mRNA in the brain. C, D: Axon tracts in the head.[41] E, F: Brain territories and axon tracts in the brain (redrawn from MacDonald *et al.*[41]).

Orientation of embryos is dorsal up and anterior to the left. Abbreviations: ac, anterior commissure; DLF, dorsal longitudinal fascicle; dvdt, dorsoventral diencephalic tract; e, epiphysis; ey, eye; MLF, medial longitudinal fascicle; nMLF, nucleus of the medial longitudinal fascicle; nPC, nucleus of the posterior commissure; poc, postoptic commissure; sot, supraoptic tract; t, telencephalon; tpc, tract of the posterior commissure, tpoc, tract of the postoptic commissure.

The pattern of neurogenesis is more complex in the developing brain than in the spinal cord. Many different cell groups express *ngn1* (Fig. 2A) and *islet-1* (Fig. 2B) in the fore- and midbrain. By the end of the first day of development, the primary neurons have extended processes to form a scaffold of neuronal connections (Fig. 1I, 2C to F).[6,7,9] Several commissures connect left and right halves of the brain and longitudinal tracts relay information along the anteroposterior axis of the CNS (Fig. 2C to F). By this stage, the trigeminal ganglion, which has differentiated lateral to the midbrain-hindbrain boundary (Fig. 2D), extends its highly branched dendrites under the epidermis of the yolk and head.[12] The trigeminal ganglia like the RB sensory neurons mediate touch response and project to the Mauthner neurons, the two largest reticulospinal interneurons of the hindbrain residing in rhombomere 4.[46,47] The axon of the Mauthner neuron crosses the midline to connect via motor neurons to the muscles of the contralateral side. The trigeminal ganglia like the RB sensory neurons in the trunk and tail mediate touch response.[12] When touched on the side of the head, the trigeminal ganglion activates the Mauthner neuron, which leads to contraction of the musculature on the contralateral side.[46] As a consequence, the animal bends away from the source of the tactile stimulus. There is evidence that other reticulospinal interneurons are also involved in the neuronal control of the startle response.[48,49]

Primary Neurons Form Distinct Domains in the Neural Plate

The organization of the primary nervous system is pre-figured on the neural plate. The first territories that express neuronal marker genes appear during gastrulation. The bHLH transcription factor Neurogenin1 (Ngn1) and the signaling molecule Delta-A are expressed in all areas of primary neurogenesis in the neural plate (Fig. 3), where they define neuroblasts or undifferentiated precursor cells.[15–17,50,51] Transcripts of other genes such as the homeobox transcription factor Islet-1 or the bHLH factor NeuroD appear slightly later in development.[15] Their expression is confined to cells that have already initiated neuronal differentiation and have been selected from the *ngn1*-positive precursor pool by lateral inhibition as will be discussed below.[3,13,15,17] The medial to lateral coordinates of the neural plate are transformed into the dorsoventral axis of the neural tube.[4,52] Accordingly, cells of the dorsal neural tube such as neural crest and RB sensory neurons are derived from the lateral edges of the neural plate. Precursors of motor neurons reside close to the midline on each side of the future floor plate and progenitors of interneurons occupy intermediate positions in the neural plate.[52]

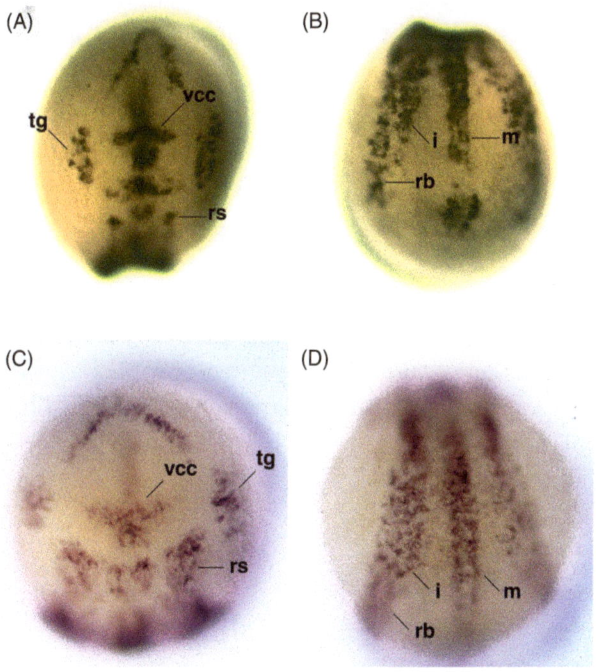

Fig. 3 *ngn1* and *delta-A* are expressed in an overlapping pattern and mark the sites of primary neurogenesis in the neural plate. A, B: *ngn1* expression in the neural plate of three-somite stage embryos.[15] C, D: Expression of *delta-A* in the neural plate of three-somite stage embryos.[51] View onto the anterior (A, C) and posterior neural plate (B, D) giving rise to the spinal cord. Anterior is up. Abbreviations: i, interneuron, m, motor neuron; rb, Rohon–Beard sensory neuron; rs, reticulospinal interneuron; tg, trigeminal ganglion; vcc, ventrocaudal cluster.

The precursors of reticulospinal interneurons of the hindbrain are also already recognizable by distinct *ngn1* expression domains in the hindbrain primordium of the neural plate.[15–17] Expression of *ngn1* is detected in medial regions of the midbrain anlage giving rise to the so-called ventrocaudal cluster (vcc), which forms the nucleus of the medial longitudinal fascicle (nMLF) later in development. *ngn1* is also expressed weakly along the anterior edge of the neural plate and in two clusters of cells lateral to the prospective MHB which will become the trigeminal ganglia (Fig. 3A). *delta-A* expression is strikingly similar to that of *ngn1* (Fig. 3C, D)[50,51]. In comparison to the posterior neural plate comprising the prospective hindbrain and spinal cord, the anterior neural plate is relatively devoid of *ngn1* and *delta-A* expression (Fig. 3A, C).

External Signals in the Spatial Control of Neurogenesis

It is obvious that neurogenesis does not occur uniformly throughout the neural plate, but is confined to defined regions (Fig. 3). An important question is therefore how the complex spatial pattern of neurogenic regions in the neural plate is controlled. The timing and the pattern of neurogenesis suggest a control by the signaling systems regulating the development of the dorsoventral and anteroposterior body coordinates during gastrulation.

Bone morphogenetic proteins (BMPs) that belong to the transforming growth factor β family of signaling molecules control differentiation of the dorsoventral body axis during gastrulation.[53,54] At early gastrulation stages, *swirl/bmp2b* expression becomes restricted to the ventral-most region of the gastrula forming a gradient of BMP expression along the dorsoventral axis.[55,56] Loss-of-function mutations in *swirl/bmp2b*, *snailhouse/bmp7* and *somitobun/smad5* impair dorsoventral pattern formation of the gastrula and perturb the pattern of neuronal differentiation.[57,58] The concentration of BMP protein appears to be relevant in the control of neuronal differentiation: embryos, in which

BMP signaling is most severely disrupted, fail to develop neural crest, RB sensory neurons and interneurons, whilst the medial motor neurons are expanded. Mutants with less severe phenotypes fail to form neural crest and RB sensory neurons but differentiate instead a hugely expanded array of interneurons.[57,58] These results suggest that BMPs act during gastrulation on the entire neural plate.[57,58] Moreover, they indicate that BMPs are morphogens in the zebrafish that trigger distinct neuronal programs depending on the amount of BMP signal that a neuronal precursor encounters.

In addition to BMPs, development of primary motor neurons has been shown to depend on Hedgehog (Hh) signaling.[59–61] Three members of the Hh family of secreted molecules are expressed in the midline of the zebrafish gastrula and neurula: *sonic hedgehog* is expressed in the floor plate and notochord while *echidna* and *tiggywinkle hedgehog* are expressed in the notochord and floor plate, respectively.[62–64] The Hhs bind to Patched, a multi-pass transmembrane protein on the surface of target cells. Patched is a repressor of a second transmembrane protein Smoothened (for a review of the Hh-signaling pathway see Ingham and McMahon[65]). Removal of maternal and zygotic activity of *smoothened* blocks signaling by all three Hhs in the embryo. These mutant embryos fail to form primary and secondary motor neurons demonstrating a requirement for Hh signaling in the differentiation of motor neurons.[60,61,66] Moreover, forced Hh signaling by mis-expression of Hhs or constitutively active components of the Hh-signaling pathway causes the differentiation of ectopic motor neurons in the spinal cord.[15,59]

Further signaling is required for induction of hindbrain and spinal cord identities. Mesoderm of the blastoderm margin was shown by grafting experiments to be the source of the posteriorizing signals.[67,68] As predicted by the ectodermal fate maps of gastrula stage embryos, the marginal mesoderm is in close juxtaposition with ectodermal regions fated to become hindbrain and spinal cord.[69] A candidate molecule for the posteriorizing signal is fibroblast growth factor 3.[70] While these posteriorizing signals from the blastoderm margin mediate the rough subdivision of the neural plate in anterior and posterior regions, they fail to explain the complex pattern of neurogenic regions in the neural plate. Very little is understood about the underlying mechanisms. It is envisaged that the cross-talk between dorsoventral patterning systems such as BMPs and Hhs and the anterior-posterior signaling systems will generate distinct positional cues in the neural plate.[71,72] Moreover, secondary signaling centers within the neural plate such as the floor plate, the midbrain/hindbrain boundary or the anterior neural ridge lead to an elaboration of the spatial pattern of neurogenesis.[73–75]

Pre-Pattern Genes in the Neural Plate

The positional information conveyed by these signals is believed to be interpreted by expression of so-called pre-pattern genes. These genes act upstream of the neural determination or proneural genes such as *ngn1*. Subsequently, neural differentiation genes act downstream of the proneural genes forming a hierarchy of gene switches that ultimately determine the specific differentiation of the post-mitotic neuron (Fig. 4). The area of expression of many pre-pattern genes cover wide areas of the neural plate within which proneural genes are expressed in a spatially much more restricted fashion.[41,76–78] This suggests that these pre-pattern genes act in combination rather than individually as regulators of neurogenesis.

The *ngn1* expression in the midbrain anlage and in the trigeminal ganglia is dependent on the homeobox transcription factors Iroquois1 and 7 (*iro1, iro7*).[79] Curiously, *iro* genes were initially identified as pre-pattern genes in *Drosophila* defining the sites of expression of the proneural genes *achaete* and *scute*.[80] Thus, aspects of the spatial control of neurogenesis appear to be conserved between insects and teleosts. In agreement with an ancient function in neural patterning, the expression of the known *iroquois* genes is restricted to the midbrain anlage and the more posterior neural plate, while the evolutionary novel forebrain does not express these genes.[76–78]

Yet another parallel to *Drosophila* is indicated by the action of the hairy/enhancer-of-split related factor Her5, which, like its *Drosophila* homolog Hairy, is a negative regulator of neurogenesis.[81,82] The midbrain-hindbrain boundary region (MHB) is normally devoid of neural differentiation.[82] When, however, the activity of Her5 is blocked

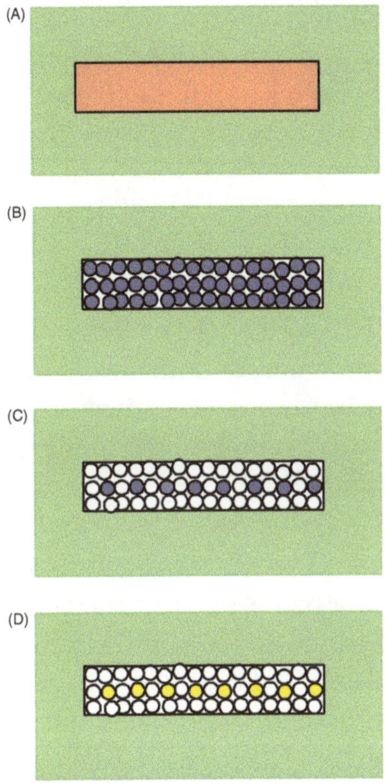

Fig. 4 The gene hierarchy controlling primary neurogenesis. A: Pre-pattern genes define region of neurogenic potential. Examples for positively acting factors are *iro1*, *iro7* and *narrowminded* (red) which define the position of proneural (neural determination) gene expression.[79,85]. In surrounding regions (green), neurogenesis is suppressed by negatively acting factors. Examples are the bHLH-WRPW factor Her5[82] or the zinc finger transcription factor Zic 2.[83] B: Proneural (neural determination) genes (blue) such as *ngn1*[15] and possibly also *olig2*[115] are expressed in regions of neurogenic potential. C: Neural progenitors are selected by lateral inhibition. Proneural gene expression is suppressed in surrounding cells. These cells are a pool of progenitors for subsequent neurogenesis and gliogenesis to form, for example, secondary motor neurons,[98] neural crest[86] or glial cells.[104] D: Neural differentiation genes (yellow) are expressed in committed neural precursors. These genes control specific aspects of neural differentiation and render the cells resistant to lateral inhibition. Examples for such genes are the transcription factors Collier[128,129] and MyT1,[130] the bHLH factors NeuroD[15,17] and LIM-homeodomain factors such as Lim3[14] and Islet1.[3]

by antisense morpholinos, ectopic *ngn1* expression and neuronal differentiation is evident in this region.[82] Moreover, forced expression of *her5* suppresses *ngn1* expression in the regions where *ngn1* is normally expressed. Thus, Her5 is a negative regulator that prevents neurogenesis in the MHB region.[82]

There are also indications for restrictions of neurogenesis in the posterior neural plate. Embryos overexpressing *ngn1* form a large number of ectopic neurons in the non-neural ectoderm that express neuronal markers such as *islet-1*.[15] The number of *islet-1* positive cells in the neural plate of mis-expressing embryos remained, however, unchanged. In particular, the regions separating the stripes of differentiating primary neurons did not show any ectopic *islet-1* positive cells, suggesting that these regions of the neural plate, in contrast to the non-neural ectoderm, are refractory to *ngn1*-controlled neuronal differentiation. The corresponding regions in *Xenopus laevis* embryos express the zinc finger factor Zic 2 that acts as a repressor of neurogenesis when overexpressed, suggesting that neurogenesis in these intervening regions of the neural plate is actively repressed in the *Xenopus* embryo.[83] It remains to be seen whether zebrafish have *zic2* homologs and whether they act in the same way as proposed for Zic2 in *Xenopus* embryos. Curiously, in zebrafish *ngn1* can trigger ectopic *delta-A* and *-B* gene expression in these neuron-free intervening regions even though it fails to induce *islet-1*-positive cells (Blader P and Strähe U unpublished). This suggests that not all aspects of neuronal gene expression are blocked in these neuron-free regions of the zebrafish neural plate. The differential response may be due to the fact that Ngn1 is a direct regulator of *delta* gene expression. Indeed, an upstream regulatory sequence was recently identified in the *delta-D* gene that contains E-boxes, the binding sites of Ngn1. This regulatory element is activated by *ngn1* expression and this response depends on intact E-boxes.[84]

In principle, negatively acting factors would be sufficient to paint the entire pattern of neurogenesis on the neural plate. There is, however, also evidence for positively acting factors that promote neurogenesis. We have already mentioned the *iro1* and *7* genes. Mutations in the *narrowminded* locus abolish the development of RB sensory neurons and early differentiating neural crest cells.[85] RB sensory neurons are selected from a progenitor pool by lateral inhibition.[86] Progenitors, in which the Notch pathway is activated, develop into neural crest. Cell transplantation indicated a cell-autonomous function of *narrowminded* in neural plate cells.[85] As both neural crest and RB sensory neurons are affected, the *narrowminded* gene appears thus to control the development of the common progenitor cells. The *narrowminded* function is dispensable for later developing neural crest cells (but not for RB sensory neurons), indicating temporally distinct mechanisms of neural crest specification.[85]

Primary Neurons are Selected from Pools of Precursors by Lateral Inhibition

The molecular mechanisms controlling neurogenesis in the zebrafish neural plate bear resemblance to those employed in neurogenesis of *Drosophila melanogaster*.[87–90] The areas in the neural plate, from which the primary neurons develop, comprise initially a lot more cells than eventually develop into post-mitotic primary neurons.[15] Thus, as in *Drosophila*, primary neurons appear to be selected from a pool of precursor cells. The transmembrane receptor Notch and its membrane-bound ligand Delta are involved in this selection process, which is also referred to as lateral inhibition.[90,91] Ngn1 expression delineates the areas of neural precursors, some of which express higher levels of Delta.[15,50,51] As a consequence Notch is activated in neighboring cells leading to expression of repressors of the hairy/enhancer of split class (named her in the zebrafish). Her proteins harbor a bHLH domain and a WRPW motif that binds the co-repressor Groucho. They suppress expression of *ngn1* and *delta* genes and prevent neural differentiation. In particular, *her4* transcripts are abundantly present in the neural plate in a pattern very similar to that of *ngn1*.[92]

The neural plate expresses several *delta* and *notch* genes.[50,93–95] which act in a redundant manner that complicates functional analysis. At neural plate stages, the *delta-A* and *delta-D* genes are expressed in a pattern very similar to that of *ngn1*[50,51,94] (Fig. 3). The overlapping expression patterns of *ngn1* and *delta* genes suggests that the genes are functionally connected. Mis-expression of *ngn1* induces ectopic *delta* gene expression, while inhibition of Ngn1 caused down-regulation of *delta-A*.[15,96] Moreover, it was recently shown that *ngn1* is a direct regulator of transgenes containing regulatory sequences of the *delta-D* gene.[84] Thus, *ngn1* appears to be structurally and functionally related to proneural genes of *Drosophila*.[97] Also as predicted from the neurogenic phenotype of *delta* mutations in *Drosophila*, mutations in the *delta-A* gene increase the number of primary neurons.[98] Similarly, mutations in the *delta-D* gene (also called *after eight*) have a neurogenic phenotype.[99] The panel of zebrafish mutations contains other, in molecular terms, uncharacterized mutants that also increases the number of primary neurons such as *deadly seven*.[100] Mutations in zebrafish homologs of Notch genes have not been identified so far. However, an increase of primary neurons is observed, when Delta-Notch signaling is blocked by an inhibitor of γ-Secretase, the protease required to process Notch upon Delta binding.[101] A similar neurogenic phenotype has been observed in another mutant *mind bomb*.[102] This locus encodes a RING ubiquitin ligase, which interacts with the intracellular domain of Delta to promote its ubiquitination and degradation. *Mind bomb* function is essential in the signaling cell for efficient activation of Notch in neighboring cells.[102] In summary, the bHLH transcription factor *ngn1* defines neurogenic regions in the neural plate that are functionally similar to proneural domains in the imaginal disks of the fruitfly. Primary neurons are selected from these regions by Delta-Notch-mediated lateral inhibition—a process that requires most likely the repressor Her4 as downstream effector (Fig. 5).

What is the fate of the cells that are prevented from becoming primary neurons? *delta-A* mutants as well as embryos overexpressing dominant negative variants of *delta* genes show a reduction of secondary motor neurons, suggesting that the increase in the number of primary neurons in these embryos occurred at the expense of the secondary neurons.[98] Delta-Notch signaling thus appears to maintain progenitor cells so that they can be available for later, secondary neurogenesis.[98]

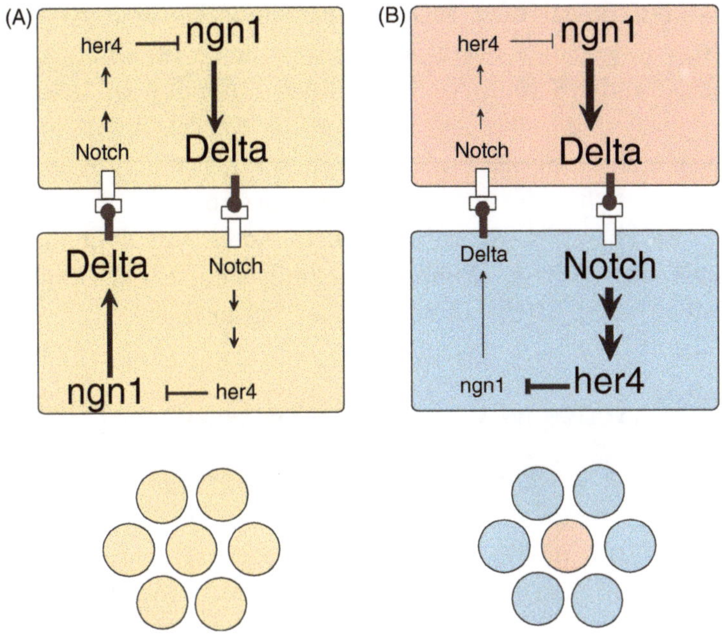

Fig. 5 The principle of lateral inhibition. Initially, all the cells within the proneural cluster or neurogenic domain in the neural plate express *ngn1* that controls expression of *delta-A* and *delta-D* (panel A, brown) By analogy with the cells of the proneural clusters in the imaginal disks of *Drosophila melanogaster*[90], these cells appear to form an equivalence group where each cell has the potential to develop into a neuron. Indeed, when lateral inhibition is blocked by mutation or forced expression of dominant negative *delta* genes, many more neurons develop from these regions.[15,98] Moreover ablation of committed motor neurons leads to recruitment of neighboring cells.[98] In the normal embryo, a few cells start to express higher levels of Delta-A or Delta-D (Panel B, pink cell). This leads to increased activation of the Delta receptor Notch in the surrounding cells and suppression of the neuronal program (Panel B, blue cells). Even though it has not directly been demonstrated in the zebrafish embryo, it is assumed that subsequent steps of Delta-Notch signaling (indicated by two consecutive, short arrows) are similar to those shown in other systems.[88,89,101,131] Upon ligand binding, the intracellular domain of Notch is released from the membrane by proteolysis and forms a complex with homologs of *supressor-of-hairless* proteins. This complex activates the expression of the bHLH-WRPW repressor *her4*,[92] which in turn blocks expression of *ngn1* and *delta* genes (Panel B, blue cells). As a consequence, these cells are prevented from immediate entry into the neuronal differentiation program. The mechanism by which cells are endowed with the ability to express higher levels of *delta* genes is not well understood. Stochastic

In the dorsal neural tube another cell fate decision is controlled by Delta-Notch signaling. There, it regulates the partitioning of cells into neural crest and RB sensory neuron fates: inhibition or lack of Delta-Notch signaling shifts the cell fate towards RB cell differentiation and depletes neural crest cells. Hence, Delta-Notch signaling represses *ngn1* expression and thereby prevents progenitors to enter the RB sensory fate.[86,96] Another possible role of Delta-Notch signaling may be the control of glial cell development as suggested by studies in higher vertebrates[103] and in the retina of the zebrafish.[104]

bHLH Genes as Regulators of Neuronal Subtype

Another central question is what controls differentiation of distinct neurons at different locations. The bHLH gene *ngn1* is broadly expressed in the neural plate, which suggests that it may control development of many distinct neurons.[15,17] Overexpression of *ngn1* induces ectopic *delta* gene expression and leads to aberrant differentiation of neurons in non-neural ectoderm.[15,16] The ectopic neurons, however, expressed most abundantly markers specific for RB sensory neurons and trigeminal ganglia.[15] Neurons expressing interneuron marker were restricted to certain sites in the embryo and ectopic motor neurons were only scored when the Hh pathway was activated artificially in embryos mis-expressing *ngn1*.[15] In agreement with a role in sensory neuron development, lack of *ngn1* activity impairs development of RB sensory neurons, the trigeminal and dorsal root ganglia.[86,105–107] Even though *ngn1* is also expressed at the other sites of primary neurogenesis it does not seem to be required for neuronal

---

Fig. 5 Continued
fluctuations in *delta* expression levels have been proposed to generate initially small differences in *delta* expression, which are then amplified by the negative feedback loop between neighboring cells. Alternatively certain cells may be endowed with higher *delta* expression by extrinsic signals from the very beginning and Delta-Notch signaling would then lead to a refinement and reinforcement of this initial bias.[90]

differentiation in these regions. The mouse has several *ngn* genes that are partially redundant.[108–113] The existence of a related *ngn* gene with overlapping functions in the motor neuron and interneuron precursors remains to be demonstrated in the zebrafish. The zebrafish *ngn3* gene is expressed in the ventral hypothalamus in post-somitogenesis stage embryos and may have a role as proneural gene during secondary neurogenesis in the ventral forebrain.[114]

Motor neuron differentiation depends on Olig2, another bHLH protein.[115] The Olig2 gene plays a role not only in motor neuron but also in oligodendrocyte differentiation suggesting a functional link between these two distinct cell types characteristic for the ventrolateral aspects of the neural tube. Indeed, differentiation of both cell types depends on hedgehog signals. *olig2* is expressed in the neural plate from mid-gastrula stages onwards.[115] When its protein expression is knocked-down by a morpholino-antisense approach, primary motor neurons and oligodendrocytes fail to differentiate.[115] Expression of *olig2* in the medial neural plate requires an intact Hh-signaling pathway suggesting that *olig2* acts downstream of Hhs.[115] Mis-expressed *olig2* triggers the ectopic differentiation of oligodendrocytes and motor neurons in wild-type embryos. However, the ectopic differentiation of motoneurons and oligodendrocytes does not occur in embryos that have a compromised Hh signaling pathway.[115] This suggests that Olig2 is necessary but not sufficient, and that other Hh-dependent factors are required in addition for motor neuron and oligodendrocyte differentiation.

Zash1a and b, two homologs of the mouse Ash1 and the *Drosophila* Achaete and Scute proneural bHLH factors may be other zebrafish neural determination genes. *zash1b* is broadly expressed in interneuron progenitors of the neural plate.[15,24] This expression is delayed with respect to that of *ngn1* and it remains to be tested whether *zash1b* has proneural function in the neural plate. The function of bHLH genes within the gene hierarchies controlling neurogenesis can vary. For example in the mouse, *ngn1* can act as a neural determination (proneural) gene in a certain context, such as the development of epibranchial placodes or the dorsal root ganglia.[111–113] In a different context, such as the neurons of the olfactory epithelium, *ngn1* acts downstream of the Achaete–Scute homolog Mash1 and can thus be classified as a later acting neural differentiation gene.[116] In the two-day old brain of the zebrafish embryo, *zash1b* is widely expressed in the ventricular zone, a region where neurons are born before they migrate to the pial surface.[117] In contrast, *ngn1* expressing cells are located in subventricular regions at these stages suggesting that *ngn1* expression is confined to differentiating post-mitotic neurons.[117] This pattern of expression suggests that *zash1b* acts upstream of *ngn1* in the two-day-old zebrafish brain and that *ngn1* may have a downstream neuronal differentiation function in this context.

The Regulatory Elements Controlling Gene Expression in Primary Neurons are Conserved among Vertebrates

Several cis-regulatory regions were identified upstream of the zebrafish *ngn1* coding sequence.[118] These regions control distinct and overlapping aspects of the *ngn1* expression pattern in the neural plate. One regulatory region (named Lateral Stripe Element or LSE), located between –6702 and –6490 upstream of the ATG, drives expression in the lateral neural plate in the precursors of the RB sensory neurons. A second enhancer (ANPE, between –3116 and –3122) controls aspects of *ngn1* expression in the anterior neural plate. The activities of these regulatory elements precede that of a third regulatory region, which is located at more proximal positions and has an overlapping activity with the ANPE and LSE in older embryos.[118] Thus multiple regulatory regions appear to drive *ngn1* expression in primary neurons in the zebrafish.

When compared to the 5' region of the human and mouse homolog of *ngn1*, the regulatory regions map to islands of conserved sequence[116] (Fig. 6). This remarkable conservation is not only restricted to the regulatory architecture of the *ngn1* gene but it was also noted for the regulatory elements of the zebrafish and mouse *delta* gene homologs.[84,119] This suggests that both lower and higher vertebrates utilize conserved mechanisms to build their nervous systems. In agreement, transgenes carrying reporters under control of zebrafish *ngn1* regulatory elements drive expression in the mouse neural tube,[120] (Blader P, Scardigli S, Guillemot F and Strähle U unpublished). This suggests that primary and secondary neurogenesis differ in timing but share regulatory mechanisms. Neurogenesis in vertebrates thus probably relies on a similar set of regulatory factors. A central but so far unresolved question is what determines the timing of neurogenesis in lower and

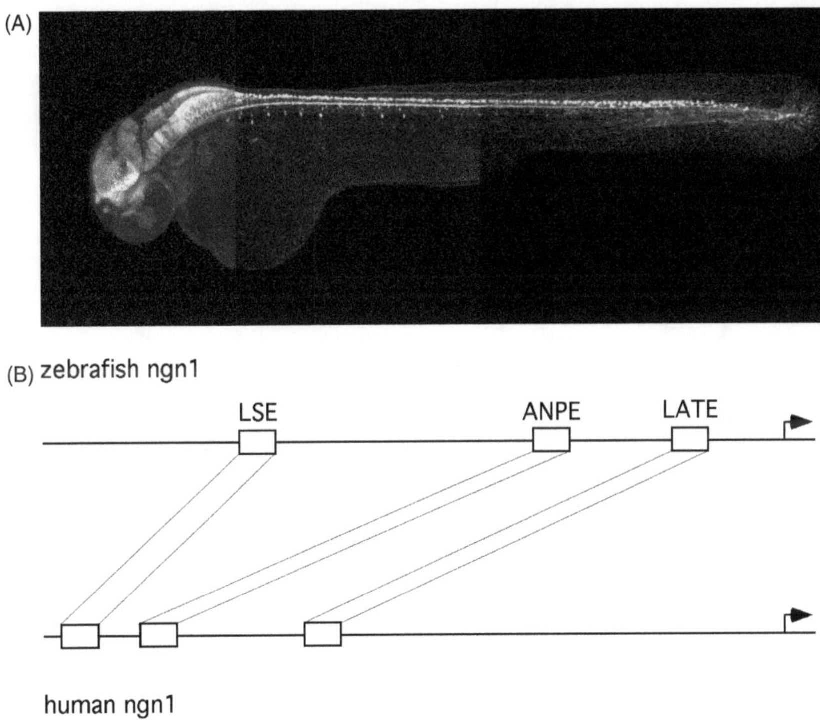

Fig. 6 Multiple regulatory elements control the expression of *ngn1* in transgenic zebrafish embryos. A: Transgenic zebrafish embryo (two days) harbouring the green fluorescent protein under control of a 3.4 kb *ngn1* upstream sequence.[118] B: Summary of regulatory elements mapped by deletion analysis in the 8.4 kb upstream region of the zebrafish *ngn1* locus. The diagram at the bottom indicates the regions of homology shared with the human *ngn1* locus. Note that both the order and the orientation of the regulatory elements are conserved.[118]

higher vertebrates. This question is linked to the question of the evolutionary origin of the primary and secondary nervous systems. Is the primary nervous system the more ancient nervous system? Has it been lost during evolution of higher vertebrates as it is not necessary for embryos developing in an amnion?

Irrespectively, the conservation of regulatory mechanisms indicates that analysis of the development of the simple primary nervous system of the zebrafish will prove extremely helpful in unraveling the principles of vertebrate neurogenesis in general. In particular, the experimental virtues of the zebrafish system such as simplicity of the primary nervous system, transgenesis,[121–124] forward genetics[125,126] and antisense technologies[127] in an optically translucent embryo will provide an efficient way to analyze development of a functional vertebrate nervous system.

Acknowledgments

We are grateful to Martine Behra, Charles Plessy, Dominique Biellmann and Sandra Metz for artwork and to Thomas Dickmeis, Charles Plessy and Eric Chen Sok Lam for comments and discussion. US was supported by CNRS/INSERM/ULP/HUS, AFM, ARC, Volkswagen Stiftung and AICR. VK received support from the Agency for Science, Technology and Research of Singapore.

References

1. Eaton RC, Bombardieri RA and Meyer DL (1977). The Mauthner-initiated startle response in teleost fish. *J Exp Biol* **66**(1): 65–81.
2. Mendelson B (1986). Development of reticulospinal neurons of the zebrafish. I. Early axonal outgrowth and cell body position. *J Comp Neurol* **251**(2): 172–184.
3. Korzh V, Edlund T and Thor S (1993). Zebrafish primary neurons initiate expression of the LIM homeodomain protein Isl-1 at the end of gastrulation. *Development* **118**(2): 417–425.
4. Schmitz B, Papan C and Campos–Ortega JA (1993). Neurulation in the anterior trunk region of the zebrafish Brachydanio rerio. *Roux's Arch Dev Biol* **202**: 250–259.

5. Papan C and Campos–Ortega JA (1994). On the formation of the neural keel and the neural tube in the zebrafish Danio (Brachidanio) rerio. *Roux's Archive Dev Biol* **203**: 178–186.
6. Wilson SW, *et al.* (1990). The development of a simple scaffold of axon tracts in the brain of the embryonic zebrafish, Brachydanio rerio. *Development* **108**(1): 121–145.
7. Bernhardt RR, *et al.* (1990). Identification of spinal neurons in the embryonic and larval zebrafish. *J Comp Neurol* **302**(3): 603–616.
8. Bernhardt RR, *et al.* (1992). Axonal trajectories and distribution of GABAergic spinal neurons in wildtype and mutant zebrafish lacking floor plate cells. *J Comp Neurol* **326**(2): 263–272.
9. Kimmel CB, (1993). Patterning the brain of the zebrafish embryo. *Annu Rev Neurosci* **16**: 707–732.
10. Wullimann MF and Knipp S (2000). Proliferation pattern changes in the zebrafish brain from embryonic through early postembryonic stages. *Anat Embryol (Berl)* **202**(5): 385–400.
11. Kimmel CB and Westerfield M (1989). Primary neurons of the zebrafish. In Edelman GM and Cowan MW (eds.), *Signals and Sense*, pp. 561–588.
12. Metcalfe WK, *et al.* (1990). Primary neurons that express the L2/HNK-1 carbohydrate during early development in the zebrafish. *Development* **110**(2): 491–504.
13. Inoue A, *et al.* (1994). Developmental regulation of islet-1 mRNA expression during neuronal differentiation in embryonic zebrafish. *Dev Dyn* **199**(1): 1–11.
14. Appel B, *et al.* (1995). Motoneuron fate specification revealed by patterned LIM homeobox gene expression in embryonic zebrafish. *Development* **121**(12): 4117–4125.
15. Blader P, *et al.* (1997). The activity of neurogenin1 is controlled by local cues in the zebrafish embryo. *Development* **124**(22): 4557–4569.
16. Kim CH, *et al.* (1997). Overexpression of neurogenin induces ectopic expression of HuC in zebrafish. *Neurosci Lett* **239**(2–3): 113–116.
17. Korzh V, *et al.* (1998). Expression of zebrafish bHLH genes ngn1 and nrd defines distinct stages of neural differentiation. *Dev Dyn* **213**(1): 92–104.
18. Dheen T, *et al.* (1999). Zebrafish tbx-c functions during formation of midline structures. *Development* **126**(12): 2703–2713.

19. Williams JA, *et al.* (2000). Programmed cell death in zebrafish Rohon Beard neurons is influenced by TrkC1/NT-3 signaling. *Dev Biol* **226**(2): 220–230.
20. Hale ME, Ritter DA and Fetcho JR (2001). A confocal study of spinal interneurons in living larval zebrafish. *J Comp Neurol* **437**(1): 1–16.
21. Granato M, *et al.* (1996). Genes controlling and mediating locomotion behavior of the zebrafish embryo and larva. *Development* **123**: 399–413.
22. Drapeau P, *et al.* (2002). Development of the locomotor network in zebrafish. *Prog Neurobiol* **68**(2): 85–111.
23. Matise M (2002). A dorsal elaboration in the spinal cord. *Neuron* **34**(4): 491–493.
24. Allende ML and Weinberg ES (1994). The expression pattern of two zebrafish achaete-scute homolog (ash) genes is altered in the embryonic brain of the cyclops mutant. *Dev Biol* **166**(2): 509–530.
25. Tamme R, *et al.* (2002). The identity and distribution of neural cells expressing the mesodermal determinant spadetail. BMC *Dev Biol* **2**(1): 9.
26. Mikkola I, *et al.* (1992). The paired domain-containing nuclear factor pax[b] is expressed in specific commissural interneurons in zebrafish embryos. *J Neurobiol* **23**(8): 933–946.
27. Myers PZ, Eisen JS and Westerfield M (1986). Development and axonal outgrowth of identified motoneurons in the zebrafish. *J Neurosci* **6**(8): 2278–2289.
28. Eisen JS and Melancon E (2001). Interactions with identified muscle cells break motoneuron equivalence in embryonic zebrafish. *Nat Neurosci* **4**(11): 1065–1070.
29. Fashena D and Westerfield M (1999). Secondary motoneuron axons localize DM-GRASP on their fasciculated segments. *J Comp Neurol* **406**(3): 415–424.
30. Pike SH, Melancon EF and Eisen JS (1992). Pathfinding by zebrafish motoneurons in the absence of normal pioneer axons. *Development* **114**(4): 825–831.
31. Beattie CE, Melancon E and Eisen JS (2000). Mutations in the stumpy gene reveal intermediate targets for zebrafish motor axons. *Development* **127**(12): 2653–2662.
32. Chiba A and Keshishian H (1996). Neuronal pathfinding and recognition: roles of cell adhesion molecules. *Dev Biol* **180**(2): 424–432.

33. Metcalfe WK, Mendelson B and Kimmel CB (1986). Segmental homologies among reticulospinal neurons in the hindbrain of the zebrafish larva. *J Comp Neurol* **251**(2): 147–159.
34. Hanneman E, *et al.* (1988). Segmental pattern of development of the hindbrain and spinal cord of the zebrafish embryo. *Development* **103**(1): 49–58.
35. Trevarrow B, Marks DL and Kimmel CB (1990). Organization of hindbrain segments in the zebrafish embryo. *Neuron* **4**(5): 669–679.
36. Kimmel CB, Sepich DS and Trevarrow B (1988). Development of segmentation in zebrafish. *Development* **104**(Suppl): 197–207.
37. Mendelson B (1985). Soma position is correlated with time of development in three types of identified reticulospinal neurons. *Dev Biol* **112**(2): 489–493.
38. Mendelson B and Kimmel CB (1986). Identified vertebrate neurons that differ in axonal projection develop together. *Dev Biol* **118**(1): 309–313.
39. Mendelson B (1986). Development of reticulospinal neurons of the zebrafish. I. Time of origin. *J Comp Neurol* **251**(2): 160–171.
40. Chandrasekhar A, *et al.* (1997). Development of branchiomotor neurons in zebrafish. *Development* **124**(13): 2633–2644.
41. Macdonald R, *et al.* (1994). Regulatory gene expression boundaries demarcate sites of neuronal differentiation in the embryonic zebrafish forebrain. *Neuron* **13**: 1039–1053.
42. Wullimann MF and Puelles L (1999). Postembryonic neural proliferation in the zebrafish forebrain and its relationship to prosomeric domains. *Anat Embryol (Berl)* **199**(4): 329–348.
43. Hauptmann G and Gerster T (2000). Regulatory gene expression patterns reveal transverse and longitudinal subdivisions of the embryonic zebrafish forebrain. *Mech Dev* **91**(1–2): 105–118.
44. Oxtoby E and Jowett T (1993). Cloning of the zebrafish krox-20 gene (krx-20) and its expression during hindbrain development. *Nucl Acids Res* **21**: 1087–1095.
45. Prince VE, Price AL and Ho RK (1998). Hox gene expression reveals regionalization along the anteroposterior axis of the zebrafish notochord. *Dev Genes Evol* **208**(9): 517–522.
46. Eaton RC, *et al.* (1977). Functional development in the Mauthner cell system of embryos and larvae of the zebra fish. *J Neurobiol* **8**(2): 151–172.
47. Kimmel CB, Hatta K and Metcalfe WK (1990). Early axonal contacts during development of an identified dendrite in the brain of the zebrafish. *Neuron* **4**(4): 535–545.

48. Liu KS and Fetcho JR (1999). Laser ablations reveal functional relationships of segmental hindbrain neurons in zebrafish. *Neuron* **23**(2): 325–335.
49. Gahtan E, *et al.* (2002). Evidence for a widespread brain stem escape network in larval zebrafish. *J Neurophysiol* **87**(1): 608–614.
50. Appel B and Eisen JS (1998). Regulation of neuronal specification in the zebrafish spinal cord by Delta function. *Development* **125**(3): 371–380.
51. Haddon C, *et al.* (1998). Multiple delta genes and lateral inhibition in zebrafish primary neurogenesis. *Development* **125**(3): 359–370.
52. Kimmel CB, Warga RM and Kane DA (1994). Cell cycles and clonal strings during formation of the zebrafish central nervous system. *Development* **120**: 265–276.
53. Mullins MC, *et al.* (1996). Genes establishing dorsoventral pattern formation in the zebrafish embryo: the ventral specifying genes. *Development* **123**: 81–93.
54. Hammerschmidt M, Serbedzija GN and McMahon AP (1996). Genetic analysis of dorsoventral pattern formation in the zebrafish: requirement of a BMP-like ventralizing activity and its dorsal repressor. *Genes Dev* **10**: 2452–2461.
55. Kishimoto Y, *et al.* (1997). The molecular nature of zebrafish swirl: BMP2 function is essential during early dorsoventral patterning. *Development* **124**(22): 4457–4466.
56. Nguyen VH, *et al.* (1998). Ventral and lateral regions of the zebrafish gastrula, including the neural crest progenitors, are established by a bmp2b/swirl pathway of genes. *Dev Biol* **199**(1): 93–110.
57. Barth KA, *et al.* (1999). Bmp activity establishes a gradient of positional information throughout the entire neural plate. *Development* **126**(22): 4977–4987.
58. Nguyen VH, *et al.* (2000). Dorsal and intermediate neuronal cell types of the spinal cord are established by a BMP signaling pathway. *Development* **127**(6): 1209–1220.
59. Hammerschmidt M, Bitgood MJ and McMahon AP (1996). Protein kinase A is a common negative regulator of Hedgehog signaling in the vertebrate embryo. *Genes Dev* **10**(6): 647–658.
60. Chen W, Burgess S and Hopkins N (2001). Analysis of the zebrafish smoothened mutant reveals conserved and divergent functions of hedgehog activity. *Development* **128**(12): 2385–2396.

61. Varga ZM, *et al.* (2001). Zebrafish smoothened functions in ventral neural tube specification and axon tract formation. *Development* **128**(18): 3497–3509.
62. Krauss S, Concordet J-P and Ingham PW (1993). A functionally conserved homolog of the Drosophila segment polarity gene hedgehog is expressed in tissues with polarizing activity in zebrafish embryos. *Cell* **75**: 1431–1444.
63. Ekker SC, *et al.* (1995). Patterning activities of vertebrate hedgehog proteins in the developing eye and brain. *Curr. Biol.* **5**: 944–955.
64. Currie PD and Ingham PW (1996). Induction of a specific muscle cell type by a hedgehog-like protein in zebrafish. *Nature* **382**: 452–455.
65. Ingham PW and McMahon AP (2001). Hedgehog signaling in animal development: paradigms and principles. *Genes Dev* **15**(23): 3059–3087.
66. Lewis KE and Eisen JS (2001). Hedgehog signaling is required for primary motoneuron induction in zebrafish. *Development* **128**(18): 3485–3495.
67. Woo K and Fraser SE (1997). Specification of the zebrafish nervous system by nonaxial signals. *Science* **277**(5323): 254–257.
68. Woo K and Fraser SE (1998). Specification of the hindbrain fate in the zebrafish. *Dev Biol* **197**(2): 283–296.
69. Woo K and Fraser SE (1995). Order and coherence in the fate map of the zebrafish nervous system. *Development* **121**: 2595–2609.
70. Koshida S, *et al.* (2002). Inhibition of BMP activity by the FGF signal promotes posterior neural development in zebrafish. *Dev Biol* **244**(1): 9–20.
71. Houart C, *et al.* (2002). Establishment of the telencephalon during gastrulation by local antagonism of Wnt signaling. *Neuron* **35**(2): 255–265.
72. Lagutin OV, *et al.* (2003). Six3 repression of Wnt signaling in the anterior neuroectoderm is essential for vertebrate forebrain development. *Genes Dev* **17**(3): 368–379.
73. Litingtung Y and Chiang C (2000). Control of Shh activity and signaling in the neural tube. *Dev Dyn* **219**(2): 143–154.
74. Rhinn M and Brand M (2001). The midbrain hindbrain boundary organizer. *Curr Opin Neurobiol* **11**(1): 34–42.
75. Wilson SW, Brand M and Eisen JS (2002). Patterning the zebrafish central nervous system. *Results Probl Cell Differ* **40**: 181–215.
76. Tan JT, Korzh V and Gong Z (1999). Expression of a zebrafish iroquois homeobox gene, Ziro3, in the midline axial structures and central nervous system. *Mech Dev* **87**(1–2): 165–168.
77. Cheng CW, *et al.* (2001). Identification and expression of zebrafish Iroquois homeobox gene irx1. *Dev Genes Evol* **211**(8–9): 442–444.

78. Wang X, *et al.* (2001). Expression of two novel zebrafish iroquois homologues (ziro1 and ziro5) during early development of axial structures and central nervous system. *Mech Dev* **105**(1–2): 191–195.
79. Itoh M, *et al.* (2002). A role for iro1 and iro7 in the establishment of an anteroposterior compartment of the ectoderm adjacent to the midbrain-hindbrain boundary. *Development* **129**(10): 2317–2327.
80. Calleja M, *et al.* (2002). How to pattern an epithelium: lessons from achaete-scute regulation on the notum of Drosophila. *Gene* **292**(1–2): 1–12.
81. Fisher A and Caudy M (1998). The function of hairy-related bHLH repressor proteins in cell fate decisions. Bioessays **20**(4): 298–306.
82. Geling A, *et al.* (2003). bHLH transcrition factor Her5 links patterning to regional inhibition of neurogenesis at the midbrain-hindbrain boundary. *Development* **130**(8): 1591–1604.
83. Brewster R, Lee J and Ruiz i Altaba A (1998). Gli/Zic factors pattern the neural plate by defining domains of cell differentiation. *Nature* **393**(6685): 579–583.
84. Hans S and Campos–Ortega JA (2002). On the organisation of the regulatory region of the zebrafish deltaD gene. *Development* **129**(20): 4773–4784.
85. Artinger KB, *et al.* (1999). Zebrafish narrowminded suggests a genetic link between formation of neural crest and primary sensory neurons. *Development* **126**(18): 3969–3979.
86. Cornell RA and Eisen JS (2000). Delta signaling mediates segregation of neural crest and spinal sensory neurons from zebrafish lateral neural plate. *Development* **127**(13): 2873–2882.
87. Artavanis–Tsakonas S and Simpson P (1991). Choosing a cell fate: a view from the Notch locus. Trends Genet, 7(11–12): 403–408.
88. Artavanis–Tsakonas S, Matsuno K and Fortini ME (1995). Notch signaling. *Science* **268**(5208): 225–232.
89. Artavanis–Tsakonas S, Rand MD and Lake RJ (1999). Notch signaling: cell fate control and signal integration in development. *Science* **284**(5415): 770–776.
90. Simpson P (1997). Notch signaling in development: on equivalence groups and asymmetric developmental potential. *Curr Opin Genet Dev* 7(4): 537–542.
91. Heitzler P and Simpson P (1991). The choice of cell fate in the epidermis of Drosophila. *Cell* **64**(6): 1083–1092.

92. Takke C, *et al.* (1999). her4, a zebrafish homologue of the Drosophila neurogenic gene E(spl), is a target of NOTCH signaling. *Development* **126**(9): 1811–1821.
93. Kortschak RD, Tamme R and Lardelli M (2001). Evolutionary analysis of vertebrate Notch genes. *Dev Genes Evol* **211**(7): 350–354.
94. Dornseifer P, Takke C and Campos-Ortega JA (1997). Overexpression of a zebrafish homologue of the Drosophila neurogenic gene Delta perturbs differentiation of primary neurons and somite development. *Mech Dev* **63**(2): 159–171.
95. Bierkamp C and Campos-Ortega JA (1993). A zebrafish homologue of the Drosophila neurogenic gene Notch and its pattern of transcription during early embryogenesis. *Mech Dev* **43**(2–3): 87–100.
96. Cornell RA and Eisen JS (2002). Delta-Notch signaling promotes formation of zebrafish neural crest by repressing Neurogenin 1 function. *Development* **129**(11): 2639–2648.
97. Hassan BA and Bellen HJ (2000). Doing the MATH: is the mouse a good model for fly development? *Genes Dev* **14**(15): 1852–1865.
98. Appel B, Givan LA and Eisen JS (2001). Delta-Notch signaling and lateral inhibition in zebrafish spinal cord development. BMC *Dev Biol* **1**(1): 13.
99. Holley SA, Geisler R and Nusslein-Volhard C (2000). Control of her1 expression during zebrafish somitogenesis by a delta-dependent oscillator and an independent wave-front activity. *Genes Dev* **14**(13): 1678–1690.
100. Gray M, *et al.* (2001). Zebrafish deadly seven functions in neurogenesis. *Dev Biol* **237**(2): 306–323.
101. Geling A, *et al.* (2002). A gamma-secretase inhibitor blocks Notch signaling in vivo and causes a severe neurogenic phenotype in zebrafish. *EMBO Rep* **3**(7): 688–694.
102. Itoh M, *et al.* (2003). Mind Bomb is a ubiquitin ligase that is essential for efficient activation of Notch signaling by Delta. *Dev Cell* **4**(1): 67–82.
103. Lundkvist J and Lendahl U (2001). Notch and the birth of glial cells. *Trends Neurosci* **24**(9): 492–494.
104. Scheer N, *et al.* (2001). An instructive function for Notch in promoting gliogenesis in the zebrafish retina. *Development* **128**(7): 1099–1107.
105. Andermann P, Ungos J and Raible DW (2002). Neurogenin1 defines zebrafish cranial sensory ganglia precursors. *Dev Biol* **251**(1): 45–58.

106. Golling G, *et al.* (2002). Insertional mutagenesis in zebrafish rapidly identifies genes essential for early vertebrate development. Nat Genet, **31**(2): 135–140.
107. Korzh V and Strähle U (2002). Proneural, prosensory, antiglial: the many faces of neurogenins. *Trends Neurosci* **25**(12): 603–605.
108. Gradwohl G, Fode C and Guillemot F (1996). Restricted expression of a novel murine atonal-related bHLH protein in undifferentiated neural precursors. *Dev Biol* **180**(1): 227–241.
109. Ma Q, Kintner C and Anderson DJ (1996). Identification of neurogenin, a vertebrate neuronal determination gene. *Cell* **87**(1): 43–52.
110. Sommer L, Ma Q and Anderson DJ (1996). neurogenins, a novel family of atonal-related bHLH transcription factors, are putative mammalian neuronal determination genes that reveal progenitor cell heterogeneity in the developing CNS and PNS. *Mol Cell Neurosci* **8**(4): 221–241.
111. Fode C, *et al.* (1998). The bHLH protein NEUROGENIN 2 is a determination factor for epibranchial placode-derived sensory neurons. *Neuron* **20**(3): 483–494.
112. Ma Q, *et al.* (1998). Neurogenin1 is essential for the determination of neuronal precursors for proximal cranial sensory ganglia. *Neuron* **20**(3): 469–482.
113. Ma Q, *et al.* (1999). Neurogenin1 and neurogenin2 control two distinct waves of neurogenesis in developing dorsal root ganglia. *Genes Dev* **13**(13): 1717–1728.
114. Wang X, *et al.* (2001). A novel zebrafish bHLH gene, neurogenin3, is expressed in the hypothalamus. *Gene* **275**(1): 47–55.
115. Park HC, *et al.* (2002). olig2 is required for zebrafish primary motor neuron and oligodendrocyte development. *Dev Biol* **248**(2): 356–368.
116. Cau E, *et al.* (2000). Hes genes regulate sequential stages of neurogenesis in the olfactory epithelium. *Development* **127**(11): 2323–2332.
117. Mueller T and Wullimann MF (2003). Anatomy of neurogenesis in the early zebrafish brain. *Brain Res Dev Brain Res* **140**(1): 137–155.
118. Blader P, Plessy C and Strähle U (2003). Multiple regulatory elements with spatially and temporally distinct activities control neurogenin1 expression in primary neurons of the zebrafish embryo. *Mech Dev* **120**(2): 211–218.
119. Beckers J, *et al.* (2000). Distinct regulatory elements direct delta1 expression in the nervous system and paraxial mesoderm of transgenic mice. *Mech Dev* **95**(1–2): 23–34.

120. Strähle U and Blader P (2003) The basic helix-loop-helix proteins in vertebrate and invertebrate neurogenesis, in Modularity in evolution and development, Schlosser G and Wagner G Editors. Chicago University Press: Chicago. in press.
121. Ju B, *et al.* (1999). Faithful expression of green fluorescent protein (GFP) in transgenic zebrafish embryos under control of zebrafish gene promoters. *Dev Genet* **25**(2): 158–167.
122. Higashijima S, Hotta Y and Okamoto H (2000). Visualization of cranial motor neurons in live transgenic zebrafish expressing green fluorescent protein under the control of the islet-1 promoter/enhancer. *J Neurosci* **20**(1): 206–218.
123. Koster RW and Fraser SE (2001). Direct imaging of in vivo neuronal migration in the developing cerebellum. *Curr Biol* **11**(23): 1858–1863.
124. Gilmour DT, Maischein HM and Nusslein–Volhard C (2002). Migration and function of a glial subtype in the vertebrate peripheral nervous system. *Neuron* **34**(4): 577–588.
125. Haffter P, *et al.* (1996). The identification of genes with unique and essential functions in the development of the zebrafish, Danio rerio. *Development* **123**: 1–36.
126. Driever W, *et al.* (1996). A genetic screen for mutations affecting embryogenesis in zebrafish. *Development* **123**: 37–46.
127. Nasevicius A and Ekker SC (2000). Effective targeted gene 'knockdown' in zebrafish. *Nat Genet* **26**(2): 216–220.
128. Bally–Cuif L, Dubois L and Vincent A (1998). Molecular cloning of Zcoe2, the zebrafish homolog of Xenopus Xcoe2 and mouse EBF-2, and its expression during primary neurogenesis. *Mech Dev* 77(1): 85–90.
129. Dubois L, *et al.* (1998). XCoe2, a transcription factor of the Col/Olf-1/EBF family involved in the specification of primary neurons in Xenopus. *Curr Biol* **8**(4): 199–209.
130. Bellefroid EJ, *et al.* (1996). X-MyT1, a Xenopus C2HC-type zinc finger protein with a regulatory function in neuronal differentiation. *Cell* **87**(7): 1191–1202.
131. de la Pompa JL, *et al.* (1997). Conservation of the Notch signaling pathway in mammalian neurogenesis. *Development* **124**(6): 1139–1148.

Chapter 7

Making Scents: Development and Function of the Olfactory Sensory System

Kathleen E. Whitlock
Department of Molecular Biology and Genetics
Cornell University

> "I should think we might fairly gauge the future of biological science, centuries ahead, by estimating the time it will take to reach a complete, comprehensive understanding of odor. It may not seem a profound enough problem to dominate all the life sciences, but it contains, piece by piece, all the mysteries."
>
> — Lewis Thomas

As indicated by Lewis Thomas, the olfactory sensory system is a spectacular sensory system holding all the mysteries of the biological world. The olfactory organ arises as a highly orchestrated interaction between the placodally derived tissues forming the sensory epithelium and the neural crest derived cells that will form the structural elements encasing the olfactory epithelium. Once formed, the sensory neurons must forge axonal connections with the developing olfactory bulbs and maintain the fidelity of these connections throughout life. This is a particular challenge in the olfactory system for the sensory neurons undergo constant regeneration throughout life. The olfactory sensory neurons express odorant receptors in order to interact with the world. Recently it has been shown that the genes encoding the olfactory receptors number around 100 in fishes and 1000 in mammals, making up a large part of their respective genomes. Thus the developing sensory neurons must choose to express a given receptor type and maintain the representation of the receptors in a constantly regenerating epithelium. Fish have long been a classic model for the study of neuroethology; we can now couple this knowledge with cellular and molecular tools to unravel the development of the amazing olfactory sensory system using the world's largest group of vertebrate animals, the fishes.

Correspondence to: Genetics and Development, Neurobiology and Behavior, Department of Molecular Biology and Genetics, 445 Biotechnology Building, Cornell University, Ithaca, New York 14850, USA. E-mail: kew13@cornell.edu, phone: 607-255-4289.

1. Amazing Conservation: Organization of Olfactory Sensory Systems

1.1. Fish as a Model System

The zebrafish has emerged over the last decade as a pre-eminent model system for the study of developmental biology and genetics in vertebrate animals.[1] But fish have been a popular group of animals for the study of developmental biology, as well as neurobiology and behavior, long before the emergence of zebrafish as a model system. This is evidenced in the early developmental work on killifish[2] and behavioral work by von Frisch using schooling fishes.[3,4] As a group, the fishes make up approximately 50% of all vertebrates. The Actinopterygiian fishes (this group includes teleosts) comprise 96% of the fishes, and the hagfishes (Myxini), lampreys (Cephalaspidomorphi), cartilaginous fishes (Chondrichthyes), and lungfishes and Coelacanth (Sarcopterygii) comprise the final 4% (see Pough *et al.*[5] for a review). Fishes as a group share a common ancestor with the tetrapods and thus share common features of cellular and molecular mechanisms controlling developmental pathways (Fig. 1A). In addition, the fishes show a spectacular diversification and adaptation to a variety of environments. Along with this diversification, fish have become specialists for any given system, be it a physiological specialization or a sensory specialization. As a result, a number of fishes have become popular models for various aspects of specialization including, but not limited to, evolutionary changes (cichlids), behavior (salmon), physiology (catfish, goldfish), and genomics (*Fugu*, zebrafish) (Fig. 1B). Therefore, the fishes are an excellent model system not only for the study of nervous system development and function, but also for the study of molecular and genetic pathways controlling development. In addition, the fishes are a unique group of animals to study when investigating the developmental and genetic changes that lead to the appearance of specialized systems.

While fish are an excellent model system for the study of developmental biology and the specialization of sensory systems, the olfactory sensory system is notably spectacular in its conservation of

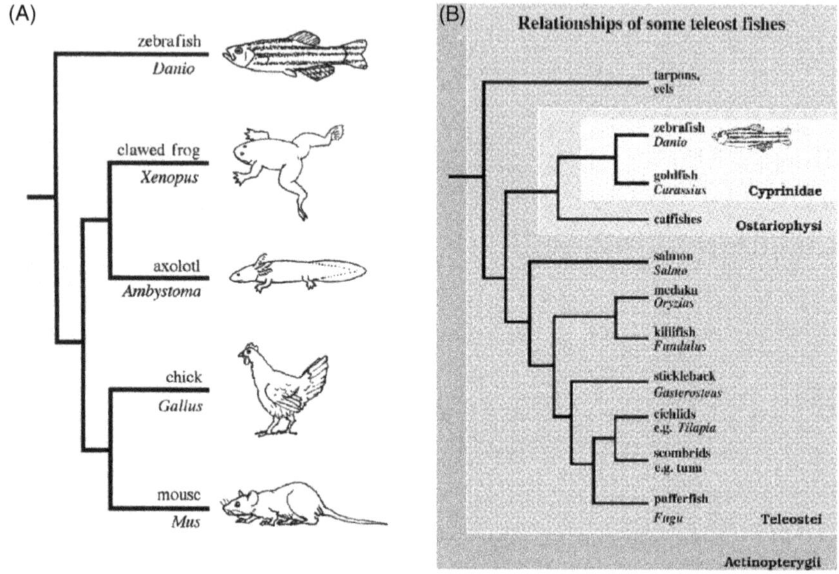

Fig. 1 Phylogenic relationship among common model systems (A) and among teleost fishes (B). (A) Zebrafish is an advanced fish whose ancestor arose before the appearance of tetrapods. (B) Zebrafish is a cyprinid most closely related to the goldfish, a common experimental model in the study of the olfactory system (from Metscher and Ahlberg[131]).

structure and function across invertebrate and vertebrate animals. The basic structure of having olfactory sensory neurons with their cell bodies located in a peripheral epithelium and their axons terminating in a highly ordered olfactory bulb, is an organization maintained across animals as diverse as insects,[6] rodents,[7,8] and fish.[9]

1.2. Sensory Epithelium

The olfactory sensory system transduces signals from the outside world through a group of sensory neurons having a direct connection with the central nervous system (CNS). These olfactory sensory neurons (OSNs) have cell bodies that arise in the peripherally located olfactory placode and their axons grow into the CNS (Fig. 2A). In the periphery odorants bind to the odorant receptors located on the dendrites of the OSNs and this stimulation conveys information about food, conspecifics

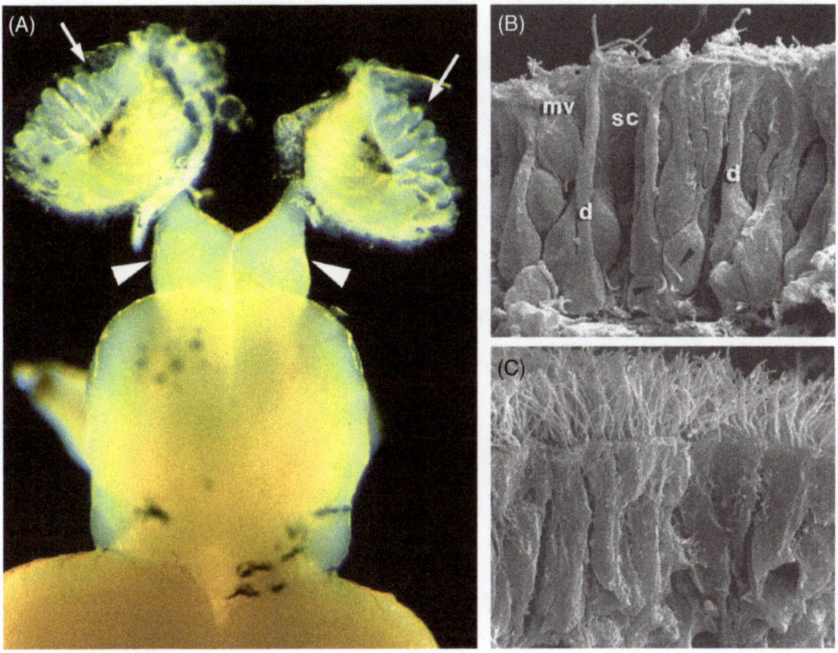

Fig. 2 Olfactory system of the adult zebrafish. (A) Intact brain showing olfactory rosettes (arrows) connected to the olfactory bulbs (arrowheads) via the olfactory nerve. (B) Electron micrograph of sensory epithelium showing microvillar (mv), support cells (sc) and ciliated sensory neurons with elongated dendrites (d). (C) Ciliated non-sensory cells in the olfactory epithelium (B, C) from Hansen and Zeiske[14]).

and mates. In vertebrates the primary sensory neurons regenerate throughout life where the new OSNs are generated by basal cells located within the olfactory epithelium. In addition, there are various types of non-sensory cells within the olfactory epithelium such as sustentacular cells or support cells that surround the sensory receptors within the epithelium, secretory cells, and ciliated non-sensory cells (Fig. 2B,C). Regeneration is proposed to occur as a result of neurons "sensing" OSN density through a lateral inhibition type mechanism where new OSNs are generated in response to a decrease in neuronal density (see Murray and Calof[10] for a review).

The olfactory epithelium of fishes is contained in a cavity with anterior and posterior nares through which the water flows across the epithelium. The sensory receptors are contained within the highly folded olfactory epithelium and these structures are called the olfactory rosettes (Fig. 2A, arrows). In general fish have around 10^6 OSNs while mammals have 10–50 times more.[7] Fish, unlike mammals, have several types of OSNs appearing together within the olfactory epithelium (see Laberge and Hara[9] for a review). In zebrafish the sensory epithelium of the adult animal has been described through scanning and electron micrograph analysis[11,12] as well as histological and immunocytochemical analysis.[13] Within the epithelium there appear to be two commonly seen OSNs, one type has dendrites bearing cilia containing microtubules (ciliated OSN), and the second type has dendrites bearing microvilli, with actin (microvillar OSN) (Fig. 2B). A third recently described sensory cell type is the crypt cell, which has both cilia and microvilli, and has an axon exiting the epithelium.[11,14] These different sensory cell types appear to have different receptors, regenerative properties and functional roles, which will be discussed later in the chapter.

1.3. Olfactory Bulb

The axons of the sensory neurons must grow into the olfactory bulb (Fig. 2A) where they terminate in characteristic clusters called glomeruli, structures that have been observed in the olfactory bulb (by Ramon y Cajal.[15]) It is within the glomeruli that the axons contact second order mitral and tufted cells, these cell types are present across vertebrates, although the organization varies.[7] Zebrafish, like other vertebrates, have a laminated bulb with olfactory nerve, glomerular, mitral/plexiform, and granule cell layers, and these layers are immunocytochemically positive for neurotransmitters observed in the olfactory bulb of other vertebrate animals.[13] In zebrafish, there is a distinct glomerular pattern that is highly reproducible across individuals where any given fish has approximately 80 glomeruli per olfactory bulb.[13,16,17]

1.4. Terminal Nerve

In vertebrate animals the olfactory nerve is associated with a cranial nerve called the terminal nerve. In mammals this nerve is mixed with the olfactory and vomeronasal nerves and terminates in the telencephalon.

In fish, in contrast, the terminal nerve is separated from the olfactory nerve and while it too terminates in the telencephalon, it also has direct projections to the retina (Fig. 3A).[18] Whether the terminal nerve in fish is equivalent to or different from the nucleus olfactoretinalis is a matter for discussion for which there is no space in this chapter (see Laberge and Hara[9] for a review). The neurons that give rise to the terminal nerve have cell bodies lying adjacent to the olfactory capsule and extend their axons centrally, although the position of the cell bodies of this cranial nerve vary greatly across fishes. The cell bodies can be uni-, bi-, or multipolar and the termination sites for these axons are variable depending upon the animals studied. Additionally, terminal nerve processes have

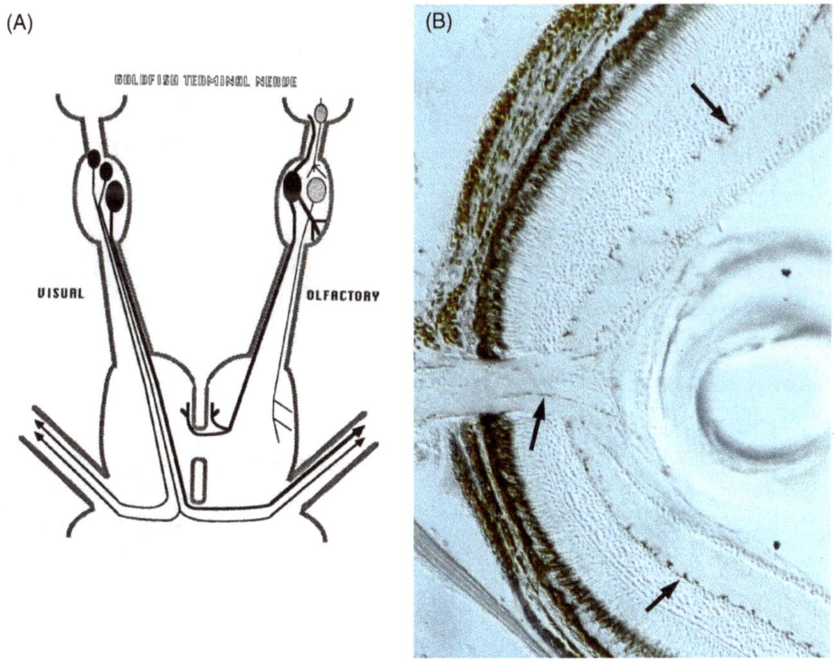

Fig. 3 Terminal nerve system associated with the olfactory sensory system. (A) Diagram of the terminal nerve showing the projections within the olfactory sensory system (right) and extending to the retina system (left). (B) Cross-section of the zebrafish retina showing GnRH immunoreactivity in the optic nerve and retina (arrows). (A) from Stell *et al.*[18]).

been described in the olfactory epithelium and a subset of terminal nerve axons terminate in the retina of the eye (Fig. 3B), where they release either FMRFamide or gonadotropin-releasing hormone (GnRH) (Fig. 3B, arrows). Previously, and possibly still, the terminal nerve was thought to have a function transducing olfactory signals via the olfactory epithelium. But at this point in time it is generally agreed that the terminal nerve has a neuromodulatory function in the nervous system. This nerve is thought to play a role in modulation of the olfactory epithelium (dwarf gourami[19,20]), the retina (white perch[21]) and targets within the CNS.

2. The Olfactory Placode Arises from the Edge of the Developing Neural Plate

2.1. Origin

Many of the original experiments examining the development of the olfactory placode were done in frog and chick embryos. Unlike frog and chick, the zebrafish nervous system develops through secondary neurulation. Therefore, the anterior neural plate shows no clear morphological neural ridge, as that seen in frog and chick. As a result, when comparing zebrafish embryos with the frog and chick embryos, the "edge of the neural plate" and the "neural ridge" are equivalent.

The vertebrate olfactory organ develops from the olfactory placode, a transient structure evident at the end of neural tube formation. Originally it was thought that the olfactory placodes arose from the induction of overlying ectoderm, much like the lens of the eye. But, with more refined techniques for labeling and following cells in the early embryo, it was possible to localize the origin of the olfactory placode to the neural ridge or neural plate as opposed to the non-neural ectoderm (Figs. 4–6). Because of differences in developmental timing it is difficult to draw direct analogies between species, but to date the collective model drawn from morphological analysis in mouse,[22] quail-chick chimeras,[23] and single cell lineage tracing in zebrafish,[24] is that the olfactory placodes arise from within the edge of the neural plate of the developing embryo. But the mechanisms giving rise to the formation of the olfactory placode from the edge of the neural plate remain a mystery. One model is that the olfactory placodes develop from the edge of the neural plate by becoming isolated by differential cell movements (see Farbman[25] for a review) and this small piece of neural plate will become surrounded by non-neuronal tissue as development proceeds (Fig. 4A). Recent work in the zebrafish has shown not only that the olfactory placode develops from an olfactory field within the anterior neural plate (Fig. 6), but that it does so in the absence of cell division in the six hours preceding the appearance of the olfactory placode.[24] As a result of the lack of cell division it was proposed that the olfactory placode (and telencephalon) develop through the anterior convergence of the cells in the olfactory placode field (Fig. 4B).[24] This model has been supported by work in the chick embryo using DiI to label small groups of cells at the 4-somite stage. In these embryos it has been shown that the otic placode develops through extensive cell movements, and that these cells converge to form the otic

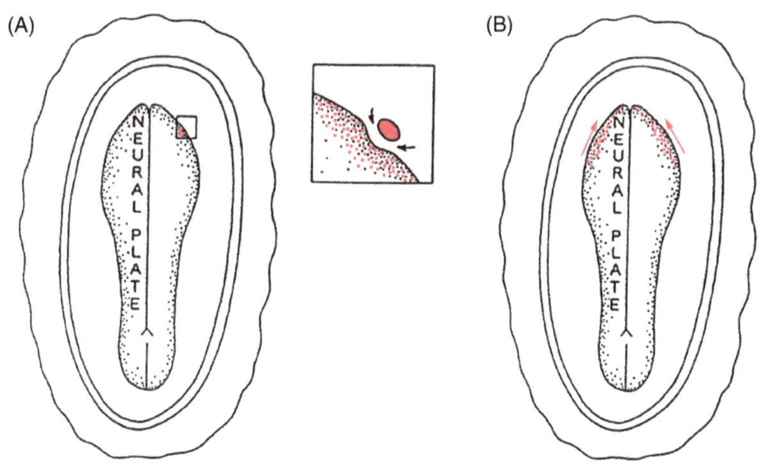

Fig. 4 Model for the development of the olfactory placode. (A) Chick neural plate with edge (red) that will dissociate from the edge of the neural plate through differential cell movements. (B) New model where olfactory placode forms through the anterior convergence of cells in the olfactory placode field (arrows) (modified from Farbman[25]).

placodes in their final position adjacent to rhombomeres 5–6.[26] Thus, cell convergence (as opposed to localized cell division) at the edge of the neural plate may be a common mechanism for placode formation across vertebrate animals.

2.2. Gene Expression in the Olfactory Field of the Anterior Neural Plate

In the zebrafish, early fate maps from 60–90% epiboly show that the olfactory placode/telencephalic regions arise from a region clustered at the animal pole[27] with the olfactory placodes arising from the lateral regions.[28] As somitogenesis and neurulation proceed, specific genes are expressed in a strip, like an upside down horseshoe, evident at the anterior end of the forming neural tube, the region from which the olfactory placodes arise. Some of the genes expressed in this region are transcription factors such as *distal-less-3, dlx3*[29]; *distal-less 7, dlx7*[30]; and *eyes absent, eya1*[31] (Fig. 5). These genes along with others[32] such as *six4.1*,[33] appear to be expressed in the same region at the anterior end of the neural plate (Fig. 5A).[34] This region of gene expression correlates with the proposed "placode domain", a region at the anterior end of the neural plate that will give rise to placodes (see Torres and Giraldez[34] for a review). The initial expression of these genes in the edge of the neural plate is maintained in the differentiated olfactory placode (Fig. 5B, C). In this model these transcription factors would work to restrict the developmental potential, thus defining the specific placodes. To truly test the model in zebrafish, experiments where cells are transplanted from this placode region at different developmental stages could be done in order to confirm the potentials of the cells at this unique border of the anterior neural plate.

2.3. Induction

While there is a general model for placode development where placodes are induced from overlying ectoderm as the neural tube forms, the olfactory placode develops from within the neural plate. At the onset of somitogenesis the genes involved in olfactory placode formation are

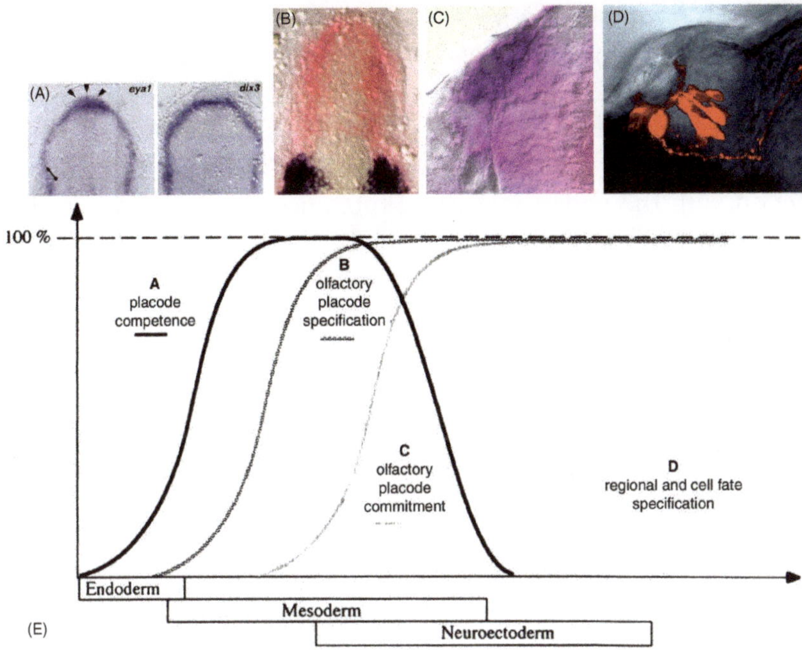

Fig. 5 Model for induction of olfactory placode. (A) Olfactory placode competence as reflected by expression of *eya1* and *dlx3* in the anterior neural plate. (B) Olfactory placode specification as reflected by *dlx3* (red) in the converging anterior neural plate. (C) Olfactory placode commitment and formation with *dlx3* expressed in the olfactory placode (blue). (D) Regional and cell fate specification as reflected by differentiation of OSNs in the olfactory organ (red). (E) Diagram depicting the sequential exposure of the placode domain to the underlying endoderm, mesoderm and neurectoderm, proposed sources of signals involved in placode formation. (A) from Sahly *et al.*[31]; (B) from Whitlock and Westerfield[24]; (D) from Whitlock and Westerfield[58]; (E) from Torres and Giraldez[34]).

already being expressed, and as somitogenesis proceeds the expression patterns of the "placode domain" genes (see above) become more restricted, reflecting the cellular changes that give rise to the olfactory placodes. But what in fact is the inducing signal(s) triggering the formation of the olfactory placodes? If one thinks of the open neural plate stage, the edge of the neural plate is at first overlying the endoderm. Then, as the neural plate forms, it comes in contact with the mesoderm and finally the neural tube (Fig. 5E) (see Whitlock and Westerfield[24], and Farbman[25] for a review). Presently there are no data defining the inductive signal(s) triggering the formation of the olfactory placode. It may be that the development of the olfactory placode is not generalizable across placodal derivatives.[35] For example, while it is agreed upon that the lens of the eye is induced from the overlying ectoderm through an interaction with the underlying neural tube, no such interaction appears to take place for the olfactory placode. In the zebrafish, the olfactory placode fate maps alongside the telencephalon and appears to develop in concert with the olfactory bulb in the zebrafish[24] (Fig. 6) rather then being induced by this tissue. A second proposed source of inductive signals for the developing olfactory placode is the anterior mesendoderm as it transiently comes in contact with the region that will give rise to the olfactory placodes (see Baker and Bronner–Fraser[32] for a review). To date, no data exists in zebrafish to support or refute this idea. A final possible source of inductive signals is the cranial neural crest that initially flanks the posterior border of the olfactory field (Fig. 6). This group of cells will migrate anteriorly to form the nasal capsule and it could be this interaction of neural crest and olfactory precursors that triggers the formation of the olfactory placode.

2.4. The Differentiating Olfactory Organ

Once the olfactory placode is formed, as evident by its morphological appearance (Fig. 7), neurons need to be generated (Fig. 5D). Based on early work in *Drosophila*, we have a clear idea of the genetic cascades determining the development of both central and peripheral sensory neurons. In general, the genes controlling neural differentiation fall into two classes, the so-called “proneural” genes belonging to the *achaete-scute* (AS-C) complex of which there are four genes, *achaete, scute, asense*, and *lethal of scute*,[36] and the neurogenic genes, *notch* and *delta*.[37] In *Drosophila* the AS-C genes are involved in the differentiation of the peripheral nervous system (PNS), while the *atonal* gene plays an analogous role in the CNS (and chordotonal organ). The vertebrate homologs of these genes have been cloned in zebrafish. Presently there are two Zebrafish achaete-scute homologues (Zash) *Zash-1a* and

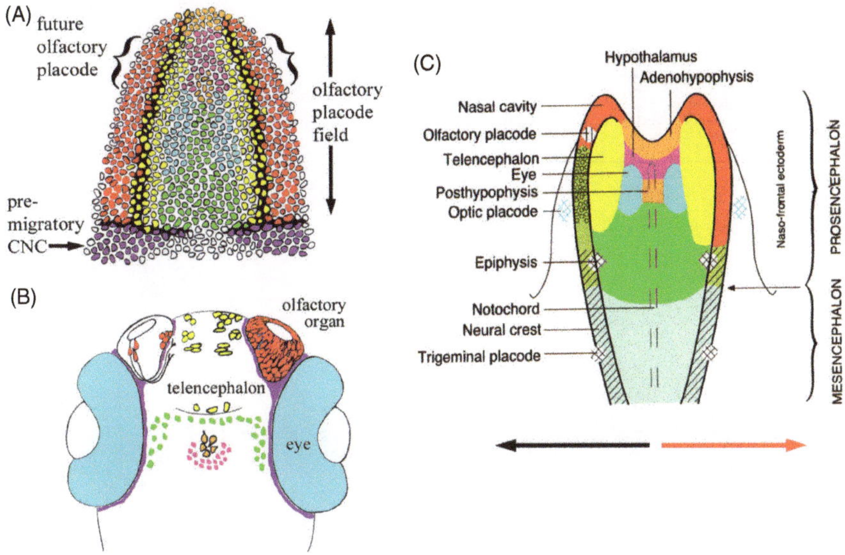

Fig. 6 Fate map of the olfactory placode field in the zebrafish and chick embryos. (A) Anterior neural plate cells are color coded for their fate when scored at two days (B). (C) Fate map for the anterior neural plate of chick. Right hand side has been color coded to reflect the findings of the zebrafish, namely by expanding the olfactory placode field back to meet anterior limit of cranial neural crest (above red arrow), ((A, B) modified from Whitlock and Westerfield[24]; (C) modified from Kalcheim and Le Douarin[51]).

Zash-1b, although there may be more *Zash* genes to be uncovered.[38] At present time there is no evidence that the Zash genes are expressed in specific PNS cells, such as the precursors of the OSNs in the developing sensory epithelium, as seen in *Drosophila*. Both *Zash-1a* and *Zash-1b* are expressed as early as 12 hours post-fertilization (hpf) in the developing telencephalic region, although their pattern is different.[38,39] Additionally, an *atonal*-like gene *neurogenin* which plays a role in lateral inhibition, has been cloned from zebrafish[40,41] and is expressed in the developing nervous system in regions where *neuroD* is expressed.[39] The *neuroD* gene is a downstream target of *neurogenin* and is expressed in the anterior neural plate "horseshoe" pattern, in the olfactory placode, and the telencephalon domains (Fig. 5).[41] Recently, a group of *neuroD*-related (*ndr*) genes was cloned and two of the genes, *ndr1a* and *ndr1b*, are expressed starting at 22 hpf and the expression pattern was limited

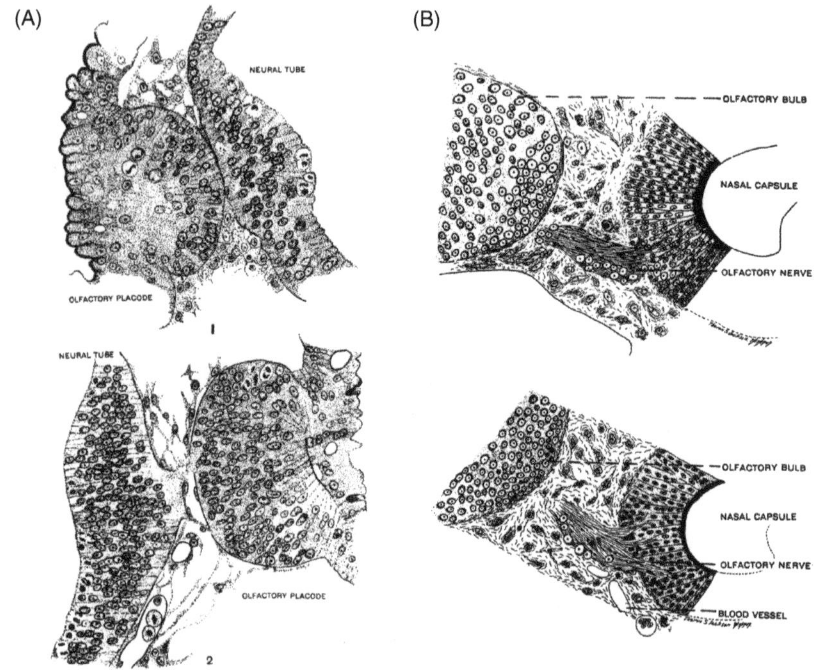

Fig. 7 The development of the olfactory placode in the gar (*Lepidosteus*) (A) and the bullhead (*Ameiurus*) B, showing transverse sections at 112 hpf (upper panel) and 120 hpf (lower panel, B). (A) The connection between the olfactory placode and neural tube is starting to form at 120 hpf. (Note the orientation of the olfactory placode/neural tube is reversed in I and II). (B) The forming olfactory nerve is evident at 120 hpf in the bullhead. The terminal nerve cells are seen associated with the olfactory nerve. (A) from Brookover[118]; (B) from Brookover and Jackson[119]).

to the olfactory system, namely the olfactory bulb and the olfactory organs.[42]

The neurogenic genes *notch* and *delta* have also been cloned in the zebrafish. In the zebrafish there are four *delta* homologues, *deltaA–D*. The *delta A, B,* and *D* genes appear to be involved in primary neurogenesis (as opposed to somitogenesis).[43] Of the four *notch* genes in the zebrafish, *notch1a, notch1b, notch5,* and *notch6* the first three are expressed in the CNS.[44] In the olfactory epithelium *notch1A* (few cells) and *deltaA* (few cells) are expressed, but rather than having cells expressing these genes distributed throughout the olfactory epithelium, the expression is in a small cluster of cells in the olfactory epithelium. The *neuroD* gene is more strongly and uniformly expressed in the developing olfactory placode.

Based on the expression of the neurogenic genes coupled with markers for cell division and neuronal differentiation, a secondary neurogenesis has been reported by Mueller and Wulliman[39] to take place at two days post-fertilization. Surprisingly, heavy labeling for proliferating nuclear cell antigen (PCNA), a marker for cell division, in the olfactory epithelium at two days is shown. This is in sharp contrast to the low level of cell division reported using an antibody against the phosphorylated histone H3 which marks cells in M-phase.[24,45] In addition, this report of high levels of cell division does not agree with small clone sizes and thus low rate of cell division reported in fate mapping of the olfactory placode[24] and eye anlage.[46]

Interestingly, the Hu antigen staining, a marker for newly differentiated neurons, is much more spatially restricted when compared to the PCNA labeling, suggesting that a large portion of the PCNA positive cells are non-neuronal. Thus, while the proneural and neurogenic genes are expressed in the developing olfactory placode there does not seem to be a pattern of expression prefiguring the differentiation of the OSNs, although more data are needed to make this statement conclusively. Additionally, the pattern of cell division in the olfactory epithelium as reported by PCNA labeling does not agree with the low level of cells division reported for the olfactory placode using lineage analysis and a different cell division marker.

2.5. Mutations Disrupting Olfactory Placode Development

Very few mutants that specifically affect the development of the olfactory organ have been uncovered. A genetic deletion that includes the *dlx3* gene was isolated and shown to be lacking the olfactory and auditory organs among other defects.[47] Further analysis showed that the deficiency also covered the *dlx7* gene, a gene whose expression pattern is almost identical to that of *dlx3*. In fact it has been shown by using morpholino technology[48] to "knock down" both *dlx3* and *dlx7*, the nose and ear can be eliminated.[49] A recently described mutant isolated in an olfactory behavior screen shows defects specific to the targeting of the OSN axons in the developing olfactory bulb.[50] As more information and markers become available for analysis of the olfactory sensory system, it will find its place in future genetic screens in zebrafish.

2.6. Structural Elements of the Olfactory System Arise from Cranial Neural Crest

An oft overlooked fact is that the olfactory epithelium is embedded in the olfactory capsule and that this structural element of the nose has a different embryonic origin from that of the olfactory epithelium. During vertebrate development the most anterior cranial neural crest cells migrate anteriorly in a route traveling dorsal to the eye (Fig. 8), and form the frontal mass, part of which is the olfactory capsule (see Le Douarin and Kalcheim[51] for a review). While there is much known about cranial neural crest development in the zebrafish as it relates to the branchial arch derivatives giving rise to jaw elements,[52] the route of crest migration passing dorsal to the eye is not well described. In Medaka, the components of the neural crest contributing to the head have been mapped out.[53] In this study the derivatives of the neural crest were assayed by recording the resulting defects in head skeletal elements after removing parts of the neural crest. Removal of cells overlying the prosencephalon did not disrupt the skeletagenic elements of the head skeleton. The next section posterior (mesencephalic) most severely disrupted the formation of the orbitonasalis or nasal capsule, the anterior orbital, and ethmoid (Fig. 8C, D). This region corresponds to the anterior limits of cranial neural crest. In zebrafish there have been intensive genetic screens that have uncovered a myriad of mutants affecting early development. Many of the mutants affect the development of neural crest derivatives of the branchial arches and pigment,[54,55] but there have not been any mutants identified that specifically affect the structural elements of the nose.

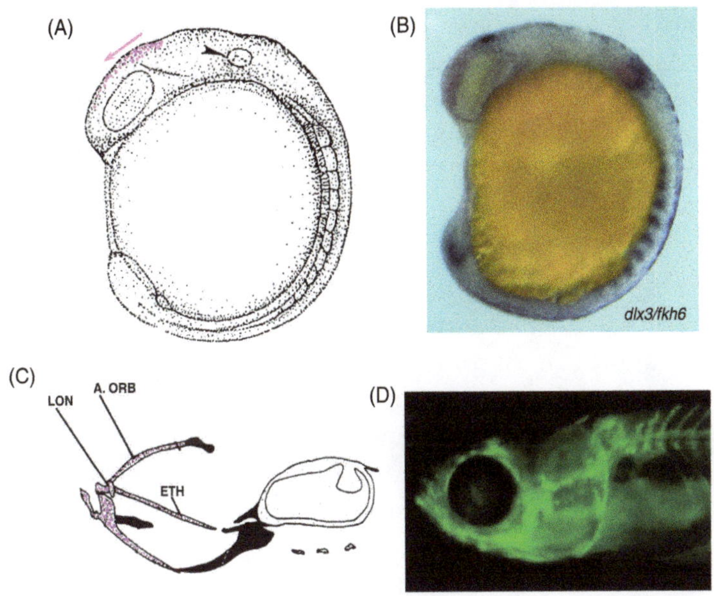

Fig. 8 Formation of the skeletal elements of the face associated with the olfactory system. (A) Cranial neural crest (purple) migrates anterior starting at 6–8 somites. (B) Pre-migratory crest expresses *fkh6* (blue) and placode field *dlx3* (red). (C) Skeletal elements derived from anterior cranial neural crest in the Medaka. (D) Skeletal elements of the head labeled with calcein. ((C) from Langille and Hall[53]; (D) courtesy of the Zebrafish Living Laboratory Resources).

3. The Olfactory Bulb Arises from the Anterior Neural Plate

3.1. Gene Expression in the Developing Olfactory Bulb

The olfactory bulb is part of the telencephalon and fate maps to the animal pole of the gastrulating zebrafish.[27,28] At the neural plate stage the telencephalon is localized anterior to the cranial neural crest domain. The telencephalic domain expresses the genes *emx1* and *emx2*,[56] with *emx1* localized to the telencephalon and *emx2* expressed in the telencephalon and other regions of the developing brain such as the diencephalon. These nested *emx1/emx2* domains are bordered laterally by the *dlx3* expression pattern marking the olfactory placode fields (Fig. 6). As has been observed in *Xenopus*,[57] the olfactory bulb domain also expresses *dlx3* in the zebrafish, but at a much lower level than the olfactory placode domain. Thus, the olfactory placode and telencephalon both express *dlx3* and appear to develop in concert.[24] These two adjacent fields of cells converge anteriorly to form their respective adult tissues. The pioneer neurons, a specialized class of neurons whose axons first make the connection between the olfactory placode and bulb, enter the developing olfactory bulb in the *emx1* expression domain.[58] As the OSNs in the placode differentiate there appears to be corresponding differentiation in the olfactory bulb as interneurons are seen in the same temporal pattern.[24] It is the interaction of the axons of the OSNs with the developing telencephalon that triggers the further development of olfactory bulb.[59] The developmental plasticity of the central olfactory processing center is directly affected by the input of the OSN axons.

3.2. Axonal Input and Olfactory Bulb Development

One of the few developmental and comparative studies of the olfactory bulb of the juvenile zebrafish reports that the olfactory bulb shows a 50% decrease in volume if deafferented at three weeks of development, whereas the adult bulb shows a 15% loss after deafferentation.[60] This is in agreement with observations from other vertebrate animals showing that the development of the olfactory bulb is dependent upon the input from the OSN axons.[61] In addition, the authors report that the juvenile zebrafish have a much reduced bulb size at three weeks when compared to the adults, even when corrected for size. This indicates that at three weeks of development the juvenile zebrafish has not reached the characteristic adult size olfactory bulb. Comparative studies indicate that there may be a metamorphosis-like event in zebrafish. Up until three weeks of age the pearl danio, *D. albolineatus*, the giant danio (*D. aquipinneatus/ D. malabaricus*), and the zebrafish *D. rerio*, appear similar in the pigment pattern after which they take on their divergent color patterns observed in the adult animal.[62] It would be interesting to see at which developmental time the severe affect of deafferentation is lost, for perhaps the severe effects of deafferentation would no longer be evident once the fish had reached its adult pigment pattern. Another interesting observation from Poling and Brunjes[60] is that the olfactory bulb of the giant danio (*Danio aequipinnatus*) is pedunculated rather then sessile, meaning that the tracts from the olfactory bulb are easily accessible, and could therefore be experimentally manipulated, much like in the goldfish. Like the olfactory epithelium, the olfactory bulb undergoes constant neurogenesis throughout life. In the adult fish the glia of the olfactory nerve and the interneurons of the internal cell layer undergo cell division as judged by BrdU labeling.[63] The rate of neurogenesis increases upon deafferentation in the developing and adult animal.[64] If the sensory axon input is removed by removing the olfactory placode during development, the olfactory bulb region of the telencephalon is substantially reduced.[61] Therefore, the olfactory placode and olfactory bulb develop in concert with one another, and it appears that the olfactory placode has an inductive influence on the development of the olfactory bulb through the axons of the OSNs.

4. Axon Guidance: The Connection of the Olfactory Sensory Axons with the Developing Olfactory Bulb

4.1. Guidance Cues

The axons of the olfactory sensory neurons make very specific and highly stereotyped connections with the developing olfactory bulb. As first suggested by work on the trout[65] and subsequently on the zebrafish,[16] the olfactory sensory axons converge on a given region of the olfactory bulb, but the cell bodies lie dispersed in the sensory epithelium showing little spatial pattern. Subsequently, work in mammals refined this original observation to show that the axons whose neurons express a given olfactory receptor converge on a single set glomeruli within the olfactory bulb.[66] The cues proposed to guide the formation of precise axonal connections during development of the olfactory sensory system are varied, ranging from cell surface molecules, to olfactory receptors, physiological activity, glia, and pioneer neurons (see Lin and Ngai[67] for a review). There is a large data set showing that the axons of the olfactory sensory system express a variety of cell-surface molecules and that these molecules can to be localized to subsets of the olfactory axons in the mouse.[68] Additionally, it has been proposed that the olfactory receptors may play a role in axon guidance in targeting the OSN axons to their specific glomeruli.[69]

4.2. Cell Surface Molecules

It has long been known that cell surface molecules are important for axon guidance during the development of the nervous system, and there are specific classes of cell surface adhesion molecules found in the developing olfactory sensory system.[68,70] A cell adhesion molecule whose expression in the olfactory system has been well studied is the neural cell adhesion molecule (NCAM). In addition to the olfactory sensory system, the NCAM protein is expressed on many neurons and is important for axon guidance in the developing nervous system. NCAM is in the immunoglobulin superfamily of adhesion molecules and this superfamily includes molecules such as the olfactory cell adhesion molecule (OCAM), found in mouse,[71] and the fasciclin II molecule found in *Drosophila*.[72] This family of molecules can be differentially glycosylated, creating greater diversity in function. For example the embryonic form of NCAM in mammals is polysialated whereas the adult form is not. NCAM is expressed in the developing olfactory system of mouse,[73] and knock-outs of NCAM have a general effect on axon fasciculation and glomerulus formation.[74] In a recent set of experiments in mouse, overexpression of two different isoforms of OCAM had specific effects on the segregation of identified sensory axons in the olfactory bulb.[75] In the zebrafish, three cell adhesion molecules from this family have been cloned, including homologous forms of NCAM and OCAM, and a novel form named zPCAM.[76] By *in situ* hybridization, NCAM is highly expressed in the developing neural tube and zPCAM also shows neural tube expression though to a lesser degree, although zPCAM does not appear to be expressed in the olfactory epithelium. The zebrafish OCAM is expressed in the developing olfactory bulb, like in mouse, and shows a highly restricted pattern in the anterior bulb in the region of first OSN axonal projections.[58,76] Additionally, polysialic acid (PSA), a carbohydrate attached to NCAM, is detectable in the olfactory bulb at 30 hpf using an immunocytochemical analysis.[77] Thus, NCAM, its carbohydrate epitope PSA, and OCAM, are all found in the developing olfactory system of the zebrafish as has been reported in other vertebrates.

4.3. Olfactory Receptors

The recent finding that the mRNA of olfactory receptors can be detected in axon terminals within the glomeruli[78] has led to the proposal that the olfactory receptors have a dual role: they transduce odorant signals in the dendrites and control axon guidance in the terminals.[79] Work in mouse has shown that if the promoter region of one receptor is paired with the coding region of another (driving the lacZ reporter gene), the axons are misrouted in the olfactory bulb. The hypothesis is that the coding region of the receptor should provide an address within the glomeruli so that even though they were being driven by a different promoter, the axons should terminate in their normal target glomerulus. Curiously, the axons are not mis-directed to the glomerulus that is appropriate for that coding sequence, rather the axons terminate in an intermediate region.[69] Therefore it is difficult to rule out whether the loss of specificity of guidance is due to a role for the receptor in axon guidance or to the chimeric construct. Thus the receptors may play a role in axon guidance. This is an intriguing idea, but there has yet to be a demonstration of protein expression in the axon terminals. While this level of genetic manipulation is not yet possible in the zebrafish, recently a transgenic manipulation of protein kinase A has led to the suggestion that axon targeting is in part due to protein kinase A (pkA) signaling. In experiments where pkA signaling was either constituitively activated or impaired under the olfactory marker protein (OMP) promoter (a OSN specific promoter), the constituitively active pkA signaling affected axon growth in the olfactory bulb and a decrease in pkA signaling affected the ability of axons to exit the developing olfactory placode.[80]

4.4. Pioneer Neurons

In zebrafish, the first connection between the olfactory placode and the bulb is forged by a set of specialized neurons called pioneer neurons.[58] Pioneer neurons were originally defined based on a finding made in the developing peripheral nervous system of the grasshopper,[81] but have subsequently been found in the developing nervous system of vertebrates.[82] Pioneer neurons are unique in that they establish a pathway and then undergo programmed cell death once the adult axons, which follow them, have made connections to the target site. Thus, they are a transient cell type involved in axon guidance. In the zebrafish, pioneer neurons for the olfactory sensory system appear at 20 hpf as large cell bodies in the basal part of the olfactory placode juxtaposed to the telencephalon. These pioneers extend their processes onto the developing telencephalon in the region expressing *emx1*[58] where they branch and form a glomerular-like pattern (Fig. 9A, B). The ablation of the pioneer neurons results in the misrouting of following OSN axons into the anterior commissure, thus bypassing their normal route into the developing olfactory bulb.[58] In a complementary study involving transgenic zebrafish expressing GFP driven by the *C. elegans unc-76* gene, which allows the GFP to be expressed throughout the axon, development of the OSN axons was visualized in the living embryo. The olfactory neurons expressing GFP were imaged during development and showed that the axons make directed growth and do not extend filopodia in an exploratory manner.[83] This is consistent with the finding that only the axons of the pioneer neurons show filopodial, exploratory-like behaviors when extending into the CNS (Fig. 9C, arrows).[24,58] The pioneers are also unique in that they arise from the specific region of the anterior neural plate and do not divide in the time between labeling (12 hpf) and scoring (28 hpf).[58] Because the olfactory receptors have been implicated in axon guidance,[79] and the olfactory receptors have been cloned in the zebrafish,[84,85] the association between receptors and

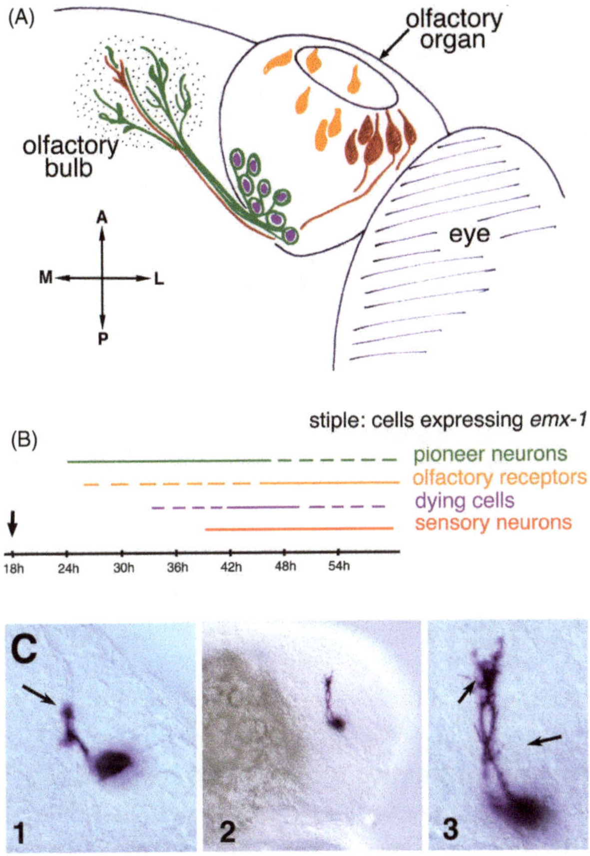

Fig. 9 Pioneer neurons in the olfactory organ of the zebrafish. (A) Olfactory sensory neurons (red) follow the pioneer neurons (green) into the developing olfactory bulb which expresses *emx1* (stippled). Pioneer neurons do not express olfactory receptors (orange) and undergo programmed cell death (purple nuclei). (B) Developmental time course of the appearance of the different cell types and receptors shown in A. (C) 1: Pioneer neurons' initially show simple growth cone (arrow). 2: Two labeled pioneer neurons in the developing olfactory placode. 3: As the axons enter the olfactory bulb they show numerous fine filopodial branches (arrows), unlike the following sensory neurons. ((A, B) from Whitlock and Westerfield[58]; (C) from Whitlock and Westerfield[24]).

targeting was also examined. It was shown that the expression of olfactory receptors was not spatially correlated with the pioneer neurons in the developing olfactory epithelium.[58]

5. Olfactory Receptors: The Interface with the Outside World

5.1. Main Olfactory Epithelium

The olfactory sensory neurons detect a wide variety of odorants and transmit this information to the central nervous system via their axonal connections with the olfactory bulb. In order to detect the odors there must be receptors that bind odorants and transduce a signal. Prior experiments indicated that olfactory responses are transduced by G-protein coupled receptors[86] and this class of receptors has seven trans-membrane domains. Based on these observations, degenerate primers were used to clone and characterize a large family of odorant receptors in the mouse.[87] At this time there are approximately 1000 olfactory receptors in mammals. These receptors fall into families all having the seven transmembrane G-protein coupled receptor motif. Subsequently olfactory receptors have been cloned from catfish,[88,89] Medaka,[90] goldfish,[91,92] *Fugu*,[93] *Drosophila*[94] and even an avian species that purportedly does not use olfaction.[95] The olfactory receptors fall into two broad categories, class I "fish-like" and class II "mammal-like" olfactory receptors. In animals whose life history is both terrestrial and aquatic, such as *Xenopus*, it has been shown that the class I receptors are expressed in the part of the nasal cavity exposed to water and class II in the part exposed to air.[96] This observation has led to the suggestion that class I receptors bind water borne molecules and class II receptors bind air borne molecules. But the class I receptors are found in mammals, and in humans the class I receptors account for 10% of the olfactory receptor genes and they are under positive selection (they have proportionally fewer pseudogenes).[97] Thus, the current theory is that class I receptors can bind odorants in either air or water and these types of odorants are important across vertebrate animals.

Using *in situ* hybridization and transgenic techniques in the mouse, cells expressing particular receptors were shown to be scattered in the olfactory epithelium, which is in agreement with previous studies primarily from fishes.[65] Nevertheless, the expression pattern is not entirely random in mammals, for olfactory receptors are expressed in four broad spatial zones. In fishes there appear to be no such zonal organization, although the olfactory receptors are expressed in concentric domains within the olfactory epithelium of the adult zebrafish.[98] In general, a given OSN will express only a single olfactory receptor, but this is not true for all systems. Given the paucity of extensive *in situ* hybridization experiments, one cannot rule out the possibility that several receptors are expressed in a single OSN. We do know that in rat and goldfish there are reports of olfactory receptor types that are expressed throughout the olfactory epithelium (see Laberge and Hara[9] for a review). In spite of this apparent lack of detailed spatial organization in the olfactory epithelium, the OSN axons target specific glomeruli in the olfactory bulb (see Mombaerts[8] for a review).

5.2. Onset of Olfactory Receptor Expression during Development

Fishes have approximately 100 olfactory receptor genes while mammals have 1000 — a difference of an order of magnitude.[97] While the overall number of receptors is less, the diversity of the family remains high. The time of onset of olfactory receptor expression has been studied in the developing zebrafish using *in situ* hybridization techniques. Initially the idea was that the olfactory epithelium would be much like the peripheral nervous system of *Drosophila* where there would be a highly ordered spatial array of differentiating sensory neurons and that this highly ordered array would result from the clear patterning of the proneural and neurogenic genes. The cloning of the zebrafish proneural and neurogenic genes in zebrafish[38,43] and the subsequent examination of their expression pattern within the olfactory epithelium has not revealed a highly ordered spatial array. Unlike what is seen in the ear of the zebrafish,[99] the neurogenic genes in the olfactory placode do not appear to show the prefiguring of sensory neuron differentiation. Additionally, the examination of the onset of olfactory receptor expression also appears to show no regulated pattern, and receptor expression is asynchronous across the olfactory epithelium.[84] Examination of the expression of olfactory receptors during sensory neuron regeneration in the catfish has shown that the olfactory receptors are expressed in the differentiated OSNs and not the basal, mitotic precursor cells. In addition, the olfactory receptors were expressed before the OSNs made axonal contact with the olfactory bulb, supporting the idea that receptor expression is independent of direct influences from the olfactory bulb.[100] Therefore, the mechanisms governing the differentiation of a given OSN and the expression of a given receptor are unclear except that they appear to be independent of the olfactory bulb.

5.3. Do Fish Have a Vomeronasal System?

Terrestrial vertebrates have a "second" olfactory system, the vomeronasal organ, that appears to be specialized for pheromones, although the main olfactory epithelium also senses pheromones.[101] The vomeronasal epithelium in mouse is distinct from the main olfactory epithelium, and the axons of the vomeronasal sensory neurons project to a specialized region of the olfactory bulb called the accessory olfactory bulb (AOB). Within the AOB, the axonal terminations do not show the clear glomerular segregation seen in the axon projections from the MOB. Additionally, the regenerative properties of the vomeronasal epithelium appear to be different from that of the sensory neurons in the main olfactory epithelium in that after nerve transection the reinnervation into the AOB is much slower and less complete in rodents.[102] The vomeronasal receptors have been cloned in mammals and are highly divergent from those found in the main olfactory epithelium. Further analysis of vomeronasal receptors has shown that they fall into two families based on their sequence, V1R and V2R.[103–105]

In fishes there is no obvious division of the epithelium into the equivalent of a main and vomeronasal epithelium. But, as stated earlier, the OSNs within the olfactory epithelium generally fall into two classes: the ciliated and the microvillar sensilla. In goldfish it has been shown that the goldfish homologues of the putative V2R vomeronasal receptors are expressed in the region of the microvillar OSNs.[91,92] The microvillar OSNs appear to respond to pheromones while the ciliated OSNs are known to respond to amino acids (see Laberge and Hara[9] for a review). This is based on the observation that after transection of the olfactory nerve in goldfish the recovering epithelium responds first to food odors, a response mediated by the ciliated OSNs, and eventually the response to pheromones recovers as the microvillar OSNs regenerate.[106,107] A further examination of the olfactory pathway exiting the olfactory bulb revealed a division into a medial (MOT) and lateral (LOT) olfactory tract. In fishes with a pedunculated bulb such as the goldfish, it is possible to selectively cut the MOT or the LOT. When the MOT is cut the animal loses its response to pheromones, and when the LOT is cut the fish loses its feeding behavioral response.[108] Thus, the microvillar OSNs within the olfactory epithelium of fishes appear to express receptors homologous to the mammalian vomeronasal receptors and have similar functions to the vomeronasal epithelium of mammals.

The vomeronasal receptors have been proposed to be pheromone receptors based on the tissue from which they were cloned and their localization to the vomeronasal epithelium by *in situ* hybridization. It has now been shown in mouse that the expression of the V2R class of vomeronasal receptors co-localizes with expression of the class 1b molecules of the major histocompatibility complex (MHC) molecules in the vomeronasal organ neurons.[109,110] It has been proposed that the expression of these MHC molecules may play a role in pheromone detection in the vomeronasal organ. In zebrafish, genes involved in MHC function have also been implicated in playing a role in olfaction and the recombination activating genes (RAG1, RAG2) of the immune system are expressed in the olfactory sensory neurons early in development.[111] This observation is curious in that the olfactory receptor genes do not contain sequence that would be indicative of the recombination events witnessed in the immune system. Yet, with the recent discovery of MHC molecules being expressed in the vomeronasal organ of mammals, there may also be a similar connection in the olfactory epithelium of the zebrafish. Furthermore, zebrafish also have a pheromonal system where they show behavioral response to an "aggregation pheromone" thought to be involved in schooling and spawning,[112] attraction to the female zebrafish by the male zebrafish by steroid glucuronides released by the ovaries,[113] and a physiological response at the level of the olfactory bulb to prostaglandin F_2-alpha and 17-alpha, 20ß-dihydroxy-4-pregnene-3-one-20-sulfate.[114] For these reasons zebrafish are well suited for studying the development of the social behaviors and the underlying cell and molecular mechanisms controlling them.

5.4. Development

There is, to date, little specific information on the development of the microvillar versus the ciliated receptors. In examining the olfactory placode fate map in the zebrafish[24] one could argue that the OSNs in the clones appear to be microvillar or ciliated based on their position in the developing epithelium. This argument is problematic in that the olfactory epithelium is not as clearly striated in the juvenile as observed in the adult, and depending upon the differentiated state of the OSNs, they may still be in the process of forming dendrites and extending axons. In general, it appears that ciliated and microvillar OSNs are both present in the developing olfactory epithelium based on the appearance of the dendritic length.[24,58,83]

In mammals it has been reported that when the olfactory nerve is transected, the OSNs from the main olfactory epithelium regenerate first and form a more complete connection in the olfactory bulb than do the OSNs of the vomeronasal organ.[102] Like the difference between the main and vomeronasal epithelium in rodents, the ciliated sensory receptors (associated with feeding behaviors) are replaced more quickly after nerve transection[106] when compared to the microvillar OSNs (associated with pheromonal response) in the goldfish. Given that these two classes of sensory neurons have different regeneration dynamics, it might be expected that they would not show equivalent differentiation dynamics during development. The need to have OSNs that respond to pheromones, such as those involved in reproduction, may be such that they develop later in comparison to the OSNs involved in the detection of food odors. In fact, juvenile zebrafish respond to amino acids earlier than they respond to pheromones such as alarm pheromones.[50]

6. Other Derivatives of the Olfactory Placode: Neuroendocrine Cells

6.1. The Terminal Nerve

The olfactory placode develops in association with the terminal nerve, a cranial nerve found together with the main olfactory nerve (see Fig. 3). Based on its close proximity to the developing olfactory placode it has been assumed that the terminal nerve arises in part or in whole from the olfactory placode. The terminal nerve is reported to have a neuromodulatory function in the nervous system and, in zebrafish as in other fishes, contains gonadotropin-releasing hormone (GnRH).[45,115] In addition, the terminal nerve contains a variety of neuroactive peptides. In zebrafish, for example, Neuropeptide Y (NPY) and FMRFamide are localized to the terminal nerve.[116,117] Because NPY and FMRFamide are localized to the cells in the "olfactory placode", their origin is often attributed to the olfactory placode. But upon closer inspection the cells appear alongside the olfactory placode. Thus, the terminal nerve contains many neuroactive peptides and has extensive axonal projections throughout the forebrain including the retina. But, with the exception of GnRH cells (see below), the developmental origin of the cell types found in the terminal nerve remains unclear.

6.2. Cells Appear to Exit the Olfactory Placode

The observation that cells appear to exit the olfactory placode during embryonic development in vertebrate animals was first reported by Brookover in the analysis of the development of the olfactory sensory system in fishes (Fig. 7B).[118,119] In these early studies using prepared sections of the developing olfactory organ, cells were seen associated with the developing olfactory nerve leaving the region of the olfactory placode.

Late in 1980 there were several reports suggesting that the olfactory placode generates not only OSNs and support cells, but also neuroendocrine cells containing the decapeptide gonadotropin-releasing hormone (GnRH).[120,121] GnRH has several forms based on amino acid sequence, and is known to exert an endocrine effect on the pituitary as well as a neuromodulatory effect within the CNS. In developing zebrafish GnRH is clearly localized to the terminal nerve, hypothalamus and the midbrain[45] (unpublished observations). The model was that the terminal nerve and GnRH cells of the hypothalamic regions had their origin in the olfactory placode, while the midbrain cells arose locally. Recent work in both Medaka and the zebrafish called this model into question.

6.3. The Neural Crest

In Medaka, a species-specific form of GnRH was cloned.[122] Its developmental expression pattern led to the proposal that the terminal nerve and hypothalamic GnRH cells did not share a common origin in the olfactory placode. Rather, it was more likely that the terminal nerve cells were placodally derived while the hypothalamic cells arose from the ventral forebrain.[122,123] In a subsequent paper it was shown through lineage tracing and analysis using mutants, that neither the terminal nerve nor hypothalamic GnRH cells have their origin in the olfactory placode.[45] The terminal nerve GnRH cells arise from the anterior cranial neural crest, a source consistent with its being a cranial nerve (Fig. 10, purple).[45] This also leads to the possibility that other cell types ascribed to the terminal nerve such as FMRFamide and NPY containing cells also arise from the cranial neural crest, a source of neuroendocrine cells in the trunk of vertebrates. This is especially plausible in the case of NPY since the neural crest derived autonomic nervous system contains NPY.[51] Therefore, cell types originally proposed to have their origins in the olfactory placode may in fact be neural crest derived. Because of the accessibility of the early embryo, the zebrafish presents the perfect model system to unravel the developmental relationships between the olfactory field and neural crest.

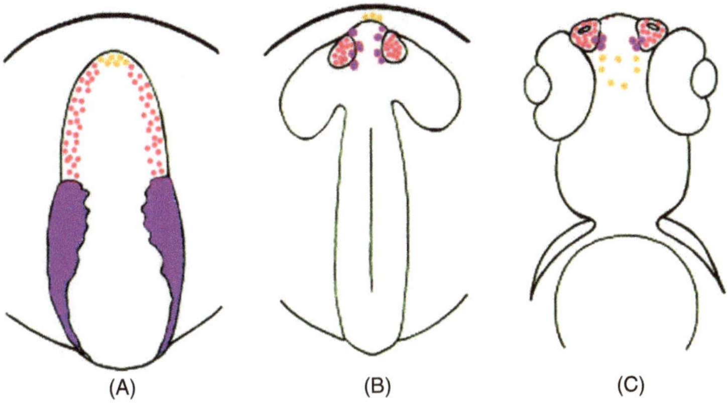

Fig. 10 Development of the GnRH cells associated with the differentiating olfactory placode. (A) GnRH cells of the terminal nerve arise in cranial neural crest (purple) while GnRH cells of the hypothalamus arise from the anterior pituitary field (orange). (B) As the placode forms GnRH cells are associated with it through cell movements of the forming neural crest and anterior pituitary field. (C) GnRH cells of the terminal nerve remain associated with the olfactory nerve and GnRH cells of the hypothalamus continue on their migration to the target in the CNS (from Whitlock *et al.*[45]).

6.4. The Adenohypophysis

The original reports of GnRH cells migrating from the olfactory placode done in the developing mouse embryo concentrated on the GnRH cells of the hypothalamus.[120,121] These analyses were done after the formation of the olfactory placode. An advantage of the zebrafish is that the embryos are accessible from the time of fertilization. It is therefore possible to label cells early in development, before the formation of the olfactory placode, and to look for embryonic origins before extensive cell migration has taken place. In the analysis of the origin of GnRH cells,[45] the origin of the hypothalamic cells was correlated with development of the adenohypophysis (anterior pituitary). The regions that give rise to the olfactory placodes in zebrafish lie at the edge of the anterior neural plate, and flank the region of the future adenohypophysis[24,32] which arises from a field of cells located on the midline at the anterior end of the neural plate (Fig. 6).[124–126] This adenohypophyseal region is also flanked by the hypothalamic precursors (see Fig. 6), and is a source of endocrine tissue. In the work of Whitlock *et al.* (2003), it was demonstrated that the loss of the pituitary using the *you-too* and *detour* mutants[125] results in the loss of the GnRH cells of the hypothalamus but not the cells of the terminal nerve. Additionally, the olfactory organs develop normally in these mutants. Thus the loss of GnRH cannot be due to olfactory placode loss. At first glance this may be a surprising finding. However, upon closer examination it was a curious observation that the olfactory placode should generate endocrine cells of the hypothalamus since no other sensory system generates endocrine tissue. Furthermore, the olfactory placodes are intimately associated with the developing adenohypophysis and hypothalamus, making the mistaken assignment of GnRH cell origin to the olfactory placodes easily understandable. A recent study examining the development of the precursors of the melanotrophs and corticotrophs used the pro-opiomelanocortin gene promoter linked to green fluorescent protein (GFP) to visualize their development.[127] In this study there is a cluster of GFP positive cells at 22 hpf closely associated with the olfactory placode, demonstrating that the olfactory placode is clearly flanked by endocrine tissue after placode formation. Thus, the GnRH cells of the hypothalamus arise from the adenohypophyseal region of the neural plate (Fig. 10, orange), while those of the terminal nerve arise from cranial neural crest (Fig. 10, purple).

7. Olfactory Physiology and Behavior: The Key to Survival and Reproduction

7.1. Fishy Smells

It has long been known that fishes respond behaviorally and physiologically to amino acids, bile acids, gonadal steroids and prostaglandins.[4,9,128] Fish use these olfactory cues for finding food sources, recognizing conspecifics, and coordinating mating behaviors. The concentration of odorants required to elicit a response in fish is exceedingly small, in the range of $1\times10^{8-12}$, or lower for certain substances.[129] Zebrafish also have a sensitive olfactory sensory system and respond to the same types of odorants as described for other fish. These responses have been measured in the adult olfactory system using electro-olfactograms[130] and optical imaging.[114] Being Cyprinids, zebrafish are close to the goldfish in terms of phylogeny (Fig. 1B),[131] and thus might be expected to show similar olfactory driven behaviors. It is well known that goldfish use hormones as pheromones to coordinate reproductive physiology and behavior.[128] In zebrafish, it is known that certain pheromonal hormones elicit physiological responses as measured by voltage-sensitive dyes.[114] Zebrafish also show a behavioral response to an as of yet chemically undefined "aggregation pheromone" that has a behavioral effect that is dependent on the density of the shoal of fish.[112] Zebrafish also remember odors experienced as juveniles, although certain odors elicit a response regardless of exposure as juveniles.[132] Male zebrafish show a marked attractive response to ovarian extracts that can be eliminated by cauterizing the olfactory epithelium.[113] Thus, zebrafish have both a behavioral and a physiological response to pheromonal substances which in the future can be coupled to the developmental expression of putative pheromone receptors.[91,92]

Unlike goldfish, zebrafish are a schooling fish, and thus have specialized behaviors that befit a group, such as response to alarm pheromone. Alarm pheromone, or Schreckstoff, was first isolated by Karl von Frisch[3] from the European minnow *Phoxinus laevis*, a schooling fish. This substance is produced by glands in the skin and is released when the skin is damaged, eliciting a rapid swimming and clustering on the bottom of the tank.[3,4,133] It has been shown that this response is present in the giant danio.[134–136] While attempts have been made to chemically reconstitute the alarm pheromone, the behavioral and physiological response does not perfectly mimic the naturally derived substance. The zebrafish *Danio rerio* shows a behavioral response to alarm pheromone, which is not surprising given that they are a close relative of the giant danio.[50]

Amino acids and bile acids are important odorants to all fishes studied thus far, and zebrafish are no exception.[4] Amino acids elicit well-defined physiological responses based on electro-olfactogram recording, optical imaging, and receptor binding assays in a variety of fishes including catfish, goldfish, salmon, and zebrafish to name a few. Zebrafish clearly respond physiologically to amino acids and bile acids,[130] and have a clear behavioral response to certain amino acids such as alanine.[137,138] Recently the developmental onset of behavioral responses to a specific amino acid, L-cysteine, has been characterized.[50, 139] Knowing the ontogeny of well characterized behavior is useful for the development of genetic screens,[140,141] where defects in olfactory behaviors can lead to the discovery of mutants with defects specific to the olfactory system. One such pilot screen has been carried out in zebrafish and a mutant defective in the formation of OSN axonal contacts with the developing olfactory bulb has been isolated.[50,139] Thus it appears that a behavioral genetic approach will be useful in uncovering mutants specific to the olfactory sensory system.

8. Evolution of the Olfactory Sensory System

8.1. Is there a Unifying Placode Theme?

In considering the evolutionary origins of the olfactory organ, the tendency is to group all placodal derivatives and look for common mechanisms driving their appearance over evolutionary time. But as cellular and molecular mechanisms governing the formation of the various placodes contributing to the vertebrate head are uncovered,[35,142,143] it is becoming increasingly apparent that the various placodes appear to march to the beat of different drummers. For example the lens placode clearly arises from non-neural ectoderm and the adenohypophyseal placode arises on the midline and can be eliminated by interfering within midline signaling.[125] Yet, the olfactory placodes arise from within the neural plate in zebrafish and appear to develop in concert with the olfactory bulbs rather than being induced by them. Additionally, the anterior neural plate placodes all give rise to rather distinct cellular derivatives such as the crystalline containing cells of the lens, the endocrine cells of the adenohypophysis and the regenerating sensory neurons of the olfactory organ. Therefore, perhaps it is more useful to consider the function of the olfactory system and the various cell types that make up this spectacular sensory system, and to look for functional equivalents of each sensory system outside of a unifying placodal doctrine.

8.2. Olfactory and Adenohypophysis

One idea in the literature that has shaped our thinking on the evolution of the olfactory placode is the idea that the olfactory placode gives rise to neuroendocrine cells. However, an extensive fate map of the olfactory organ in the zebrafish demarcated an olfactory field whose posterior and anterior borders aligned with the cranial neural crest and the adenohypophyseal region respectively (Fig. 6). Yet, GnRH precursors in this olfactory field were never uncovered.[24] This suggests that the olfactory placode does not give rise to the GnRH neuroendocrine cells in the zebrafish.[45] The idea that the OSNs and GnRH cells arose from the same placode led to the idea that there is an ancestral link between chemoreception and GnRH. Cephalochordates such as amphioxus (*Branchiostoma*), a proposed representative of an ancestral state of the chordates, and Urochordates have been examined for just such an association. But, having a chemosensory system that is also endocrine in nature is perhaps the wrong trip to be on. This author once barked up that tree too, looking for the ancestral relationship between neuroendocrine and chemosensory systems. In fact, vital dye labeling showed a label in the buccal cirri of the oral hood suggesting possible chemoreceptors (Fig. 11A, B, arrows), and GnRH-like immunoreactivity was localized to clusters of cells in the anterior neural tube of amphioxus (Fig. 11C). Nevertheless, the fact that the olfactory sensory system does not give rise to the GnRH cells does not make the problem any less interesting. Rather the independent origins of the GnRH cells in the zebrafish reinforce our thinking about neural crest derivatives and expose our thinking to the relationship between the endocrine producing cells in the anterior neural plate and the olfactory placode fields. In amphioxus, Hatschek's pit has been suggested as being the homologue to the adenohypophysis of modern day vertebrates. This organ lies ventral to the notochord and appears to contain endocrine substances;[144]

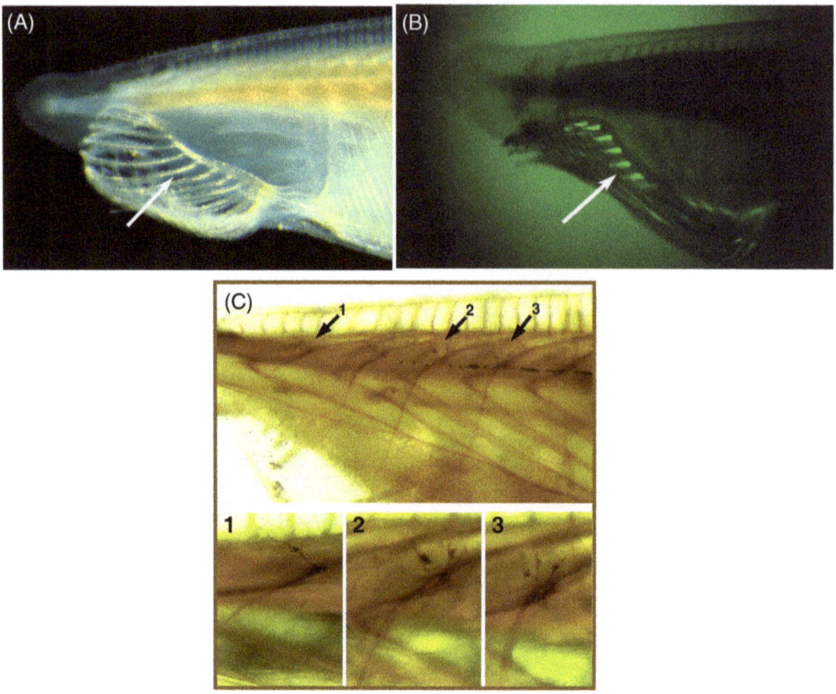

Fig. 11 Amphioxus neuroendocrine and sensory cells. (A) Head of living amphioxus with buccal cirri indicated (arrow). (B) Same head showing vital dye label at base of buccal cirri (arrow). (C) GnRH-like immunoreactivity in five segmentally repeated cell clusters. Arrows 1–3 are shown at higher magnification enlargements in panels 1, 2, and 3 below).

thus one might look for a common region giving rise to Hatschek's pit and the type II chemosensory cells (see below).

Many of the genes expressed in the anterior end of the forming neural tube are expressed in both the olfactory placode field and the adenohypophysis of modern vertebrates, and homologues of these genes are found in amphioxus. For example, *Pax6* is expressed in the developing eyes, nose, and pituitary of modern day vertebrates. This gene has been cloned in amphioxus and is expressed in the neurula stage animal in the rostral ectoderm.[145] Amphioxus is known to have sensory cells in the rostral end of the animal. The type II class of sensory cells, which have both a cilium and microvilli, are suggested to be chemosensory cells, which is interesting because these cells also show a cycle of degeneration and regeneration.[146] The assignment of the type II cells as chemosensory is intriguing for zebrafish have crypt cells, also containing central cilia surrounded by a ring of microvilli;[11] perhaps these are the remnants of the ancestral chemoreceptors. In amphioxus the type II sensory neurons do not develop until after metamorphosis and may use the pioneer neuron tracts, which develop prior to metamorphosis, to develop connections with the central nerves.[147,148] In analyzing the nerve tracts in the rostral end of the nervous system, the dorsal and ventral tracts have been proposed as possible olfactory and terminal nerve equivalents, although no clear telencephalic structure analogous to the olfactory bulb has been identified.[148] Therefore, in amphioxus there are cells that appear to function as peripheral chemosensory receptors (type II cells), and these cells, like the modern day olfactory sensory neurons, undergo regeneration. Additional similarities lie in the fact that amphioxus has a class of pioneer neurons,[147] as has been reported for the olfactory system of the zebrafish,[58] and it may have structural homologues of the olfactory and terminal nerves.

8.3. Olfactory Placode Field and Neural Crest

The olfactory organ is comprised of a mixture of placode (sensory epithelium) and neural crest (structure of frontal mass) derived cells. While amphioxus has no apparent neural crest or sensory placodes, it is intriguing that the amphioxus *Hox* cluster does have elements that can drive reporter gene expression in the sensory placodes and neural crest derivatives of transgenic chick and mouse.[149] When examining the expression pattern of genes such as *dlx3* in the zebrafish, the *dlx3* gene is expressed in a continuous region at the edge of the neural plate flanking the premigratory neural crest.[24] Perhaps the *dlx3* is an ancestral "pre-crest" gene, for it has been shown that in amphioxus *dlx* is expressed at the border of the neural plate. These *dlx* expressing cells in amphioxus show motile behaviors extending across the midline as the neural tube forms.[150] This cellular behavior has been interpreted as being neural-crest-like because the cells extend processes across the edge of the forming neural tube.[142] Therefore there is an ancestral relationship between the first hints of cells migration and the expression of the *dlx* gene. Perhaps the cell movements that have been proposed to give rise to the placodal structures in vertebrate animals[24,26] are also remnants of the movement that we so associate with the neural crest. In this respect the olfactory placode is like the neural crest in that the cells of the olfactory placode field delaminate from their shared border with the developing telencephalon as they migrate to form the olfactory placode.

A World of Questions Remain

What are the signals that induce the onset of expression of the genes involved in olfactory placode formation? Clearly, genes involved in patterning the midline affect development of the nose in that cyclopic fish also have a fused nose. Additionally, anterior patterning defects such as those seen in the *masterblind* mutant[151] result in the loss of the nose although this phenotype varies with genetic background.[152] A genetic screen focusing on the developing olfactory system may help lead us to genes involved in the induction of the olfactory sensory system.

How do the axons target the bulb? The rules governing the development of the specific axonal connections in the CNS are complex. The olfactory sensory system is a perfect model system to investigate these rules for the neurons are readily identifiable and accessible to experimental manipulation and physiological recording.

What are the mechanisms governing the initial expression and maintenance of the olfactory receptors? As described here, the factors controlling the decision to express a single or a few olfactory receptors are poorly understood. The fact that the olfactory sensory neurons regenerate throughout life presents the more challenging problem of the mechanism used to maintain faithful, invariant receptor expression in a regenerating system.

What is the role of the olfactory sensory neurons in determining behaviors? The nose is the interface with the odorous world — how does this world interact with the nose? There are clear and dramatic examples of olfactory imprinting in fishes[153] and zebrafish present a tractable model system to better understand the cellular and molecular processes controlling olfactory memory formation.[132]

What evolutionary forces gave rise to the nose and its close association with the adenohypophysis? Recent developmental analysis has uncovered the adenohypophyseal origin of the GnRH cells[45] which develops in close association with the olfactory placode[126] and express common genes in the anterior neural plate (see Fig. 5). Perhaps there is an ancestral tissue from which the nose and adenohypophysis diverged.

Acknowledgments

The author would like to thank J. Ewer, A. Gopinath, D. Lin, and M. Rivard for their helpful critiques of the manuscript. The author's work is supported by a Basil O'Connor March of Dimes Grant, NYS Hatch Grant, NIH R01 DC0421801A1.

References

1. Grunwald DJ and Eisen JS (2002). *Nature Reviews Genetics* **3:** 717–724.
2. Trinkaus JP (1985). *Journal of Neuroscience Research* **13:** 1–19.
3. von Frisch K (1938). *Naturwissenschaften* **26:** 601–606.
4. Hara TJ (1971). In: Hoar WS and Randall DJ (eds.), *Fish Physiology, Vol. V.* Academic Press, New York, pp. 79–114.
5. Pough FH, Janis CM and Heiser JB (1999). In: *Vertebrate Life*, Prentice Hall, Upper Saddle River, New Jersey.
6. Tolbert LP (1998). *Annals of the New York Academy of Science* **855:** 95–103.
7. Hildebrand JG and Shepherd GM (1997). *Annual Review of Neuroscience* **20:** 595–631.
8. Mombaerts P (2001). *Nature Neuroscience* **4 Supplement:** 1192–1198.
9. Laberge F and Hara TJ (2001). *Brain Reserach Reviews* **36:** 46–59.
10. Murray RC and Calof AL (1999). *Seminars in Cell and Developmental Biology* **10:** 421–431.

11. Hansen A and Finger TE (2000). *Brain Behavior and Evolution* **55:** 100–110.
12. Hansen A and Zeiske E (1993). *Journal of Comparative Neurology* **333:** 289–300.
13. Byrd CA and Brunjes PC (1995). *Journal of Comparative Neurology* **358:** 247–259.
14. Hansen A and Zeiske E (1998). *Chemical Senses* **23:** 39–48.
15. Ramon y Cajal S (1995). In: *Histology of the Nervous system, Vol II.* Oxford University Press, New York, pp. 532–554.
16. Baier H and Korsching S (1994). *Journal of Neuroscience* **14:** 219–230.
17. Baier H, Rotter S and Korsching S (1994). *Proceedings of the National Academy of Sciences USA* **91:** 11646–11650.
18. Stell WK, Walker SE and Ball AK (1987). *Annals of the New York Academy of Science* **519:** 80–96.
19. Wirsig-Wiechmann CR and Oka Y (2002). *Neuroscience Research* **44:** 337–341.
20. Wirsig-Wiechmann CR (2001). *Keio Journal of Medicine* **50:** 81–85.
21. Umino O and Dowling JE (1991). *Journal of Neuroscience* **11:** 3034–3046.
22. Verwoerd CDA and Oostrom CGV (1979). *Advances in Anatomy Embryology and Cell Biology* **58:** 1–75.
23. Couly GF and Le Douarin NM (1985). *Developmental Biology* **110:** 422–439.
24. Whitlock KE and Westerfield M (2000). *Development* **127:** 3645–3653.
25. Farbman AI (1992). In: Barlow PW, Bray D, Green PB and Slack JMW (eds.), *Developmental and Cell Biology Series,* Cell Biology of Olfaction. Cambridge University Press, Cambridge.
26. Streit A (2002). *Developmental Biology* **249:** 237–254.
27. Woo K and Fraser SE (1995). *Development* **121:** 2595–2609.
28. Kozlowski DJ, Murakami T, Ho RK and Weinberg ES (1997). *Biochemical Cell Biology* **75:** 551–562.
29. Akimenko MA, Ekker M, Wegner J, Lin W and Westerfield M (1994). *The Journal of Neuroscience* **14:** 3475–3486.
30. Ellies DL *et al.* (1997). *Genomics* **45:** 580–590.
31. Sahly I, Andermann P and Petit C (1999). *Development, Genes and Evolution* **209:** 399–410.

32. Baker CV and Bronner–Fraser M (2001). *Developmental Biology* **232:** 1–61.
33. Kobayashi M, Osanai H, Kawakami K and Yamamoto M (2000). *Mechanisms of Development* **98:** 151–155.
34. Torres M and Giraldez F (1998). *Mechanisms of Development* **71:** 5–21.
35. Graham A and Begbie J (2000). *Trends in Neuroscience* **23:** 313–316.
36. Campuzano S and Modolell J (1992). *Trends in Genetics* **8:** 202–208.
37. Campos–Ortega JA (1995). *Molecular Neurobiology* **10:** 75–89.
38. Allende ML and Weinberg ES (1994). *Developmental Biology* **166:** 509–530.
39. Mueller T and Wullimann MF (2003). *Brain Research Developmental Brain Research* **140:** 137–155.
40. Blader P, Fischer N, Gradwohl G, Guillemont F and Strähle U (1997). *Development* **124:** 4557–4569.
41. Korzh V, Sleptsova I, Liao J, He J and Gong Z (1998). *Developmental Dynamics* **213:** 92–104.
42. Liao J, He J, Yan T, Korzh V and Gong Z (1999). *DNA Cell Biology* **18:** 333–344.
43. Haddon C *et al.* (1998). *Development* **125:** 359–370.
44. Kortschak RD, Tamme R and Lardelli M (2001). *Development, Genes and Evolution* **211:** 350–354.
45. Whitlock KE, Wolf CD and Boyce ML (2003). *Developmental Biology.*
46. Varga ZM, Wegner J and Westerfield M (1999). *Development* **126:** 5533–5546.
47. Fritz A, Rozowski M, Walker C and Westerfield M (1996). *Genetics* **144:** 1735–1745.
48. Nasevicius A and Ekker SC (2000). *Nature Genetics* **26:** 216–220.
49. Solomon KS and Fritz A (2002). *Development* **129:** 3127–3136.
50. Reyes R, Vitebsky A and Whitlock KE (2002). *5th International Conference on Zebrafish Genetics and Development*, University of Wisconsin, Madison, WI.
51. Le Douarin NM and Kalcheim C (1999). *The Neural Crest*, Developmental and Cell Biology Series Cambridge University Press, Cambridge, ed. Second.
52. Schilling TF and Kimmel CB (1994). *Development* **120:** 483–494.
53. Langille RM and Hall BK (1988). *Journal of Anatomy and Embryology* **177:** 297–305.

54. Neuhauss SC *et al.* (1996). *Development* **123:** 357–367.
55. Kelsh RN and Eisen JS (2000). *Development* **127:** 515–525.
56. Morita T, Nitta H, Kiyama Y, Mori H and Mishina M (1995). *Neuroscience Letters* **198:** 131–134.
57. Papalopulu N and Kintner C (1993). *Development* **117:** 961–975.
58. Whitlock KE and Westerfield M (1998). *Journal of Neuroscience* **18:** 8919–8927.
59. Gong Q and Shipley MT (1995). *Neuron* **14:** 91–101.
60. Poling KR and Brunjes PC (2000). *Brain Research* **856:** 135–141.
61. Dryer L and Graziadei PP (1994). *Perspectives on Developmental Neurobiology* **2:** 163–174.
62. McClure M (1999). *Journal of Morphology* **241:** 83–105.
63. Byrd CA and Brunjes PC (2001). *Neuroscience Abstracts* **104:** 793–801.
64. Byrd CA (2000). *Brain Research* **866:** 92–100.
65. Riddle DR and Oakley B (1992). *The Journal of Comparative Neurology* **324:** 575–589.
66. Mombaerts P *et al.* (1996). *Cell* **87:** 675–686.
67. Lin DM and Ngai J (1999). *Current Opinion in Neurobiology* **9:** 74–78.
68. Mori K, Nagao H and Yoshihara Y (1999). *Science* **286:** 711–715.
69. Wang F, Nemes A, Mendelsohn M and Axel R (1998). *Cell* **93:** 47–60.
70. Yu TW and Bargmann CI (2001). *Nature Neuroscience* **4:** 1169–1176.
71. Yoshihara Y *et al.* (1997). *Journal of Neuroscience* **17:** 5830–5842.
72. Harrelson AL and Goodman CS (1988). *Science* **242:** 700–708.
73. Au WW, Treloar HB and Greer CA (2002). *Journal of Comparative Neurology* **446:** 68–80.
74. Treloar HB, Tomasiewicz H, Magnuson T and Key B (1997). *Journal of Neurobiology* **32:** 643–658.
75. Alenius M and Bohm S (2003). *Development* **130:** 917–927.
76. Mizuno T *et al.* (2001). *Molecular and Cellular Neuroscience* **18:** 119–130.
77. Marx M, Rutishauser U and Bastmeyer M (2001). *Development* **128:** 4949–4958.
78. Ressler KJ, Sullivan SL and Buck LB (1994). *Cell* **79:** 1245–1255.
79. Singer MS, Shepherd GM and Greer CA (1995). *Nature* **377:** 19–20.
80. Yoshida T, Ito A, Matsuda N and Mishina M (2002). *Journal of Neuroscience* **22:** 4964–4972.
81. Klose M and Bentley D (1989). *Science* **245:** 982–984.

82. McConnell SK, Ghosh A and Shatz CJ (1994). *Journal of Neuroscience* **14:** 1892–1907.
83. Dynes JL and Ngai J (1998). *Neuron* **20:** 1081–1091.
84. Barth AL, Justice NJ and Ngai J (1996). *Neuron* **16:** 23–34.
85. Barth AL, Dugas JC and Ngai J (1997). *Neuron* **19:** 359–369.
86. Sklar PB, Anholt RR and Snyder SH (1986). *Journal of Biological Chemistry* **261:** 15538–15543.
87. Buck L and Axel R (1991). *Cell* **65:** 175–187.
88. Ngai J *et al.* (1993b). *Cell* **72:** 667–680.
89. Ngai J, Dowling MM, Buck L, Axel R and Chess A (1993a). *Cell* **72:** 657–666.
90. Kondo R, Kaneko S, Sun H, Sakaizumi M and Chigusa SI (2002). *Gene* **282:** 113–120.
91. Speca DJ *et al.* (1999). *Neuron* **23:** 487–498.
92. Cao Y, Oh BC and Stryer L (1998). *Proceedings of the National Academy of Sciences USA* **95:** 11987–11992.
93. Asano–Miyoshi M *et al.* (2000). *Journal of Biochemistry (Tokyo)* **127:** 915–924.
94. Clyne PJ *et al.* (1999). *Neuron* **22:** 327–338.
95. Leibovici M, Lapointe F, Aletta P and Ayer-Le Lievre C (1996). *Developmental Biology* **175:** 118–131.
96. Freitag J, Krieger J, Strotmann J and Breer H (1995). *Neuron* **15:** 1383–1392.
97. Kratz E, Dugas JC and Ngai J (2002). *Trends in Genetics* **18:** 29–34.
98. Weth F, Nadler W and Korsching S (1996). *Proceedings National Academy of Sciences USA* **93:** 13321–13326.
99. Haddon C, Jiang YJ, Smithers L and Lewis J (1998). *Development* **125:** 4637–4644.
100. Fan J and Ngai J (2001). *Developmental Biology* **229:** 119–127.
101. Eisthen HL (1997). *Brain Behavior and Evolution* **50:** 222–233.
102. Matsuoka M *et al.* (2002). *Brain Research* **946:** 52–63.
103. Dulac C and Axel R (1995). *Cell* **83:** 195–206.
104. Matsunami H and Buck LB (1997). *Cell* **90:** 775–784.
105. Dulac C (2000). *Current Opinion in Neurobiology* **10:** 511–518.
106. Hansen A, Zippel HP, Sorensen PW and Caprio J (1999). *Microscope and Research Techniques* **45:** 325–338.

107. Zippel HP, Sorensen PW and Hansen A (1997). *Journal of Comparative Physiology* **180:** 39–52.
108. Stacey NE and Kyle AL (1983). *Physiology and Behavior* **30:** 621–628.
109. Loconto J *et al.* (2003). *Cell* **112:** 607–618.
110. Ishii T, Hirota J and Mombaerts P (2003). *Current Biology* **13:** 394–400.
111. Jessen JR, Jessen TN, Vogel SS and Lin S (2001). *Genesis* **29:** 156–162.
112. Bloom HD and Perlmutter A (1977). *Journal of Experimental Zoology* **199:** 215–226.
113. van den Hurk R and Lambert JGD (1983). *Canadian Journal of Zoology* **61:** 2381–2387.
114. Friedrich RW and Korsching SI (1998). *Journal of Neuroscience* **18:** 9977–9988.
115. Oehlmann VD, Korte H, Sterner C and Korsching SI (2002).*Mechanisms of Development* **117:** 357–361.
116. Mathieu M, Tagliafierro G, Bruzzone F and Vallarino M (2002). *Brain Research Developmental Brain Research* **139:** 255–265.
117. Pinelli C *et al.* (2000). *Brain Research Developmental Brain Research* **119:** 195–208.
118. Brookover C (1914). *Journal of Comparative Neurology* **24:** 113–130.
119. Brookover C and Jackson TS (1911). *Journal of Comparative Neurology* **21:** 237–259.
120. Wray S, Grant P and Gainer H (1989). *Proceedings National Academy of Sciences USA* **86:** 8132–8136.
121. Schwanzel-Fukuda M and Pfaff DW (1989). *Nature* **338:** 161–163.
122. Parhar IS, Soga T, Ishikawa Y, Nagahama Y and Sakuma Y (1998). *Journal of Comparative Neurology* **401:** 217–226.
123. Dubois EA, Zandbergen MA, Peute J and Goos HJ (2002). *Brain Research Bulletin* **57:** 413–418.
124. Sbrogna JL, Barresi MJ and Karlstrom RO (2003). *Developmental Biology* **254:** 19–35.
125. Karlstrom RO, Talbot WS and Schier AF (1999). *Genes & Development* **13:** 388–393.
126. Herzog W *et al.* (2003). *Developmental Biology* **254:** 36–49.
127. Liu NA *et al.* (2003). *Molecular Endocrinology* **17:** 959–966.
128. Sorensen PW and Caprio J (1997). In: Evans DH (ed.), *The Physiology of Fishes.* CRC Press, Boca Raton, pp. 375–406.

129. Kleerekoper H (1969). *Olfaction in Fishes* Indiana University Press, Bloomington.
130. Michel WC and Lubomudrov LM (1995). *Journal of Comparative Physiology A* **177:** 191–199.
131. Metscher BD and Ahlberg PE (1999). *Developmental Biology* **210:** 1–14.
132. Whitlock KE, Newton LA and Boyce ML (2002). *5th International Conference on Zebrafish Genetics and Development*, University of Wisconsin, Madison, WI.
133. von Frisch K (1941b). *Z. Vergl. Physiol.* **29:** 46–145.
134. Pfeiffer W (1978). *Journal of Chemical Ecology* **4:** 665–673.
135. Pfeiffer W and Lamour D (1976). *Revue Suisse de Zoologie* **83:** 861–873.
136. Pfeiffer W and Lemke J (1973). *Journal of Comparative Physiology* **82:** 407–410.
137. Steele CW, Ownes DW and Scarfe AD (1990). *Journal of Fish Biology* **36:** 341–352.
138. Steele CW, Scarfe AD and Owens DW (1991). *Journal of Fish Biology* **38:** 553–564.
139. Reyes R, Vitebsky A and Whitlock KE (2001). *Society for Neuroscience Abstracts* **27:** 365.
140. Bargmann CI and Kaplan JM (1998). *Annual Review of Neuroscience* **21:** 279–308.
141. Carlson JR (1996). *Trends in Genetics* **12:** 175–180.
142. Baker CV and Bronner–Fraser M (1997). *Mechanisms of Development* **69:** 13–29.
143. Shimeld SM and Holland PW (2000). *Proceedings of the National Academy of Sciences USA* **97:** 4449–4452.
144. Nieuwenhuys R (1998). In: Nieuwenhuys R, ten Donkelaar HJ and Nicholson C (eds.), *The Central Nervous System of Vertebrates, Vol. I.* Springer–Verlag, Berlin, pp. 365–396.
145. Glardon S, Holland LZ, Gehring WJ and Holland ND (1998). *Development* **125:** 2701–2710.
146. Lacalli TC (1999). *Acta Zoologica* **80:** 125–134.
147. Yasui K, Tabata S, Ueki T, Uemura M and Zhang SC (1998). *Journal of Comparative Neurology* **393:** 415–425.
148. Lacalli TC (2002). *Acta Zoologica* **83:** 149–166.

149. Manzanares M *et al.* (2000). *Nature* **408:** 854–857.
150. Holland ND, Panganiban G, Henyey EL and Holland LZ (1996). *Development* **122:** 2911–2920.
151. Heisenberg CP *et al.* (1996). *Development* **123:** 191–203.
152. Sanders LH and Whitlock KE (2003). *Developmental Dynamics.*
153. Hasler AD (1971). In: Hoar WS and Randall DJ (eds.), *Fish Physiology, Vol VI.* Academic Press, New York, pp. 429–456.

Chapter 8

Somite Segmentation: A View from Fish

Hiroyuki Takeda* and Yumiko Saga†

**Department of Biological Sciences, Graduate School of Sciences, University of Tokyo*

†Division of Mammalian Development, National Institute of Genetics

Somite formation, a process in which reiterated epithelial structures are progressively demarcated from the mesenchymal presomitic mesoderm (PSM) in a anterior-posterior sequence, is the earliest manifestation of segmentation and is a feature shared by all vertebrate embryos. The temporal and spatial regulation of this process requires a molecular oscillator, the segmentation clock. The mechanisms driving and regulating the oscillation in PSM cells have been actively studied in zebrafish, chick and mouse. The oscillator is comprised of genetic circuit involving the Notch signaling pathway and its target genes *her1* and *her7* in zebrafish. Converting clock oscillation into the periodic arrangement of segment boundaries is achieved at the 'wavefront' located in the anterior PSM. The level of Fgf/MAPK activation (highest in the posterior PSM) serves as a positional cue within the PSM to restrict the wavefront to the anterior PSM. Once the level of Fgf/MAPK signaling declines in the anterior PSM, the wavefront activity mediated by a transcription factor, Fss/Tbx24, arrests the oscillation and leads to activation of a number of key genes required for subsequent sequences of somite formation. In the anterior PSM or wavefront, a complicated gene network centered on Mesp, a bHLH transcription factor, finally establishes a rostrocaudal subdivisions within somite primordium, which is prerequisite for formation of morphological distinct somite boundaries.

1. Introduction

Somites are transient segments of the paraxial mesoderm that are present in developing cephalochordates and vertebrates. In many vertebrate species, such as frog, chick, mouse and zebrafish, somites form as blocks of cells, which bud off in a highly coordinated fashion from the anterior end of the unsegmented presomitic mesoderm (PSM). Although the somites are transient structure during early embryogenesis, they create the metamerism of all somite-derived tissues (axial skeleton, the dermis of the back, and all striated muscle of the adult body,[1] and the segmental organization of somites imposes a segmental property on the spinal cord[2] and spinal ganglia.[3] Because of its strict periodicity (30 min for zebrafish, 90 min for chick and 120 min for mouse) and beautifully coordinated morphogenesis, somite segmentation has attracted many developmental and theoretical biologists.[4,5] However, until recently, the mechanisms that create the periodicity of somites were largely unknown. One recent breakthrough is a finding of cyclic genes linked to periodic somite formation.[6] Since then, genetic and experimental evidence has accumulated to unravel how the periodicity is generated, how the positions of segment border are determined and how rostrocaudal polarity within the somite primordum is generated. Based on these findings, we now know that the Notch/Delta signaling pathway plays crucial roles in establishment of temporal periodicity in the PSM and rostrocaudal polarity within segments. Intriguingly, segmentation in *Drosophila melanogaster*, one of the well-studied embryonic segmentation, does not appear to use the Notch/Delta pathway. Indeed, unlike vertebrate segmentation, segments in the fly (a long germ insect) form simultaneously in the syncytial blastoderm. Thus, the genetic cascade leading to fly and vertebrate segmentation may not be conserved, raising the question as to whether segmentation evolved independently in invertebrates and vertebrates.

1.1. General View of Vertebrate Somite Segmentation

During vertebrate embryonic development, the paraxial mesoderm is subdivided into metameric subunits called somites. The somites are epithelial spheres of paraxial mesoderm and the first segmented structures to form during embryogenesis. Individual pairs of somites, located symmetrically on either side of the neural tube, are formed in a anterior-posterior progression within the PSM. It is believed that the process of somitogenesis can be divided into four distinct stages, which may be regulated by different genetic mechanisms (Fig. 1E):[7]

(1) *Specification as paraxial mesoderm*: the mesoderm derived from the primitive streak in mice and chick, the marginal zone in amphibians,

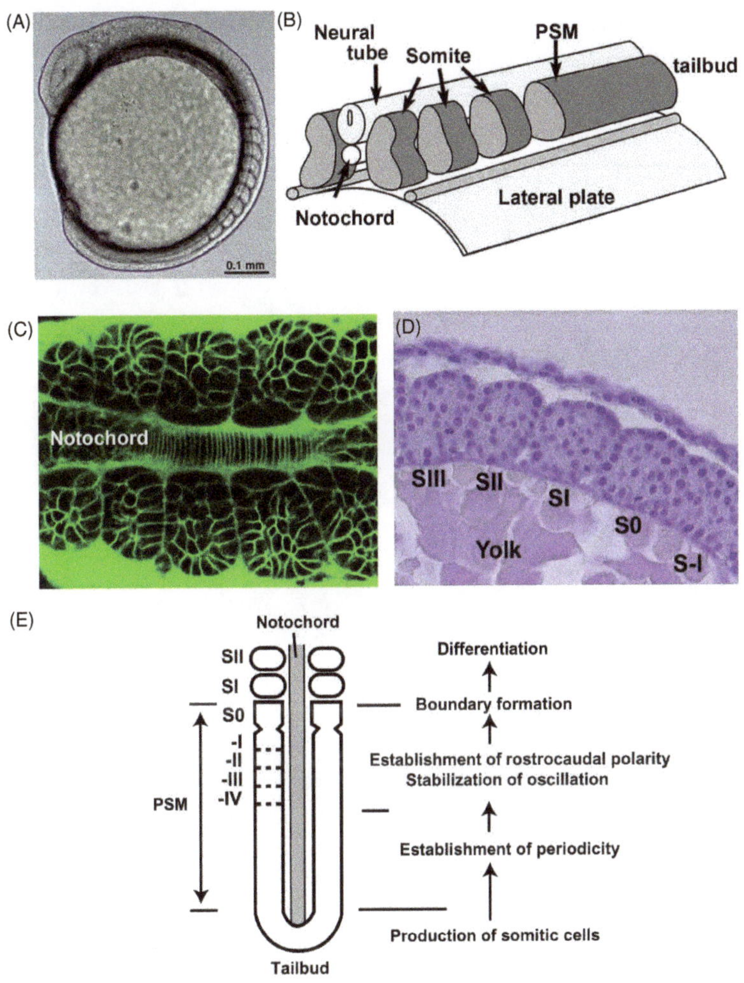

Fig. 1. Overview of somitogenesis. A–D Morphological aspects of zebrafish somitogenesis. (A) Lateral view of a live embryo at 11-somite stages (~14qhpf). (B) Schematic drawing of zebrafish somitogenesis. (C) A live embryo (~10-somite stage) stained with BODPY-ceramide. A dorsal view at the level of the notochord. Anterior to the left. (D) Sagittal histological section of the somites and PSM at 8-somite stage. In (C) and (D), epithelial cells surrounding loosely packed central cells are visible in several of the somites. (E) A schematic drawing showing sequential steps of vertebrate somitogenesis. In this chapter, nomenclature of the segmented and presumptive somites follows that proposed by Pouquie and Tam,[86] e.g. SI labels the most recently formed somite, S0 represents forming somite and S-I is the most anterior presumptive somite.

and germ ring in fish or the tailbud, is arranged on both sides of the neural tube as the paraxial mesoderm.

(2) *Establishment of periodicity*: through a molecular oscillator, the segmentation clock, the PSM acquires a periodicity that will later be translated into regular-spaced boundary formation.

(3) *Boundary formation*: the paraxial (somitic) mesoderm generates each segmental border and is divided into the so-called epithelial somites. The establishment of rostrocaudal polarity within somite primordium is prerequisite for formation and maintenance of segment border.

(4) *Differentiation*: Soon after their formation, epithelial somites become patterned in response to local signals derived from the surrounding tissues.

Signaling molecules such as Sonic hegehog, BMPs, Wnts and Noggin have been identified and implicated in patterning and differentiation within the somites. In general, the dorsal part of a somite differentiates into the dermomyotome, which mainly gives rise to all trunk and lib skeletal muscle and the dermis of the back, and the ventral part into a mesenchymal compartment called the sclerotome that gives rise to axial skeleton. In zebrafish, contribution of somitic cells to dermis has not been established.

Somitogenesis in amniotes and discussion of general mechanisms of somite formation and differentiation have been extensively reviewed elsewhere.[8–12] Here we focus on the processes that control establishment of periodicity and boundary formation in the PSM, and try to integrate zebrafish and amniote data to draw a general scheme of vertebrate somite segmentation. We start by introducing morphological events of zebrafish somitogenesis.

1.2. Zebrafish Somite Formation and Mutants

In the zebrafish, the first somites appear approximately 10.5 hours after fertilization. Cells in the extreme anterior region of the PSM alter their adhesive properties and undergo mesenchymal-epithelial transitions, forming epithelial cells around loosely organized mesenchymal cells

(Fig. 1A–D). The earliest few somites in zebrafish seem to form more quickly than later ones: 3 per hour for the first six, and 2 per hour thereafter,[13] in a bilaterally symmetric, anterior to posterior wave until a total of about 30 somites pairs is formed. It had been thought that the mechanical forces arising from compaction of the presumptive internal mesenchymal cells of prospective somite cause them to detach from the unsegmented PSM. However, zebrafish doubly mutant for *knypek* and *trilobite* (both are characterized as a convergent extension mutant) can make somites without internal mesenchymal cells or compaction.[14] Thus, two rows of presumptive border cells are sufficient to make segment borders in zebrafish, suggesting that PSM cells can be segmented into somites by local cell-to-cell interactions (or cell behaviors), without mechanical aids of internal mesenchymal cells. Unlike amniote, muscle differentiation takes place prior to furrow formation in a special population of PMS cells adjacent to the notochord, adaxial cells (see Stickney *et al.*[15] for a review).

In a large-scale screening for mutations affecting development of zebrafish embryo, two groups of genes, *fss*-type and *you*-type genes, were identified that play an important role in somite formation. The *you*-type mutants, *you, sonic-you, you-too, chameleon* and *u-boot*, do not exhibit obvious defects during somite segmentation but do have defects in somite patterning such as muscle differentiation.[16] Recent work has shown that *you*-type genes encode components of Sonic hedgehog pathway that is required for slow muscle differentiation in the somite and the ventral neural tube. Indeed, *sonic you* and *you-too* were found to be *sonic hedgehog* and *gli2*, respectively (see Currie and Ingham[10] for a review). Another group of somite mutants is *fss*-type genes, containing 5 complementation groups, *fused somites* (*fss*), *beamter* (*bea*), *deadly seven* (*des*), *after eight* (*aei*) and *mind bomb* (*mib*), and exhibits a defect in somite boundary formation. However, the spatial distribution of the defects is different among mutants: *fss* controls the formation of all somites while the other four only govern the formation of posterior somites, the first several somites remain intact in these mutants.[16,17] Most of the genes were already identified, which made a significant contribution to understanding of vertebrate somite segmentation.[18]

2. Gene Expression and Segmental Property in the PMS

In zebrafish, like in other vertebrates, somite formation is presaged by stripes of gene expression within the morphologically unsegmented PSM. Metameric patterns of paired bilateral stripes of transcripts have been observed for a number of zebrafish genes (Fig. 2), clearly indicating that cells in the anterior PSM are allocated to specific somites or acquire a segmental property before epithelial segment boundaries become evident. Interestingly, some genes such as *mesp, ephA4* and *notch5* show a preferential expression within prospective rostral or caudal parts of somites, indicating rostrocaudal specification of the somite primordia prior to boundary formation. As described below, segmental prepattern and rostrocaudal specification are established through the Notch signaling pathway. Indeed, loss-of-function and gain-of-function experiments in the chick, *Xenopus*,[19,20] mouse[21–23] and zebrafish[24] have demonstrated that the Notch pathway is involved in creating the regular somitic pattern (see below).

The Notch pathway is an intercellular signaling cascade consisting of the transmembrane receptor Notch and its transmembrane ligands Delta and Serrate. Upon activation, the extracellular domain of Notch is released by Furin-protease and the membrane-bound intracellular domain of Notch (NICD) is further processed by Presenilin. NICD interacts with Suppressor-of-Hairless/RPBJk and enters the nucleus to activate bHLH genes in the hairy-Enhancer of split (E(spl)) family, such as *c-hairy1* in chick, *Hes* in mouse and *Her* in zebrafish. Lunatic fringe is known to act in the Go lgi as a glycosyltransferase enzyme that modifies the extracellular domain of Notch and regulates the activity of the Notch receptor.[25,26] A number of homologues of components in the Notch pathway have been identified in zebrafish,[27] and some of them show segmental or specific expression in the PMS and segmented somites (Fig. 2A).

3. A Molecular Clock Functions in the PSM

Existence of the clock in the PSM has been predicted by theoretical models. According to the “Clock and Wavefront” model, a widely

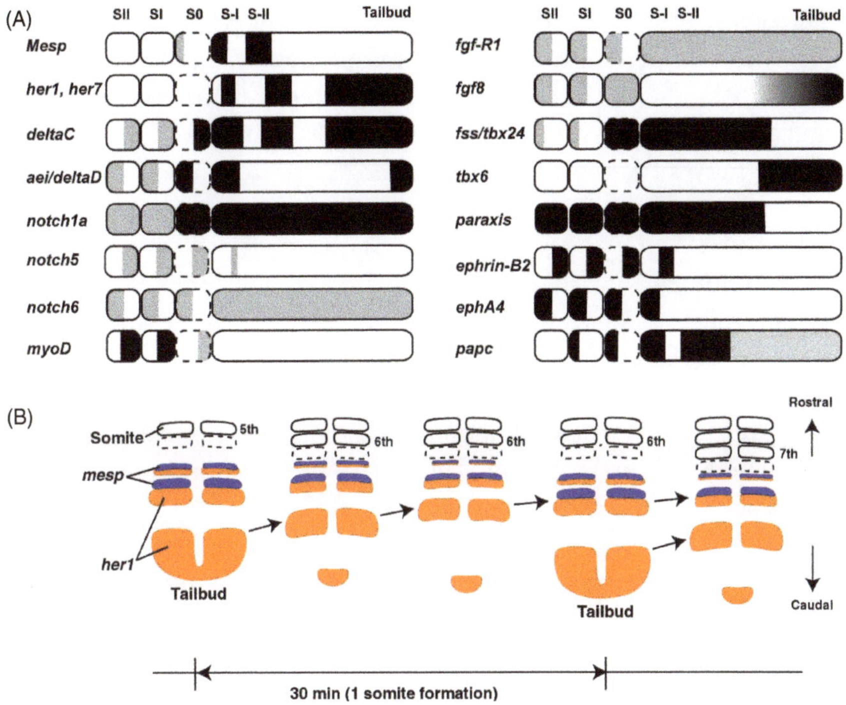

Fig. 2. Patterns of gene expression in zebrafish segmented and presomitic mesoderm (A) Expression patterns of genes though to be involved in somite segmentation, boundary formation and differentiation. The two most recently formed somites (SI, SII), a forming somite (S0, dotted line) and the presomitic mesoderm (PSM) are shown. Drawings are oriented with the anterior to the left. Color density (from black to gray) largely corresponds to the level of expression. *her1*, *her7* and *deltaC* expression oscillates in the posterior to intermediate PSM and becomes stabilized in the anterior PSM. Thus, the drawings depict their expression pattern at one moment. (B) Schematic diagrams illustrating expression profile of *her1* (red) and *mesp-a* (blue). Solid lines represent formed somites while dotted lines represent successively forming somite. The *her1* expression domain appears around the tailbud, and moves anteriorly until it finally overlaps with the most anterior stripe of *mesp-a*. Both stripes disappear near the point of furrow formation. The posterior tip of the tailbud is always positive for *her1* transcripts. A stripe of *her1* appears every 30 minutes in the tailbud region, and persists for about 1.5 hours (three somite cycles in zebrafish).

accepted model,[28,29] the clock creates a temporal periodicity, such as a cyclic wave of gene expression in the PSM, which would later be interpreted by the wavefront (or determination front) to generate spatial periodicity of the somites. The wavefront that exists in the anterior PSM gradually moves back at a constant speed as somitogenesis proceeds and the tailbud retreats. Thus, the stepwise interaction between the clock and the wavefront (or periodic entry of the wave into the wavefront) leads to regularly spaced furrow formation in the anterior PSM. The first molecular evidence for the existence of a somitogenesis oscillator came from the discovery of the oscillating expression of the chick *hairy* homologue, *c-hairy1*.[6] The expression domain appears in the tailbud and sweeps up the PSM once per somite formation. Subsequently, additional *hairy*-related genes have been shown to oscillate in both the chick (*c-hairy2* and *c-Hey2*) and mouse (*hes1* and *hes7*) PSM.[30–32] The expression of *lunatic fringe* (*lfng*) also oscillates within the chick and mouse PSM.[33–35] Their cycling behavior in the PSM is regulated at transcriptional level. A vital connection between the oscillator and somitogenesis has been established by knockout mice in which the oscillating gene, *lfng* or *hes7*, was mutated: in both cases, somite formation and rostrocaudal patterning within segments are disrupted.[32,36,37]

In the zebrafish, the expression of *hairy*-related genes, *her1* and *her7* have been shown to oscillate.[38–40] As shown in Fig. 2B, *her1* expression usually appears as three stripes in the PSM. A new wave of *her1* expression appears in the tailbud every 30 minutes (the duration of one-somite formation in zebrafish), becomes narrower as it moves anteriorly, and finally stabilizes at the future segmentation point in the anterior PSM before decaying. The one known zebrafish *lfng* homologue does not appear to exhibit an oscillating pattern of expression.[41] Instead, the expression of the Delta homologue, *deltaC* oscillates in a pattern overlapping that of *her1* and *her7*.[40,42]

4. Molecular Circuit in the Clock

The zebrafish *fss*-type mutants have defects in somite segmentation. Except for *fss*, four mutants exhibit relatively similar phenotype: several anterior somites are formed and irregular somitic borders are formed in the posterior paraxial mesoderm. Furthermore, extra defects, that can be interpreted as being deaf to Notch signaling, are observed. For example, the *aei* mutant has neuronal hyperplasia and *des* shows neurogenic abnormalities in the neural plate.[43] Indeed, subsequent work has shown that *aei*, *des* and *mib* encode DeltaD, Notch1 and RING E3 ubiquitin ligase (a novel Notch component), respectively.[38,44,45]

The expression of the oscillating genes, *her1, her7* and *deltaC*, is perturbed in zebrafish Notch pathway mutants.[17,38,42] In *aei/deltaD* and *des/notch1* embryos, *her1, her7* and *deltaC* expression are absent or greatly reduced in the posterior PSM, but are expressed in a disorganized "salt and pepper" pattern in the rostral PSM. Based on the assumption that cells of the mutant PSM are simply uncoordinated in their expression of these genes, Jiang *et al.*[42] proposed that Notch signaling is crucial for synchronization of oscillation between neighboring cells. However, recent work in zebrafish, chick and mouse supported the notion that the Notch pathway functions within the oscillator itself. The transcription of *hairy* genes in PSM cells depends on the activation of Notch signaling; misexpression of Notch1a receptor causes ubiquitous expression of *her1* in zebrafish PSM,[46] and *her1* and *her7* expression is down-regulated in Notch pathway mutants.[38,40,44] Her1 and Her7, being transcriptional repressors, then inactivate their transcription as well as *delta* genes. Overexpression of Her1 leads to a decreased level of *deltaD* and *deltaC*[46] while morpholino antisense (MO)-mediated elimination of Her1, or both Her1 and Her7 function causes the widespread, elevated expression of *deltaC* and *her1*, resulting in elimination of the oscillation of *her* and *deltaC* expression.[38,40,44] Thus, the transcriptional response to Notch signaling in zebrafish PSM cells can be negatively regulated by the action of induced Her proteins. The expression of all Notch components (*delta, notch*, and *her*) is initially activated at the tip of the tailbud (Fig. 2A) probably by a certain factor perhaps controlling mesoderm fate, when cells are allocated to the paraxial mesoderm. The subsequent activities of these proteins could then establish a negative feedback loop to create oscillation in gene expressions.

The model proposed by Oates and Ho[40] predicts the following events within one oscillation cycle (Fig. 4B).

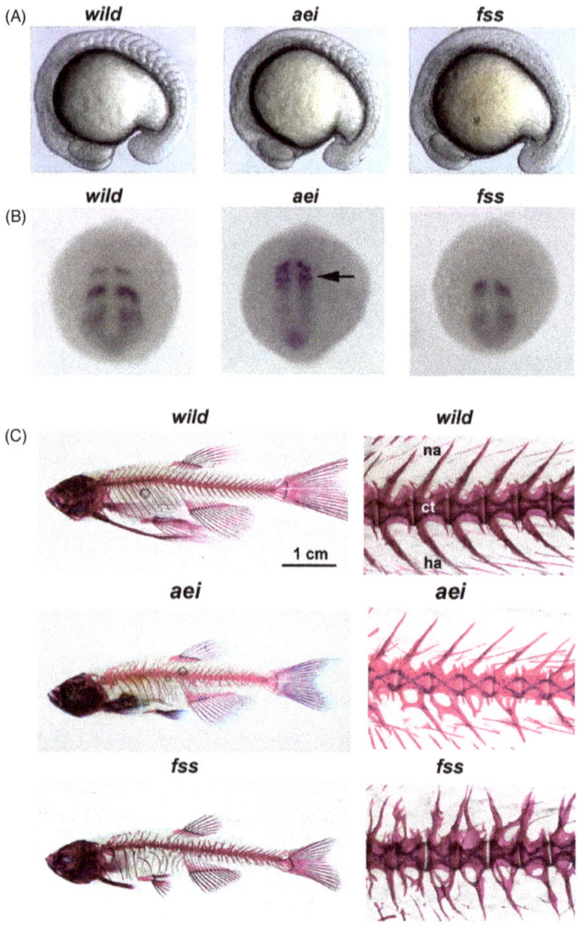

Fig. 3. Zebrafish segmentation mutants (A–B) Phenotypic appearance of wild-type (~16-somite stage), *aei/deltaD* and *fss/tbx24*. Lateral views (anterior is left) of live embryos (A) and dorsal views (anterior is top) of embryos stained with *her1* probes (B) are shown. *aei*: the first eight somites appear normal while no somite is formed beyond this point. *fss*: no somite formation is observed. In a *fss* embryo, most anterior *her1* stripe is missing while the posterior two are normal. By contrast, in *aei* embryo, only broad expression of *her1* is detectable in the anterior PSM (arrow). (C) Skeletal phenotype of zebrafish segmentation mutants. When carefully maintained, *aei* and *fss* homozygous mutants are viable and fertile. Low (left) and high (right) magnification views of wild-type and mutant skeletons are shown. Severe defects are seen in vertebra in *fss* mutant. The centra (ct) in *fss* are almost normal in shape but the length of individual centrum is slightly more variable. In contrast, *fss* mutant shows irregular formation of the arches (neural arch, na on the dorsal and hemal arch on the ventral). As compared with *fss*, the phenotype of *aei* is less severe.

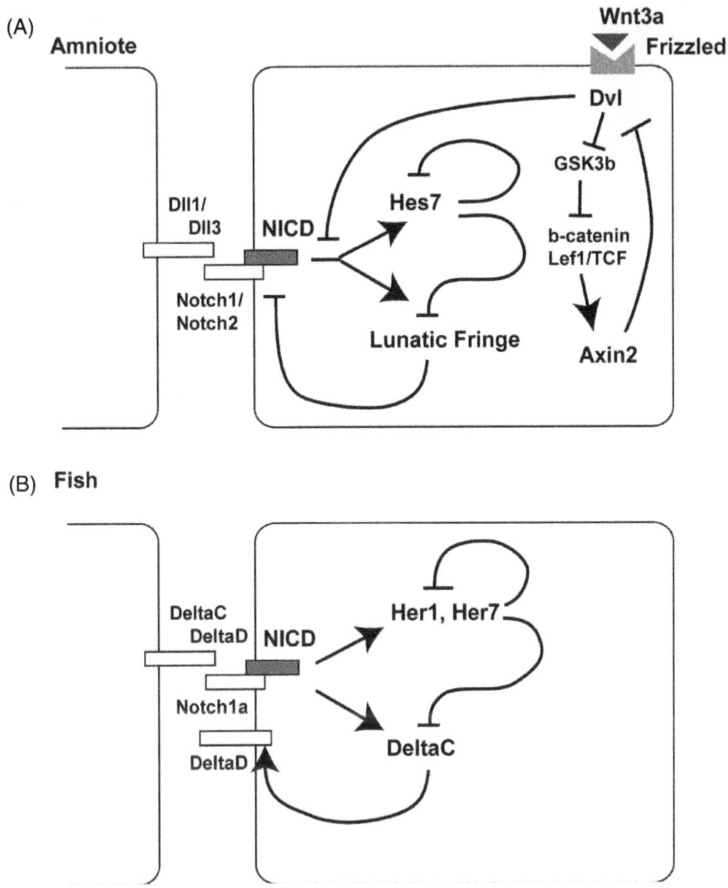

Fig. 4. Possible mechanisms for the segmentation clock based on the negative feedback loop of Notch signal. In both amniotes and fish, the negative feedback loop of Hairy proteins (Hes7 in mouse and Her1/Her7 in zebrafish) may lie in the core of the oscillator, and deltaC and Lunatic fringe are differently coupled with the oscillator (for detailed, see text). Involvement of Wnt singal in the segmentation clock has been shown in mouse. It was shown that Axin2 and Lfng transcription oscillate out of phase, suggesting a inhibitory interaction between the two pathways.[84] Involvement of Wnt signal has not been tested in fish. Dll1, Delta-like 1; Dll3, Delta-like 3; Dvl, disheveled.

(1) Delta activates Notch signaling. The *deltaC* gene could be an immediate target of Notch signaling, providing a rapid amplification of basal *deltaC* expression.
(2) After a short lag time, the transcription of *her* genes is activated and Her proteins accumulate. Finally, Her proteins act on promoters of their own and *deltaC*, switching off the loop.
(3) When the level of Her proteins drops below a specific threshold, activated Notch signaling starts the cycle again.

Reactivation of Notch signaling could depend on the low and/or constant level of Delta proteins present throughout the PSM. In addition to a clock generator, Delta proteins may have a non-cell-autonomous function with which to synchronize oscillations in adjacent cells and could serve as a clock synchonizer. Therefore, it is likely that the generation and coordination of the oscillation cannot be separated to each other but rather they are one and the same. In any case, it would be essential to examine whether the segmentation clock is cell-autonomous or whether the clock still runs in Notch pathway mutants.

As described above, in chick and mouse, *lunatic fringe* (*Lfng*), another important target of Notch signaling, displays a cyclic oscillation in the PSM.[33–35] In chick embryo, Lfng is activated by Notch signaling and negatively regulate Notch signaling, forming a negative feedback loop in PSM cells (Fig. 4A).[47] Consistent with this, RBPJk (a cofactor of NICD) binding sites are located in mouse *Lfng* promoter that is responsible for its own cyclic expression.[48,49] In mammals, Hes7 that shows cyclic expression of mRNA in the PSM appears to play a central role in the segmentation clock because its knockout mutation results in somite phenotypes and up-regulation of Hes7 transcripts.[32] Spatial comparison, using the antibody specific to Hes7 protein, revealed that *Hes7* and *Lfng* transcription occurs in the Hes7 protein-negative domains in the PSM. Thus, periodic repression by Hes7 protein is critical for the cyclic transcription of Hes7 and Lfng.[50]

Taken together, in all vertebrates examined so far, a negative feedback loop mediated by the Notch pathway seems to lie at the heart of the oscillator (Fig. 4). The model, however, requires a rapid degradation of Hairy and Lfng proteins. Indeed, those proteins were shown to be highly unstable in mouse and chick embryos.[47,50]

5. Wavefront

5.1. Fused Somites/Tbx24 Regulate the Wavefront Activity

As described above, four of *fss*-type mutants show essentially the same phenotype, with posterior somite defects and neuronal hyperplasia, and encode components of the Notch signaling cascade. The fifth mutant, *fss*, however, exhibits a different phenotype, characterized by complete lack of somite formation along the entire anterior-posterior axis. Expression analyses with the *fss*-type mutants have demonstrated that the Fss and Notch pathways are functionally distinct.[38] As it rostrally travels in wild-type PSM, *her1* expression domains becomes narrower, slows down and finally arrests in the anterior PSM before decaying. The gradual slowing and stabilization of the oscillation are thought to occur in the wavefront that is established through the activity of Fss. In *fss* mutants, the anterior-most stripe of *her1* is always missing, while the posterior two normally appear in the PSM. By contrast, the posterior *her1* stripes are disorganized or abolished, while the anterior one is detectable in *aei/deltaD* mutants. This indicates that Fss does not affect the oscillation and is sufficient to induce or maintain *her1* expression in the anterior PSM in the absence of normal Notch signaling. In addition to a defect in stabilizing the oscillating gene expression, *fss* mutation blocks the induction of a number of segmentation key genes such as *mesp* and *papc* in the anterior PSM. Therefore, Fss is required for nearly all events in the wavefront prior to segment border formation. Recently, it was found that the *fss* gene encodes a novel T-box transcription factor, Tbx24, which is specifically expressed in the PSM (Fig. 5A, B).[51] Comparing amino-acid sequences corresponding to the T-domains revealed that the new T-box protein does not belong to any other T-box clusters, and no significant similarity is found in the region outside of the T-domain.

Since the *fss/tbx24* mutation affects only the rostral-most *her1* stripe, leaving the caudal two stripes intact, Fss/Tbx24 function was thought to be required only in the rostral PSM. However, *fss/tbx24* is widely expressed in the intermediate and anterior PSM (the anterior border of the expression domain resides in the anterior of S0) (Fig. 5B). In

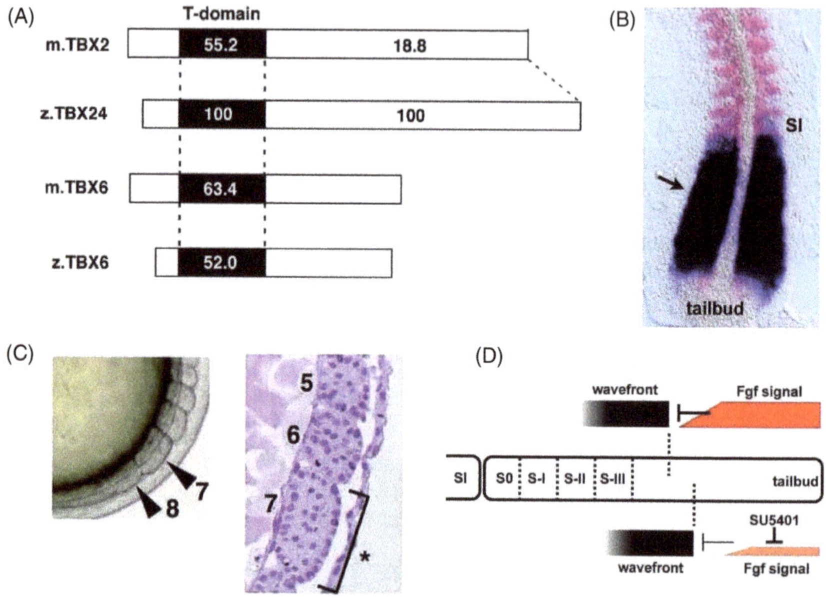

Fig. 5. Fss/Tbx24 and Fgf signaling control the wavefront (A) Amino acid sequence identity (%) between Fss/Tbx24 and other T-box genes. According to the BLASTP search, Fss/Tbx24 is most related to Tbx6 proteins, although the identity is low. (B) Expression of *fss/tbx24* at the segmentation stage. Double *in situ* hybridization of *MyoD*[87] (red) and *fss/tbx24* (blue, arrow) in a flat-mounted embryo. *fss/tbx24* is widely expressed in the intermediate to anterior PSM. The anterior is oriented to the top. SI is most recently formed somite. (C) Manipulation of Fgf signaling using a chemical FGF-R inhibitor (SU5402) alters somite size. Lateral view of a treated live embryo at 7-somite and its histological section are shown. The somite number is indicated by Arabic numerals near the somites. The embryos were incubated in SU5402-containing medium for 8 minutes at 2-somite stage, followed by intense washing. Large somites are observed at the level of 7th somites (arrowheads). Histological section confirms the formation of large somites. Note that a large somite (asterisk) contains more somitic cells that exhibit no cellular abnormality. (D) A model of how Fgf/MAPK signaling is involved in somite boundary formation. An Fgf signal activated in the intermediate and posterior PSM antagonizes maturation of PSM cells. In the wavefront, the anterior PSM, which is devoid of MAPK activation, becomes competent to initiate furrow formation in response to the oscillator. When an Fgf signal is transiently compromised by SU5402, the maturation is accelerated and the wavefront is posteriorly expanded, which leads to a posterior shift in furrow formation.

other words, PSM cells start to express *fss/tbx24* when they pass from the posterior to intermediate PSM and maintain expression until segment border formation is complete. This expression pattern provides an important clue to Fss/Tbx24 functions in somite boundary formation. It is thought that PSM cells, born in an immature state in the tailbud, mature during the process of segmentation and become competent to segment as they pass from the intermediate to the anterior PSM. When they reach the anterior PSM, cells activate segmentation genes and stabilize the expression of oscillating genes, at which point segmentation occurs. The fact that both of these events fail to occur in *fss* mutant embryos and that *fss/tbx24* is expressed in maturing cells located in the anterior and intermediate PSM supports the idea that Fss/Tbx24 is an essential factor in the maturation process leading to a segmentation-competent state. Thus, the phenotype of *fss/tbx24* homozygous embryos could be explained as a result of a defect in the maturation process.

T-box family genes have been implicated in development of the paraxial mesoderm. Especially Mouse Tbx6 protein is thought to play a crucial role in a cell fate decision between the paraxial mesoderm and neuroectoderm. In *Tbx6*-mutant mice, because of a biased cell fate decision toward the neural, additional neural tubes are formed at the expense of somite formation.[52] Thus far, no mammalian counterpart of *fss/tbx24* can be found in the databases (Y. Saga, personal communication), suggesting that *fss/tbx24* evolved uniquely in fish lineage. Indeed, Fugu ahd medaka have the *fss/tb24* gene (HT, personal communication). Furthermore, from the following observations, it is tempting to speculate that, due to subfunctionalization of duplicated genes during fish evolution, the function of Tbx24 takes over a part of Tbx6 function. First, the T-domain of Fss/Tbx24 is related to that of Tbx6, especially mouse TBX6 proteins, although the identity is not so high. Second, the expression patterns of zebrafish *tbx6* and *fss/tbx24* are likely to recapitulate the overall expression of mouse *tbx6*: mouse *tbx6* is broadly expressed throughout the PSM (from the tailbud to anterior PSM),[53] while zebrafish *tbx6* expression is confined to the tailbud region but instead, *fss/tbx24* expression is detectable in the intermediate to anterior PSM.[51,54] Finally, a recent work using a weak allele of *Tbx6* mutation has revealed that in addition to its role in the formation of paraxial mesoderm, mouse Tbx6 acts on rostrocaudal specification of the somites.[53] Partial restoration of Tbx6 expression in null mutants rescues development of the paraxial mesoderm development and somites, but the resulting somites lose the anterior characters leading to the fusion of rib and vertebral fusions. This is reminiscent of the fact that in *fss* mutants, expression of most of the rostral-specific genes (e.g. *mesp* and *ephA4*) are lost and thus somitic cells appear to be posterior in nature.[17]

5.2. Positioning the Wavefront in the Anterior PSM

The clock and wavefront model predicts the presence of positional information as well as molecular oscillator in the PSM. The positional information is required for restricting wavefront to the anterior PSM. In fact, somite boundaries are formed one by one in the anterior to posterior direction and the maturation process must be tightly regulated in a way that the wavefront is always placed in a specific region of the rostral PSM and moves caudally as development proceeds. What determines the position of the wavefront or the point of the transition from an immature to mature state in the PSM? Studies performed in both the chick and zebrafish, address this question by showing that Fgf signaling, especially mediated by Fgf8, provides positional information along the rostral-caudal axis of the PSM.[55,56]

Fgf8 is expressed in a graded fashion in the chick PSM with the high end of the gradient at the caudal end. Similarly in zebrafish, the activation level of Fgf signaling is high in the caudal PSM and drops between the intermediate and the rostral PSM, as indicated by the phosphorylation of MAPK, which is one of the major downstream targets of Fgf signaling. The activation pattern of MAPK closely resembles the expression pattern of zebrafish *fgf8* (Fig. 2A). The pattern of *fgf8* expression and activated MAPK suggests a role for Fgf signaling in the maturation of the PSM. Indeed, FGF8, when mis-expressed in the entire PSM of chick embryos, up-regulates a caudal marker, *Brachyury* (also known as an early mesodermal marker), in the rostral

PSM and suppresses segmentation indicating that Fgf8 maintains the posterior identity (immature state) of the PSM. More interestingly, a transient manipulation of Fgf signaling in chick and zebrafish embryos alters the size of the somites (Fig. 5C): transient inhibition of Fgf signaling results in the formation of larger somites, whereas transient activation results in smaller somites. Detailed analyses of gene expression in manipulated wild-type and mutant embryos revealed that Fgf/MAPK signaling regulates the position of the wavefront within the PSM. Suppression of Fgf signaling by a chemical inhibitor of Fgf-receptor mediated signaling (SU5401) caudally shifts the wavefront: this causes *her1* expression to be prematurely terminated and become Fss-dependent in the intermediate PSM instead of the rostral PSM (Fig. 5D). Accordingly, the expression of segmentation genes such as *mesp* and *papc* is induced in the intermediate PSM leading to a posterior shift in segment border formation and larger somites. These results are complementary to those obtained with transplantation of Fgf beads, strengthening the idea that an Fgf signal determines the position of segment border formation by negatively regulating the wavefront and the maturation of the PSM (Fig. 5D). Importantly, as development proceeds, the Fgf activation domain gradually moves caudally at a constant speed, suggesting that Fgf signaling functions as a constant source of positional information within the PSM. Fgf signaling appears to function independently of Notch and Fss activity because the pattern of *fgf8* expression and MAPK activation remains unchanged in these mutants. Furthermore, as in wild-type embryos, inhibition of Fgf signaling causes a posterior shift in the persistent anterior *her1* expression domain in *aei/deltaD* mutants.[56] Therefore, it is reasonable to assume that Fgf signaling directly controls the transition from immature to mature state of PSM cells, and, thereby, determines the position of the wavefront.

Another intriguing result obtained with the chemical inhibitor is that four to five somites are normally formed after the treatment even though the level of Fgf signal drops immediately after addition of the inhibitor. This indicates that the positioning of furrow formation is already specified or made Fgf-insensitive at least at the position of –IV to –V in the

PSM. Interestingly, the Fgf-sensitive region approximately corresponds to the *heat-shock* sensitive zone in zebrafish: the initial defects in the segmental pattern after *heat shock* are observed five somites posterior to the forming somite at the time of heat shock.[57] It was shown that position –IV to –V represents a position at which the level of Fgf/MAPK activation drops below a threshold allowing the cells to mature. Similarly, grafting experiments in chick revealed that the rostral-caudal polarity within the somite is irreversibly determined in the PSM around the level of somite –IV.[55]

6. Establishment of Rostrocaudal Polarity

Establishment of rostrocaudal subdivisions within each somite primordia is thought to precede somite boundary formation. It has been shown in chick by reversing the grafts of anterior PSM that rostrocaudal polarity of the somite is established in the PSM and is maintained independently of its orientation with respect to the environment.[58] Additional grafting experiments suggest that somite borders form only when rostral and caudal somite compartments are juxtaposed to each other.[59]

6.1. Gene Network Centered on *Mesp* Gene

The generation of the rostrocaudal polarity is also thought to be controlled by the molecular clock. However, in zebrafish, defects in the rostrocaudal polarity are often not distinguished from defects in the molecular clock function, because most of Notch pathway mutants in zebrafish exhibit similar phenotypes. For example, zebrafish *aei*, *des*, and *bea* mutant embryos commonly show a salt-and-pepper (randomized) expression pattern of the rostral- or caudal-half marker genes, instead of normal regular stripes.[42,44] This phenotype is virtually indistinguishable from the phenotype seen in the *her1*- and *her7*-morpholino-injected embryo, which shows disruption of cyclic gene expression.[40] Thus, there is no available Notch pathway mutant in zebrafish that enables further analysis of the mechanism of rostrocaudal patterning separately from the molecular clock. However, the expression patterns of several genes such as *deltaC, notch5* and *mesp* reveals rostrocaudal subdivisions prior to boundary formation and maintained in segmented somites. A bHLH transcription factor, *mesp-b* (zebrafish homologue of mouse Mesp-2) is shown to play a crucial role in this process. Mesp-b is expressed in the anterior parts of somite primordia (Fig. 2A). Ectopic expression of Mesp-b in embryos causes a loss of the posterior identity within the somite primordium, leading to a defect in segment border formation. These injected embryos show a reduction in expression of the posterior genes, *myoD* and *notch5*, with uniform expression in the anterior genes, *FGFR1, papc* and *notch6*. Thus, Mesp-b act upstream of the Notch pathway to confer the anterior identity to the presumptive somites, by regulating the essential signaling pathways mediated by Notch-Delta and FGFR.[39]

In addition to Mesp2, Notch pathway mutants in mouse exhibit various patterns of phenotypes regarding the rostrocaudal polarity of somites. For example, in *Delta-like 1* (*Dll1*)- and *RBPjk*-null embryos, somites show neither rostral nor caudal property,[59] whereas *Delta-like3* (*Dll3*), *lfng* and *Hes7*-null embryos show a salt-and-pepper expression pattern of caudal marker genes.[32,36,37,60,61] Among them, *Mesp2*-null and *Presenilin1* (*Psen1*)-null embryos show opposite phenotypes with respect to the rostrocaudal polarity of somites.[62] The *Mesp2*-null embryo exhibits caudalized somites, i.e., the somite loses the rostral-half property, and the whole somite acquires the caudal-half characteristics. The reverse is true for the *Psen1*-null embryo. The rostrocaudal polarity of somites well correlates with the spatial pattern and the level of expression of the Notch ligand *Dll1*. Genetic analyses of *Mesp2*-null, and *Psen1*-null mice, and mice carrying an activated Notch1 in the *Mesp2* locus have led to a model for rostrocaudal patterning, in which two Notch pathways can be active in the anterior PSM.[62] One is the Psen1-dependent Notch pathway for inducing expression of *Dll1*, and the other is the Psen1-independent Notch pathway for suppressing expression of *Dll1* (Fig. 6A). In mouse and zebrafish, initial expression of *mesp* genes are observed over the length of one somite and become localized to the rostral half, which is crucial for creating rostrocaudal polarity in somite primordia. Mesp2 normally suppresses the *Dll1*-inducing pathway and potentiates the *Dll1*-suppressing pathway in a region corresponding to one presumptive somite. When *Mesp2* expression becomes restricted to the presumptive rostral half, expression of *Dll1* is induced in the presumptive caudal half by the Psen1-dependent Notch pathway. In both zebrafish and mouse embryos, at least two Notch ligands (*deltaC* and *deltaD*, and *Dll1* and *Dll3*, respectively) are co-expressed in the PSM, and their expression domains are finally segregated into the rostral or caudal half of formed somites.[63–65] These expression patterns imply that these ligands do not have merely redundant functions, but also have distinct roles in somite patterning and boundary formation. Further genetic analyses revealed that Dll1- and Dll3-Notch signaling and Mesp2 constitute a complex signaling network for stripe formation in the anterior PSM.[66] Feedback loops of Dll1 and Mesp2 are essential for establishment of the rostrocaudal polarity, while Dll3 is necessary for localization and integration of expression of *Dll1* and *Mesp2*. In addition, Dll3-Notch signaling is shown to counteract Psen1-dependent Dll1-Notch signaling (Fig. 6B).

Another important factor that regulates rostrocaudal polarity is Foxc1a, a winged helix transcription factor. In *foxc1a*-morpholino injected embryos, somite segmentation is severely disturbed.[67] Detailed expression analysis further revealed that knock-down of Foxc1a function does not affect the oscillating expression of *her1* and *delataC* but that it abolishes or reduce segmental expression of genes normally transcribed in either caudal or rostra half of the segments, such as *mesp-b*, *ephrinB2*, *ephHA4*, *notch5* and *notch6*. Similar results were obtained in mutant mice that are compound null mutants for Foxc1 and the closely related Foxc2,[68] indicating an essential and conserved function of Foxc1 family gene in rostrocaudal patterning of the somites.

6.2. Rostrocaudal Patterning and Clock Mechanism

As discussed above, expression of some genes are considered to reflect the molecular clock, such as chick *c-hairy1*, oscillates as a "traveling wave" from the posterior PSM, stabilizes at the anterior PSM and finally forms a half-a-somite stripe retained in somites.[6] Therefore, the rostrocaudal patterning, i.e. formation of half-a-somite stripe pattern

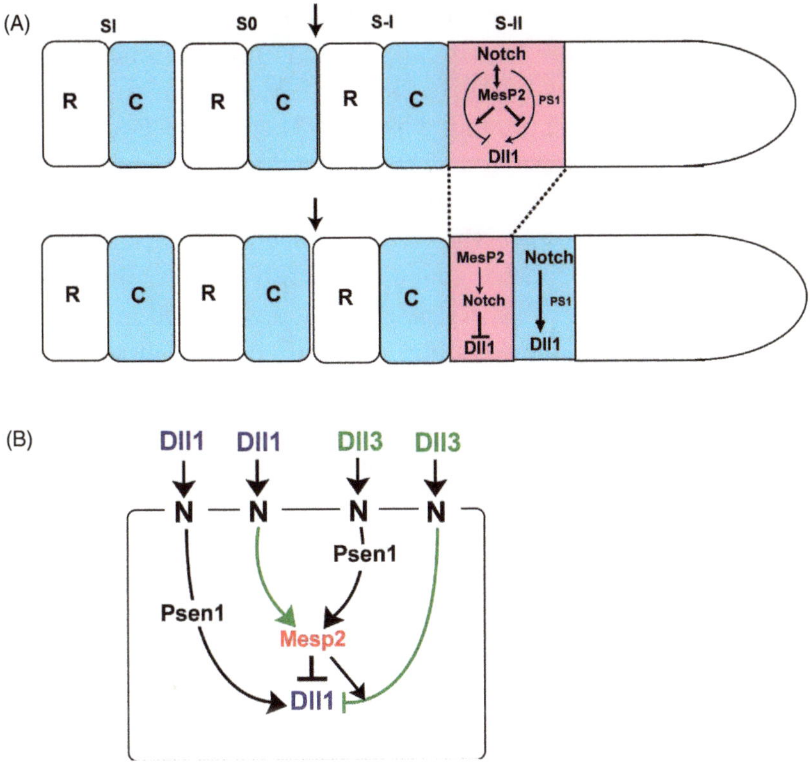

Fig. 6. Genetic cascade leading to the establishment of rostrocaudal polarity of mouse somites (A) Establishing rostrocaudal polarity in a somite primordium. *Dll1* expression is regulated through two Notch signaling pathways. When *Mesp2* is initially expressed both in prospective rostral and caudal regions at S-II, Mesp2 suppresses *Dll1* in the entire somite primordium by suppressing the Psen1-dependent Notch pathway and by activating the Psen1-independent Notch signaling pathway. When *Mesp2* expression becomes localized to the presumptive rostral half of the somite after 40–60 min, *Dll1* expression is induced in the caudal half via Psen1. Pink shading indicates *Mesp2* expression, while blue indicates *Dll1* expression. Vertical arrow indicates the position of new segmental border. (B) A putative feedback signaling cascade in the anterior PSM. Dll1-Notch signaling results in induction of both *Dll1* itself and *Mesp2*. The positive feedback of Dll1 is mediated by the Psen1-dependent signal. Induction of *Mesp2* is mediated via Psen1-independent Dll1-Notch signaling and Psen1-dependent Dll3-Notch signaling. Mesp2 negatively regulates *Dll1* expression. In contrast to Dll1, Dll3 has roles in up-regulation of *Mesp2* and suppression of *Dll1*. The Psen1-independent pathways are shown with green arrows.

of gene expression, has been regarded as a result of stabilization of oscillating expression in the posterior PSM. However, none of the half-a-somite stripe patterns of Dll1, Dll3 and Mesp2 are formed in the absence of Mesp2 function and at present, there is no evidence of the half-a-somite stripe prepattern upstream of Mesp2 in mouse.

There is another example that the oscillation in the posterior PSM seems to be separated from the stripe formation. As described above, Holley *et al.*[44] have reported that in zebrafish embryos injected with *her1*-MO, a normal stripe of *deltaC* expression is formed in the anterior PSM, in the absence of oscillation of *deltaC* or *her1* in the posterior PSM. In this case, the *deltaC* stripe at the anterior PSM is not a result of simple stabilization of oscillating expression in the posterior PSM, but is likely to be formed by another mechanism. This stripe formation also appears to be mediated by Notch signaling, because the additional loss of DeltaD function disrupts stripe formation. In addition, injection of *her1/her7* double-MO completely abolishes stripe formation.[40] Holley *et al.*[44] suggested that Notch signaling acts in oscillator of cyclic gene expression in the posterior PSM as well as in stripe formation (refinement of the stripe) at the anterior PSM. Therefore it could be possible to assume that the narrowing stripe formed at the anterior PSM of mouse embryo, by the positive and negative feedback loops among Dll1, Dll3 and Mesp2 (Fig. 6B). These feedback loops may constitute a kind of cellular oscillator in the anterior PSM, which is distinct from the oscillator in the posterior PSM. This process may be normally linked with the oscillation process in the posterior PSM.

7. Formation of Morphologically Distinct Somite Boundary

The final step in somite formation is the creation of morphologically distinct boundaries. For this, the segregation of PSM cells and mesenchymal-epithelial transition along the forming epithelial boundary could be essential. The Eph intercellular signaling system has been implicated in somitogenesis. The receptor *EphA4* is expressed in the anterior while its ligand, *ephrinB2*, is in the posterior half of the presumptive and developing somites. EphrinB2 is a transmembrane protein capable of transducing a signal into the expressing cell. Boundary formation seems to take place at the site of interaction between EphA4-expressing in the posterior of the forming somite and EphringB2-expressing cells in the anterior of the presumptive somite. Interruption of Eph signaling leads to abnormal somite boundary formation in zebrafish.[69] Given that Eph receptors and Ephrins are thought to function intercellulary as repulsive factors by influencing cytoskeletal architecture and cell adhesion,[70] the presence of EphrinB2 on one side of the boundary and EphA4 in cells on the other could mediate changes in cell shape and adhesion required for furrow formation. In zebrafish *fss* mutants, the expression of EphA4 is lost, while ephrin-B2 is expressed throughout this region. Surprisingly, transplanting of cells ectopically expressing EphA4 into the paraxial mesoderm of *fss* mutant rescued boundary formation, indicating that the Eph system is sufficient to cause morphological boundary formation in the paraxial mesoderm.[71] Interestingly, ectopic boundaries were sometimes only visible on one side of a group of transplanted Eph4-expressing cells, suggesting that the cells may become polarized (a sign of epithelialization) during boundary formation.

Cadherin, one of the major cell adhesion molecules, has been implicated in somite boundary formation in amniotes. Cadherins cluster on the cell surface and bind to cadherins on adjacent cells through a Ca^{2+}-dependent homotypic interaction. N-cadherin, which is the primary cadherin associated with somitogenesis, is expressed in the anterior presomitic mesoderm and segmented somites. Inhibition of N-cadherin with an antibody or by genetic inactivation disrupts the formation of epithelial somites.[72–74] Interestingly, N-cadherin deficient somite tends to be cleaved into the anterior and posterior halves, and the cleaved halves maintains the cluster state with epithelial morphlogy.[75] Thus, one role of N-cadherin is to connect the two halves into a single unit, a somite. Recently, zebrafish *parachute* (*pac*) mutant was found to carry a potential null mutation in zebrafish N-cadherin homologue (*pac/ncad*).[76,77] In spite of defects in neural tube and eye development, no gross somite defects were apparent in *pac* mutant, probably due to a gene redundancy, and thus, a role of N-cadherin in zebrafish somitogenesis remains unclear. Recent work has also identified the protocadherin, *papc*, another member of the cadherin superfamily as an important regulator of somitogenesis in *Xenopus*, zebrafish and mouse. *papc* is expressed in the anterior of the presumptive somites and its inactivation leads to the abnormal formation of the somites.[78,79] Although mice deficient for *papc* do not display no clear segmentation defect, *in vitro* analysis with a soluble form of Papc that could act in a dominant-negative fashion demonstrated that Papc is an important regulator of somite epithelialization at the intersomitic border.[80]

Finally, the bHLH gene, *paraxis* may also be involved in the final step of somite formation. *paraxis* is expressed in the PSM and segmented somites, whose expression pattern is largely conserved in mouse, chicken and zebrafish (Fig. 2A).[81] A targeted null mutation of *paraxis* caused a deficiency in somite epithelialization.[82] Interestingly, somitic cells in the mutant segment at a regular interval, just like in wild-type embryos, but there is no sign of epithelialization. This indicates that the two processes, segmentation and epithelialization, can be genetically separated. Intriguingly, unlike other segmentation key genes, *paraxis* expression in zebrafish is independent of Fss activity, but depends on Foxc1a.[67] Thought loss-of-function analysis is yet to be done, it is reasonable to speculate that the same might be true for the function of zebrafish *paraxis*.

8. Unanswered Questions

Accumulating experimental and genetic data lead us to conclude that the molecular clock and wavefront activity establish the periodic pattern of the PSM. However, there are many questions yet to be answered.

Although a number of components of the segmentation clock have been isolated, their detailed interactions within the oscillator still remain unclear. Recently, Hirata *et al.*[83] have reported that serum treatment of various cultured mouse cell lines induces cyclic expression of both mRNA and protein of Hes1, a repressor of Notch signaling, with a periodicity (2 hours) similar to that seen in mouse segmentation. While we do not know how much this *in vitro* system reflects the segmentation clock working in the PSM, this *in vitro* system could greatly facilitate the analysis of complicated interaction among clock components. Furthermore, a novel link between Wnt signaling and the segmentation clock has recently been established.[84] Axin2, a negative regulator of the Wnt pathway, shows oscillating expression in the PSM, even when Notch signaling is impaired. Moreover, Wnt3a is required for oscillating Notch signaling activity in the PSM. These results suggest that Wnt signaling functions upstream of the Notch pathway (Fig. 4A). The interaction between the two signaling system awaits the further biochemical studies.

The interaction between the oscillation wave and wavefront in the anterior PSM should be analyzed molecularly. The formation of the wavefront is thought to depend on Fss/Tbx24, and one of the key molecules involved in wavefront activity is likely to be Mesp2. Mesp acts both downstream and upstream of the Notch pathway to establish rostrocaudal subdivisions within the somite primordium. As described above, Fss/Tbx24 regulates expression of *mesp* and *ephA4*, a possible key molecule in epithelial boundary formation, while Foxca1, but not Fss/Tbx24, regulates the expression of *paraxis*, which is implicated in epithelialization of somitic cells. Therefore, a complicated genetic network is working in the anterior PSM. Positional information in the PSM also should interact with this network. In mouse, Wnt3a was proposed to act upstream of or together with Fgf8 in the PSM.[84] However, this is yet to be confirmed in other vertebrates. At least in zebrafish *wnt5* mutant (*pipetail*), no obvious segmentation defect has been reported.[85] Ultimately, determining how the PSM cells mature and how this leads to segmentation and establishment of segment polarity will require integrating the function of many downstream genes, such as Notch, ephA4, Mesp and foxc1a, into a complex succession of positive and negative feedback loops.

Acknowledgment

We thank Mr. Akihito Omori and Miss. Kana Ishimatsu for providing us pictures of Fig. 3C, and Dr. Kazuki Horikawa for Fig. 1C. We also thank Drs. Y. Bessho and R. Kageyama for our citing their exciting data before publication. Work in author's laboratories is funded by the Organized Research Combination System Grant and Grant-in-Aids of the Ministry of Education, Culture, Sports, Science and Technology Japan.

References

1. Christ B and Ordahl CP (1995). Early stages of chick somite development. *Anat. Embryol.* **191**: 381–396.
2. Eisen JS and Pike SH (1991). The spt-1 mutation alters the segmental arrangement and axonal development of identified neurons in the spinal cord of the embryonic zebrafish. *Neuron* **6**: 767–776.
3. Keynes R and Stern C (1988). Mechanisms of vertebrate segmentation. *Development* **103**: 413–429.
4. Meinhardt H (1982). *Models of Biological Pattern Formation*, Academic Press.
5. Meinhardt H (1986). Models of Segmentation. In: Bellairs R, Ede DA and Lash JW (eds.), *Somites in Developing Embryos.* Plenum Press.
6. Palmeirim I, Henrique D, Ish Horowicz D and Pourquié O (1997). Avian hairy gene expression identifies a molecular clock linked to vertebrate segmentation and somitogenesis. *Cell* **91**: 639–648.
7. Tam PPL and Trainor PA (1994). Specification and segmentation of the paraxial mesoderm. *Anat. Embryol.* **189**: 275–305.
8. Pourquié O (2000). Segmentation of the paraxial mesoderm and vertebrate somitogenesis. *Curr. Top. Dev. Biol.* **47**: 81–105.
9. Rawls A, Wilson–Rawls J and Olson EN (2000). Genetic regulation of somite formation. *Curr. Top. Dev. Bio.* **47**: 131–154.
10. Currie PD and Ingham PW (1998). The generation and interpretation of positional information within the vertebrate myotome. *Mech. Dev.* **73**: 3–21.
11. Saga Y and Takeda H (2001). The making of the somite: molecular events in vertebrate segmentation. *Nat. Rev. Genet.* **2**: 835–845.
12. Brennan C, Ammarcher SL and Currie PD (2002). Somitogenesis. In: Solnica–Krezel L (ed.), *Pattern Formation in Zebrafish. Results and Problems in Cell Differentiation,* vol. 40. Springer-Verlag, Berlin, Heidelberg, pp. 271–297.
13. Kimmel CB, Ballard WM, R. KS, Ullmann B and Schilling TF (1995). Stages of Embryonic Development of the Zebrafish. *Dev. Dynam.* **203**: 253–310.

14. Henry CA, Hall LA, Burr Hille M, Solnica–Krezel L and Cooper MS (2000). Somites in zebrafish doubly mutant for knypek and trilobite form without internal mesenchymal cells or compaction. *Curr. Biol.* **10**: 1063–1066.
15. Stickney HL, Barresi MJ and Devoto SH (2000). Somite development in zebrafish. *Dev. Dyn.* **219**: 287–303.
16. van Eeden FJM, Granato M, Schach U, Brand M, Furutani–Seiki M, Haffter P, Hammerschmidt M, Heisenberg C-P, Jiang Y-J, Kane DA, Kelsh RN, Mullins MC, Odenthal J, Warga RM, Allende ML, Weinberg ES and Nüsslein–Volhard C (1996). Mutations affecting somite formation and patterning in the zebrafish *Danio rerio*. *Development* **123**: 153–164.
17. van Eeden FJM, Holley SA, Haffter P, Campos–Ortega J and Nüsslein–Volhard C (1998). Zebrafish segmentation and pair-rule patterning. *Dev. Genet.* **23**: 65–76.
18. Holley SA and Takeda H (2002). Catching a wave: the oscillator and wavefront that create the zebrafish somite. *Seminars Cell Dev. Biol.* **13**: 481–488.
19. Jen W-C, Wettstein D, Turner D, Chitnis A and Kintner C (1997). The Notch ligand, X-Delta-2, mediates segmentation of the paraxial mesoderm in Xenopus embryos. *Development* **124**: 1169–1178.
20. Jen WC, Gawantka V, Pollet N, Niehrs C and Kintner C (1999). Periodic repression of Notch pathway genes governs the segmentation of Xenopus embryos. *Genes Dev.* **13**: 1486–1499.
21. Oka C, Nakano T, Wakeham A, de la Pompa JL, Mori C, Sakai T, Okazaki S, Kawaichi M, Shiota K, Mak TW and Honjo T (1995). Disruption of the mouse RBP-J Kappa results in early embryonic death. *Development* **121**: 3291–3301.
22. Conlon RA, Reaume AG and Rossant J (1995). Notch1 is required for the coordinate segmentation of somites. *Development* **121**: 1533–1545.
23. Hrabé Angelis M, McIntyre J and Gossler A (1997). Maintenance of somite borders in mice requires the Delta homologue Dll1. *Nature* **386**: 717–721.
24. Dornseifer P, Takke C and Campos–Ortega JA (1997). Overexpression of a zebrafish homologue of the Drosophila neurogenic gene Delta perturbs differentiation of primary neurons and somite development. *Mech. Dev.* **63**: 159–171.
25. Greenwald I (1998). LIN-12/Notch signaling: lessons from worms and flies. *Genes Dev.* **12**: 1751–1762.

26. Artavanis–Tsakonas S, Rand MD and Lake RJ (1999). Notch signaling: Cell fate control and signal integration in development. *Science* **284**: 770–776.
27. Bierkamp C and Campos–Ortega JA (1993). A zebrafish homologue of the Drosophila neurogenic gene Notch and its pattern of transcription during early embryogenesis. *Mech. Dev.* **43**: 87–100.
28. Cooke J and Zeeman EC (1976). A clock and wavefront model for control of the number of repeated structures during animal morphogenesis. *J. Theor. Biol.* **58**: 455–476.
29. Dale KJ and Pourquie O (2000). A clock-work somite. *Bioessays* **22**: 72–83.
30. Jouve C, Palmeirim I, Henrique D, Beckers J, Gossler A, Ish Horowicz D and Pourquié O (2000). Notch signalling is required for cyclic expression of the hairy-like gene HES1 in the presomitic mesoderm. *Development* **127**: 1421–1429.
31. Leimeister C, Dale K, Fischer A, Klamt B, Hrabe de Angelis M, Radtke F, McGrew MJ, Pourquie O and Gessler M (2000). Oscillating expression of c-Hey2 in the presomitic mesoderm suggests that the segmentation clock may use combinatorial signaling through multiple interacting bHLH factors. *Dev. Biol.* **227**: 91–103.
32. Bessho Y, Sakata R, Komatsu S, Shiota K, Yamada S and Kageyama R (2001). Dynamic expression and essential functions of Hes7 in somite segmentation. *Genes Dev.* **15**: 2642–2647.
33. McGrew MJ, Dale JK, Fraboulet S and Pourquié O (1998). The Lunatic Fringe gene is a target of the molecular clock linked to segmentation in avian embryos. *Curr. Biol.* **8**: 979–982.
34. Forsberg H, Crozet F and Brown NA (1998). Waves of mouse Lunatic fringe expression, in four-hour cycles at two-hour intervals, precede somite boundary formation. *Curr. Biol.* **8**: 1027–1030.
35. Aulehla A and Johnson RL (1999). Dynamic expression of lunatic fringe suggests a link between notch signaling and an automomous cellular oscillator driving somite segmentation. *Dev. Biol.* **207**: 49–61.
36. Zhang N and Gridley T (1998). Defects in somite formation in lunatic fringe-deficient mice. *Nature* **394**: 374–377.
37. Evrard YA, Lun Y, Aulehla A, Gan L and Johnson RL (1998). Lunatic fringe is an essential mediator of somite segmentation and patterning. *Nature* **394**: 377–381.

38. Holley SA, Geisler R and Nüsslein–Volhard C (2000). Control of her1 expression during zebrafish somitogenesis by a Delta-dependent oscillator and an independent wave-front activity. *Genes Dev.* **14**: 1678–1690.
39. Sawada A, Fritz A, Jiang Y, Yamamoto A, Yamasu K, Kuroiwa A, Saga Y and Takeda H (2000). Zebrafish Mesp family genes, mesp a and mesp b are segmentally expressed in the presomitic mesoderm, Mesp b confers the anterior identity to the developing somites. *Development* **127**: 1691–1702.
40. Oates AC and Ho RK (2002). Hairy/E(spl)-related (Her) genes are central components of the segmentation oscillator and display redundancy with the Delta/Notch signaling pathway in the formation of anterior segmental boundaries in the zebrafish. *Development* **129**: 2929–2946.
41. Prince VE, Holley SA, Bally–Cuif L, Prabhakaran B, Oates AC, Ho RK and Vogt TF (2001). Zebrafish lunatic fringe demarcates segmental boundaries. *Mech. Dev.* **105**: 175–180.
42. Jiang YJ, Aerne BL, Smithers L, Haddon C, Ish Horowicz D and Lewis J (2000). Notch signaling and the synchronization of the somite segmentation clock. *Nature* **408**: 475–479.
43. Jiang Y-J, Brand M, Heisenberg C-P, Beuchle D, Furutani–Seiki M, Kelsh RN, Warga RM, Granato M, Haffter P, Hammerschmidt M, Kane DA, Mullins MC, Odenthal J, van Eeden FJM and Nüsslein–Volhard C (1996). Mutations affecting neurogenesis and brain morphology in the zebrafish, *Danio rerio. Development* **123**: 205–216.
44. Holley SA, Julich D, Rauch GJ, Geisler R and Nüsslein–Volhard C (2002). her1 and the notch pathway function within the oscillator mechanism that regulates zebrafish somitogenesis. *Development* **129**: 1175–1183.
45. Itoh M, Kim CH, Palardy G, Oda T, Jiang Y-J, Maust D, Yeo SY, Lorick K, Wright GJ, Ariza–McNaughton L, Weissman AM, Lewis J, Chandrasekharappa SC and Chitnis AB (2003). Mind bomb is a ubiquitin ligase that is essential for efficient activation of notch signaling by delta. *Dev. Cell* **4**: 67–82.
46. Takke C and Campos–Ortega JA (1999). her1, a zebrafish pair-rule gene, acts downstream of notch signaling to control somite development. *Development* **126**: 3005–3014.
47. Dale KJ, Maroto M, Dequeant M-L, Malapert P, McGrew M and Pourquié O (2003). Periodic Notch inhibition by lunatic fringe underlies the chick segmentation clock. *Nature* **421**: 275–278.

48. Morales AV, Yasuda Y and Ish–Horowicz D (2002). Periodic Lunatic fringe expression is controlled during segmentation by a cyclic transcriptional enhancer responsive to notch signaling. *Dev. Cell* **3**: 63–74.
49. Cole SE, Levorse JM, Tilghman SM and Vogt TF (2002). Clock regulatory elements control cyclic expression of Lunatic fringe during somitogenesis. *Dev. Cell* **3**: 75–84.
50. Bessho Y, Hirata H, Masamizu Y and Kageyama R (2003). Periodic repression by the bHLH factor Hes7 is an essential mechanism for the somite segmentation clock. *Gene Dev.* **17**: 1451–1456.
51. Nikaido M, Kawakami A, Sawada A, Furutani–Seiki M, Takeda H and Araki K (2002). Tbx24, encoding a T-box protein, is mutated in the zebrafish somite-segmentation mutant fused somites. *Nat. Genet.* **31**: 195–199.
52. Chapman DL and Papaioannou VE (1998). Three neural tubes in mouse embryos with mutations in the T-box gene Tbx6. *Nature* **391**: 695–697.
53. White PH, Farkas DR, McFadden EE and Chapman DL (2003). Defective somite patterning in mouse embryos with reduced levels of Tbx6. *Development* **130**: 1681–1690.
54. Hug B, Walter V and Grunwald DJ (1997). tbx6, a Brachyury-related gene expressed by ventral mesendodermal precursors in the zebrafish embryo. *Dev. Biol.* **183**: 61–73.
55. Dubrulle J, McGrew MJ and Pourquie O (2001). FGF signaling controls somite boundary position and regulates segmentation clock control of spatiotemporal Hox gene activation. *Cell* **106**: 219–232.
56. Sawada A, Shinya M, Jiang YJ, Kawakami A, Kuroiwa A and Takeda H (2001). Fgf/MAPK signalling is a crucial positional cue in somite boundary formation. *Development* **128**: 4873–4880.
57. Roy MN, Prince VE and Ho RK (1999). Heat shock produces periodic somitic disturbances in the zebrafish embryo. *Mech. Dev.* **85**: 27–34.
58. Aoyama H and Asamoto K (1988). Determination of somite cells: independence of cell differentiation and morphogenesis. *Development* **104**: 15–28.
59. del Barco Barrantes I, Elia A, Wunnch K, Hrabe De Angelis M, Mak T, Rossant J, Conlon R, Gossler A and Luis de la Pompa J (1999). Interaction between Notch signaling and Lunatic Fringe during somite boundary formation in the mouse. *Curr. Biol.* **9**: 470–480.
59. Stern CD and Keynes RJ (1987). Interactions between somite cells: the formation and maintenance of segment boundaries in the chick embryo. *Development* **99**: 261–272.

60. Kusumi K, Sun ES, Kerrebrock AW, Bronson RT, Chi DC, Bulotsky MS, Spencer JB, Birren BW, Frankel WN and Lander ES (1998). The mouse pudgy mutation disrupts Delta homologue Dll3 and initiation of early somite boundaries. *Nat. Genet.* **19**: 274–278.
61. Dunwoodie SL, Clements M, Sparrow DB, Sa X, Conlon RA and Beddington RS (2002). Axial skeletal defects caused by mutation in the spondylocostal dysplasia/pudgy gene Dll3 are associated with disruption of the segmentation clock within the presomitic mesoderm. *Development* **129**: 1795–1806.
62. Takahashi Y, Koizumi K, Takagi A, Kitajima, Inoue T, Koseki H and Saga Y (2000). Mesp2 initiates somite segmentation through the Notch signalling pathway. *Nat. Genet.* **25**: 390–396.
63. Bettenhausen B, Hrabe de Angelis M, Simon D, Guenet JL and Gossler A (1995). Transient and restricted expression during mouse embryogenesis of Dll1, a murine gene closely related to Drosophila Delta. *Development* **121**: 2407–2418.
64. Dunwoodie SL, Henrique D, Harrison SM and Beddington RS (1997). Mouse Dll3: a novel divergent Delta gene which may complement the function of other Delta homologues during early pattern formation in the mouse embryo. *Development* **124**: 3065–3076.
65. Haddon C, Smithers L, Schneider-Maunoury S, Coche T, Henrique D and Lewis J (1998). Multiple delta genes and lateral inhibition in zebrafish primary neurogenesis. *Development* **125**: 359–370.
66. Takahashi Y, Inoue T, Gossler A and Saga Y (2003). Feedback loops comprising Dll1, Dll3 and Mesp2, and differential involvement of Psen1 are essential for rostrocaudal patterning of somites. *Development*, in press.
67. Topczewska JM, J. T, Shostak A, Kume T, Solnica-Krezel L and Hogan BL (2001). The winged helix transcription factor Foxc1a is essential for somitogenesis in zebrafish. *Genes Dev.* **15**: 2483–2493.
68. Kume T, Jiang H, Topczewska JM and Hogan BL (2001). The murine winged helix transcription factors, Foxc1 and Foxc2, are both required for cardiovascular development and somitogenesis. *Genes Dev.* **15**: 2470–2482.
69. Durbin L, Brennan C, Shiomi K, Cooke J, Barrios A, Shanmugalingam S, Guthrie B, Lindberg R and Holder N (1998). Eph signaling is required for segmentation and differentiation of the somites. *Genes Dev.* **12**: 3096–3109.
70. Wilkinson DG (2001). Multiple roles of EPH receptors and ephrins in neural development. *Nat. Rev. Neurosci.* **2**: 155–164.

71. Durbin L, Sordino P, Barrios A, Gering M, Thisse C, Thisse B, Brennan C, Green A, Wilson S and Holder N (2000). Anteroposterior patterning is required within segments for somite boundary formation in developing zebrafish. *Development* **127**: 1703–1713.
72. Duband JL, Dufour S, Hatta K, Takeichi M, Edelman GM and Thiery JP (1987). Adhesion molecules during somitogenesis in the avian embryo. *J. Cell Biol.* **104**: 1361–1374.
73. Linask KK, Ludwig C, Han MD, Liu X, Radice GL and Knudsen KA (1998). N-cadherin/catenin-mediated morphoregulation of somite formation. *Dev. Biol.* **202**: 85–102.
74. Radice GL, Rayburn H, Matsunami H, Knudsen KA, Takeichi M and Hynes RO (1997). Developmental defects in mouse embryos lacking N-cadherin. *Dev. Biol.* **181**: 64–78.
75. Horikawa K, Radice G, Takeichi M, Chisaka O. (1999). Adhesive subdivisions intrinsic to the epithelial somites. *Dev. Biol.* **215**: 182–189.
76. Lele Z, Folchert A, Concha M, Rauch GJ, Geisler R, Rosa F, Wilson SW, Hammerschmidt M and Bally–Cuif L (2002). Parachute/n-cadherin is required for morphogenesis and maintained integrity of the zebrafish neural tube. *Development* **129**: 3281–3294.
77. Masai I, Lele Z, Yamaguchi M, Komori A, Nakata A, Nishiwaki Y, Wada H, Tanaka H, Nojima Y, Hammerschmidt M, Wilson SW and Okamoto H (2003). N-cadherin mediates retinal lamination, maintenance of forebrain compartments and patterning of retinal neurites. *Development* **130**: 2479–2494.
78. Yamamoto A, Amacher SL, Kim SH, Geissert D, Kimmel CB and De Robertis EM (1998). Zebrafish paraxial protocadherin is a downstream target of spadetail involved in morphogenesis of gastrula mesoderm. *Development* **125**: 3389–3397.
79. Kim SH, Yamamoto A, Bouwmeester T, Agius E and Robertis EM (1998).The role of paraxial protocadherin in selective adhesion and cell movements of the mesoderm during Xenopus gastrulation. *Development* **125**(23): 4681–4690.
80. Rhee J, Takahashi Y, Saga Y, Wilson–Rawls J, Rawls A (2003). The protocadherin papc is involved in the organization of the epithelium along the segmental border during mouse somitogenesis. *Dev. Biol.* **25**: 248–261.
81. Shanmugalingam S and Wilson SW (1998). Isolation, expression and regulation of a zebrafish paraxis homologue. *Mech. Dev.* **78**: 85–89.

82. Burgess R, Rawls A, Brown D, Bradley A and Olson EN (1996). Requirement of the paraxis gene for somite formation and musculoskeletal patterning. *Nature* **384**: 570–573.
83. Hirata H, Yoshiura S, Ohtsuka T, Bessho Y, Harada T, Yoshikawa K, Kageyama R (2002). Oscillatory expression of the bHLH factor Hes1 regulated by a negative feedback loop. *Science* **298**: 840–843.
84. Aulehla A, Whrie C, Brand–Saberi B, Kemler R, Gossler A, Kanzler B and Herrmann BG (2003). Wnt3a plays a major role in the segmentation clock controlling somitogenesis. *Dev. Cell* **4**: 395–406.
85. Rauch GJ, Hammerschmidt M, Blader P, Schauerte HE, Strahle U, Ingham PW, McMahon AP and Haffter P (1997). Wnt5 is required for tail formation in the zebrafish embryo. *Cold Spring Harb. Symp. Quant. Biol.* **62**: 227–234.
86. Pourquie O and Tam PP (2001). A nomenclature for prospective somites and phases of cyclic gene expression in the presomitic mesoderm. *Dev. Cell.* **1**: 619–620.
87. Weinberg ES, Allende ML, Kelly CS, Abdelhamid A, Murakami T, Anderman P, Doerre OG, Grunwald DJ and Riggleman B (1996). Developmental regulation of zebrafish MyoD in wild-type, no tail and spadetail embryos. *Development* **122**: 271–280.

Chapter 9

Vertebrate Somite Development, Notch Signaling and Others

Yun-Jin Jiang

Laboratory of Developmental Signalling and Patterning
Institute of Molecular and Cell Biology
1 Research Link
National University of Singapore
Singapore 117604

The molecular basis of somite development: the periodic generation of somites, rostrocaudal (RC) polarization in formed somites, somite furrow formation and somite differentiation has been substantially explored among different vertebrates for last few decades, enabling us to understand it from a more mechanistic way. The work on chicken *c-hairy1* cycling, mouse knock-outs of Notch components and zebrafish somite mutants has demonstrated a vital role of Notch signaling in somite segmentation. A mechanism involving cyclical activation of transcription and delayed negative feedback regulation is emerging. Fgf8 and Wnt3a gradients are important in positioning somite boundaries and, probably, in coordinating tail growth and segmentation. In addition to segmentation, Notch signaling is also essential for RC polarity and boundary formation in collaboration with a variety of genes, including *Mesp, Eph, ephrin, Protocadherin* (*Papc*), *Foxc* and T-box genes. Zebrafish has played an indispensable role in recent progress. Studies of other species will also be discussed in a comparative and complementary way.

1. Introduction

Vertebrate somites are the most obviously segmental structures in the early vertebrate embryos. They are formed out of the unsegmented posterior mesenchyme, the presomitic mesoderm (PSM) in a head-to-tail sequence and at a regular tempo — 2 hours per somite for mice;

Correspondence to: Laboratory of Developmental Signalling and Patterning, Institute of Molecular and Cell Biology, Proteos Building (Rm 8-01B), 61 Biopolis Drive, Singapore 138673. E-mail: yjjiang@imcb.a-star.edu.sg; Tel: (65) 65869718; Fax: (65) 67791117

90 minutes for chicken; 30 minutes for zebrafish. Eventually, each somite will differentiate and form skeletal muscle, dermis, vertebrae, etc.

In contrast to the syncytial and simultaneous body segmentation of long-germ band insects such as *Drosophila* and discrete rhombomere segregation in vertebrate hindbrain, where growth and segmentation are uncoupled, vertebrates use a cellular and sequential mechanism to build up their somite metameres. Only vertebrates and other chordates have a tail bud. After gastrulation, the vertebrate tail bud continues to produce somites.[1–4] Thus, vertebrate segmentation is theoretically an open-ended system, where growth and segmentation are connected.

In general, the vertebrate somitogenesis begins with the recruitment of prospective mesodermal cells early from the primitive streak (in mouse and chicken); or its equivalents, the marginal zone in amphibians and the germ ring in fish. Later on, cells are recruited from the tail bud to the caudal end of the PSM. The mesenchymal cells are then prepatterned in the PSM and subsequently mature by increasing cell number and packing density and by accumulating extracellular matrix. Somite segmentation is accomplished symmetrically along the midline by the compaction of cells in the anterior-most PSM and the formation of intersomitic furrows. After the mesenchymal-to-epithelial transition, the newly formed somites remain as a partially epithelial structure, where mesenchymal cells are surrounded by epithelial cells, but later they differentiate into a dorsolateral dermomyotome and a ventromedial sclerotome.[5–10] In summary, there are several interrelated processes happening in the PSM and they are independent of subsequent differentiation into skeletal muscle, vertebrae, etc. They consist of the determination of paraxial mesoderm, the generation of a metameric pattern, the genetic specification of compartments within somites, the coordination of somite formation in either sides of the midline and the somite epithelialization.

Several theoretical models have been proposed, attempting to incorporate all different aspects of somitogenesis (reviewed by Keynes and Stern,[5] and Schnell and Maini[11]). The important assumption of some models is a clock or oscillator (segmentation clock) in the PSM, generating a temporal periodicity, which is then translated into the spatial metameric somite pattern. The identification of a chicken *hairy* homolog is a breakthrough for the somite segmentation, which demonstrated the existence of a clock suggested by experimental manipulations long ago.[12]

The evolutionarily conserved Notch pathway is an important cell–cell communication signaling and plays an essential role in the development of many tissues. There are several different ways to use the Notch pathway in cell differentiation (reviewed by Artavanis-Tsakonas *et al.*,[13] and Bray[14]). First, during lateral inhibition Notch activity inhibits Delta expression and neighboring cells are driven to be unlike. Known examples can be found in neurogenesis of insects and vertebrates. Second, during lateral induction Notch activity stimulates Delta expression and neighboring cells are driven to be alike. A known example is boundary formation in the *Drosophila* wing disk (reviewed in Irvine and Vogt[15]). In addition, Notch signaling can be used for purposes other than the control of cell differentiation. It is involved in control of neurite outgrowth[16–18] and, as have recently been shown, somitogenesis (reviewed in Jiang *et al.*,[19] and Pourquié[20]).

This review will focus on the molecular and genetic aspects of somitogenesis, particularly the segmentation clock and somite compartmentation, where Notch signaling is involved. As for the morphological and cellular aspects of somitogenesis, refer to other reviews.[5–10,21–23] For discussions about the clock mechanism and comparison to circadian clock, refer to Bessho and Kageyama,[24] and Rida *et al.*[25]

2. The Way to a Segmentation Clock

For some decades, developmental biologists have been fascinated by the problem of how vertebrate embryos regulate somite number, size and temporal pattern. Many experiments have been done to address this problem. For example, it has been shown in several vertebrates, including amphibian, chicken and mouse, that somite segmentation is an autonomous property of PSM cells, independent of surrounding tissues and PSM continuity.[12,26–30] Based on the results of teratology, embryology and experimental manipulations, several models have been proposed (reviewed by Keynes and Stern[5]).

Here I would like to single out two sets of outstanding experiments, which not only lead to the proposal of two models — the clock and wavefront model, and the cell cycle model — but also await satisfactory molecular explanations in terms of a segmentation clock (see below).

2.1. Heat-Shock Experiment

Heat-shock treatment affects somite formation in amphibians, chicken and zebrafish: the abnormal somites appear after a species-specific time interval and recover gradually afterwards.[31–37] For example, zebrafish embryos exhibit a somite anomaly of 1–2 somites wide in 4–5 somite-forming time after heat shock, which corresponds to 2–3 hours of zebrafish development. One thing is worthy of notice: perturbations of somite formation induced by a short while of heat shock lead to pattern abnormalities that are broader than that expected from the duration of the heat shock itself. This indicated that an irregularity formed in one somite induces another irregularity in the next succeeding somite. This observation suggested that a cell–cell communication mechanism is involved. It was also observed that within each zone of abnormality, the defect was most severe at the anterior border and gradually became less severe near the caudal margin.[32]

These anomalies have been explained by assuming that heat shock alters the action of the wave at a single critical point in its passage, resulting in somite abnormalities a short interval later. The delay between the time of the heat shock and the appearance of the anomalous segments reflects the time interval between the determination of a group of cells to segment together and the somite boundary formation.[31–33]

2.2. Cell Cycle

It has been observed that a certain degree of synchrony in the cell cycle exists in the chicken PSM cells.[38] Moreover, the heat-shock experiment done in chicken and other results (e.g. treatment of antimitotic agents) related to cell cycle favor the idea that there is an oscillatory event, perhaps linked to the cell cycle, which plays a role in gating those cells predetermined to segment together and is susceptible to heat-shock treatment. It was found that in chicken these somite anomalies appeared at multiple positions from head to tail, with a constant and reliable interval of 6–7 somites in between. Since a pair of somites in chicken forms every 90 minutes, this interval corresponds significantly to the cell cycle time, which is 9–10 hours in chicken PSM.[35,39] The repeated somite abnormalities can also be seen in heat-shock-treated *Xenopus* and zebrafish embryos, though rarely.[31,37] Consistent with this view, the expression pattern of CS131, a member of the group of small acidic proteins, including myd118 and GADD45, that arrests cell cycle, is affected in zebrafish somite mutants.[40]

A cell cycle model has been proposed,[39] and subsequently expressed in a mathematical form.[41] This model is the only one that can account for the multiple but discrete somite anomalies found in chicken heat-shock experiments. However, the finding of a cycling *c-hairy1* gene[12] and the discrepancy between chicken and zebrafish in respect of the relationship between duration of cell cycle and interval of heat-shock anomalies[37] argue against a direct role for the cell cycle in somite formation.

2.3 Previous Models

In addition to the clock and wavefront model[42] and the cell cycle model[39] mentioned above, many different models have been proposed to account for the experimental observations: the wave gradient model[43]; a positional information (reaction-diffusion-type) model[44] and the cell polarization model.[45] More recently, after the emergence of the molecular basis of a segmentation clock, there was a boom of many models with mathematical simulations.[11,41,46–48] They successfully explained some aspects of somitogenesis, but failed to elucidate, or even contradict other observations (reviewed by Schnell and Maini[11]). Moreover, they are devoid of molecular details.

The clock-and-wavefront model which Cooke and Zeeman proposed few decades ago[42] can explicate many aspects of the periodic generation of somites and gain support from recent findings (see below). Their idea was that there is an underlying **oscillator** governing the behavior of the cells that are destined to segment together and form a somite. While cells are in the PSM, their clock runs — that is, they oscillate. As a **wavefront** (a front of rapid cell change, controlled by a rate or timing gradient[42]) sweeps rostrocaudally, the cells mature anteriorly, their clock slows down and is finally arrested. Cells in different positions mature at different times and are frozen in different phases of the oscillation cycle. This creates the oscillating spatial pattern of the somites. It is worthy of mentioning that Slack has suggested to call this as "clock and gradient" model, since a wavefront is difficult to visualize and sometimes misleading.[49,50] We will come back to the gradient issue later.

3. Molecular Era — Oscillation and Clock

3.1. Cyclic Genes — Expression Level Changing with Time and Space

Cyclic and dynamic expression pattern of *c-hairy1* is the first molecular evidence of such a clock.[12] Palmeirim *et al.* have shown that expression of *c-hairy1* mRNA is oscillating in the PSM and this process is correlated with the somite formation. Furthermore, its periodic expression does not require signals from the surrounding tissues and is independent of cell movement and *de novo* protein synthesis.[12] The cyclic *c-hairy1* mRNA indicates a molecular clock to the establishment of periodic metameres and a link to somitogenesis. It appears that *c-hairy1* is more likely to correspond to a clock output than a clock component, since its progression is insensitive to blockage of protein synthesis.

After the finding of oscillating *c-hairy1* expression in the PSM, many cycling genes have been found in mice: *Lunatic fringe* (*Lfng*),[51,52] *Hes1*[53] and *Hes7*[54]; in chicken: *Lfng*,[55] *c-hairy2*[53] and *c-Hey2*[56]; in zebrafish: *her1*,[57,58] *deltaC*[59] and *her7*[60,61] and in *Xenopus*: *esr9* and *esr10*.[62] They are either downstream target genes of Notch, such as *Hes1, Hes7, c-hairy1, c-hairy2, c-Hey2, her1, her7, esr9* and *esr10* from *hairy/Enhancer of Split* family or Notch regulators, such as *deltaC* and *Lfng*. Recently, the first non-Notch-related cycling gene, *Axin2*, has been found in mice and it encodes a Wnt negative regulator (see the following section, and Aulehla *et al.*[63]). Cycling expression of *c-Hey2* and *esr9* in the PSM is insensitive to cycloheximide treatment,[56,62] similar to *c-hairy1*. On the contrary, *Lfng* expressing oscillation depends on *de novo* protein synthesis.[55]

3.2. Notch Signaling and Mutants

Notch is a transmembrane receptor that interacts with Delta and Serrate, two alternative ligands. When Notch is activated, the signal is transduced through intracellular components, such as *Su(H)* (*Drosophila* counterpart for *RBPjκ*), down to the target genes, such as *E(spl)* gene complex, which contain basic helix-loop-helix (bHLH) domains and behave as DNA-binding transcriptional repressors (reviewed in Davis and Turner[64]). Modulation by Fringe and processing by Presenilin are essential for the proper activation of the Notch pathway.[65,66] In addition to the nervous system, Notch and/or its ligands Delta and Serrate, are expressed in a variety of different tissues, including somites, gut, testis, epidermis, thymus, muscle, limb buds, kidney, lung, vasculature, spleen, etc. At least four human disorders, including a T-cell acute lymphoblastic leukemia/lymphoma, a late onset neurological disease (CADASIL), a developmental genetic disease (Alagille syndrome) and a vertebral-segmentation defect (spondylocostal dysostosis), are associated with mutations in the *NOTCH1*, *NOTCH3*, *JAGGED1* (*Serrate* homolog) and *DLL3* genes, respectively.[67–71]

The expression patterns suggest that Notch signaling is important for somite formation. Injecting mRNA of a wild-type or a dominant negative *Delta* construct, or of a *Su(H)* homolog resulted in a similar abnormality in the *Xenopus* somitogenesis[72,73] as well as in that of zebrafish.[74,75] The most decisive evidence is that mutations in many different Notch pathway components from both mice and zebrafish lead to a similar somite abnormality. All the knock-out mice where the components and modulators of Delta-Notch signaling, such as *Notch1*,[76] *RBPκ*,[77] *Dll1*,[78] *Presenilin1*,[79] *Lfng*,[80,81] *Hes7*,[82] *Dll3*[83] are mutated and the spontaneous *pudgy* mice, which is caused by a small deletion in *Dll3* gene,[84] exhibit a common segmentation phenotype: irregular somite formation and disrupted segment polarity. Most of the zebrafish somite mutants have been shown to have a Notch component, including *after eight* (*aei*), *deadly seven* (*des*) and *mind bomb* (*mib*), which encodes *deltaD*, *notch1a* and a novel E3 ligase endocytosing Delta, respectively.[58,85,86] A zebrafish mutant with a chromosome deletion covering *her1* and *her7* also exhibits a similar somite phenotype.[87] In spite of the defects in somites, the basic metameric somite pattern is nevertheless established in these mutants. Interestingly, in all the mutants of Notch components, both from mouse and zebrafish, the anterior somites are normal or mildly affected by these mutations (see below). In addition, the human Alagille syndrome, an autosomal dominant developmental disorder and spondylocostal dysostosis, with both autosomal dominant and autosomal recessive modes of inheritance reported, are caused by mutations in Notch ligands, *JAGGED1*[69,70] and *DLL3*,[71] respectively. The most common skeletal anomalies are due to a weaker segmentation defect, 'butterfly' vertebrae or hemivertebrae, resulting from clefting abnormalities of the vertebral bodies.

3.3. The Clockwork of the Segmentation Clock

Although recent progress has demonstrated the existence of a Notch-dependent biochemical clock that drives somite segmentation, a fundamental question is how the segmentation clock works. The nature of an oscillator is a system that regularly swerves from and boomerangs to equilibrium. Two major ways to make an oscillator:[241] (i) positive feedback, a deviation-amplifying process, in which threshold is a common phenomenon, e.g. Ca^{2+} oscillations; and (ii) negative feedback (autoinhibition), a deviation-counteracting process, which is necessary but not sufficient for homeostasis, e.g. circadian rhythms. What is needed to achieve autoinhibitory oscillation is a process whose product feeds back to decrease the rate of the process itself (negative element) and a delay in the enactment of the feedback. A further requirement for sustaining a biological oscillation is a source of activation or excitation (positive element) that keeps the oscillator from damping out. A familiar example of this type is a grandfather clock. While the angular momentum moves the pendulum through and away from the still position (equilibrium) toward one side to a greater height, the weight of pendulum (negative element) generates a gradually (time delay) increased potential energy until it becomes strong enough to counteract the straying momentum and draws the pendulum back to equilibrium. The same process repeats again toward the other side and then it completes a cycle. The mainspring (positive element) is a crucial component, consistently supplying energy to oppose friction and to keep pendulum swaying.

Work done so far has highlighted four significant characters of the segmentation clock: (i) autoinhibitory loops; (ii) two transcriptional factors driving interconnected loops; (iii) dual function of Notch signaling pathway; and (iv) post-translational regulations, and is summarized in Fig. 1.

3.3.1. *Autoinhibitory loops — the essence of the segmentation clock*

The cycling expression of a putative transcriptional repressor, *c-hairy1*, triggered the possibility that unstable components and negative feedback regulation, found in circadian clock control mechanism, may be responsible for its oscillation. The cycloheximide experiment on *c-hairy1* cycling expression, however, argued against such a mechanism.[12] Recent data on *Hes1* gene, a *c-hairy2* homolog, demonstrated that *Hes1* gene and its protein are actually cycling out-phasedly in cell culture and a negative feedback regulation and protein degradation are indeed responsible for mRNA and protein oscillation shown both in cell culture and in embryos.[88] The deficiency of mouse *Hes1*, however, did not give rise to any detectable somite phenotypes.[89] Furthermore, the segmentation clock remained functional in such mutants and cycling *Hes1* expression is lost in *Dll1* deficient embryos,[53] which suggests that *Hes1*, a *hairy*-like gene, is a readout or output of the segmentation clock, though genetic redundancy cannot be entirely ruled out at present.

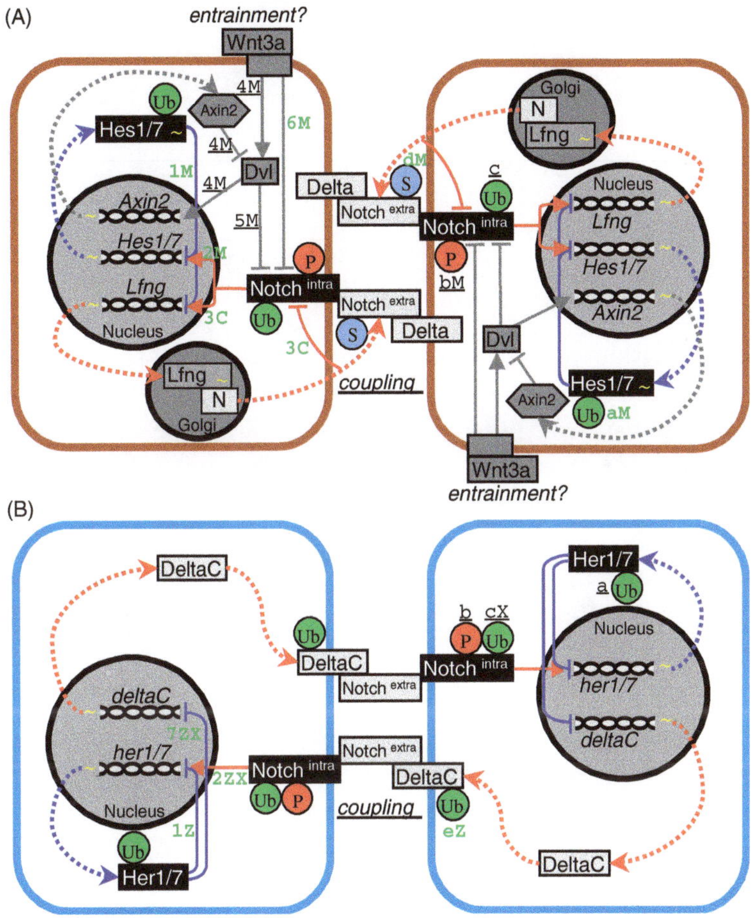

Fig. 1 Summary of current knowledge in vertebrate segmentation clock of mouse (mainly) and chick (a); and of zebrafish (mainly) and *Xenopus* (b). Numbered footnotes indicate known activation (solid arrow line) or inhibition (solid blunt line); lettered footnotes indicate known post-translational modifications. Ub, P and S stand for ubiquitylation (in green circle), phosphorylation (in red circle) and glycosylation (in blue circle), respectively. M, C, Z and X mean that results obtained from mouse, chick, zebrafish and *Xenopus*, respectively. Two transcriptional factors, NICD and Hes/Her proteins, are highlighted in reverse black-white background. Underlined items stand for those regulations not yet shown to be related to somite segmentaion but probably would be; the rest in green have been otherwise shown to be linked to somite segmentation. Dotted arrow lines mean time delay due to transcription, translation, post-translational

Different from *hairy*-like bHLH genes, *E(spl)*-like bHLH genes (for phylogeny, see Davis and Turner[64]), including mouse *Hes7* and zebrafish *her1* and *her7*, are not only cycling in the PSM but they also play a vital role in vertebrate somite segmentation: mutations in *Hes7* and *her1/her7* or gene expression knockdown of *her1* and/or *her7* cause somite phenotypes seen in other mutants deficient in Notch components.[60,61,82,85,87,90] Moreover, it has been shown that their transcripts are negatively regulated by their proteins both in mouse and zebrafish.[60,82,85,90] This is very likely to be the biochemical basis for the oscillation of segmentation clock (see below).

In mice and chicken, another essential target of Notch signaling is the glycosyltransferase, Lfng, whose mRNA shows periodic oscillations in the PSM.[51,52,55] It was observed that mis-expression of *N*otch *i*ntra*c*ellular *d*omain (NICD) caused ectopic expression of *Lfng*. Conversely, mis-expression of *dnRBPJκ* abolished the oscillatory expression of the chicken *Lfng*.[91] These results are consistent with the findings that *RBPJκ*-binding sites are located in *Lfng* promoter and it responds to Notch activation.[92] Moreover, there was rapid turnover of Lfng protein in the PSM, probably via ubiquitin-dependent protein degradation (see below). The mis-expression of *Lfng* resulted in an

---

Fig. 1 Continued
modification, translocation and protein turnover; yellow tilde sign indicates the cycling of transcripts or proteins. Hes1/7 and Her1/7 feedback loops (in blue) are mainly responsible for intracellular oscillation; on the contrary, Lfng/Notch and DeltaC/Notch loops (in red) can additionally **couple** the oscillations between neighboring cells. Wnt3a/Axin2 loop (in gray) could, in theory, behave as an input signal to entrain the oscillation. However, it needs to be investigated further. There are several feedback loops and post-translational modifications in these two broad systems, which can be taken to formulate an accessible and meaningful model either as a whole or partially (e.g. a Hes-dependent oscillation only or plus a Lfng-dependent feedback loop) for a simple two-celled coupled oscillator (see Lewis [103], simple as it is but there are some interesting findings and possible mechanistic ways beyond intuition), for a cluster of cells in 2-D region, or even for a group of cells in 3-D space—a more realistic situation. **1**[60,82,85,90], **2**[54,73,75,101], **3**[91], **4**[234,235], **5**[97], **6**[63], **7**[60,72,75,85], **a**[88,90], **b**[95], **c**[94], **d**[65,104], **e**[86]. Modified from Fig. 3 of Rida *et al.*[25].

inhibition of Notch signaling, destruction of cyclic gene expression and irregular positioning of somite boundaries, indicating a pronounced disruption of the segmentation clock in the chicken PSM. This result demonstrated that Lfng can negatively regulate Notch signaling and this feedback loop could potentially underlie the periodic inhibition of Notch signaling during segmentation, at least in chicken.[91]

3.3.2 *Two transcription factors drive interconnected loops of segmentation clock*

Many experiments have demonstrated that transcriptional feedback regulation is an essential feature of Notch signaling. Among Notch components, NICD and the Hairy/E(spl) proteins are the two key transcription factors that constitute the prime driving force of the segmentation clock and have the following features in common. First, they both manifest that there is an intracellular cyclic Notch activation (see above, though this has not been shown directly for NICD). Second, they negatively modulate their own transcript levels directly or indirectly (see below). Third, they are transient and likely degraded via the ubiquitin-proteasome pathway.[88,90,93–95,242] Furthermore, Notch is a membrane-bound transcriptional factor, whose maturation and activation are rigorously controlled within the Notch signaling pathway and by other pathways as well.[13,96–100]

The NICD loop and the Hairy/E(spl) loop of the segmentation clock are coupled. Compared to the Hairy/E(spl) loop, the NICD loop is more intricate and the details may vary among species (see below). Notch activation can induce expression of Hairy/E(spl)-related genes, such as *Hes1*, *Hes7*, *her1* and *ESR4*, in a Su(H)/RBPjκ-dependent way.[54,73,75,101] It has been demonstrated that Her1 and Her7 can repress expression of *deltaC* and *deltaD* in zebrafish, although it is not yet certain whether this repression is direct or indirect.[60,61,75,85] Similar results were found in *Xenopus* for ESR4 and ESR5.[73] Therefore, Hairy/E(spl) proteins are both effectors and upstream regulators (as repressors of ligand expression) of the Notch signaling cascade, forming an auto-inhibitory loop.

3.3.3. *Dual role of Notch signaling in somite segmentation*

The *c-hairy1* oscillator and the Notch signaling pathway both seem to be key parts of the somite-formation machinery. Somehow they must be linked, and one of the big questions about somite formation is how this linkage is implemented.[19,20] In Cooke and Zeeman's landmark paper, they described this segmentation clock as "an oscillator, shared by all the pre-somite cells, with respect to which they are an entrained and closely phase-organized population, because of intercellular communication".[42] In other words, the PSM cells are coupled oscillators. We have proposed that Notch signaling is required for the synchronization of the segmentation clock.[59,102,103] However, the work done by others[58,82,85,91] argued for a more direct role of Notch signaling — the generation of periodic rhythms — in the segmentation clock. Nevertheless, a reconciled view is emerging: Hairy/E(spl)-dependent Delta/Notch signaling is the oscillator with dual functions — a clock generator as well as a clock synchronizer.[60] Interestingly, the data for arguing the role of Notch signaling in signal synchronization and generation are mostly related to NICD and Hes/Her loops, respectively. Lewis has demonstrated and discussed the effect of different wiring within segmentation clock by mathematical modeling.[103]

At least in zebrafish, it seems that Notch signaling pathway performs both functions. This unique capacity of Notch signaling could be due to its essential character as a module that allows cells to communicate to each other and adjust their behavior accordingly. A perturbation of one function would likely lead to a perturbation of the other to some extent — an important trait of coupled oscillators. We are confined by the degree to which these functions are genetically separable due to technical limitations. Moreover, the circuitry that consists of the clock mechanism may be wired dissimilarly in different organisms and mutations could affect the two Notch-dependent functions to different degrees (see below). Recent evidence from cell culture experiments has shown that serum shock can induce oscillatory expression of *Hes1* in several cell lines.[88] Interestingly, similar periodic *Hes1* expression was observed when the cells were mixed with Delta-expressing S2 cells. This phenomenon once again indicates the possibility of dual function of Notch signaling — signal induction and/or signal synchronization — among cells in culture.

3.3.4. *Post-translational regulation in Notch signaling and others*

The first direct evidence for the involvement of a post-translational regulation of Notch signaling in somite segmentation is the glycosylation of Notch by Lfng.[65,80,81,91,104] This is then followed the characterization of a somite mutant, *mib*, which harbors a mutation in a gene that encodes a RING E3 ligase.[86,105] Though it has not yet been shown to be responsible for the degradation of any Notch component, Mib has been demonstrated to ubiquitylate Delta and result in its endocytosis.[86] There is indirect evidence suggesting a role for regulated protein turnover in somite segmentation. First, Hes1 and Hes7 proteins have been shown to have a short *in vivo* half-life due to ubiquitin-proteasome-mediated degradation[88,90] and Hes1 level oscillates every two hours in cultured cells, matching the time for a somite to form in mouse.[88] Second, Lfng protein behaves likewise.[91] Third, when it coexists with Nrarp, NICD is short-lived.[94,106]

In consistency, other proteins have been shown to regulate Notch signaling by post-translational regulation, although it is not clear whether these modifications are genuinely indispensable for somite segmentation. Sel-10, an F-box/WD-40 repeat protein, and Itch, a HECT domain-containing mammalian Su(dx) homologue, can target NICD for ubiquitin-dependent degradation.[107–111] Additionally, Deltex, encoding a RING-H_2 E3 ligase (Takeyama *et al.*,[112] and unpublished data) has been shown to positively or negatively regulate Notch signaling.[113,114] Similar to Mib, Neuralized also contains a RING domain and can ubiquitylate Delta leading to its endocytosis.[115–117] Table 1 summarizes the known ubiquitylation of Notch components.

Other forms of post-translational modification also exist. It has been shown that only a specifically phosphorylated form of NICD interacts with Sel-10.[108,118] There is also evidence that the glycogen synthase kinase-3β (GSK-3β) phosphorylates NICD and thereby protects it from degradation by the proteasome.[95] Hes1 is phosphorylated on the bHLH

Table 1 Ubiquitylation of Notch components.

E3 ligase	positive/negative	E3 domain	substrate	mechanism
Su(dx)[a]	negative	HECT	NotchIC	degradation
Sel-10[b]	negative	F-box, WD-40	NotchIC	degradation
c-Cbl[c]	negative	RING	NotchTM	lysosomal deg.
LNX[d]	positive	RING	Numb	degradation
Siah-1[e]	positive	RING	Numb	degradation
Mdm2[f]	positive	RING	Numb	degradation
Deltex[g]	positive/negative	RING-H_2	NotchIC	degradation
Neuralized[h]	positive	RING	Delta	endocytosis
Mib[i]	positive	RING	Delta	endocytosis

At least, nine ubiquitin E3 ligases have been found in regulating Notch activation.
[a]ref. 109, 110, 236.
[b]ref. 108, 118, 107.
[c]ref. 237.
[d]ref. 238.
[e]ref. 239.
[f]ref. 240.
[g]ref. 112 and our unpublished data.
[h]ref. 115, 116, 117.
[i]ref. 86.

domain and this phosphorylation inhibits its transactivation function.[119] It is also known that both Hairy and Hey proteins can form homo- and heterodimers, raising the possibility of combinatorial action and additional levels of regulation.[56] It has been documented that bHLH proteins can be regulated by short-lived HLH proteins of the Id family in mammals[120,121] and Extramacrochaetae (Emc) in *Drosophila*.[122,123] These HLH proteins bind to and titrate out the bHLH proteins and, in effect, inhibit the DNA-binding ability of the latter.

Hes7 transcription as well as *Hes7* mRNA have been demonstrated to be cyclically activated and regulated, respectively,[90] suggesting an RNA decay regulation on top of the Hes7 autoinhibition on *Hes7* mRNA.[82] Consistently, it has been suggested that there is a specific degradation signal for *her1* mRNA residing in the 5′UTR,[61] which is different from that of *Xenopus hairy2* gene, where the 3′UTR is essential for modulating *hairy2* RNA level.[124] These and other unidentified mechanisms may be responsible for stabilizing the nexus of interactions around the core segmentation clock machinery and for ensuring that the oscillations are robust and resistant to perturbations. The resiliency, adaptability and fine-tuning of the segmentation clock, therefore, could be attributed to regulations of NICD and Hairy/E(spl) proteins at multiple levels.

4. Molecular Era — Gradients and Clock Output

Both a small amphibian embryo size-reduced at blastula and a haploid embryo with cells half the volume of a diploid counterpart develop with a species-typical somite number at corresponding stages compared to a wild-type control.[125,126] These results suggest that there is a global system of positional information, which regulates the completeness and proportions of the body as a whole, acting to normalize the somite number at the expense of somite size and cell number within a somite. The early embryological manipulations done in amphibians have clearly demonstrated that separation of an embryo into two parts right caudal to formed somites will not perturb normal somite segmentation in the posterior half of embryos.[26] Furthermore, a quail node graft can induce PSM tissue in the chicken host to develop a secondary axis with segmented somites and the pattern of the somites depends on the medial-lateral position of the quail node graft, which led the authors to propose that there is a morphogen gradient originated from the node.[127] What could the posterior signal(s) be? Will it (they) affect somite segmentation? Though *Wnt-3a*, *Fgfr-1* and *Fgf8* have been shown to express in the PSM and tail bud, early loss-of-function studies suggested that they are essential for morphogenesis and mesoderm specification but not somite segmentation.[128–131]

4.1. FGF Signaling — A Gradient Positioning the Boundary

Dubrulle *et al.* have shown that the expression of *Fgfr1* in rostral PSM and the graded expression of *Fgf8* in caudal PSM including tail bud overlap in the somite–IV, the determination front, roughly where the anterior determined region and posterior undetermined region abut.[132] When the Fgf signaling is compromised by drug treatment or Fgf8 protein level is ectopically increased by grafting Fgf8-soaked beads, the somite size will become bigger or smaller, respectively.[132,133] Interestingly, the response of somite size change is not immediate but with a time delay, while few new normal-looking somites form prior to the appearance of size change.[132,133] This corresponds approximately to the sensitivity zone of heat shock — where interactions between wavefront (a timing gradient[42] or prior wave[32]) and oscillator take place some hours before segmentation. In summary, a transient surge or sink in Fgf signaling will lead to a somite size change in those PSM cells located in the vicinity of determination front.

4.2. WNT Signaling — A Gradient Harmonizing Tail Growth with Segmentation and, Therefore, the Cellular Oscillator

Recently, Aulehla *et al.* have demonstrated that a negative regulator of the Wnt pathway, *Axin2*, is strongly expressed in the tail bud and has a graded distribution in PSM. More excitingly, the authors found that its expression level in PSM is indeed oscillating up and down, just as *c-hairy1* or other cycling genes do — the first cycling gene uncovered outside the Notch signaling pathway.[63] *Axin2* expression out-phases *Lfng* transcript. While *Axin2* is still cycling in *Dll1* deficient mutant, the cycling expression of *Lfng* is abolished in a hypomorphic *Wnt3a* mutant, which suggests that Notch signaling acts downstream of Wnt signaling. Indeed, there is no lack of biochemical linkages between these two signaling pathways in addition to the one through Dishevelled binding to Notch.[97–99] Similar to Fgf signaling, an increase or decrease on Wnt signaling will result in smaller or bigger somites, respectively.[63]

Fgf8 is highly down-regulated in hypomorphic *Wnt3a* mutants,[63] which manifests that *Fgf8* is controlled by *Wnt3a*. The phenotypes of null mutants for *Fgf8*, *Fgfr1* and *Wnt3a*, however, do not suggest a similar regulation for anterior somites, since *Wnt3a* null mutants are able to form anterior somites, while *Fgf8* and *Fgfr1* mutants have a reduced mesoderm and do not form any somites.[128–131] Interestingly, mutants lacking of Fgfr1α isoform can still form anterior somites.[134] This observation suggests that other signaling pathway and/or regulation may involve and interplay with *Fgf8* and *Wnt3a* to establish a proper gradient for patterning anterior somites during gastrulation.

The work on *Fgf8* and *Wnt3a* signaling also signify a less-noticed feature of vertebrate segmentation: the growth of tail bud. Only a small portion of PSM is laid down during gastrulation, maybe 10–20 somites, depending on species; the rest is derived from the tail bud when the embryo grows. Therefore, the coordination between somite segmentation and tail outgrowth is essential for proper reiterated pattern. Consistent with this view, Notch and Xwnt3a have been shown to be vital for *Xenopus* tail outgrowth.[135]

4.3. *Hox* Genes are Clock-Controlled

Vertebrate *Hox* genes are essential for organizing structures from head to tail for many tissues, including somitic derivatives, which become regionalized into different morphological domains later, such as cervical, thoracic, lumbar and sacral regions.[136] They are located in the genome as clusters in a constitutively repressed state and activated sequentially by the order of the genes along the clusters.[137] While genes located at the 3' end are activated early on and thus they function in anterior structures, genes located near 5' end are activated later and hence operate in more posterior structures. This correlation is known as colinearity.[138] It is conceivable that there are mechanisms ensuring tight coordination between serial segment production and progressive anteroposterior (AP) identification.

Zákány *et al.* have shown that, indeed, there is a connection between sequential somites and AP patterning.[139] After re-examination of the expression pattern of *Hoxd1* and *Hoxd3*, they found that they display a temporally cyclic expression in the PSM, which is controlled transcriptionally. Interestingly, this dynamic expression of *Hox* genes is abolished in *RBPjκ* mutants and *HoxD* null mice do not have segmentation phenotypes, which indicate that *Hox* genes are very likely to be transcriptionally regulated by the segmentation clock to achieve the temporal pace of body axis specification, though the regulatory element has not been identified.[139] It has been also demonstrated that *Hox* gene expression is not dictated by the absolute AP axial position but rather a spatiotemporal activation by the segmentation clock.[132]

The results not only signify the coordination between somite segmentation and AP axis specification but also probably the way how an oscillator impinge a temporal pace on its target genes: one or several clock-regulated enhancers govern the expression of a vital region of the chromatin. The latter point is echoed by recent studies done in circadian clock, where many *Drosophila* circadian clock or clock-controlled genes are found to be clustered and the promoter regions of mouse circadian clock genes exhibit periodicity in H3 acetylation and RNA polymerase II binding that corresponds synchronously to mRNA rhythms.[140,141]

5. Molecular Era — Compartmentation, Notch Signaling and Others

After maturation, somites can be further subdivided into rostral and caudal compartments, which exhibit different properties with respect to neural crest cell and motorneuron axon migration.[5] This subdivision also provides a scaffold for the future vertebrae, which are formed by a process, called resegmentation — the fusion of the caudal part of anterior somite with the rostral part of the posterior one.[142] While dorsoventral and mediolateral somite patterning occur after segmentation, the determination of rostral and caudal compartments in the somites occurs before and during segmentation at the level of the PSM.[5] Acquisition of these rostral and caudal identities by anterior PSM cells is seen by the striped expression of several genes, which are later expressed either in the rostral or caudal compartments of the formed somites. Embryological experiments in chicken and analysis of zebrafish somite mutants show that establishment of RC polarity is required for formation and maintenance of the somite boundary — presumably as a result of different cell surface properties of rostral and caudal somitic cells.[40,143] Notch signaling is essential, working together with other genes such as *Mesp*, *Papc*, and *Foxc*, for establishing and maintaining somite boundaries by setting up differences between rostral and caudal halves of somites (reviewed in Saga and Takeda,[23] and Holley and Takeda[144]). All the important developmental mechanisms in vertebrate somitogenesis are summarized in Fig. 2.

5.1. Gene Expression Patterns — Transition to Fixed Fates

Many genes are expressed in the anterior PSM of vertebrates in domains that have appropriate positions to represent nascent rostral- or caudal-half of somites. In mouse, *Mesp2*, *Dll3*, *FGFR1*, *Cer1* and *EphA4* are expressed in the rostral halves of somites, while *Hes1, Jagged1, Lfng, Dll1*, *ephrinB2* and *Uncx4.1* in the caudal halves.[53,145–154] In chicken, *Lfng*, *EphA4* and *c-hairy2* are expressed in the rostral halves, while *C-Delta-1*, *ephrinB2*, *c-hairy1*, *c-Hey1*, *c-Hey2* and *cMeso2* (a *Mesp* homologue) in the caudal halves.[12,29,53,55,56,155–157] In zebrafish, *notch6*, *deltaD*, *fgfr1*, *papc*, *ephA4*, *lfng*, *mesp-a* and *mesp-b* are expressed in rostral parts of nascent somites, whereas *notch5*, *deltaC*, *myoD* and *ephrinB2* in caudal parts.[57,158–163] In *Xenopus*, *X-Delta-2*, *ESR-4/-5*, *PAPC*, *Thylacine1* (a *Mesp* homologue) are expressed in the rostral half segments, whereas *Hairy2A* is expressed in the caudal half segments.[72,73,164,165]

5.2. Boundary Formation and Rostrocaudal Patterning

5.2.1. Lfng

It has been suspected that Notch signaling is involved in somite boundary formation for some time but there is no direct evidence to support this idea due to the early activity of Notch signaling in segmentation. Sato *et al.* have nicely shown in the chicken system that somite boundaries form via a Lfng- and Notch-dependent induction from posterior border cells located at B-1 (for the border between S0 and S-1, see Pourquié and Tam[166]).[167] The authors first demonstrated that the cells posterior to B-1 can induce boundary formation by elegant transplanation. While no ectopic boundaries formed when cells at the

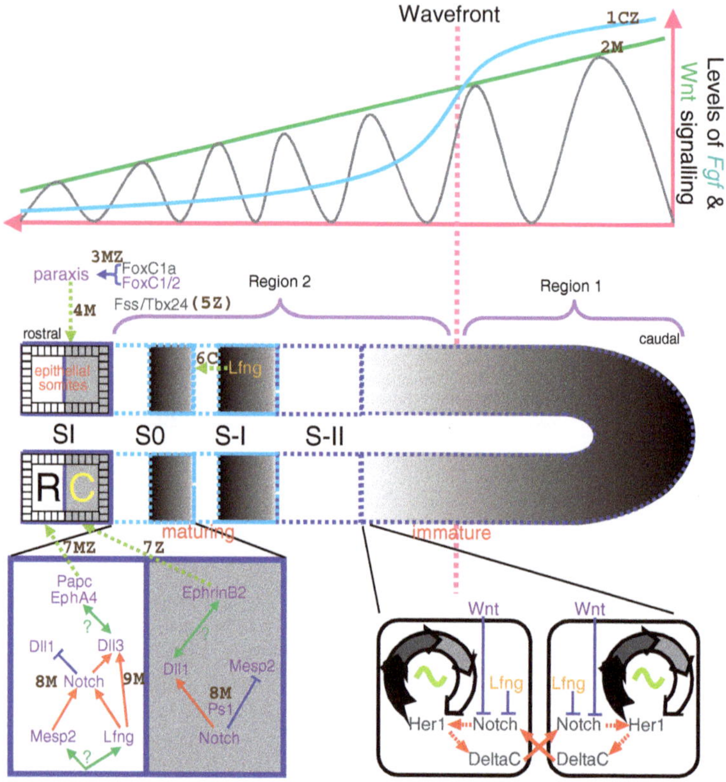

Fig. 2 Schematized presentation of the somitogenesis of an imaginary animal, mouchickish (= mouse + chicken + zebrafish). This figure is a summary of the known developmental mechanisms of somitogenesis gathering from mouse (gene names in purple), chicken (gene names in orange) and zebrafish (gene names in gray). Some homologs behave similarly, see footnotes for details. Red solid and dotted arrow lines mean gene activation; blue blunt lines mean gene repression; olive dotted lines mean morphological changes and green arrow lines depict possible interactions. The PSM can be further divided into two regions: region I, where high level of Fgf signal maintains mesenchymal cells in the PSM in an immature state and, through Notch signaling, the oscillation in each cell might be translated into periodic expression of cycling genes; and region II, where Fgf signaling reduces, the segmentation clock slows down and the PSM cells become mature until they complete the transition and end up with the epithelial somites[23]. Blue and light blue outlines represent mesenchymal cells in immature and maturing status, respectively. Refer to Fig. 1. for the detail of the segmentation clock. R and C mean rostral and caudal halves of the formed somite, respectively. S0 is the forming somite and SI is the newly segmented somite (see Pourquié and Tam[166]). M, C and Z mean that results obtained from mouse, chicken and zebrafish, respectively. **1**[132,133], **2**[63], **3**[173,175], **4**[184], **5**[177], **6**[167], **7**[160,161,188,189], **8**[170,171,243], **9**[228].

level of –1.5 or –2.5 from non-electroporated donor were transplanted to the level of –1.5 in host, an ectopic interface did occur when a Lfng-electroporated donor has been used. NICD-electroporated donor had the same effect.[167] This clearly demonstrates that a Lfng- and Notch-mediated induction is required for boundary formation. Interestingly, if the cells are from NICD-electroporated donor at the level of –4.5, no morphological effects can be seen, suggesting the more posterior PSM has not reached maturation, which requires *Mesp*, *Papc* and other genes (see below).

5.2.2. Mesp

Analysis of *Mesp* gene expression in embryos with disrupted Notch signaling strongly suggests a link between Notch signaling and the establishment of rostral somite identity. *mesp-a* and *mesp-b*, zebrafish homologues of murine *MesP1* and *MesP2*, are expressed in narrow stripes corresponding to the rostral half of anterior nascent somites.[40,57] Ectopic expression of *mesp-b* leads to an expansion of rostral somite compartment at the expense of caudal somite part, giving a slab of tissue where no boundaries form.[57] A similar result is observed when *Mesp* genes are mis-expressed in mouse, chicken and *Xenopus*.[145,165,168,169] In *aei/deltaD* mutant embryos, PSM *mesp-a* expression is lost, and *mesp-b* expression, though weak, loses its striped appearance. These findings suggest that expression of *mesp* genes and establishment of rostral somite identity are downstream to Notch-signaling in zebrafish.[57] In mouse, *Mesp2* is thought to establish RC somite identity by controlling Presenilin1-dependent and -independent Notch signaling pathways which induce *Dll1* expression in the caudal domains of nascent somites and suppress *Dll1* expression in the rostral counterparts, respectively.[170] The rostral restriction of *Mesp2* transcription fail to occur in *Presenilin1* deficient embryos, suggesting that Notch signaling is responsible for reducing *Mesp2* transcription in the caudal halves.[171] The lack of the initial segment border and the loss of rostral properties of the somite result in the formation of a caudalized vertebrae.[145] In *mesp-b* knock-in mice, the RC polarity was disrupted, as shown by expression of *Uncx4.1* and *Dll1*; in contrast, the expression of *EphA4*, *Lunatic fringe* and *Papc*, thought to be involved in segment border formation, was fairly normal in hypomorphic mutant embryos.[172] These results suggest that the Mesp family of transcription factors are involved in both segment border formation and establishment of RC polarity through different genetic cascades.

5.2.3. Foxc

Members of the forkhead/winged helix family of transcription factors have also been implicated in somite formation in both zebrafish and mouse. *Foxc1* and *Foxc2* in mouse and *foxc1a* and *foxc1b* in zebrafish, are expressed in the PSM and nascent somites during segmentation.[173,174] Inactivation of both *Foxc1* and *Foxc2* in mouse or of *foxc1a* alone in zebrafish leads to a disruption of RC somite identity and failure in forming an epithelial somite.[173,175] In zebrafish embryos void of Foxc1a protein by morpholino knockdown, *her1* and *deltaC* stripes are still cycling as in wild-type embryos. Thus, neither the operation of the clock nor the rate of progression of the wavefront is affected. However, the anterior PSM is affected, as *mesp-a*, *ephA4* and *ephrinB2* are no longer expressed. Moreover, expression of *deltaC* and *deltaD* is not maintained in formed somites. Thus, Foxc1a seems to be required for stabilizing RC somite identity and boundary formation. Expression of *notch5* and *notch6* is also severely reduced in embryos lacking Foxc1a, suggesting a pathway by which Foxc1a may regulate RC somite identity. Similarly, analysis of mouse double homozygotes shows that Foxc1 and Foxc2 are both required for transcription in the anterior presomitic mesoderm of *paraxis*, *Mesp1*, *Mesp2*, *Hes5* and *Notch1*, and for the formation of sharp boundaries of *Dll1*, *Lfng* and *ephrinB2* expression, suggesting that these two genes interact with the Notch signaling pathway and are required for the prepatterning of rostral and caudal domains in the presumptive somites through a putative Notch/Delta/Mesp regulatory loop.[173]

5.2.4. fss *and T-box gene*

The zebrafish *fused somites (fss)* mutant shows a complete lack of somite boundaries along the entire body axis even though the cycling gene expression is normal in the posterior PSM.[58,59,176] The *fss* mutant is therefore instrumental in showing that the process of somite boundary formation can be uncoupled from prepattern implemented via a Notch-dependent segmentation clock. The *fss* gene, which encodes a T-box protein Tbx24, is expressed in maturing cells in the intermediate to anterior PSM.[177] Tbx24 is required to stabilize the pattern of oscillating gene expression in the anterior PSM and is also essential for the expression of genes such as *mesp* and *papc* in the anterior PSM.[57,59] Genetic analyses of zebrafish somite mutants have shown that the Fss and Notch pathways are functionally distinct and perhaps independent of each other.[58,59] Transcriptional regulation of *tbx24* is also independent of the Notch pathway.[177] Additionally, it is well documented that Fgf signaling can activate T-box genes and the T-box proteins can interplay among themselves.[178,179] Since Activin can activate and suppress *Xenopus Brachyury* promoter at low and high concentration, respectively,[178] it will be particularly intriguing to examine the regulation of *fss/tbx24* in the anterior PSM, where the concentration of Fgf8 is low.

The murine Tbx6 has been shown to be essential for the formation of posterior somites.[180] Interestingly, *Tbx6* genetically interacts with *Dll1*, whose gene expression is completely lost in *Tbx6* null mice and restored in Tg46 rescued embryos, suggesting that *Dll1* could be a target of Tbx6.[180–182] *Dll3* expression in *Tbx6* mutants is indistinguishable from wild-type, indicating that this is not a simple reduction in expression of all PSM-specific genes in these mutant embryos.[182] Consistent with this, no genetic interaction between *Dll3* and *rib-vertebrae* (*rv*, a weak *Tbx6* allele) was detected.[181] *Dll1* is required for proper RC patterning of the somites and for epithelialization of the somites.[78] In *Dll1*-null mutants, somites appear to be rostralized,[183] in contrast to their caudalization in embryos with reduced levels of *Tbx6*.[182] A complete explanation awaits the identification of additional Tbx6 targets.

5.3. Epithelialization and Cell Adhesion

Striped gene expression occurs in mouse and chicken PSM explants lacking ectoderm and cultured *in vitro*, indicating that establishment of this segmental pattern is an intrinsic property of the PSM. In these explants, boundary formation and epithelialization do not occur, indicating that these morphological processes require a signal derived from the ectoderm and can be uncoupled from the genetic determination of the anterior and posterior compartments.[29,30,184,185] Moreover, intercellular signaling through cell–matrix and cell–cell interactions is one of the key processes that underlie the final events of somitogenesis. There is clear evidence for the importance of local adhesive interactions, which comprise cell–matrix (Fibronectin and Integrin) and cell–cell interactions (Cadherin), during epithelialization and somite formation.[6,186] The connection between periodic Notch signaling activity and these intercellular interactions, however, is not yet clear.

5.3.1. Papc

Papc is a very potent homotypic cell adhesion molecule.[187] *Xenopus PAPC* may link the process of RC patterning of segments with the generation of AP boundaries required for segmental morphogenesis (for more on adhesive differences, see Kim *et al.*[188]) *X-Delta-2*, *Thylacine* and *ESR-5* are segmentally expressed one somitomere advanced that of *PAPC.*[188] Cycloheximide treatment disrupted segmental expression of *X-Delta-2*, *Thylacine* and *ESR-5* in somitomere 1 after one hour, as compared to two hours for *PAPC*[188]: segmental *PAPC* expression is likely to be a downstream consequence of these patterning events that establish segmental identity within the PSM. These results suggest the model in which segmental identity is established by a mechanism that produces a segmental expression of the selector gene, *Thylacine* (a *Mesp* homolog), in somitomere 1, which then establishes the segmental expression of *PAPC* in somitomere 2.[188] The same scenario is likely to apply to zebrafish embryos, where the *papc* is expressed in rostral half segments in a pattern that overlaps with, but is downstream of, the segmental expression of *mesp.*[57,160] Moreover, ectopic expression of *mesp-b* in zebrafish embryos induces ectopic expression of *papc.*[57] In mice, *Mesp2* transcription precedes *Papc* transcription in S-I and when applied a soluble form of dominant-negative Papc, the treated embryos showed a somite epithelialization phenotype similar to those seen in *Mesp2-/-* and *Lfng-/-* mutants.[189] Thus, regulation of segmental expression of *Papc* by the Mesp proteins may be an evolutionarily conserved mechanism for linking segmental identity to segmental differences in cell adhesion. Redundancy of Papc function might explain why somitogenesis occurs normally in mice with a targeted inactivation of a mouse *Papc* homolog.[190]

5.3.2. Paraxis *and epithelialization*

Paraxis is a bHLH transcription factor that is expressed in the anterior PSM and newly formed somites. As somite matures, *paraxis* expression becomes restricted to the dermomyotome in mouse and chicken.[191–193] The importance of Paraxis during the process of epithelialization was revealed through the analysis of *paraxis* deficient mice that fail to form epithelial somites.[184] The mutant embryos exhibit a normal segmented pattern of paraxial mesoderm derivatives, such as axial skeleton, skeletal muscle and dermis. The implication of Paraxis in somite formation is corroborated by experiments using antisense oligonucleotide treatment against *paraxis* mRNA in chicken embryos, which also inhibited normal somite epithelialization.[193] These results clearly indicate that Paraxis function is required for the formation of epithelial somites and the process of epithelialization can be uncoupled from the processes of metamerism. However, it has since been reported that isolated chicken PSM is able to maintain *paraxis* expression after four hours of incubation, though the explants form no somite borders,[29] suggesting that Paraxis may be a necessary but not a sufficient component in the generation of somitic boundaries. Interestingly, *paraxis* expression seems to be affected when Eph or Notch signaling is compromized. Disruption of Eph signaling in zebrafish affects the normal down-regulation of the zebrafish *paraxis* homologue, *par1*, in the rostral halves of somites.[161] In principle, it is possible that *par1* down-regulation is a direct effect of Eph signaling. Moreover, the expression of *paraxis* is reduced in *Dll1* mutant mice,[78] raising the possibility that normal *paraxis* expression is controlled by Notch signaling.

5.3.3. Eph *signaling*

The Eph receptor tyrosine kinases and their transmembrane ligands, the ephrins, are a large family of surface molecules, which together mediate a bi-directional cell–cell signaling,[194–196] involved in several developmental processes, such as boundary formation, cell migration, axon guidance, synaptogenesis and angiogenesis/vasculogenesis (reviewed in Holder and Klein,[197] and Wilkinson[198]) Some of the receptors and ligands are expressed in the anterior PSM and in the somites of a number of vertebrates. In particular, the specific PSM expression profile of the *ephrinB2* ligand and the *EphA4* receptor in the PSM is conserved among chicken, mouse and zebrafish.[149,153,161,199,200] Experiments done in zebrafish show that blockage of Eph signaling results in abnormal somite boundary formation, though the RC polarity within the somites is not affected. Furthermore, this disruption of Eph signaling in zebrafish affects the modulation and normal down-regulation of *deltaD* and *her1* in the anterior PSM.[161] Interestingly, expression of *EphA4* and *ephrinB2* is severely down-regulated in several Notch pathway null mutant mice, such as *Dll1* and *RBPJκ*, suggesting the existence of another feedback loop between Eph and Notch signaling pathways.[183,201] However, *EphA4* and *ephrinB2* null mice exhibit no somite abnormalities,[202,203] which may be due to genetic redundancy of other Eph and ephrin homologs.

Given that Eph signaling regulates cytoskeletal organization and adhesion in a number of *in vitro* systems,[198] the striped expression of *EphA4* and *ephrinB2* in the anterior PSM suggests that these molecules mediate changes in cell adhesion and cell shape associated with boundary formation. Morphological observations of boundary formation in zebrafish are consistent with this hypothesis[204,205]: zebrafish somite furrow formation first appears as local de-adhesions in the PSM along the line of the nascent boundary accompanied by a mesenchymal-to-epithelial transition that spreads laterally. The site and timing of this de-adhesion correlates with EphA4 and ephrinB2 expression, and loss of Eph signaling results in loss of somite boundary formation.[161] Similarly, it has been shown in chicken embryos via *in vivo* time-lapse microscopy that it takes a series of repeatable steps for an embryo to sculpt a furrow between somites.[206] While the somite border cells exhibit ball-and-socket separation, the straight-line expression of *EphA4* remains unchanged, suggesting a rapid change of *EphA4* expression in the cells moving across the presumptive somite boundary. Though the mechanism of plasticity of gene expression is unclear, it is reminiscent of that sharpening rhombomere boundaries.[207]

Analysis of zebrafish embryos of double mutant for *knypek* and *trilobite* shows that somite boundary formation occurs in the absence of any central mesenchymal cells.[205] Thus, the force that drives boundary formation must come from interactions among border cells themselves, not from an increase in compaction of central cells as proposed in chicken.[208] Moreover, transplantation of EphA4-expressing cells into the PSM of *fss* mutant, which express *ephrinB2* evenly instead in caudal halves, is sufficient to induce a furrow between transplanted and host cells.[40] In the mouse, reduced Mesp activity leaves the expression pattern of *EphA4* unaffected and the initial somitic segregation normal, but soon somites fuse again.[172] These results imply that Eph signaling is necessary and sufficient for intersomitic furrow formation but not for its maintenance. The similar expression patterns of *EphA4* and *ephrinB2* in mouse, chicken and zebrafish suggest that such a role for Eph signaling in somite formation may be conserved.[161,209]

5.3.4. *Fibronectin and Integrin*

Fibronectin is an adhesion molecule that interacts with its cell surface receptor, Integrin, to mediate cell–extracellular matrix adhesions.[210–213] As evidenced by the phenotype of the null mutant mice, the functions of both Fibronectin and Integrin proteins are required during somitogenesis. The *Fibronectin* deficient embryos die at 8.5 days post coitus (dpc) and their phenotypes suggest that there is a quantitative deficit in mesoderm formation, as manifested by the lack of notochord and somites.[213,214] A less severe phenotype is observed in the *α5-Integrin* deficient embryos, which fail to produce epithelial somites even if the paraxial mesoderm shows segmented blocks of mesenchymal cells.[215]

This milder phenotype for the null mutant of *α5-Integrin* compared to that of *Fibronectin* could be due to functional redundancy provided by another Fibronectin receptor expressed in the PSM, such as a different homolog of the Integrin family.

5.3.5. *Cadherins*

Likewise, there is also evidence that Cadherins play a role in modulating cell–cell adhesion during the course of somite formation.[208,216] Cadherins are the central components of major sites of cell–cell adhesion, the adherens junctions. They cluster on the cell surface and bind to Cadherins on adjacent cells through a Ca^{2+}-dependent homotypic interaction. The cytoplasmic domain interacts with β-Catenin, which anchors the Cadherins to F-actin and the cytoskeleton. β-Catenin is also an integral component of the Wnt signal transduction pathway, raising the possibility that Cadherins are able to modulate gene expression.[217,218] During mesenchymal-to-epithelial transitions, adherens junctions are redistributed to the apical and lateral cell membranes. Within chicken somitic mesoderm, the redistribution of adherens junctions, as determined by localization of β-Catenin, occurs just prior to visualization of somite boundaries and has been proposed to be important for boundary formation.[208] N-cadherin is the primary Cadherin associated with somitogenesis and is expressed in the anterior PSM and epithelial somites. Cadherin mutants have somite abnormalities, both for *N-cadherin* single mutant and more severe in double one of *N-cadherin* and *Cadherin11*.[219,220] This phenotype resembles those observed after anti-N-cadherin antibody treatment in chicken embryos,[208] implying that Cadherin-mediated cell–cell adhesion is required to maintain the epithelial somite by keeping rostral- and caudal-half cohesive. Zebrafish *N-cadherin* mutants also show a similar phenotype, though it was not analyzed in detail.[221] Furthermore, application of RGD peptides that contain the minimal specific adhesion recognition signal of Fibronectin stimulates N-cadherin synthesis during somitogenesis.[208] This result implies that N-cadherin mediated cell–cell events are coordinated with Fibronectin-associated cell–substratum adhesion. It is also intriguing to indicate that isolated chicken PSM explants treated with RGD peptides are able to segment and make somites,[211] similar to what is observed when the PSM is cultured together with overlying ectoderm in chicken and mouse.[29,30] Interestingly, E-cadherin-mediated cell–cell adhesion has been shown to regulate expression of *Eph* receptors and *ephrin* ligands.[222]

6. Differences among Vertebrates

Despite the similarities of somitogenesis, e.g. Notch signaling as the core of the segmentation clock and the role of Eph signaling in somite boundary formation, the details are not all alike among mammals, birds, amphibians and fish, which exist at least in three aspects: morphology; gene expression pattern; and transcriptional regulation. *Xenopus* somite segmentation is very different from those of mouse, chicken and zebrafish: when a group of myotomal cells mature and become somites, they will segregate, rotate ninety degrees and orient parallel to the AP axis.[126] Even more, within amphibians the somite development varies (reviewed by Keller[21]). There are variations in respect of somite furrow formation: in chicken, the posterior border cells ephithelialize first and gradually extend to medial then anterior ones (See Sato *et al.*[167]; Fig. 1); in zebrafish, the border cells of adjacent somites become ephithelialized simultaneously (See Jiang *et al.*[59]; Supp. Fig.).

As for gene expression patterns: *Mesp* is expressed in rostral halves of somites in mouse (*Mesp2*) and zebrafish (*mesp-a*) but in caudal halves in chicken (*cMeso2*)[57,145,157]; *Lfng* is expressed in rostral halves of somites in chicken and zebrafish but in caudal counterparts in mice[55,151,162]; mouse *Jagged1* is expressed in a thin stripe of cells in the forming somite boundary, a feature not observed in chicken homolog, *C-Serrate-1*[223, 224]; and *Pax2* is segmentally regulated during chicken somitogenesis but not expressed in mouse PSM.[225]

The transcriptional regulation of *Hes7* and its zebrafish homolog, *her1*,[64] seems to be conserved: a 0.9-kb *Hes7* promoter, which consists of a pair of putative RBPJκ binding sites, two E-boxes and one N-box — target sequences for Hes7 protein, and the sequence between 2.3 and 8.6 kb upstream of the *her1* transcriptional start can regulate their cyclic expression.[54,61,90] Coexpression of *NICD* and *Hes7* can up-regulate and down-regulate *Hes7* promoter activity, respectively.[54,90] Furthermore, Hes7 can override the Notch-induced transcription from the *Hes7* promoter.[90]

While the expression of chicken and mouse *Lfng* has been shown to oscillate during somite formation, *Xenopus* and zebrafish counterparts do not cycle in the posterior PSM.[51,52,55,162,226,227] A conserved 2.3 kb region in the promoter of the mouse *Lfng*, which includes *cis*-acting elements for both enhancing and repressing factors, governs the cyclic expression in PSM cells. Moreover, Notch signaling acts directly via RBPJκ-binding sites to activate *Lfng* expression.[92] Mutation or deletion of E-boxes in the A/2 region of *Lfng* promoter (region A in ref. 92, region 2 in ref. 228) has been shown to eliminate *Lfng* periodic expression in posterior PSM, suggesting a direct regulation by the cyclically expressed Hes proteins.[90] Interestingly, in such transgenic mutants, *Lfng* is still expressed in anterior PSM and formed somites in a manner similar to that of zebrafish counterpart.[92,227] This observation implies that the A/2 region is responsible for *Lfng* cycling in posterior PSM, whereas the rest of 2.3 kb region could be an ancestral promoter shared by all vertebrates that controls the expression of *Lfng* in anterior PSM and formed somites.

On the other hand, zebrafish *deltaC* is cycling in posterior PSM, whereas chicken *Delta1* and mouse *Dll1* are non-cyclically expressed in PSM.[29,59,78] Though not cyclically expressed, *X-Delta-2* has been shown to be dynamically expressed within the PSM and mediates somite segmentation, reminiscent of zebrafish *deltaC* and *deltaD*.[72] Furthermore, mouse *Dll1* and its closest homolog, zebrafish *deltaD* are expressed in the posterior parts and in the anterior parts of somites, respectively. The promoter analysis of zebrafish *deltaC* is not yet available, but the analyses done in mouse *Dll1* and zebrafish *deltaD* have shown that their mesodermal elements are more divergent than neural elements during evolution,[229,230] suggesting that the corresponding transcription factors and hence the regulatory circuit are dissimilar as well.

The difference in the expression dynamics of *Fringe* and *Delta* genes among species suggests a different wiring for NICD regulation. In chicken and likely in mouse as well, this loop is more "intrinsic", since the feedback loop can occur in individual cells with minimal interactions with their neighbors. In zebrafish, the NICD loop, if not entirely "extrinsic", exploits mutual interactions to certain degree. In other words, the coupling strength between individual PSM oscillators is stronger in zebrafish than that in amniotes. This may explain the observed differences in expression patterns of key genes among species. The finding that mouse *Axin2* is cycling suggests another possible mechanism in the entrainment of individual PSM oscillators, probably from Wnt3a (extrinsic factor) through Dishevelled (intrinsic factor), which then binds and antagonizes Notch.[97] It would be intriguing to know whether Wnt signaling plays an indispensable role in zebrafish somite segmentation. However, *pipetail* (*ppt*) mutants, which is devoid of functional Wnt5, show no obvious segmentation phenotypes except defective tail outgrowth.[231]

7. Perspectives

Notch signaling is involved in several steps in somitogenesis: generation/ synchronization of oscillation in the posterior PSM; suppression and activation of Delta expression in the presumptive rostral half and caudal half of the somite primordia in the anterior PSM; and re-shaping the Mesp2 expression domain. One challenge will be to identify molecules that modulate the Notch signaling pathway to distinguish among these Notch activities. Furthermore, it would be necessary to identify genes that lie immediately downstream of the segmentation clock (clock outputs) and to study their expression patterns and functions to understand the interface between dynamic gene expression, cellular differentiation and morphogenesis at the organismal level.

Latest progress in *in vitro* studies can provide valuable insights into the mechanisms underlying gene oscillation, as it has been done for *Hes1*,[88] but final conclusions could only be achieved with the establishment of *in vivo* reporter transgenic lines. Another interesting question is whether segmentation clock is cell-autonomous. Although indirect observations coming from zebrafish and mouse[59, 88] support cell autonomy, direct evidence is still missing.

The role of Fgf8 and Wnt3a gradients in somite segmentation are not yet clear. These gradients could solely be an input signal to entrain the segmentation clock. Alternatively, Fgf8 and Wnt3a may biochemically interact with Notch signaling to maintain the tempo and coherence of somite formation. The *Axin2* knock-out mouse should be able to answer this question to some extent.

Promoter analysis of cycling genes in different species will provide valuable insight into transcriptional regulation underlying the segmentation clock. It can answer the question of the differences in circuit wiring among different vertebrates. It can also help understand the differences between clock output and clock component, for example, whether *Hes1* (a *hairy* homolog) and *Hes7* (an *E(spl)* homolog) are an output and a component, respectively, and whether *Hey-2* (*c-Hey-2*) has similar transcriptional regulation as *Hes1*.

We still owe a mechanistic explanation to heat-shock experiments, particularly with the current knowledge of segmentation clock. Both experimental and theoretical analyses shall shed light on its mechanism. The evidence from the effect of treating chicken embryos with cell-cycle inhibitors and other observations suggest a connection between the cell-cycle control machinery and the segmentation clock.[39,232] The nature of this link, however, remains to be elucidated — the work done by Zákány *et al.* might bridge the gap.[139]

The nexus of interactions that surrounds the basic segmentation clock is indeed multiplex and complex. Dissecting this gamut of pathways in order to go beyond simple feedback loops into the realm of molecular networks of astonishing complexity poses a considerable challenge. Intuition is a poor guide for understanding coupled oscillators — the nature of the segmentation clock, whose dynamics can be very complicated in the real embryos (e.g. see Lewis,[103] and Monk[233]). Mathematical modeling is well known in making assumptions explicit, compressing information, clarifying essential features of system, describing dynamical processes, handling complexity, allowing predictions and uncovering general principles. For such a dynamic and complicated system, mathematical modeling and simulation will definitely complement the experimental methods and facilitate the progress in understanding the clockwork, the perturbation consequences and the evolutionary constraints of the segmentation clock.

Acknowledgments

I am grateful to Kate Osborne and Karuna Sampath for comments on early versions, to Nguyet Le Minh for critical reading and suggestions and to lab members for discussions. The work on the Notch signaling pathway and zebrafish somitogenesis in my lab was funded by the Agency of Science, Technology and Research (A*STAR), Singapore.

References

1. Gont LK, Steinbeisser H, Blumberg B and De Robertis EM (1993). *Development* **119**: 991.
2. Tucker AS and Slack JMW (1995). *Development* **121**: 249.
3. Kanki JP and Ho RK (1997). *Development* **124**: 881.
4. Knezevic V, De Santo R and Mackem S (1998). *Development* **125**: 1791.
5. Keynes RJ and Stern CD (1988). *Development* **103**: 413.
6. Tam PPL and Trainor PA (1994). *Anat. Embryol. (Berl)* **189**: 275.
7. Gossler A and Hrabe de Angelis M (1998). *Curr. Top. Dev. Biol.* **38**: 225.
8. Stockdale FE, Nikovits W, Jr. and Christ B (2000). *Dev. Dyn.* **219**: 304.
9. Pourquié O (2001). *Annu. Rev. Cell Dev. Biol.* **17**: 311.
10. Brennan C, Amacher SL and Currie PD (2002). Somitogenesis. In Solnica–Krezel L (ed.). *Results and Problems in Cell Differentiation: Pattern Formation in Zebrafish*, vol. 40. Springer–Verlag, Berlin, pp. 271.
11. Schnell S and Maini PK (2000). *Dev. Dyn.* **217**: 415.
12. Palmeirim I, Henrique D, Ish–Horowicz D and Pourquié O (1997). *Cell* **91**: 639.
13. Artavanis–Tsakonas S, Rand MD and Lake RJ (1999). *Science* **284**: 770.
14. Bray S (1998). *Semin. Cell Dev. Biol.* **9**: 591.
15. Irvine KD and Vogt TF (1997). *Curr. Opin. Cell Biol.* **9**: 867.
16. Franklin JL, Berechid BE, Cutting FB, Presente A, Chambers CB, Foltz DR, Ferreira A and Nye JS (1999). *Curr. Biol.* **9**: 1448.
17. Sestan N, Artavanis–Tsakonas S and Rakic P (1999). *Science* **286**: 741.
18. Redmond L, Oh SR, Hicks C, Weinmaster G and Ghosh A (2000). *Nat. Neurosci.* **3**: 30.
19. Jiang Y-J, Smithers L and Lewis J (1998). *Curr. Biol.* **8**: R868.
20. Pourquié O (1999). *Curr. Opin. Genet. Dev.* **9**: 559.

21. Keller R (2000). *Curr. Top. Dev. Biol.* **47**: 183.
22. Stickney HL, Barresi MJ and Devoto SH (2000). *Dev. Dyn.* **219**: 287.
23. Saga Y and Takeda H (2001). *Nat. Rev. Genet.* **2**: 835.
24. Bessho Y and Kageyama R (2003). *Curr. Opin. Genet. Dev.* **13**: 379.
25. Rida PCG, Le Minh N and Jiang Y-J (2004). *Dev. Biol.* **265**: 2.
26. Deuchar EM and Burgess AMC (1967). *J. Embryol. Exp. Morphol.* **17**: 349.
27. Packard DS, Jr. (1978). *J. Exp. Zool.* **203**: 295.
28. Aoyama H and Asamoto K (1988). *Development* **104**: 15.
29. Palmeirim I, Dubrulle J, Henrique D, Ish–Horowicz D and Pourquié O (1998). *Dev. Genet.* **23**: 77.
30. Correia KM and Conlon RA (2000). *Mech. Dev.* **91**: 19.
31. Elsdale T, Pearson M and Whitehead M (1976). *J. Embryol. Exp. Morphol.* **35**: 625.
32. Pearson M and Elsdale T (1979). *J. Embryol. Exp. Morphol.* **51**: 27.
33. Elsdale T and Pearson M (1979). *J. Embryol. Exp. Morphol.* **53**: 245.
34. Cooke J and Elsdale T (1980). *J. Embryol. Exp. Morphol.* **58**: 107.
35. Primmett DRN, Stern CD and Keynes RJ (1988). *Development* **104**: 331.
36. Kimmel CB, Sepich DS and Trevarrow B (1988). *Development* **104** (Suppl.), 197.
37. Roy MN, Prince VE and Ho RK (1999). *Mech. Dev.* **85**: 27.
38. Stern CD and Bellairs R (1984). *Anat. Embryol.* **169**: 97.
39. Primmett DRN, Norris WE, Carlson GJ, Keynes RJ and Stern CD (1989). *Development* **105**: 119.
40. Durbin L, Sordino P, Barrios A, Gering M, Thisse C, Thisse B, Brennan C, Green A, Wilson S and Holder N (2000). *Development* **127**: 1703.
41. Collier JR, McInerney D, Schnell S, Maini PK, Gavaghan DJ, Houston P and Stern CD (2000). *J. Theor. Biol.* **207**: 305.
42. Cooke J and Zeeman EC (1976). *J. Theor. Biol.* **58**: 455.
43. Flint OP, Ede DA, Wilby OK and Proctor J (1978). *J. Embryol. Exp. Morphol.* **45**: 189.
44. Meinhardt H (1986). Models of segmentation. In Bellairs R, Ede DA and Lash JW (eds.). *Proceedings of the NATO Advanced Research Workshop on Somites in Developing Embryos*, vol. 118. Plenum Press, New York, pp. 179.
45. Polezhaev AA (1992). *J. Theor. Biol.* **156**: 169.
46. Kaern M, Menzinger M and Hunding A (2000). *J. Theor. Biol.* **207**: 473.
47. Kerszberg M and Wolpert L (2000). *J. Theor. Biol.* **205**: 505.

48. Jaeger J and Goodwin BC (2001). *J. Theor. Biol.* **213**: 171.
49. Slack JMW (1983). Chapter 7, The problems of early development and the means for their solution. In *From Egg to Embryo: Determinative Events in Early Development*, Cambridge University Press, Cambridge, pp. 214.
50. Slack JMW (1991). Chapter 3, Theoretical embryology. In *From Egg to Embryo: Regional Specification in Early Development*, Cambridge University Press, Cambridge, pp. 65.
51. Forsberg H, Crozet F and Brown NA (1998). *Curr. Biol.* **8**: 1027.
52. Aulehla A and Johnson RL (1999). *Dev. Biol.* **207**: 49.
53. Jouve C, Palmeirim I, Henrique D, Beckers J, Gossler A, Ish–Horowicz D and Pourquié O (2000). *Development* **127**: 1421.
54. Bessho Y, Miyoshi G, Sakata R and Kageyama R (2001). *Genes Cells* **6**: 175.
55. McGrew MJ, Dale JK, Fraboulet S and Pourquié O (1998). *Curr. Biol.* **8**: 979.
56. Leimeister C, Dale K, Fischer A, Klamt B, Hrabe de Angelis M, Radtke F, McGrew MJ, Pourquié O and Gessler M (2000). *Dev. Biol.* **227**: 91.
57. Sawada A, Fritz A, Jiang Y-J, Yamamoto A, Yamasu K, Kuroiwa A, Saga Y and Takeda H (2000). *Development* **127**: 1691.
58. Holley SA, Geisler R and Nüsslein–Volhard C (2000). *Genes Dev.* **14**: 1678.
59. Jiang Y-J, Aerne BL, Smithers L, Haddon C, Ish–Horowicz D and Lewis J (2000). *Nature* **408**: 475.
60. Oates AC and Ho RK (2002). *Development* **129**: 2929.
61. Gajewski M, Sieger D, Alt B, Leve C, Hans S, Wolff C, Rohr KB and Tautz D (2003). *Development* **130**: 4269.
62. Li Y, Fenger U, Niehrs C and Pollet N (2003). *Differentiation* **71**: 83.
63. Aulehla A, Wehrle C, Brand–Saberi B, Kemler R, Gossler A, Kanzler B and Herrmann BG (2003). *Dev. Cell* **4**: 395.
64. Davis RL and Turner DL (2001). *Oncogene* **20**: 8342.
65. Moloney DJ, Panin VM, Johnston SH, Chen J, Shao L, Wilson R, Wang Y, Stanley P, Irvine KD, Haltiwanger RS and Vogt TF (2000). *Nature* **406**: 369.
66. Fortini ME (2001). *Curr. Opin. Cell Biol.* **13**: 627.
67. Ellisen LW, Bird J, West DC, Soreng AL, Reynolds TC, Smith SD and Sklar J (1991). *Cell* **66**: 639.
68. Joutel A, Corpechot C, Ducros A, Vahedi K, Chabriat H, Mouton P, Alamowitch S, Domenga V, Cecillion M, Marechal E, Maciazek J, Vayssiere C, Cruaud C, Cabanis EA, Ruchoux MM, Weissenbach J, Bach JF, Bousser MG and Tournier–Lasserve E (1996). *Nature* **383**: 707.

69. Oda T, Elkahloun AG, Pike BL, Okajima K, Krantz ID, Genin A, Piccoli DA, Meltzer PS, Spinner NB, Collins FS and Chandrasekharappa SC (1997). *Nat. Genet.* **16**: 235.
70. Li L, Krantz ID, Deng Y, Genin A, Banta AB, Collins CC, Qi M, Trask BJ, Kuo WL, Cochran J, Costa T, Pierpont MEM, Rand EB, Piccoli DA, Hood L and Spinner NB (1997). *Nat. Genet.* **16**: 243.
71. Bulman MP, Kusumi K, Frayling TM, McKeown C, Garrett C, Lander ES, Krumlauf R, Hattersley AT, Ellard S and Turnpenny PD (2000). *Nat. Genet.* **24**: 438.
72. Jen W-C, Wettstein D, Turner D, Chitnis A and Kintner C (1997). *Development* **124**: 1169.
73. Jen W-C, Gawantka V, Pollet N, Niehrs C and Kintner C (1999). *Genes Dev.* **13**: 1486.
74. Dornseifer P, Takke C and Campos–Ortega JA (1997). *Mech. Dev.* **63**: 159.
75. Takke C and Campos–Ortega JA (1999). *Development* **126**: 3005.
76. Conlon RA, Reaume AG and Rossant J (1995). *Development* **121**: 1533.
77. Oka C, Nakano T, Wakeham A, de la Pompa JL, Mori C, Sakai T, Okazaki S, Kawaichi M, Shiota K, Mak TW and Honjo T (1995). *Development* **121**: 3291.
78. Hrabe de Angelis M, McIntyre II J and Gossler A (1997). *Nature* **386**: 717.
79. Wong PC, Zheng H, Chen H, Becher MW, Sirinathsinghji DJ, Trumbauer ME, Chen HY, Price DL, Van der Ploeg LH and Sisodia SS (1997). *Nature* **387**: 288.
80. Zhang N and Gridley T (1998). *Nature* **394**: 374.
81. Evrard YA, Lun Y, Aulehla A, Gan L and Johnson RL (1998). *Nature* **394**: 377.
82. Bessho Y, Sakata R, Komatsu S, Shiota K, Yamada S and Kageyama R (2001). *Genes Dev.* **15**: 2642.
83. Dunwoodie SL, Clements M, Sparrow DB, Sa X, Conlon RA and Beddington RSP (2002). *Development* **129**: 1795.
84. Kusumi K, Sun ES, Kerrebrock AW, Bronson RT, Chi DC, Bulotsky MS, Spencer JB, Birren BW, Frankel WN and Lander ES (1998). *Nat. Genet.* **19**: 274.
85. Holley SA, Jülich D, Rauch GJ, Geisler R and Nüsslein–Volhard C (2002). *Development* **129**: 1175.

86. Itoh M, Kim CH, Palardy G, Oda T, Jiang Y-J, Maust D, Yeo SY, Lorick K, Wright GJ, Ariza–McNaughton L, Weissman AM, Lewis J, Chandrasekharappa SC and Chitnis AB (2003). *Dev. Cell* **4**: 67.
87. Henry CA, Urban MK, Dill KK, Merlie JP, Page MF, Kimmel CB and Amacher SL (2002). *Development* **129**: 3693.
88. Hirata H, Yoshiura S, Ohtsuka T, Bessho Y, Harada T, Yoshikawa K and Kageyama R (2002). *Science* **298**: 840.
89. Ishibashi M, Ang SL, Shiota K, Nakanishi S, Kageyama R and Guillemot F (1995). *Genes Dev.* **9**: 3136.
90. Bessho Y, Hirata H, Masamizu Y and Kageyama R (2003). *Genes Dev.* **17**: 1451.
91. Dale JK, Maroto M, Dequeant ML, Malapert P, McGrew M and Pourquié O (2003). *Nature* **421**: 275.
92. Morales AV, Yasuda Y and Ish–Horowicz D (2002). *Dev. Cell* **3**: 63.
93. Schroeter EH, Kisslinger JA and Kopan R (1998). *Nature* **393**: 382.
94. Lamar E, Deblandre G, Wettstein D, Gawantka V, Pollet N, Niehrs C and Kintner C (2001). *Genes Dev.* **15**: 1885.
95. Foltz DR, Santiago MC, Berechid BE and Nye JS (2002). *Curr. Biol.* **12**: 1006.
96. Kopan R (2002). *J. Cell Sci.* **115**: 1095.
97. Axelrod JD, Matsuno K, Artavanis–Tsakonas S and Perrimon N (1996). *Science* **271**: 1826.
98. Soriano S, Kang DE, Fu M, Pestell R, Chevallier N, Zheng H and Koo EH (2001). *J. Cell Biol.* **152**: 785.
99. Ross DA and Kadesch T (2001). *Mol. Cell. Biol.* **21**: 7537.
100. Martinez Arias A, Zecchini V and Brennan K (2002). *Curr. Opin. Genet. Dev.* **12**: 524.
101. Takebayashi K, Sasai Y, Sakai Y, Watanabe T, Nakanishi S and Kageyama R (1994). *J. Biol. Chem.* **269**: 5150.
102. Jiang Y-J and Lewis J (2001). Notch as a synchroniser in somite segmentation. In Sanders EJ, Lash JW and Ordahl CP (eds.). *Proceedings of the NATO Advanced Research Workshop on The Origin and Fate of Somites*, vol. 329. IOS Press, Amsterdam, pp. 71.
103. Lewis J (2003). *Curr. Biol.* **13**: 1398.
104. Brückner K, Perez L, Clausen H and Cohen S (2000). *Nature* **406**: 411.

105. Jiang Y-J, Brand M, Heisenberg C-P, Beuchle D, Furutani–Seiki M, Kelsh RN, Warga RM, Granato M, Haffter P, Hammerschmidt M, Kane DA, Mullins MC, Odenthal J, van Eeden FJM and Nüsslein–Volhard C (1996). *Development* **123**: 205.
106. Krebs LT, Deftos ML, Bevan MJ and Gridley T (2001). *Dev. Biol.* **238**: 110.
107. Öberg C, Li J, Pauley A, Wolf E, Gurney M and Lendahl U (2001). *J. Biol. Chem.* **276**: 35847.
108. Wu G, Lyapina S, Das I, Li J, Gurney M, Pauley A, Chui I, Deshaies RJ and Kitajewski J (2001). *Mol. Cell. Biol.* **21**: 7403.
109. Cornell M, Evans DAP, Mann R, Fostier M, Flasza M, Monthatong M, Artavanis–Tsakonas S and Baron M (1999). *Genetics* **152**: 567.
110. Qiu L, Joazeiro C, Fang N, Wang HY, Elly C, Altman Y, Fang D, Hunter T and Liu YC (2000). *J. Biol. Chem.* **275**: 35734.
111. McGill MA and McGlade CJ (2003). *J. Biol. Chem.* **278**: 23196.
112. Takeyama K, Aguiar RCT, Gu L, He C, Freeman GJ, Kutok JL, Aster JC and Shipp MA (2003). *J. Biol. Chem.* **278**: 21930.
113. Matsuno K, Diederich RJ, Go MJ, Blaumueller CM and Artavanis–Tsakonas S (1995). *Development* **121**: 2633.
114. Izon DJ, Aster JC, He Y, Weng A, Karnell FG, Patriub V, Xu L, Bakkour S, Rodriguez C, Allman D and Pear WS (2002). *Immunity* **16**: 231.
115. Yeh E, Dermer M, Commisso C, Zhou L, McGlade CJ and Boulianne GL (2001). *Curr. Biol.* **11**: 1675.
116. Lai EC, Deblandre GA, Kintner C and Rubin GM (2001). *Dev. Cell* **1**: 783.
117. Deblandre GA, Lai EC and Kintner C (2001) *Dev. Cell* **1**: 795.
118. Gupta–Rossi N, Le Bail O, Gonen H, Brou C, Logeat F, Six E, Ciechanover A and Israël A (2001). *J. Biol. Chem.* **276**: 34371.
119. Strom A, Castella P, Rockwood J, Wagner J and Caudy M (1997). *Genes Dev.* **11**: 3168.
120. Bounpheng MA, Dimas JJ, Dodds SG and Christy BA (1999). *FASEB J.* **13**: 2257.
121. Jogi A, Persson P, Grynfeld A, Pahlman S and Axelson H (2002). *J. Biol. Chem.* **277**: 9118.
122. Baonza A, de Celis JF and Garcia–Bellido A (2000). *Development* **127**: 2383.
123. Campuzano S (2001). *Oncogene* **20**: 8299.
124. Davis RL, Turner DL, Evans LM and Kirschner MW (2001). *Dev. Cell.* **1**: 553.

125. Cooke J (1975). *Nature* **254**: 196.
126. Hamilton L (1969). *J. Embryol. Exp. Morphol.* **22**: 253.
127. Hornbruch A, Summerbell D and Wolpert L (1979). *J. Embryol. Exp. Morphol.* **51**: 51.
128. Takada S, Stark KL, Shea MJ, Vassileva G, McMahon JA and McMahon AP (1994). *Genes Dev.* **8**: 174.
129. Sun X, Meyers EN, Lewandoski M and Martin GR (1999). *Genes Dev.* **13**: 1834.
130. Yamaguchi TP, Harpal K, Henkemeyer M and Rossant J (1994). *Genes Dev.* **8**: 3032.
131. Deng CX, Wynshaw–Boris A, Shen MM, Daugherty C, Ornitz DM and Leder P (1994). *Genes Dev.* **8**: 3045.
132. Dubrulle J, McGrew MJ and Pourquié O (2001). *Cell* **106**: 219.
133. Sawada A, Shinya M, Jiang Y-J, Kawakami A, Kuroiwa A and Takeda H (2001). *Development* **128**: 4873.
134. Xu X, Li C, Takahashi K, Slavkin HC, Shum L and Deng CX (1999). *Dev. Biol.* **208**: 293.
135. Beck CW and Slack JMW (1999). *Development* **126**: 1611.
136. Krumlauf R (1994). *Cell* **78**: 191.
137. van der Hoeven F, Zákány J and Duboule D (1996). *Cell* **85**: 1025.
138. Duboule D (1994). *Development* **120 Suppl.**, 135.
139. Zákány J, Kmita M, Alarcon P, de la Pompa J-L and Duboule D (2001). *Cell* **106**: 207.
140. McDonald MJ and Rosbash M (2001). *Cell* **107**: 567.
141. Etchegaray JP, Lee C, Wade PA and Reppert SM (2003). *Nature* **421**: 177.
142. Christ B and Ordahl CP (1995). *Anat. Embryol. (Berl.)* **191**: 381.
143. Stern CD and Keynes RJ (1987). *Development* **99**: 261.
144. Holley SA and Takeda H (2002). *Semin. Cell Dev. Biol.* **13**: 481.
145. Saga Y, Hata N, Koseki H and Taketo MM (1997). *Genes Dev.* **11**: 1827.
146. Dunwoodie SL, Henrique D, Harrison SM and Beddington RSP (1997). *Development* **124**: 3065.
147. Yamaguchi TP, Conlon RA and Rossant J (1992). *Dev. Biol.* **152**: 75.
148. Biben C, Stanley E, Fabri L, Kotecha S, Rhinn M, Drinkwater C, Lah M, Wang CC, Nash A, Hilton D, Ang SL, Mohun T and Harvey RP (1998). *Dev. Biol.* **194**: 135.

149. Nieto MA, Gilardi–Hebenstreit P, Charnay P and Wilkinson DG (1992). *Development* **116**: 1137.
150. Cohen B, Bashirullah A, Dagnino L, Campbell C, Fisher WW, Leow CC, Whiting E, Ryan D, Zinyk D, Boulianne G, Hui CC, Gallie B, Phillips RA, Lipshitz HD and Egan SE (1997). *Nat. Genet.* **16**: 283.
151. Johnston SH, Rauskolb C, Wilson R, Prabhakaran B, Irvine KD and Vogt TF (1997). *Development* **124**: 2245.
152. Bettenhausen B, Hrabe de Angelis M, Simon D, Guenet JL and Gossler A (1995). *Development* **121**: 2407.
153. Bergemann AD, Cheng HJ, Brambilla R, Klein R and Flanagan JG (1995). *Mol. Cell. Biol.* **15**: 4921.
154. Mansouri A, Yokota Y, Wehr R, Copeland NG, Jenkins NA and Gruss P (1997). *Dev. Dyn.* **210**: 53.
155. Schmidt C, Christ B, Maden M, Brand–Saberi B and Patel K (2001) *Dev. Dyn.* **220**: 377.
156. Baker RK and Antin PB (2003). *Dev. Dyn.* **228**: 128.
157. Buchberger A, Bonneick S, Klein C and Arnold HH (2002). *Dev. Dyn.* **223**: 108.
158. Westin J and Lardelli M (1997) *Dev. Genes Evol.* **207**: 51.
159. Haddon C, Smithers L, Schneider–Maunoury S, Coche T, Henrique D and Lewis J (1998). *Development* **125**: 359.
160. Yamamoto A, Amacher SL, Kim S-H, Geissert D, Kimmel CB and De Robertis EM (1998). *Development* **125**: 3389.
161. Durbin L, Brennan C, Shiomi K, Cooke J, Barrios A, Shanmugalingam S, Guthrie B, Lindberg R and Holder N (1998). *Genes Dev.* **12**: 3096.
162. Leve C, Gajewski M, Rohr KB and Tautz D (2001). *Dev. Genes Evol.* **211**: 493.
163. Weinberg ES, Allende ML, Kelly CS, Abdelhamid A, Murakami T, Andermann P, Doerre OG, Grunwald DJ and Riggleman B (1996) *Development* **122**: 271.
164. Kim S-H, Yamamoto A, Bouwmeester T, Agius E and De Robertis EM (1998). *Development* **125**: 4681.
165. Sparrow DB, Jen WC, Kotecha S, Towers N, Kintner C and Mohun TJ (1998). *Development* **125**: 2041.
166. Pourquié O and Tam PPL (2001). *Dev. Cell* **1**: 619.
167. Sato Y, Yasuda K and Takahashi Y (2002). *Development* **129**: 3633.
168. Saga Y, Hata N, Kobayashi S, Magnuson T, Seldin MF and Taketo MM (1996). *Development* **122**: 2769.

169. Buchberger A, Seidl K, Klein C, Eberhardt H and Arnold HH (1998). *Dev. Biol.* **199**: 201.
170. Takahashi Y, Koizumi K, Takagi A, Kitajima S, Inoue T, Koseki H and Saga Y (2000). *Nat. Genet.* **25**: 390.
171. Koizumi K, Nakajima M, Yuasa S, Saga Y, Sakai T, Kuriyama T, Shirasawa T and Koseki H (2001). *Development* **128**: 1391.
172. Nomura-Kitabayashi A, Takahashi Y, Kitajima S, Inoue T, Takeda H and Saga Y (2002). *Development* **129**: 2473.
173. Kume T, Jiang H, Topczewska JM and Hogan BLM (2001). *Genes Dev.* **15**: 2470.
174. Topczewska JM, Topczewski J, Solnica-Krezel L and Hogan BLM (2001). *Mech. Dev.* **100**: 343.
175. Topczewska JM, Topczewski J, Shostak A, Kume T, Solnica-Krezel L and Hogan BLM (2001). *Genes Dev.* **15**: 2483.
176. van Eeden FJM, Granato M, Schach U, Brand M, Furutani-Seiki M, Haffter P, Hammerschmidt M, Heisenberg C-P, Jiang Y-J, Kane DA, Kelsh RN, Mullins MC, Odenthal J, Warga RM, Allende ML, Weinberg ES and Nüsslein-Volhard C (1996). *Development* **123**: 153.
177. Nikaido M, Kawakami A, Sawada A, Furutani-Seiki M, Takeda H and Araki K (2002). *Nat. Genet.* **31**: 195.
178. Latinkic BV, Umbhauer M, Neal KA, Lerchner W, Smith JC and Cunliffe V (1997). *Genes Dev.* **11**: 3265.
179. Griffin KJP, Amacher SL, Kimmel CB and Kimelman D (1998). *Development* **125**: 3379.
180. Chapman DL and Papaioannou VE (1998). *Nature* **391**: 695.
181. Beckers J, Schlautmann N and Gossler A (2000). *Mech. Dev.* **95**: 35.
182. White PH, Farkas DR, McFadden EE and Chapman DL (2003). *Development* **130**: 1681.
183. del Barco Barrantes I, Elia AJ, Wünsch K, Hrabe de Angelis M, Mak TW, Rossant J, Conlon RA, Gossler A and de la Pompa JL (1999). *Curr. Biol.* **9**: 470.
184. Burgess R, Alan R, Brown D, Bradley A and Olson EN (1996) *Nature* **384**: 570.
185. Sosic D, Brand-Saberi B, Schmidt C, Christ B and Olson EN (1997). *Dev. Biol.* **185**: 229.
186. Duband JL, Dufour S, Hatta K, Takeichi M, Edelman GM and Thiery JP (1987). *J. Cell Biol.* **104**: 1361.

187. Obata S, Sago H, Mori N, Rochelle JM, Seldin MF, Davidson M, St John T, Taketani S and Suzuki ST (1995). *J. Cell Sci.* **108**: 3765.
188. Kim S-H, Jen W-C, De Robertis EM and Kintner C (2000). *Curr. Biol.* **10**: 821.
189. Rhee J, Takahashi Y, Saga Y, Wilson–Rawls J and Rawls A (2003). *Dev. Biol.* **254**: 248.
190. Yamamoto A, Kemp C, Bachiller D, Geissert D and De Robertis EM (2000). *Genesis* **27**: 49.
191. Quertermous EE, Hidai H, Blanar MA and Quertermous T (1994). *Proc. Natl. Acad. Sci. USA* **91**: 7066.
192. Burgess R, Cserjesi P, Ligon KL and Olson EN (1995). *Dev. Biol.* **168**: 296.
193. Barnes GL, Alexander PG, Hsu CW, Mariani BD and Tuan RS (1997). *Dev. Biol.* **189**: 95.
194. Holland SJ, Gale NW, Mbamalu G, Yancopoulos GD, Henkemeyer M and Pawson T (1996). *Nature* **383**: 722.
195. Xu Q, Mellitzer G, Robinson V and Wilkinson DG (1999). *Nature* **399**: 267.
196. Mellitzer G, Xu Q and Wilkinson DG (1999). *Nature* **400**: 77.
197. Holder N and Klein R (1999). *Development* **126**: 2033.
198. Wilkinson DG (2001). *Nat. Rev. Neurosci.* **2**: 155.
199. Flenniken AM, Gale NW, Yancopoulos GD and Wilkinson DG (1996). *Dev. Biol.* **179**: 382.
200. Gale NW, Holland SJ, Valenzuela DM, Flenniken A, Pan L, Ryan TE, Henkemeyer M, Strebhardt K, Hirai H, Wilkinson DG, Pawson T, Davis S and Yancopoulos GD (1996). *Neuron* **17**: 9.
201. De Bellard ME, Ching W, Gossler A and Bronner–Fraser M (2002). *Dev. Biol.* **249**: 121.
202. Dottori M, Hartley L, Galea M, Paxinos G, Polizzotto M, Kilpatrick T, Bartlett PF, Murphy M, Kontgen F and Boyd AW (1998). *Proc. Natl. Acad. Sci. USA* **95**: 13248.
203. Wang HU, Chen ZF and Anderson DJ (1998). *Cell* **93**: 741.
204. Cooper MS and Kimmel CB (1998). Morphogenetic cell behaviors and specification of cell fate during early teleost development. In Soll DR (ed.). *Motion analysis of living cells*, Wiley–Liss, Inc., New York, pp. 177.
205. Henry CA, Hall LA, Hille MB, Solnica–Krezel L and Cooper MS (2000). *Curr. Biol.* **10**: 1063.

206. Kulesa PM and Fraser SE (2002). *Science* **298**: 991.
207. Trainor PA and Krumlauf R (2000). *Nat. Rev. Neurosci.* **1**: 116.
208. Linask KK, Ludwig C, Han MD, Liu X, Radice GL and Knudsen KA (1998). *Dev. Biol.* **202**: 85.
209. Rawls A, Wilson–Rawls J and Olson EN (2000). *Curr. Top. Dev. Biol.* **47**: 131.
210. Ostrovsky D, Cheney CM, Seitz AW and Lash JW (1983). *Cell Differ.* **13**: 217.
211. Lash JW and Yamada KM (1986).The adhesion recognition signal of fibronectin: a possible trigger mechanism for compaction during somitogenesis. In Bellairs R, Ede DA and Lash JW (ed.). *Proceedings of the NATO Advanced Research Workshop on Somites in Developing Embryos*, vol. 118. Plenum Press, New York, pp. 201.
212. Lash JW, Linask KK and Yamada KM (1987). *Dev. Biol.* **123**: 411.
213. George EL, Georges–Labouesse EN, Patel–King RS, Rayburn H and Hynes RO (1993). *Development* **119**: 1079.
214. Georges–Labouesse EN, George EL, Rayburn H and Hynes RO (1996). *Dev. Dyn.* **207**: 145.
215. Yang JT, Bader BL, Kreidberg JA, Ullman–Cullere M, Trevithick JE and Hynes RO (1999). *Dev. Biol.* **215**: 264.
216. Shin CS, Lecanda F, Sheikh S, Weitzmann L, Cheng SL and Civitelli R (2000) *J. Cell. Biochem.* **78**: 566.
217. Marambaud P, Shioi J, Serban G, Georgakopoulos A, Sarner S, Nagy V, Baki L, Wen P, Efthimiopoulos S, Shao Z, Wisniewski T and Robakis NK (2002). *EMBO J.* **21**: 1948.
218. Marambaud P, Wen PH, Dutt A, Shioi J, Takashima A, Siman R and Robakis NK (2003). *Cell* **114**: 635.
219. Radice GL, Rayburn H, Matsunami H, Knudsen KA, Takeichi M and Hynes RO (1997). *Dev. Biol.* **181**: 64.
220. Horikawa K, Radice G, Takeichi M and Chisaka O (1999). *Dev. Biol.* **215**: 182.
221. Malicki J, Jo H and Pujic Z (2003). *Dev. Biol.* **259**: 95.
222. Orsulic S and Kemler R (2000). *J. Cell Sci.* **113**: 1793.
223. Mitsiadis TA, Henrique D, Thesleff I and Lendahl U (1997). *Development* **124**: 1473.
224. Myat A, Henrique D, Ish–Horowicz D and Lewis J (1996). *Dev. Biol.* **174**: 233.

225. Suetsugu R, Sato Y and Takahashi Y (2002). *Gene Expr. Patterns* **2**: 157.
226. Wu JY, Wen L, Zhang WJ and Rao Y (1996). *Science* **273**: 355.
227. Prince VE, Holley SA, Bally–Cuif L, Prabhakaran B, Oates AC, Ho RK and Vogt TF (2001). *Mech. Dev.* **105**: 175.
228. Cole SE, Levorse JM, Tilghman SM and Vogt TF (2002) *Dev. Cell* **3**: 75.
229. Beckers J, Caron A, Hrabe de Angelis M, Hans S, Campos–Ortega JA and Gossler A (2000). *Mech. Dev.* **95**: 23.
230. Hans S and Campos–Ortega JA (2002). *Development* **129**: 4773.
231. Rauch G-J, Hammerschmidt M, Blader P, Schauerte HE, Strähle U, Ingham PW, McMahon AP and Haffter P (1997). *Cold Spring Harb. Symp. Quant. Biol.* **62**: 227.
232. Gorodilov YN (1992). *Zool. Sci.* **9**: 1101.
233. Monk NA (2003). *Curr. Biol.* **13**: 1409.
234. Jho EH, Zhang T, Domon C, Joo CK, Freund JN and Costantini F (2002). *Mol. Cell. Biol.* **22**: 1172.
235. Lustig B, Jerchow B, Sachs M, Weiler S, Pietsch T, Karsten U, van de Wetering M, Clevers H, Schlag PM, Birchmeier W and Behrens J (2002). *Mol. Cell. Biol.* **22**: 1184.
236. Mazaleyrat SL, Fostier M, Wilkin MB, Aslam H, Evans DA, Cornell M and Baron M (2003). *Dev. Biol.* **255**: 363.
237. Jehn BM, Dittert I, Beyer S, von der Mark K and Bielke W (2002). *J. Biol. Chem.* **277**: 8033.
238. Nie J, McGill MA, Dermer M, Dho SE, Wolting CD and McGlade CJ (2002). *EMBO J.* **21**: 93.
239. Susini L, Passer BJ, Amzallag–Elbaz N, Juven-Gershon T, Prieur S, Privat N, Tuynder M, Gendron M-C, Israël A, Amson R, Oren M and Telerman A (2001). *Proc. Natl. Acad. Sci. USA* **98**: 15067.
240. Juven–Gershon T, Shifman O, Unger T, Elkeles A, Haupt Y and Oren M (1998). *Mol. Cell. Biol.* **18**: 3974.
241. Goldbeter A (2002). *Nature* **420**: 238.
242. Fryer CJ, Lamar E, Tubachova I, Kinther C and Jones KA (2002). *Genes Dev.* **16**: 1397.
243. Takahashi Y, Inoue T, Gossler A and Saga Y (2003). *Development* **130**: 4259.

Chapter 10

Molecular Regulation of Fish Muscle Development and Growth

Shao Jun Du
Center of Marine Biotechnology
University of Maryland Biotechnology Institute

The formation of skeletal muscles in vertebrate embryos involves a series of events including induction, specification and differentiation. The multi-potential mesoderm cells in gastrula stage embryos are first induced to become paraxial mesoderm that subsequently form segmented somites. Somitic cells are specified into osteoblasts, myoblasts and dermal mesenchyme that ultimately differentiate into axial skeleton, skeletal muscles, and dermis of the skin, respectively. Somite formation and subsequent specification and differentiation of myoblasts are regulated by many extracellular signaling molecules and intracellular transcription factors. Signaling molecules of the Notch and FGF families are involved in somitogenesis, whereas Hedgehog (Hh), Wnt and TGF-β families play critical roles in myoblast specification and differentiation. These signaling molecules bind to their receptors and activate or repress intracellular transcription factors, such as members of the MyoD family and the paired transcription factors Pax3 and Pax7 that directly regulate muscle-specific gene expression, and muscle cell differentiation. The challenging task at present is to understand how these signaling cascades coordinate with each other and how they control the myogenic transcription network to allow for precise changes in gene expression during myoblast specification and differentiation.

Introduction

Fish, especially zebrafish, have become the excellent model for muscle research because fish skeletal muscles have several unique features that offer many advantages over other systems. First, fish embryos develop

---

Correspondence to: Shao Jun Du, Center of Marine Biotechnology, University of Maryland Biotechnology Institute, 701 E. Pratt St., Baltimore, MD 21202, USA. Tel: 410-234-8854; Fax: 410-234-8896; E-mail: dus@umbi.umd.edu

externally, allowing detailed patterns of somite formation and myofiber differentiation to be easily observed during myogenesis. Second, fish embryos have a relatively simple muscle system in which fast and slow muscles as well as their respective precursors are clearly separated from each other, thus providing an ideal model for tracing their development and studying their differentiation.[1,2] Third, myogenesis in zebrafish embryos begins relatively early during development, the first spontaneous muscle contractions occur at 18 hours post-fertilization (hpf). By 24 hpf, functional embryonic myofibers are well developed, greatly reducing the time required to carry out the study.[3] Fourth, growing collections of zebrafish mutants with defects in early muscle development provide a rich resource for identifying new regulatory factors in muscle development.[4,5] Moreover, the transparent nature of the zebrafish embryos and reliable transgenic technology provides a powerful tool to analyze the regulation of muscle-specific gene expression in developing embryos.[6] Finally, the established protocol for expressing gene products through RNA or DNA injection or knocking down gene expression by morpholino antisense, complement very well with the genetic approach in delineating the *in vivo* functions of genes that regulate muscle development.[7] In the past few years, fish have been used as models to facilitate many studies that were difficult or impractical in other vertebrates. Results from these studies have made several important contributions to the field of muscle developmental biology.[8] In this review, I will discuss the recent progress in this fast developing field. The primary focus of this review will be on somitogenesis, signaling in muscle type specification, regulation of myogenic gene expression, muscle growth regulation, and fish as a model for muscular dystrophy research.

1. Somitogenesis in Fish Embryos

As in other vertebrates, fish skeletal muscles are derived from cells in the somites. Zebrafish has been used as a model to study the molecular mechanisms regulating somitogenesis (see reviews in Stickney *et al.*[9]; Holley and Nusslein-Volhard[10]; Pourquie[11]; Brennan *et al.*[12]). Somites are formed by somitogenesis which is segmentation of the paraxial mesoderm in the trunk and tail of vertebrate embryos. The somitic cells give rise to the axial skeleton, the dermis of the back and skeletal muscles of the body. In contrast to somitic cells in high vertebrates, most of the somitic cells in fish contribute to the myotome that gives rise to skeletal muscles representing approximately 60% of the body weight. Only a small group of somitic cells located at the ventral region of the somite are committed to sclerotome that give rise to axial skeleton.[13] Because fish are supported by the buoyancy of water and their swim bladder, they have no use for the robust skeleton required for bigger animals living on land. Instead, fish develop large muscles to locomote through water.

1.1. Notch Signaling in Somitogenesis

Better understanding of muscle development requires knowledge of the molecular regulation that underlies segmentation, somite epithelialization, and somite patterning. Both embryological and genetic studies have contributed to the current understanding of how somites form and differentiate in fish. Embryological studies indicate the ability to form segmental somites appears to be a property specific to the paraxial mesoderm. There appears to be a prepattern within the unsegmented presomitic mesoderm (PSM) acquired during early development[14] that is linked to a molecular clock that controls cells in the PSM going through cyclic expression of the Notch pathway target genes.

Notch signaling, a widely used mechanism regulating cell fate during development, has been implicated in somitogenesis regulation. Notch signaling pathway is extremely conserved during evolution. Interaction of Notch receptors (Notch) with their ligands (Delta, Jagged, or Serrate proteins) leads to cleavage of Notch intracellular domain (NICD) and nuclear translocalization. In the nucleus, NICD associates with a transcription factor, RBP-Jk, to activate expression of Notch target genes, such as hairy and *enhancer of split* (*hes*) and HES-related repressor protein (HER) transcriptional repressors.[15–17] Members of the Notch signaling pathway are expressed in PSM and developing somites. It has been demonstrated in mouse and zebrafish that Notch/Delta signaling pathway is required for synchronization of the somite segmentation clock.[18–20] Expression analysis revealed that stripes of *her1* and *her7* expression oscillate within the PSM and are central components of the segmentation oscillator.[21,22] They function together to refine alternating somite boundaries.[23] Recent studies further revealed that the Notch/Delta pathway is not simply required for the oscillator readout, but also constitutes the core components of the segmentation clock.[24]

Genetic screens in zebrafish have identified over 50 genes that affect virtually every aspect of somite development.[4] Two groups of mutants with defects in this patterning process have been isolated. The first mutant group showed defects in the anteroposterior patterning within individual somites. Mutants in this group include *fused somites* (*fss*), *beamter* (*bea*), *deadly seven* (*des*), *after eight* (*aei*) and *white tail* (*wit*). In these mutant embryos, the early boundaries between individual somites are invisible although irregular somite boundaries become visible at a later stage. The zebrafish mutation *after eight* (*aei*) required for both somitogenesis and neurogenesis has been mapped, and it encodes the Notch ligand DeltaD.[25] In the second group of mutants, formation of the horizontal myoseptum, which separates the dorsal and ventral part of the myotome, is reduced. Six genes have been defined in this group (*you-type genes*). Several of the *you-type* mutants carry mutations in the Hedgehog signaling pathway and show developmental defect in slow muscles.

1.2. FGF and Eph Signaling Pathways in Somitogenesis

In addition to Notch/Delta pathway, other signaling molecules expressed in the PSM and developing somite are also involved in specifying paraxial mesoderm fates and regulation of somite formation. It has been shown that *Fgfr1* is expressed in the PSM and anterior somites in mice and zebrafish embryos.[26,27] Fgfr1 knock-out mice show defect in the segment borders.[28] Fgf8 is also expressed in the PSM and newly formed somites in chick and zebrafish.[29,30] Molecular and embryological studies have demonstrated that FGF8 provides a crucial positional cue that controls the somite boundary formation and regulation of segmentation clock.[29,30] Mice and zebrafish *fgf8* mutants fail to develop through gastrulation making it difficult to study FGF8 function in somitogenesis.[31,32] Recent studies suggest an important role for Eph family of receptor kinases in anteroposterior patterning of the somite and somite boundary formation in zebrafish embryos.[33,34] Though it is apparent that combinations of signals are required to specify the paraxial mesoderm along the entire anterior-posterior axis during somitogenesis, how FGF and Eph signaling pathways interact with the Notch/Delta pathway is not clear at present.

1.3. Functions of Transcription Factors Mesp and Foxc in Somitogenesis

Several families of intracellular transcription factors have been implicated in anteroposterior patterning of developing somites. The Mesp family of bHLH transcription factors, *mesp-a* and *mesp-b*, are segmentally expressed in the paraxial mesoderm in the anterior parts of somite primordia during somitogenesis.[26] In *fused somites* (*fss*) embryos, in which all early somite boundary formation is blocked, the expression of *mesp-a* and *mesp-b* cannot be detected during the segmentation period. In the trunk region of zebrafish embryos, *mesp* transcripts are co-expressed with *her1* in every somite primordium posterior to the forming somites.[21] Ectopic expression of Mesp-b in zebrafish embryos causes a loss of the posterior identity within the somite primordium, leading to a segmentation defect.[26] The direct target genes of Mesp-a and Mesp-b and the molecular mechanisms by which Mesp-a and Mesp-b control their expression remain to be identified. Mespb homologue (Mesp2) plays an essential role in somite segmentation in mice.[35] Takahashi and colleagues[36] have shown that Mesp2 initiates somite segmentation through the Notch signaling pathway.

Recent studies have demonstrated that the winged helix transcription factor Foxc1 is another essential factor for somitogenesis in zebrafish.[37] Zebrafish foxc1a and foxc1b are homologous of the mouse forkhead gene, Foxc1.[38] Mouse embryos that are compound null mutants for Foxc1 and the closely related Foxc2 have no morphological somites and show abnormal expression of Notch signaling pathway genes in the anterior PSM.[39] Zebrafish *foxc1a* and *foxc1b* are expressed in the unsegmented presomitic mesoderm, newly formed somites, adaxial cells, and head mesoderm.[38] Knockdown of Foxc1a (but not Foxc1b) expression by morpholino antisense approach blocks formation of somite segment boundaries, and segmented expression of genes in anterior or posterior somites involved in somite epithelialization,[37] indicating that zebrafish Foxc1a plays an essential and conserved role in somite formation and the anteroposterior patterning of somite primordia.

2. Signaling Molecules in Myoblast Specification and Differentiation

Somite patterning following somitogenesis requires extensive tissue interactions. Virtually all tissues surrounding the somites provide signals that induce or inhibit particular differentiation pathways of somitic cells. It is well documented that the myogenic differentiation pathway is regulated by extracellular signaling molecules from the notochord, neural tube, surface ectoderm and lateral plate. Signal molecules Hh, Noggin, Wnt and Bone morphogenetic proteins (BMPs) expressed by these tissues are important regulators for myogenesis.[40–42] These signal molecules provide both positive and negative regulations on the myogenic transcriptional network. Hh, Wnt and Noggin signals, for example, activate *myf5* and *myoD* expression.[43,44] By contrast, BMP signals from the lateral plate inhibit *MyoD* expression and myogenesis.[43,45] These multiple signals coordinate together to regulate myogenesis.

The essential role of the notochord in zebrafish myogenesis was first discovered during the analysis of zebrafish mutant with notochord defect. Halpern and colleagues[46] found that *no tail* (*ntl*) mutants not only exhibited a defect in notochord formation, but also in skeletal muscles. The muscle defect in the *ntl* mutant could be rescued by transplanted wild-type notochord cells.[46] Additional support for the functional role of notochord in muscle development came from the characterization of *floating head* (*flh*) and *bozozok* (*boz*) mutants, which exhibit both notochord and muscle defects.[47–49] Recently, it was discovered that Hedgehog proteins are candidate signals from the notochord, required for maintenance of *myf5* and *myoD* expression and specification and differentiation of slow muscle cells in zebrafish embryos.[49–52]

2.1. Hedgehog Signaling in Slow Muscle Development

Vertebrate muscles are composed of several fiber types, the precise combination of which determines the muscle function. Like vertebrates, fish skeletal muscles are composed of two major types of muscle fibers, classified as fast or slow fibers. The fast muscles represent the major portion of fish muscles. Fast and slow muscles express different isoforms of myosin heavy chains. Fast muscles have a fast contraction speed and they are used during bursts of rapid swimming. Slow muscles, on the other hand, have slow contractions that are specialized for slow swimming. Slow fibers in zebrafish embryos are smaller and are mononucleated cells in contrast to the nutinucleated fast fibers.[53] Slow muscles are more heavily vasculated than fast fibers. Because of their distinct functions and patterns of gene expression, they have become a model system for studying molecular mechanisms involved in cell fate determination.[54–57] Extensive research in chick and mouse have demonstrated that although neonatal and adult fiber type is influenced by extrinsic factors, such as neural input and muscle load, there is little knowledge of how muscle cells are initially determined in the early embryos. In the past, the traditional model has suggested that all myogenic precursor cells originate as "generic" myoblasts derived from paraxial mesoderm and subsequently differentiate into either fast or slow muscle fibers.[54,56] More recent findings have shown that specification of slow and fast muscle cells occurs much earlier (Fig. 1), and that the earliest embryonic myoblasts have intrinsic properties of fast or slow fibers.[57]

In mammals and birds, the fast and slow muscles are often intermingled with each other, making it difficult to study their specification and differentiation. By contrast, fish fast and slow muscles

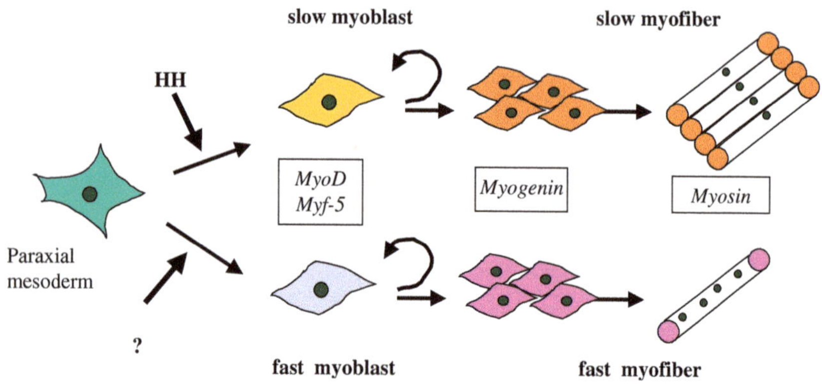

Fig. 1 Diagram showing myoblast specification and differentiation during development. In zebrafish embryos, slow muscles are mononucleate myofibers while the fast muscles are multinucleate fibers. Hedgehog signal is involved in slow muscle formation, while signals in fast muscle formation is not clear. Signal molecules such as FGFs and Wnts may be involved in fast muscle differentiation, however, their roles are yet to be determined.

are clearly separated. Fast muscles in fish are located in the deeper portion of the myotome, while slow muscles are found in wedge-shaped triangle on the lateral surface of adult myotome (Fig. 2). Precursors of slow and fast muscles can also be identified very early in development. Devoto *et al.*[1] have demonstrated that slow and fast muscle cells in zebrafish arise from two distinct populations of precursors in the paraxial mesoderm. Slow muscle precursor cells, known as adaxial cells, are derived from medial presomatic cells in the segmental plate (Fig. 3). These cells differentiate into the muscle pioneer (MP), and non-muscle pioneer (NMP) slow muscle cells which migrate to the superficial layer of the myotome. MP cells are early developing muscle fibers that differentiate adjacent to the notochord at the level of the future horizontal myoseptum. Unlike NMP, they express Engrailed proteins.[58] There are some evidence indicating that zebrafish MP cells are neither required for proper development of other fibers,[49] nor for axon guidance.[59] MP cells may instead play a role in development of the horizontal myoseptum.[46] By comparison, fast muscle precursors arise from lateral presomitic cells in the segmental plate and remain deep within the myotome of the somites (Fig. 3). The clear distribution of fast and slow muscles in fish make fish excellent models for studying

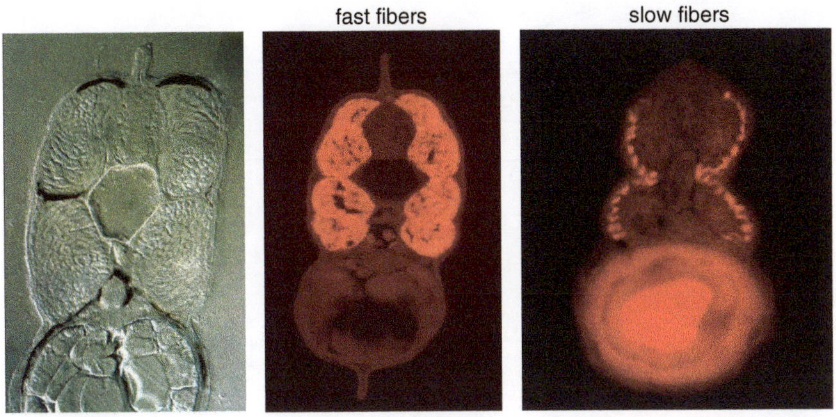

Fig. 2 The slow and fast muscles revealed by antibody staining are located in separate regions in the fish somite.

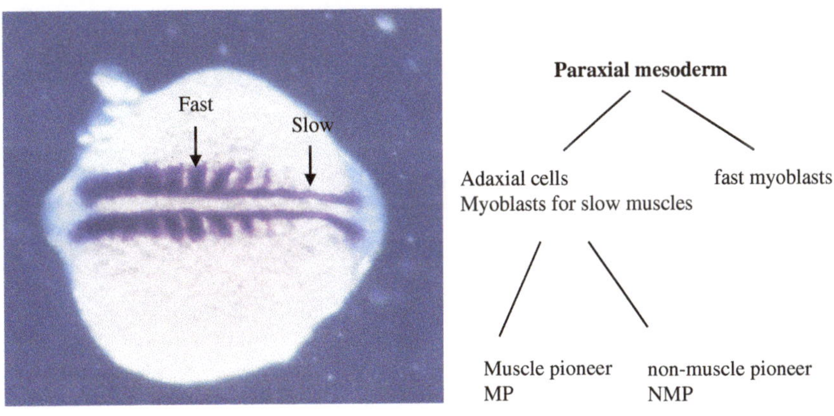

Fig. 3 MyoD expression reveals distinct localization of slow and fast muscle precursors in early stage zebrafish embryos. The slow muscle precursors also called adaxial cells, further separate into muscle pioneer and non-muscle pioneer cells.

the molecular mechanisms involved in muscle fiber specification and differentiation.[60–62]

Recent studies have demonstrated that members of the Hh family induce the formation of slow muscle fibers. *Hedgehog* gene was first cloned in *Drosophila* in 1992.[63–65] Since then several Hhs have been

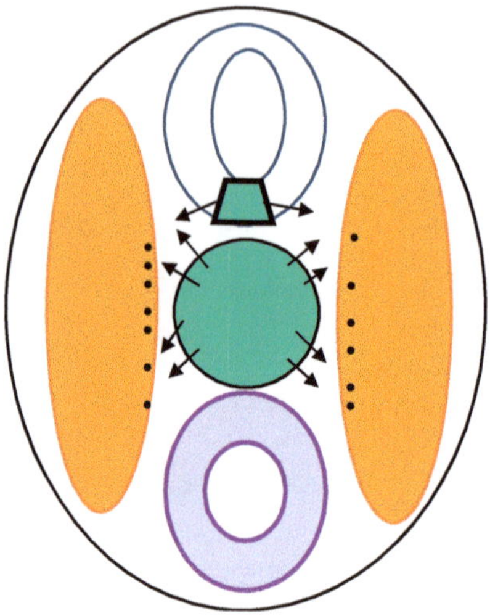

	N	F
Shh	+	+
Twhh	−	+
Ehh	+	−

Fig. 4 The overlapping expression patterns of *shh*, *twhh* and *ehh* in notochord (N) and floor plate (F) of zebrafish embryos.

identified in many invertebrate and vertebrate species. Zebrafish have at least three *Hh* genes, *sonic hedgehog* (*shh*), *tiggy-winkle hedgehog* (*twhh*), and *echidna hedgehog* (*ehh*). The three *Hhs* are expressed in an overlapping pattern in the midline tissues of zebrafish embryos (Fig. 4). *Shh* is expressed in both notochord and floor plate, whereas *twhh* or *ehh* is expressed in floor plate or notochord, respectively.[50,66–68] Shh, Ehh and Twhh have partial redundant functions in regulating slow muscle formation in zebrafish. Ectopic expression of Ehh, Shh or Twhh induces the formation of extra slow muscles (Fig. 5).[49–51] Mutation of Shh (*syu* mutant) alone, however, results in a reduction, but not complete elimination of slow muscles (Fig. 5),[69,70] suggesting that other Hh signals are able to maintain, or reinitiate, some slow muscle development in *syu* mutants. Consistent with this idea, Coutelle *et al.*[52] have shown that in the *cyclops* mutant, the absence of floor plate-derived Twhh and Shh signals has little effect on slow muscle formation. Similarly, the absence of notochord-derived Shh and Ehh signals delay slow muscle development. Removal of both notochord and floor plate-derived Shh and Twhh signals in *cyclops/syu* double mutants, however, essentially abolishes myogenesis of slow muscles. These studies indicate

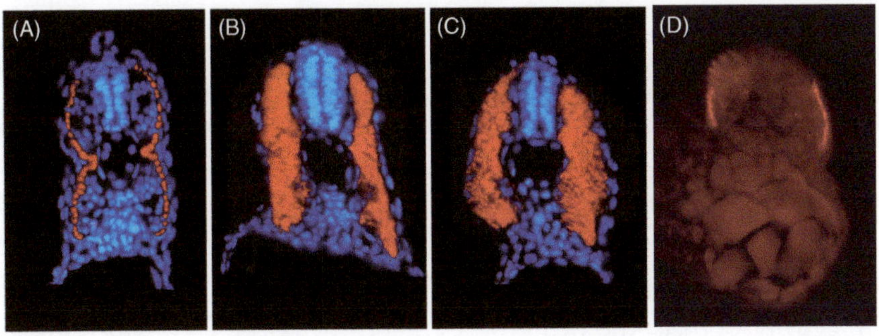

Fig. 5 Hedgehog signaling is involved in slow muscle formation in zebrafish embryos. Ectopic expression of HH induces extra slow muscles, while mutation in Shh (syu) results in slow muscle defect. A, B, C and D: Embryonic sections (dorsal to the top) showing localization of slow muscle cells labeled with F59, an anti-myosin heavy chain antibody, in 24 hpf wild-type (A), Shh mRNA (B) or Twhh mRNA (C) injected embryos, and *syu* mutant (D) embryos.

that the midline signals, likely to be various Hedgehogs, collaborate to control the myogenesis of adaxial slow muscles in the zebrafish embryo. Recent studies by Wolff *et al.*[208] showed that multiple muscle cell identities are induced by distinct levels and timing of Hedgehog activity in the zebrafish embryos.

Hedgehogs are secreted protein that function through the membrane receptors and intracellular transcription factors. Efforts to understand the mechanisms of Hh action have resulted in the identification of their receptors such as Patched and Smoothened, and several downstream intracellular components of the signal transduction pathway, including Fused, Costal-2, Protein Kinase A (PKA), and Glis/Ci.[71] Taipale and colleagues[72] have demonstrated that Patched acts catalytically to suppress the activity of Smoothened. Hh binds to receptor Patched and releases Smoothened from suppression by Patched. Smoothened encodes a seven transmembrane protein, becomes active, and transduces Hh signal into the cytoplasm. The intracellular consequence of this activation is the repression of PKA. In the absence of Hh signal, PKA phosphorylates the Gli family of transcription factors (Ci in Drosophila) and converts them into repressors that constitutively represses Hh target genes. Inactivation of PKA by Hh signal results in the nuclear accumulation of activated forms of Glis that induce Hh target gene transcription. A conserved target of Hh signal is the *patched* gene itself, as up-regulation of *patched* expression by Hh serves to restrict its signaling range.

The molecular components of the Hh signaling pathway appear to be highly conserved during evolution. Several members of the Hh signaling pathway have been identified in fish. Interestingly, two *patched* genes have been identified in zebrafish. Expression analysis revealed a striking correlation between the *patched1* and *patched2* expression and cells responding to Shh activity both in the neurectoderm and mesoderm.[73,74] *Patched 1* and *2* are strongly expressed in adaxial cells before somite formation.[73,74] Their expression is regulated by Hh and PKA activity. At present, there is no evidence for any specificity in the interactions between the various Hh proteins and the two different Patched receptors in zebrafish. Genes encoding the membrane protein, Smoothened (Smo), and the intracellular effector, Glis, of Hh have also been identified in zebrafish. *Smo* is expressed both maternally and zygotically with expression that appears to be widespread throughout the embryo.[75] Three members of the Gli family of transcription factors, namely Gli1, Gli2 and Gli3, have also been isolated in zebrafish. These three members have partial overlapping pattern of expression in Hh-responding tissues.[70,76,77]

The function of Hh receptor and their intracellular effectors in slow muscle formation have been analyzed in zebrafish. Barresi *et al.*[78] and Varga *et al.*[75] have demonstrated that mutation in Smoothened resulted in slow muscle defect. *Slow-muscle-omitted* (*smu*) mutant which is caused by mutation in Smo show a 99% reduction in the number of slow muscle fibers and a complete loss of Engrailed-expressing muscle pioneers. Consistent with the fact that Smo is a membrane receptor for Hh, Smo functions cell-autonomously in slow muscle formation. Cells from wild-type embryos can develop into slow muscle fiber when transplanted into *smu* (-/-) embryos, whereas cells from *smu* mutant embryos cannot develop into slow muscle fibers in wild-type embryos.[78] The functions of Patched were analyzed by overexpression or down-regulating of expression of the Patched1 protein in zebrafish embryos.[69,208] The results showed that overexpression of *patched1* mRNA in zebrafish embryos results in Hh loss-of-function phenotype. The embryos have both paraxial mesoderm and neural tube defects. They display U-shaped somites and

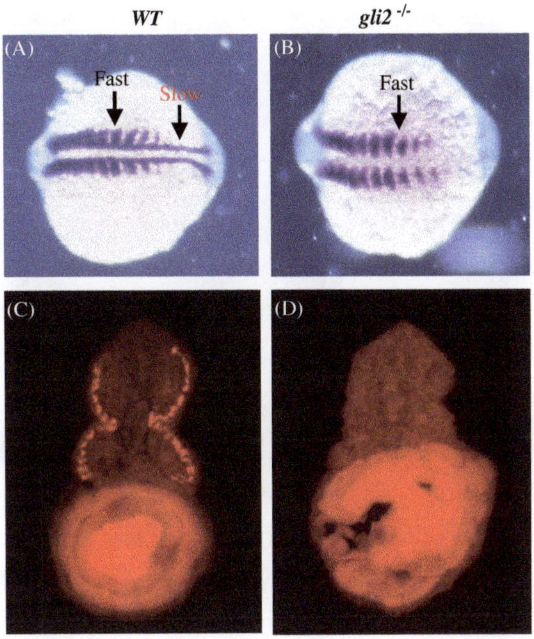

Fig. 6 Gli2 mutation affects *MyoD* expression and causes developmental defect in slow muscles in zebrafish embryos. A and B: *In situ* hybridization showing MyoD expression in wild type (A) or mutant (B) embryos at 12 somites. C and D: Sections (dorsal to the top) showing localization of slow muscle cells labeled with F59, an anti-myosin heavy chain antibody, in 24 hpf wild-type (C) and *yot* mutant (D) embryos.

suppression of slow muscle differentiation is clearly observed. Knockdown of Ptc expression resulted in marked increase in the numbers of muscle pioneer cells in zebrafish embryos.[208] These data demonstrate that Patched acts as a negative regulator of Hh signaling in zebrafish embryos.

The function of Glis in slow muscle formation has been analyzed in zebrafish mutants. In *yot* mutant that carries a mutation in Gli2, development of slow muscles was completely blocked (Fig. 6).[69,70] This defect could not be rescued by ectopic expression of any members of the Hh family in the *yot* mutant embryos. These data suggest that Gli2 play a key role in the differentiation of slow muscle cells by acting as an essential mediator of Shh, Twhh and Ehh. Unlike Gli2, recent studies in zebrafish have demonstrated that the Gli1 mutant (*detour*) exhibits defects in the development of a subset of Hh regulated nerve cells, but shows little or no defect in the development of slow muscle cells.[79,80] These data suggest that Gli1 may not be involved in differentiation of slow muscle cells, or at least it is not essential. At present, it is not clear whether Gli3 is required by Hh for the development of slow muscle cells. In chick embryos, Gli3 and Shh have been found to be mutually repressive in regulating each other's expression in the developing limb.[81] Gli3 mutant mice exhibit abnormalities including lung defects and limb polydactyly, similar to Shh gain-of-function phenotypes.[82,83] Therefore, it remains to be determined if Gli3 has a negative role in Hh-mediated induction of slow muscle cells. It should be noted that Gli2 mutation in *yot* mutant (*yot* ^{ty119}) generates a C-terminal truncated protein.[76] The truncated mutant protein appears to have a dominant repressor effect over other members of the Gli family.[77] Therefore, the possible role of other members of the Gli family in mediating Hh in slow muscle induction cannot be ruled out completely.

Several studies have demonstrated that PKA acts as an inhibitor downstream of Smoothened in the Hh signaling pathway in slow muscle formation. Ectopic expression of a constitutive active form of PKA inhibits slow muscle formation, while overexpression of a dominant negative form of PKA (dnPKA) induced extra slow muscles.[51] In addition, the *smu* mutant phenotype is phenocopied by treatment of wild-type embryos with forskolin, which inhibits the response of cells to Hh signaling by indirect activation of PKA.[78] Overexpression of dnPKA in *smu* mutant embryos rescues the developmental defects of slow muscles in *smu* mutant embryos, whereas overexpression of Shh had no effect.[78]

Recent molecular and genetic studies have identified additional genes that are involved in slow muscle differentiation controlled by Hedgehog signal. Nakano and colleagues[209] have characterized the genetic mutations in *chameleon* mutant, a "*you-type*" mutant with slow muscle defects. They showed that *chemeleon* mutant alleles are associated with premature termination in the *dispatched-1* gene. Dispatched is a multipass transmembrane protein that is dedicated to the secretion of lipid modified Hedgehog from expressing cells.[210] Wolff *et al.*[211] demostrated that *iguana* encodes the zebrafish ortholog of Dzip1, a novel zinc-finger protein with coiled-coil domains, is essential for

Hedgehog signal transduction and slow muscle differentiation in zebrafish.

All above genetic and molecular studies have demonstrated that Hh signaling is required in cell fate determination of slow muscles. There is, however, little information on whether the Hh action is direct or indirect. Using an *in vitro* culture system, Norris and colleagues[84] tested directly the ability of zebrafish myoblasts to respond to exogenous Shh peptide. They found that Shh peptide can control the binary cell fate selection of embryonic myoblasts of zebrafish *in vitro*. These investigators have also used this culture system to assay the relative activities of different members of the Hh family and to investigate the possible involvement of heterotrimeric G-proteins in Hh signal transduction. They showed that different Hh peptides exhibit varying levels of inductive activity in the *in vitro* assay. They found no evidence for involvement of heterotrimeric G-proteins in this process.

In addition to *syu* and *yot*, the *u-boot* mutant has been implicated in the Hh-regulated myogenic switch for fiber-type diversification in the zebrafish embryo.[53] Roy and colleagues showed that in the zebrafish, the *u-boot* gene acts as a myogenic switch that regulates the choice of myoblasts to adopt slow versus fast fiber developmental pathways. In *u-boot* mutant embryos, slow muscle precursors abort their developmental program and differentiate into fast muscle fibers. *u-boot* has recently been mapped to a gene named *blimp1*.[85] Blimp1 is a SET domain containing protein that is likely to be involved in chromatin remodeling through histone methylation. Chromatin remodeling plays a key role in activation or repression of gene expression during myogenic cell differentiation. Recently, we have identified a gene named *Bop* that plays a critical role in myoblast maturation in zebrafish. *Bop* encodes an intracellular protein containing SET and MYND domains involved in histone methylation and recruitment of histone deacetylases, respectively.[86] Bop is specifically expressed in skeletal and cardiac muscles. Knockdown of *Bop* expression by morpholino antisense induced malfunction in skeletal and cardiac muscles. The effected embryos could not swim and had no heartbeat.[86] Molecular and cellular analyses revealed that the initial specification of myoblasts was not affected as shown by normal expression of myogenic genes, *MyoD, myogenin* and *Myf5*. The structure of myofibers was, however, disorganized in *Bop* knockdown embryos, and formation of sarcomere was completely inhibited. Together, these data indicated that histone methylation and deacetylation may play a critical role in slow muscle development and slow myofibril assembly.[85,86]

It appears that the cellular and molecular mechanisms controlling development of slow muscles in zebrafish are similarly involved in other fish and vertebrate species. Grimaldi and colleagues[212] recently showed that Hedgehog regulates the formation of superficial slow muscles in Xenopus embryos. It has been shown that adaxial cells in trout embryos also express the slow isoform of Myosin Heavy Chain (MyHC).[87] The slow MyHC-positive cells migrate radially through the somites and differentiate into slow muscles at the superficial layer.[87] The timing of slow and fast muscle differentiation appears varied between different fish species. In contrast to zebrafish where slow muscle differentiates early, Rescan and colleagues[87] showed that in trout embryos, fast muscles express MyHC in the medial region of the somite before the migration and differentiation of the slow muscle precursors. The reason for the discrepancy is currently unknown. It should be noted that there are possible multiple MyHC isoforms expressed in slow or fast muscles at different stages of development. Without knowing the exact expression patterns of these MyHCs, one antibody staining may not provide the conclusive answer.

Although Hh signaling is required for specification and differentiation of embryonic slow muscle fibers, the development of slow muscles in larvae and adult fish is not known. Barresi and colleagues[78] have demonstrated that distinct mechanisms are involved in regulating slow muscle development and growth during slow-muscle-stratified hyperplasia in the larval period. These investigators showed that in the absence of Hh signaling, stratified hyperplastic growth of slow muscle occurs at the correct time near the dorsal and ventral extremes of the myotome.

2.2. BMP Signal Inhibits Slow Muscle Differentiation

It appears that the development of MP slow muscles in zebrafish requires both positive and negative regulations. The positive and negative regulation on MP cell fate may be determined by competing influences between Hh

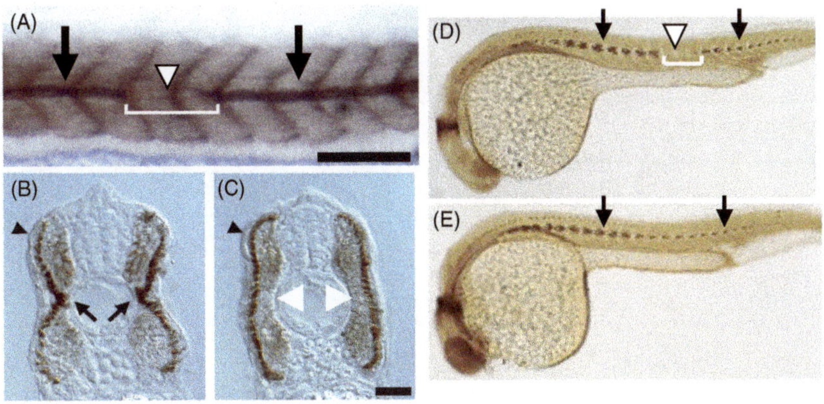

Fig. 7 BMP-like signal (Dorsalin) inhibits muscle pioneer slow muscle formation. The muscle pioneer cells are labeled with anti-myosin antibody F59 (A, B and C), or specifically with anti-Engrailed antibody (D and E). Ectopic expression of Dorsalin in notochord cells blocked differentiation of adaxial cells into muscle pioneer cells in the nearby region (indicated by the arrow).

and BMP signals.[51] The Hh signal is the positive factor for MP formation, while the BMP signal expressed in the dorsal and ventral portions of the somite may act as an inhibitory signal. This idea is based on the following two observations. First, muscle pioneer cells do not normally develop in the dorsal and ventral portions of the myotome where BMP4-related genes are expressed.[51,88] Second, ectopic expression of Dorsalin, which encodes a BMP4-related protein, in notochord cells inhibits final differentiation of adaxial cells into muscle pioneers (Fig. 7).[51]

2.3. Induction of Fast and Slow Muscles by Different Signals in Zebrafish

Most of the studies in the past few years have been focused on the regulation of fish slow muscle formation. The molecular signal(s) that induces fast muscles is still unknown. Mutations in Hh signaling disrupted slow muscle formation in zebrafish, however, fast muscles could still form. These data indicate that the fate of fast muscle cells are determined by different signals. It should be, however, noted that Hh signaling appears to be required for the proliferation and/or survival of fast muscles. In *syu* and *yot* mutant, the muscle mass of fast muscle cells is clearly reduced.[70] This is probably resulting from increased cell death in *syu* and *yot* mutant embryos.[70] This is consistent with recent findings that Shh is required for survival and proliferation of myogenic cells in chick embryos,[89–91] and the fact that Gli2 is strongly expressed in fast muscles.[70] Together these data suggest that although Hh/Gli signaling is not required for specification and differentiation of fast muscles, Hh and Gli may play a role in survival and proliferation of fast muscle cells. This notion is further supported by a recent finding on the association between Hh signaling and cell cycle control.[92]

Increasing evidence demonstrate that members of the Wnt family are involved in myogenesis.[93] Wnt proteins produced by the neural tube and dorsal ectoderm have been shown to pattern the developing somites. Co-culture of presomitic mesoderm with Wnt-expressing cells promote muscle specification. β-catenin, a key mediator in the Wnt signaling pathway is essential and sufficient for inducing skeletal myogenesis in cell culture.[94] Wnt signaling has been shown to induce myogenic gene expression in presomitic mesoderm.[95] Recent studies indicate that there are two myogenic induction pathways that induce *myoD* expression in two distinct domains in mouse and chick somites that give rise to epaxial and hypaxial muscles, respectively.[96–101] Expression of *myoD* in the dorsal-medial region of the somite is induced by Hh signal(s) from the neural tube/notochord complex through Myf-5,[101] while expression of *myoD* in the lateral domain is induced by Wnt7a signal(s) from the surface ectoderm via pax3.[97–99] Wnt1 and Wnt3a expressed in the dorsal neural tube induced *myoD* expression and myogenesis in co-culture experiments with un-segmented paraxial mesoderm and somites in the chick and mouse embryos.[95,99,102] Targeted deletion of Wnt1 or Wnt3a in mice results in myotomal defects.[103] Moreover, myogenesis in presomitic mesoderm and early somites is inhibited by overexpression of soluble Frizb1 protein, a Wnt antagonist.[104] T (Brachyury) has been shown to be a direct target of Wnt3a during paraxial mesoderm specification.[105] Together, these data suggest Wnts play a crucial role in skeletal myogenesis *in vivo.*

In zebrafish, several Wnt gene homologues have been characterized.[106–111] Zebrafish Wnt 11 is expressed in presumptive mesoderm region as its homologue in other vertebrates.[109] Wnt11 expression in the somite correlates with the migration and differentiation of slow muscle precursors. These observations suggest a role for Wnt11 in patterning the somites.[109] Wnt8 is expressed in the ventral mesoderm and tail bud at gastrula and tail bud stage, respectively.[107] Knockdown Wnt8 expression results in patterning defects in both mesoderm and neural ectoderm.[112] Recent studies in zebrafish have shown that Wnt3A mutation affects *myoD* expression in the lateral region of the somite that gives rise to fast muscles. In contrast, *myoD* expression in slow muscle cells were not affected,[113] suggesting that Wnt signals may play an important role in fast muscle formation in zebrafish. It will be interesting to analyze if Wnt signals are involved in the induction of fast muscles in zebrafish.

The antagonistic actions of Hh and BMP may be involved in specifying fast muscle fate. Ectopic expression of Shh, in wild-type embryos, leads to development of extra slow muscles at the expense of fast muscle. At present, the mechanism of this antagonism is not well understood. Meng *et al.*[114] have identified a zebrafish zinc-finger protein, Terra, that is expressed in the lateral presomitic mesoderm and in the newly formed somites that give rise to fast muscles. It has been suggested that Terra may play a role in fast muscle development.[114] Supporting this idea, Meng and colleagues have shown that Hh overexpression inhibits *terra* expression and fast muscle formation. In contrast, Swirl, the zebrafish homologue of BMP2 is required for *terra* expression. It remains to be determined if BMP2 acts through Terra to induce fast muscle cell fate.

2.4. Heat Shock Protein and Somitogenesis

Members of the Hsp90 family function as molecular chaperones in the assembly, folding and activating many cellular signaling molecules and transcription factors. Several molecules that Hsp90 interacts with, such as the bHLH transcription factor MyoD, are important regulators of muscle cell differentiation. Sass *et al.*[115] have shown that *hsp90α* is expressed in developing somites in zebrafish embryos, and is expressed in adaxial cells that give rise to slow muscles shortly following *myoD* activation. Expression of *hsp90α* is down-regulated in slow muscle fibers by 24 hpf but becomes detectable in fast fibers at later stages. Inhibiting Hsp90α function by pharmacologic inhibition, geldanamycin, disrupts zebrafish somitogenesis.[116] Geldanamycin treatment inhibits formation of *eng-2*-expressing muscle pioneer cells in the somite. However, early development of adaxial cells appeared unaffected, indicating a possible function of Hsp90 during differentiation of adaxial cells to muscle pioneers. *Hsp90α* gene expression appears to be a conserved feature of vertebrate somitogenesis. Sass and Krone[117] have demonstrated *hsp90α* expression in a subset of somitic cells of chicken embryos and that the expression pattern correlates closely to that of *myoD*. Furthermore, expression of the *hsp90α* is strongly up-regulated throughout the embryo following heat shock in zebrafish and chick embryos. The data suggest that the *hsp90α* gene may play an evolutionarily conserved role during somitogenesis in vertebrates in addition to providing protection to embryonic cells following stress.

3. Myogenic Regulatory Network in Muscle Development

3.1. Characterization of MRF and Mef2 Transcription Factors

Signal molecules activate myogenesis through the myogenic transcription network. The myogenic transcription network is composed of two major families of myogenic transcription factors, the family of myogenic regulatory factor (MRF) and the family of myocyte enhancer factor 2 (Mef2).[40,41,118,119] MyoD and its related proteins, Myf-5, Myogenin and MRF4, are members of the MRF family that share the conserved basic helix-loop-helix motif required for DNA binding and protein dimerization. Members of the MRF family are specifically expressed in developing somite, limb bud and skeletal muscles, and have the remarkable property of converting a variety of cells into myoblasts and myotubes.[120] MRF proteins bind to the E-box consensus sequence (CANNTG) in the control region of muscle-specific genes, such as *myosin*, *troponin* and *muscle creatine kinase*, and activate their muscle-specific expression.[121–124] Analyses of null mutations of MRF genes have revealed that MRFs function hierarchically during myogenesis in mouse embryos. MyoD and Myf5 function in parallel and play redundant roles in establishing myoblast identity, whereas Myogenin and MRF4 are involved in terminal differentiation. Mice lacking both *MyoD* and *Myf5* have a complete absence of myoblasts,[125] whereas mice lacking *myogenin* or *MyoD* plus *MRF4* show my oblasts with the inability to differentiate into myofibers.[126–128]

The Mef2 family of transcription factors are MADS-box proteins that have been shown to bend DNA upon high-affinity binding to DNA.[119] The Mef2 family contains four members, named Mef2A, B, C and D. They are expressed in skeletal, cardiac and smooth muscles, as well as several non-muscle tissues, such as thymus and neural cells.[119] Mef2 expression in muscle cell lineages is concomitant with the activation of myoblast differentiation. Unlike members of the MyoD family, Mef2s cannot convert non-myogenic cells to myoblasts. Mef2s function cooperatively with MyoD to regulate muscle gene expression. Knock-out of individual members of the Mef2 family results in quite different phenotypes in mice. Null mutations of either Mef2A or 2B cause no significant defect to embryonic development.[119] By contrast, null mutants of Mef2C or Mef2D are embryonic lethal. Mice lacking Mef2C die at about E9.5 from cardiovascular defects,[129] whereas Mef2D mutant mouse embryos die during gastrulation.[119]

Members of the MRF and Mef2 gene families have been identified in several fish species. The MRF genes isolated include *myoD*, *myf5*, *myogenin* and the recently identified MRF4,[130–133] whereas three members of the Mef2 family including Mef2A, C and D have been isolated.[134] Expression analysis revealed that these myogenic transcription factors are expressed in highly conserved patterns compared with their homologues in high vertebrates.[52,135–140] *myoD* and *myf5* mRNA expression in zebrafish embryos first occurs near the end of gastrulation (70–80% epiboly) prior to somite formation. Their initial expression is restricted to adaxial cells that give rise to slow muscles. *myogenin* expression occurs 1–2 hours later than that of *myoD* with a similar spatial pattern. During somitogenesis, *myoD*, *myf5* and *myogenin* expression expands to the lateral region of the somite that give rise to fast muscles. *myf5*, *myoD* and *myogenin* exhibit distinct patterns of temporal and spatial expression in the developing somite. Zebrafish *myf5* transcripts are restricted mainly in the segmental plates and the newly formed somites. Its expression declines gradually to an undetectable level by 26 hpf. *myoD* and *myogenin* expression increased substantially during somitogenesis, with their transcripts clearly detectable in all the somites in 24-hour embryos. This expression pattern is consistent with the idea that Myf5 is involved in early myoblast specification whereas Myogenin is important for myoblast differentiation.

3.2. Regulation of *MyoD* and *Myf5* Gene Expression

Consistent with the fact that notochord signal is required for slow muscle development. Expression of *myf5* and *myoD* in adaxial cells depends on signals from the notochord. In *ntl* mutant embryos, the early phase of *myoD* expression in slow muscle precursors is absent.[130] However, the later phase of *myoD* expression in the lateral somite cells that give rise to fast muscles appears normal, indicating that *myoD* expression in these two different regions of the somite are controlled by distinct signals. Because Hh is a primary signaling molecule from the notochord responsible for slow muscle formation, *myoD* and *myf5* expression is also affected in the *syu*, *yot*, and *smu* mutants.[4,52,69,141] Shh signaling is found to be necessary for normal expression of both *myf5* and *myoD* in adaxial slow muscle precursors, but not in lateral paraxial mesoderm. Ectopic expression of Shh in the embryo resulted in entire paraxial mesoderm to express *myoD*. In contrast, inhibition of Hh signaling in zebrafish embryos by overexpression of the Patched receptor blocks *myoD* expression in adaxial cells.[69] It appears that Hh signal is required for maintenance rather than initiation of *myf5* and *myoD* expression in adaxial cells. Expression of both *myoD* and *myf5* is initiated normally in rostral presomitic mesoderm in *syu* and *yot* mutants.[52,69] Without Hh signaling, their expression in adaxial cells is dramatically reduced to a level that is insufficient to commit cells to adaxial myogenesis. These adaxial cells fail to differentiate into slow muscles. Recent studies through lineage tracer have shown that these adaxial cells in *smu* and *yot mutants* are converted to fast muscle lineage.[142]

3.3. Promoter Analysis of MyoD, Myf5 and Myogenin Expression in Fish

Given the essential role of MRF in establishing the skeletal muscle lineage and myoblast differentiation, a clear understanding on how *MRF* gene expression is regulated offers a powerful means to define the upstream signaling pathways that control muscle lineage determination. In the past decade, a large number of studies using transgenic mice have demonstrated that expression of *myf5*, *myoD* or *myogenin* is controlled by enhancer sequences in their promoters.[143–145] The regulation of *myf5*, *myoD* or *myogenin* expression in fish, however, is poorly understood. Recent studies have begun to use transgenic approach to analyze the regulation of myogenic gene expression in fish.

The genomic clones of *myf5*, *myoD* and *myogenin* have been isolated and characterized from several fish species.[86,139,140,146] The promoter activity and muscle-specificity have been determined by transient expression assay in zebrafish embryos by directing GFP expression. The results revealed that the muscle-specific expression of *myogenin* in zebrafish embryos is directed by a 0.8 kb *myogenin* promoter,[86] whereas *myf5* expression in zebrafish is regulated by a small 82 bp sequence in the *myf5* promoter.[139] The zebrafish *myoD* promoter of 262 bp could drive GFP expression in muscle cells of zebrafish embryos. However, non-muscle expression was also found in the injected embryos.[86] It is apparent that additional regulatory elements for muscle-specific expression of *myoD* are located outside of the short promoter sequence. It has been demonstrated in mice that the muscle-specific expression of *myoD* requires a 258 bp element that lies 20 kb upstream of the transcription start site.[143,144] A recent study by Yang *et al.*[147] demonstrated that the muscle-specificity of a 6 kb zebrafish *myoD* promoter could be dramatically improved by adding a 250 bp enhancer sequence located 20 kb upstream of the human *myoD* gene.

The core cis-regulatory elements controlling the muscle-specific expression of *myogenin* in zebrafish embryos have been identified by sequence analysis and functional studies. Sequence analysis revealed that zebrafish *myogenin* promoter contains two putative E boxes

(GCAGTTG), a MFE2 (TATATTT) and a MEF3 (TCAGGTT) binding site. Mutating these sites alone had little effect on the muscle-specific expression. However, mutating E boxes, MEF2 and MEF3 sites together nearly abolished the promoter activity, indicating that multiple regulatory mechanisms are involved in controlling the muscle-specific expression of *myogenin* in zebrafish embryos (Fig. 8).[86] The regulatory mechanism controlling *myogenin* expression appears to be conserved during evolution because similar regulatory elements have been found to regulate *myogenin* expression in mouse embryos.[148–150]

The E boxes in the *myogenin* promoter are targets for auto-regulation by muscle regulatory factors of the MyoD family. Because Myf5 and MyoD are expressed earlier than Myogenin, Myf5 and MyoD likely regulate *myogenin* gene expression through binding to the E boxes. The Mef2 and Mef3 sites are recognized by the muscle-specific enhancer factor MEF-2 or the SIX/sine oculis homeoproteins, respectively. Several *Mef2* genes have been identified in zebrafish.[134] Their transcripts are first detected in the adaxial cells of zebrafish embryos and progressing in the lateral region of developing somites.[134] The expression patterns of *Mef2s* are consistent with their potential roles in regulating *myogenin* expression in zebrafish embryos. Members of the SIX protein family that bind to the MEF3 site have been identified in mice, chick, frog and zebrafish. SIX1, 2 and 4 are expressed specifically in developing somites and skeletal muscles in mouse, chick, and frog embryos.[151–153] These different transcription factors may regulate *myogenin* expression in different types of myofibers. It has been shown that the MEF3 binding site in the human aldolase A pM promoter is required for fast fiber-specific expression.[154] Further studies are needed to determine the functions of these transcription factors in regulating *MRF* gene expression in zebrafish embryos, and to identify the upstream signals that control their activities.

3.4. Functional Analysis of Myf5 in Fish

Although the functions of MRFs are well-established in mice by gene knock-out studies, their functions in fish have not been well-analyzed.

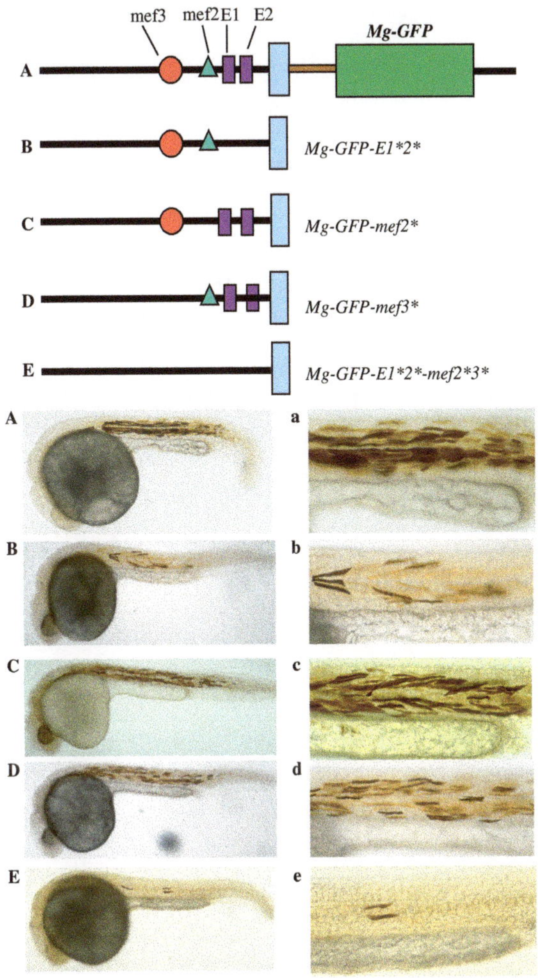

Fig. 8 Structure of *myogenin-GFP* constructs carrying various mutations in the promoter. Construct A is the native *myogenin-GFP* construct, whereas DNA constructs (B to E) carrying various mutations in the E box, MEF2 or MEF3 binding sites are generated and microinjected into zebrafish embryos. The asterisk (*) indicates the site of mutations, e.g. construct *Mg-GFP-E1*2*-mef2*3** carries mutations in the two E boxes, the mef2 and mef3 binding sites. These constructs were microinjected into zebrafish embryos. The promoter activity of these gene constructs was analyzed by GFP antibody staining. Letter labels of the zebrafish embryos correspond to the construct injected. Fish embryos with lower case letters show the trunk region of corresponding embryos with the upper case letters.

The highly conserved sequences and patterns of expression suggest that members of the MyoD family likely play a similar role in skeletal muscle development and growth in fish. Chen and Tsai[155] used the gene knockdown approach to study the biological functions of Myf5 in zebrafish myogenesis. Embryos injected with the Myf5-morpholinos antisense nucleotides displayed abnormalities in myogenesis and developmental defect in head and brain. Molecular studies demonstrated that *myod* expression was normal, however, *myogenin* expression was substantially down-regulated in whole somites; and *desmin* expression was partly inhibited in newly forming somites. These data suggest that zebrafish Myf5 may play important roles in regulation *myogenin* expression and muscle cell differentiation. This is consistent with findings from chick and mammals.

3.5. Characterization of Myogenic Regulatory Factors in Other Fish Species

In addition to zebrafish, the homologues of MRFs have been isolated in several other fish species. Members of the MRF family are expressed specifically in developing somites and skeletal muscles in trout, carp and seabream.[135,137,146] In addition to strong expression of *myoD*, *myogenin, myf5* and *mef2* in embryos, strong expression of *myogenin* and *myoD* is observed in 1-month-old carp juveniles.[137] *myoD*, *myogenin* and *mef2a* expression declines between 1 and 7 months after hatching and *myf5* expression is very weak in the older fish.[137] The relatively high levels of *MRF* mRNA in juvenile fish probably reflect the recruitment of new muscle fibers from the satellite cell population. It remains to be determined the spatial location of Myogenic gene expression in juvenile and adult fish muscles.

In some species, two distinct *myoD* genes have been identified.[146,156] The two *myoD* genes exhibited overlapping but distinct patterns of expression in slow and fast muscles of fish embryos and adult skeletal muscles.[146,156,157] Seabream *myoD1* is expressed in both slow and fast muscles in embryos and adult fish (Fig. 9).[146] By contrast, *myoD2* is initially expressed in both slow and fast muscle precursors in seabream

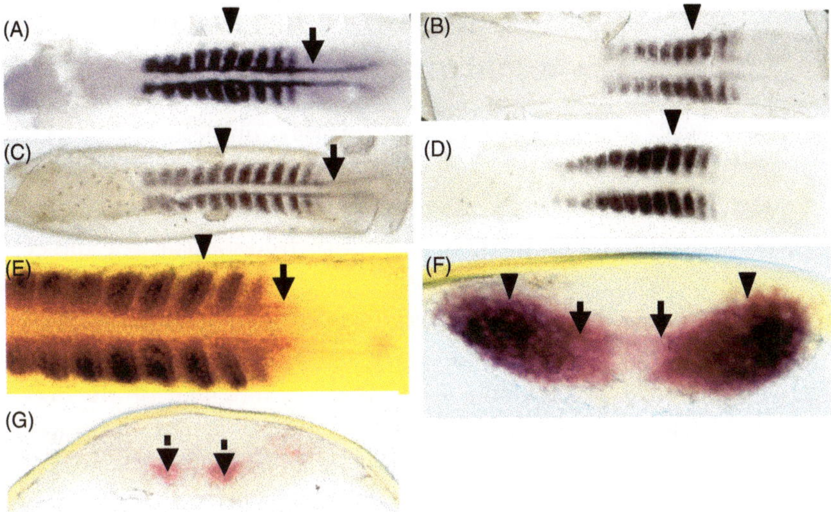

Fig. 9 Temporal and spatial expression of *MyoD1* and *MyoD2* in seabream embryos. A and C: Whole mount *in situ* hybridization (dorsal view, head to the left) showing *MyoD1* expression in 10 somites (A) and 15 somites (C) embryos. Medial (Adaxial) cells are indicated by black arrows. Lateral somitic cells giving rise to fast muscles are indicated by arrow heads. B and D: *MyoD2* expression in 10 somites (B) and 15 somites (D) embryos. E: Two color *in situ* hybridization showing *MyoD1* (red) and *MyoD2* (blue) expression. Note the adaxial cells only express are *MyoD1*. F and G: Cross-sections showing *MyoD1* and *MyoD2* expression. Note the adaxial cells express only *MyoD1* (red, arrow), whereas lateral somitic cells express both *MyoD1* and *MyoD2*.

embryos. *myoD2* expression gradually disappeared in the adaxial cells of 10- to 15-somite-stage embryos, whereas its expression in fast muscle cells is maintained (Fig. 9). Thus, in adult seabream, *myoD2* is exclusively expressed in fast muscles.[146] Hedgehog signaling is required for *myoD* expression in adaxial cells in seabream embryos. Together, these data suggest that the two non-allelic *myoD* genes are functional in seabream and their expression is regulated differently in fast and slow muscles. It remains to be determined if the two distinct *myoD* genes have evolved separately, acquiring specific roles in slow and fast muscle specification and differentiation.

4. Myostatin and Follistatin in Regulating Muscle Development and Growth

Muscle growth and patterning are tightly coupled during animal development and growth. While significant advances have been made in understanding the mechanisms in patterning of various tissues including skeletal muscles, the control of growth, especially the molecular mechanism(s) underlying size control of specific tissues and organs remains a unsolved mystery in developmental biology.[158,159] In addition to global regulators, such as Growth Hormone, recent studies have demonstrated that Myostatin (or GDF-8), a member of the TGF-β superfamily, has a specific role in inhibiting skeletal muscle growth.[160] *Myostatin* knock-out mice show a dramatic increase of skeletal muscle mass, resulting from a combination of hyperplasia and hypertrophy.[161] The "Double muscle" breeds of cattle that have significantly more muscle mass than standard breeds were found to carry natural mutations in the *myostatin* gene.[162–165] By using a conditional gene knock-out approach, Grobet *et al.*[166] have shown that postnatal inactivation of the *myostatin* gene in striated muscles is sufficient to cause a generalized muscular hypertrophy. This study clearly demonstrates that Myostatin regulates muscle mass not only during early embryogenesis but throughout development. It indicates that Myostatin antagonist could be used to treat muscle wasting and to promote muscle growth in man and animals.

In vitro studies have demonstrated that Myostatin functions by inhibiting myoblast proliferation and differentiation.[167–170] This is, in part, accomplished by down-regulating myogenic gene expression involved in myoblast specification and differentiation.[170,171] Myostatin inhibits Pax3, *Myf5* and *MyoD* expression in myoblasts.[170] Several studies have implicated Myostatin in myoblast cell cycle control. Expression of Myostatin is undetectable in proliferating myoblasts and only become expressed during myoblast differentiation that is associated with cell cycle exit.[168] Addition of Myostatin in myoblast culture up-regulates p21 expression and inhibits myoblast proliferation.[167–169]

Because Myostatin has an important function and a potential application in clinical and agricultural research, studies of Myostatin has become an active field of research in the past few years. The *myostatin* gene has been cloned from many vertebrate species ranging from fish to human.[162,172–179] Comparison of *myostatin* sequences revealed that *myostatin* gene was extremely well conserved throughout evolution. Remarkably, the murine, rat, human, porcine, chicken, and turkey Myostatin sequences are identical in the biological active C-terminal region.[162] Fish Myostatin proteins are over 85% identical to the mammalian homologue in the C-terminal region. The high sequence identity suggests that this gene might play a similar role in regulating muscle growth in non-mammalian vertebrates.

4.1. Characterization of Fish Myostatin Gene Expression

Myostatin has been cloned in over 30 species of fish. Although fish *myostatin* genes share high sequence identity with their mammalian counterparts, *myostatin* expression in fish is not identical compared with that in mammals. In mice, *myostatin* is strongly expressed in developing somite and skeletal muscles, and weakly expressed in cardiomyocytes, mammary glands and adipose tissue.[161,180,181] Fish *myostatin* is primarily expressed in skeletal muscles, but is also expressed in eyes, spleen, gill filaments, ovaries, gut, brain and testes.[173,174,177,178,182] Moreover, in contrast to strong expression in developing somites in mouse embryos, little or no *myostatin* expression could be detected in developing somites of fish embryos by whole mount *in situ* hybridization.[183] *Myostatin mRNA* expression in early stage embryos of several fish species could only be detected by RT-PCR.[174,183,184] Interestingly, Vianello and colleagues[184] showed that *myostatin* transcripts are present in fertilized eggs in zebrafish, suggesting that Myostatin may be expressed maternally. It has been shown that *myostatin* mRNA is expressed differently in red and white muscles in some fish species. Roberts and Goetz[177] reported that *myostatin* was primarily expressed in red muscles in brook trout, king mackeral, and yellow perch, but white muscles in the little tunny. Rescan *et al.*[179] showed that trout *myostatin-2* was predominantly expressed in red muscles. These data suggest that Myostatin may be involved in controlling the differential growth of red or white muscle fibers in different fish species. In some fish species, two *myostatin* genes have been found. The two *myostatin* genes are expressed differently in embryos and adult tissues in these fish species.[177–179,213]

4.2. Functions of Myostatin in Fish

Like other members of the TGF-β family, Myostatin is synthesized as a prepro-peptide that undergoes two steps of proteolytic cleavage to generate the biologically active C-terminal domain.[167,168] The bioactive C-terminal domain dimerizes and binds to membrane receptors on target cells.[167] The mature TGF-β C-terminal dimer often forms an inactive complex with the N-terminal TGF-β prodomain.[185] This observation suggested that Myostatin might also exist as a secreted latent complex with the prodomain region of the protein.[160,186] Recently, Thies and colleagues[186] showed that Myostatin prodomain was able to bind to the bioactive Myostatin molecule and inhibit its activity, presumably by preventing Myostatin from binding to its receptor on the cell surface. Thus, the Myostatin prodomain has been used as a "dominant negative" to inhibit the function of endogenous Myostatin. Transgenic mice expressing the Myostatin prodomain showed enhanced muscle growth similar to *myostatin* knock-out mice.[187,188]

To develop a fish model for functional study of Myostatin, transgenic zebrafish expressing the *myostatin* prodomain have been generated.[183] The Myostatin prodomain was specifically expressed and secreted by skeletal muscle cells using a muscle-specific promoter/enhancer derived from rat myosin light chain gene.[189] The adult transgenic fish exhibited a 10–15% increase in the number of myofibers in skeletal muscles, but no significant difference in fiber size. These data demonstrated that Myostatin may play an inhibitory role in hyperplastic muscle growth in zebrafish. Although the increase is not as dramatic as in mice, this data is consistent with a recent finding that transgenic zebrafish expressing a growth hormone transgene only show a 20% increase in growth rate.[190]

Histological analysis revealed that transgenic fish had increased stratified hyperplastic muscle growth. Stratified hyperplasia generates new fibers along a distinct germinal layer. This type of hyperplasia is found in all fish species. In addition to stratified hyperplasia, another type of hyperplasia is known as mosaic hyperplasia.[191] Mosaic hyperplasia results in large increase in total fiber number during juvenile growth, and is therefore very common in commercially important aquatic species that grow to a large size.[191] Mosaic hyperplasia is greatly reduced or entirely lacking in species such as zebrafish, guppies and other fish which remain small.[192–195] The small effect on muscle growth in zebrafish could be due to the lack of mosaic hyperplasia in zebrafish. It will be interesting to determine if blocking Myostatin function in large aquatic species will have a more dramatic effect in stimulating muscle fiber growth.

There are several possible explanations why inhibiting Myostatin function have a more dramatic effect in mammals than fish. In the *myostatin* knock-out mice, the marked increase in muscle mass is attributed to both hypertrophy and hyperplasia.[161] The different response to Myostatin in fish and mice could be due to the different types of muscle growth in postnatal or post-larval stages. Postnatal muscle growth in mammals is largely contributed by hypertrophy. In contrast, in most fish species, the muscle growth in post-embryonic life is attributed to continuous hyperplastic and hypertrophic growth (reviewed by Rowlerson and Veggetti[191]). In addition, it should be noted that Myostatin protein may be expressed differently between fish and mammals. In the adult muscle of fish, western blot analyses revealed that only the precursor protein form was detected in the adult lateral muscle.[184] The expression of Myostatin precursor in fish suggests that Myostatin function may well be repressed at the proteolysis level. This may suggest that Myostatin function is already controlled at a low level in fish. This may be another possible explanation why blocking Myostatin function in fish has less effect compared with that in mice and cattle.

The possibility that *myostatin*-related genes may be also involved in the process should not be overlooked. Recently, Lee and McPherron[187] have demonstrated that overexpression of Follistatin, a TGF-β /BMPs inhibitor, in skeletal muscles of transgenic mice induced hyperplasia and hypertrophy. Interestingly, the muscle phenotype is more dramatic than that obtained from the *myostatin* knock-out, suggesting that Follistatin may have additional function than simply blocking Myostatin activity in skeletal muscles.[187] Follistatin has been cloned in zebrafish and is expressed in developing somite and skeletal muscles.[196] It remains to be determined if overexpressing Follistatin in fish skeletal muscles will have a more dramatic effect in stimulating fish muscle growth.

4.3. Regulation of Myostatin Gene Expression

To determine the regulation of *myostatin* expression in skeletal muscles, Spiller *et al.*[197] have cloned and characterized the bovine *myostatin* gene promoter, and compared with the upstream sequences of human, porcine and mouse *myostatin* genes. They found that the *myostatin* promoter is highly conserved during evolution. Sequence analysis revealed 10 E-box motifs (E1 to E10) and a single MEF2 site in the 1.6 kb upstream region of the bovine *myostatin* gene. Deletion and mutation analysis of the *myostatin* gene promoter revealed that the E6 motif plays a significant role in its expression in muscle cells. Gel shift and chromatin immunoprecipitation assay show that the E6 E-box motif binds to MyoD *in vitro* and *in vivo.* MyoD up-regulates *myostatin* promoter activity in cell transfection assay. These investigators also showed that the *myostatin* promoter activity is regulated during the cell cycle. *Myostatin* promoter is more active at the G1 phase of the cell cycle when MyoD expression levels are high, while in quiescent "reserve" cells, which lack MyoD expression, a significant reduction in the *myostatin* promoter activity is observed. These results suggest that the *myostatin* expression may be controlled, in part, by MyoD. Since Myostatin has been implicated in controlling G1 to S progression of myoblasts, MyoD could be triggering myoblast withdrawal from the cell cycle by regulating *myostatin* gene expression.[197]

Sequence analyses of zebrafish *myostatin* gene promoter have identified several E-box sites that are conserved with that in mice.[183] To determine if the *myostatin* promoter could drive gene expression in muscle cells, the DNA construct *myostatin-GFP* was microinjected into zebrafish embryos for transient expression analysis. The results show that the zebrafish *myostatin* promoter could drive GFP expression in muscle fibers of zebrafish embryos.[183] These results indicated that expression of *myostatin* in muscle fibers is, at least in part, regulated by regulatory elements in the promoter region, possibly involving E-boxes.

5. Zebrafish as a Model for Medical Research on Muscular Disease

5.1. Characterization of Dystrophin and Its Associated Proteins in Fish

Zebrafish not only serve as an ideal model for basic research in muscle biology, they are also an excellent model for studying muscular disease, such as dystrophy. The large scale mutagenesis led to the identification of several classes of mutants with motility defects.[5] Some of the mutants are caused by defects in the nerve system, while many mutants may be due to developmental defects in skeletal muscles. These valuable mutants provide rich resource for identifying new genes in regulating muscle development. Dystrophin, the protein responsible for Duchenne Muscular Dystrophy, has been isolated from zebrafish.[198,199] Bassett *et al.*[214] have demonstrated that *sapje* mutants contain mutations in the zebrafish ortholog of the *dystrophin* gene and Dystrophin is required for the formation of stable muscle attachments in the zebrafish embryo. In addition, several members of the Dystrophin associated proteins have been identified in zebrafish.[200] Blocking the expression of Dystroglycan, a Dystrophin associated protein, produces juvenile zebrafish that are less active, which resemble the dystrophy phenotype.[200,201]

5.2. Muscle-Specific Genes and Transgenic Fish Models for Muscle Research

Final differentiation and maturation of skeletal muscle cells is characterized by expression of a group of muscle-specific genes involved in sarcomere assembly and muscle contraction. Large numbers of muscle-specific genes have been isolated from zebrafish.[202] These include skeletal muscle *α-actin*, fast skeletal muscle *α-tropomyosin*, *troponin C*, *troponin T*, *myosin heavy chain 1*, *myosin light chain 2*, *myosin light chain 3*, muscle *creatine kinase*, *parvalbumin*, and *desmin*. These muscle specific genes are expressed in both embryos and adult fish with distinct patterns of temporal and spatial expression patterns. These genes provide useful molecular tools for studying muscle cell specification, differentiation as well as muscle filament assembly during muscle development and growth in fish.

The use of zebrafish as a model for muscle research has been further enhanced by the production of transgenic fish models expressing GFP directed by muscle-specific promoters. Higashijima and colleagues[203] have isolated the zebrafish muscle-specific *α-actin* promoter and generated transgenic zebrafish that expresses GFP specifically in skeletal muscles using this promoter. Ju and colleagues[204] have generated a transgenic zebrafish line using the muscle-specific *creatine kinase* promoter (MCK). The 1.5 kb mck promoter/gfp was expressed exclusively in skeletal muscles. Recently, Ju *et al.*[205] have reported the isolation and characterization of a 1.9 kb muscle-specific promoter from the zebrafish *mylz2* gene. By using this promoter, these investigators have created stable transgenic lines with the *mylz2-gfp* construct. GFP expression in the transgenic lines accurately mimicked the expression pattern of endogenous *mylz2* mRNA transcripts in both somitic muscle and non-somitic muscles, including fin, eye, jaw and gill muscles.[205] In addition to transgenics, another powerful approach is the direct injection of DNA into fish skeletal muscles for ectopic gene expression or analysis of gene expression. Xu *et al.*[206] have used this method to study the regulation of *mlc2f* expression in zebrafish skeletal muscles. These investigators linked the promoter of the zebrafish *mlc2f* gene to the chloramphenicol acetyltransferase (CAT) reporter gene. The *MLC2f/CAT* chimeric constructs were analyzed by direct injection into the zebrafish skeletal muscle. Significant CAT activity was observed in skeletal muscles of the injected fish. This study also demonstrated that direct injection of DNA into skeletal muscle is a valid and valuable approach to analyze muscle gene promoters in the zebrafish. Together, these studies further enhance the power of zebrafish as a model for basic research and also as a model for medical research in muscular diseases.

In summary, muscle development and growth are complex processes that require precise temporal and spatial control of myoblast specification, proliferation and differentiation. A large number of signaling molecules that play critical roles in muscle development and growth have been identified. Among these signaling molecules, some exert a positive effect, while others play a negative role. These signaling molecules act on myogenic transcription factors of the MyoD and Mef2 families that form the core myogenic regulatory network directly involved in regulation of muscle-specific gene expression, and myogenesis program. The myogenic network can be activated or repressed in response to patterning signals. It is the balance between the positive and negative regulation that ultimately determines the growth rate and final size of skeletal muscles. One of the most challenging tasks at present is to understand how these signaling cascades communicate with each other to allow for rapid and precise changes in gene expression during myobalst specification and differentiation that ultimately control muscle development and growth.

Acknowledgment

Special thanks to Tracey Robinson for editorial assistance. This work was supported by a NIH grant to SJD (RO1GM58537).

References

1. Devoto SH, Melancon E, Eisen JS and Westerfield M (1996). Identification of separate slow and fast muscle precursor cells in vivo, prior to somite formation. *Development* **122**: 3371–3380.
2. Sanger AM and Stoiber W (2001). Muscle fiber diversity and plasticity. In: Johnston IA (ed.), *Muscle Development and Growth.* Academic Press.
3. Felsenfeld AL, Walker C, Westerfield M, Kimmel C and Streisinger G (1990). Mutations affecting skeletal muscle myofibril structure in the zebrafish. *Development* **108**: 443–459.
4. van Eeden FJM, Granato M, Schach U, Brand M, Furutani–Seiki M, Hafftter P, Hammerschmidt M, Heisenberg C-P, Jiang Y-J, Kane DA,

Kelsh RN, Mullins MC, Odenthal J, Warga RM, Allende ML, Weinberg ES and Nusslein–Volhard C (1996). Mutations affecting somite formation and patterning in the zebrafish and Danio rerio *Development* **123**: 153–164.

5. Granato M, van Eeden FJ, Schach U, Trowe T, Brand M, Furutani–Seiki M, Haffter P, Hammerschmidt M, Heisenberg CP, Jiang YJ, Kane DA, Kelsh RN, Mullins MC, Odenthal J and Nusslein–Volhard C (1996). Genes controlling and mediating locomotion behavior of the zebrafish embryo and larva. *Development* **123**: 399–413.
6. Udvadia AJ and Linney E (2003). Windows into development: historic, current, and future perspectives on transgenic zebrafish. *Dev. Biol.* **256**: 1–17.
7. Ekker SC and Larson JD (2001). Morphant technology in model developmental systems. *Genesis* **30**: 89–93.
8. Johnston IA (2001). *Muscle Development and Growth.* Academic Press.
9. Stickney HL, Barresi MJ and Devoto SH (2000). Somite development in zebrafish. *Dev. Dyn.* **219**: 287–303.
10. Holley SA and Nusslein–Volhard C (2000). Somitogenesis in zebrafish. *Curr. Top. Dev. Biol.* **47**: 247–277.
11. Pourquie O (2001). Vertebrate somitogenesis. *Annu. Rev. Cell. Dev. Biol.* **17**: 311–350.
12. Brennan C, Amacher SL and Currie PD (2002). Somitogenesis. *Results Probl. Cell Differ.* **40**: 271–297.
13. Morin–Kensicki EM and Eisen JS (1997). Sclerotome development and peripheral nervous system segmentation in embryonic zebrafish. *Development* **124**: 159–167.
14. Aoyama H and Asamoto K (1988). Determination of somite cells: independence of cell differentiation and morphogenesis. *Development* **104**: 15–28.
15. Schroeter EH, Kisslinger JA and Kopan R (1998). Notch-1 signaling requires ligand-induced proteolytic release of intracellular domain. *Nature* **393**: 382–386.
16. Blaschuk KL, ffrench–Constant C (1998). Developmental neurobiology: notch is tops in the developing brain. *Curr. Biol.* **8**: R334–337.
17. Iso T, Hamamori Y and Kedes L (2003). Notch signaling in vascular development. *Arterioscler Thromb. Vasc. Biol.* **23**: 543–553.
18. Pourquie O (1999). Notch around the clock. *Curr. Opin. Genet. Dev.* **9**: 559–565.
19. Jiang YJ, Smithers L and Lewis J (1998). Vertebrate segmentation: the clock is linked to Notch signaling. *Curr. Biol.* **8**: R868–871.

20. Jiang YJ, Aerne BL, Smithers L, Haddon C, Ish–Horowicz D and Lewis J (2000). Notch signaling and the synchronization of the somite segmentation clock. *Nature* **408**: 475–479.
21. Muller M, v Weizsacker E, Campos–Ortega JA (1996). Expression domains of a zebrafish homologue of the Drosophila pair-rule gene hairy correspond to primordia of alternating somites. *Development* **122**(7):2071–2078.
22. Oates AC and Ho RK (2002). Hairy/E(spl)-related (Her) genes are central components of the segmentation oscillator and display redundancy with the Delta/Notch signaling pathway in the formation of anterior segmental boundaries in the zebrafish. *Development* **129**: 2929–2946.
23. Henry CA, Urban MK, Dill KK, Merlie JP, Page MF, Kimmel CB and Amacher SL (2002). Two linked hairy/Enhancer of split-related zebrafish genes, her1 and her7, function together to refine alternating somite boundaries. *Development* **129**: 3693–3704.
24. Holley SA, Julich D, Rauch GJ, Geisler R and Nusslein–Volhard C. (2002). her1 and the notch pathway function within the oscillator mechanism that regulates zebrafish somitogenesis. *Development* **129**(5):1175–1183.
25. Holley SA, Geisler R and Nusslein–Volhard C (2000). Control of her1 expression during zebrafish somitogenesis by a delta-dependent oscillator and an independent wave-front activity. *Genes Dev.* **14**: 1678–1690.
26. Sawada A, Fritz A, Jiang Y, Yamamoto A, Yamasu K, Kuroiwa A, Saga Y and Takeda H (2000). Zebrafish Mesp family genes, mesp-a and mesp-b are segmentally expressed in the presomitic mesoderm, and Mesp-b confers the anterior identity to the developing somites. *Development* **127**: 1691–1702.
27. Yamaguchi TP, Conlon RA and Rossant J (1992). Expression of the fibroblast growth factor receptor FGFR-1/flg during gastrulation and segmentation in the mouse embryo. *Dev. Biol.* **152**: 75–88.
28. Yamaguchi TP, Harpal K, Henkemeyer M and Rossant J (1994). fgfr-1 is required for embryonic growth and mesodermal patterning during mouse gastrulation. *Genes Dev.* **8**: 3032–3044.
29. Dubrulle J, McGrew MJ and Pourquie O (2001). FGF signaling controls somite boundary position and regulates segmentation clock control of spatiotemporal Hox gene activation. *Cell* **106**: 219–232.
30. Sawada A, Shinya M, Jiang YJ, Kawakami A, Kuroiwa A and Takeda H (2001). Fgf/MAPK signaling is a crucial positional cue in somite boundary formation. *Development* **128**: 4873–4880.

31. Sun X, Meyers EN, Lewandoski M and Martin GR (1999). Targeted disruption of Fgf8 causes failure of cell migration in the gastrulating mouse embryo. *Genes Dev.* **13**: 1834–1846.
32. Reifers F, Bohli H, Walsh EC, Crossley PH, Stainier DY and Brand M (1998). Fgf8 is mutated in zebrafish acerebellar (ace) mutants and is required for maintenance of midbrain-hindbrain boundary development and somitogenesis. *Development* **125**: 2381–2395.
33. Durbin L, Brennan C, Shiomi K, Cooke J, Barrios A, Shanmugalingam S, Guthrie B, Lindberg R and Holder N (1998). Eph signaling is required for segmentation and differentiation of the somites. *Genes Dev.* **12**: 3096–3109.
34. Durbin L, Sordino P, Barrios A, Gering M, Thisse C, Thisse B, Brennan C, Green A, Wilson S and Holder N (2000). Anteroposterior patterning is required within segments for somite boundary formation in developing zebrafish. *Development* **127**: 1703–1713.
35. Saga Y, Hata N, Koseki H and Taketo MM (1997). Mesp**2**: a novel mouse gene expressed in the presegmented mesoderm and essential for segmentation initiation. *Genes Dev.* **11**: 1827–1839.
36. Takahashi Y, Koizumi K, Takagi A, Kitajima S, Inoue T, Koseki H and Saga Y (2000). Mesp2 initiates somite segmentation through the Notch signaling pathway. *Nat. Genet.* **25**: 390–396.
37. Topczewska JM, Topczewski J, Shostak A, Kume T, Solnica–Krezel L and Hogan BL (2001b). The winged helix transcription factor Foxc1a is essential for somitogenesis in zebrafish. *Genes Dev.* **15**: 2483–2493.
38. Topczewska JM, Topczewski J, Solnica–Krezel L and Hogan BL (2001a). Sequence and expression of zebrafish foxc1a and foxc1b, encoding conserved forkhead/winged helix transcription factors. *Mech. Dev.* **100**: 343–347.
39. Kume T, Jiang H, Topczewska JM and Hogan BL (2001). The murine winged helix transcription factors, Foxc1 and Foxc2, are both required for cardiovascular development and somitogenesis. *Genes Dev.* **15**: 2470–2482.
40. Buckingham M. (2001). Skeletal muscle formation in vertebrates. *Curr. Opin. Genet. Dev.* **11**: 440–448.
41. Pownall ME, Gustafsson MK and Emerson CP Jr. (2002). Myogenic regulatory factors and the specification of muscle progenitors in vertebrate embryos. *Annu. Rev. Cell. Dev. Biol.* **18**: 747–783.

42. Brent AE and Tabin CJ (2002). Developmental regulation of somite derivatives: muscle, cartilage and tendon. *Curr Opin Genet Dev.* **12**: 548–557.
43. Reshef R, Maroto M and Lassar AB (1998). Regulation of dorsal somitic cell fates: BMPs and Noggin control the timing and pattern of myogenic regulator expression. *Genes Dev.* **12**: 290–303.
44. Gustafsson MK, Pan H, Pinney DF, Liu Y, Lewandowski A, Epstein DJ and Emerson CP Jr. (2002). Myf5 is a direct target of long-range Shh signaling and Gli regulation for muscle specification. *Genes Dev.* **16**: 114–126.
45. Pourquie O, Fan CM, Coltey M, Hirsinger E, Watanabe Y, Breant C, Francis–West P, Brickell P, Tessier–Lavigne M and Le Douarin NM (1996). Lateral and axial signals involved in avian somite patterning: a role for BMP(4). *Cell* **84**: 461–471.
46. Halpern ME, Ho RK, Walker C and Kimmel, CB (1993). Induction of muscle pioneer cells and floor plate is distinguished by the zebrafish no tail mutation. *Cell* **75**: 99–111.
47. Talbot WS, Trevarrow B, Halpern ME, Melby AE, Farr G, Postlethwait JH, Jowett T, Kimmel CB and Kimelman D (1995). A homeobox gene essential for zebrafish notochord development. *Nature* **378**: 150–157.
48. Odenthal J, Haffter P, Vogelsang E, Brand M, van Eeden FJ, Furutani–Seiki M, Granato M, Hammerschmidt M, Heisenberg CP, Jiang YJ, Kane DA, Kelsh RN, Mullins MC, Warga RM, Allende ML, Weinberg ES and Nusslein–Volhard C (1996). Mutations affecting the formation of the notochord in the zebrafish and Danio rerio *Development* **123**: 103–115.
49. Blagden CS, Currie PD, Ingham PW and Hughes SM (1997). Notochord induction of zebrafish slow muscle mediated by Sonic hedgehog. *Genes Dev.* **11**: 2163–2175.
50. Currie PD and Ingham PW (1996). Induction of a specific muscle cell type by a Hedgehog-like protein in zebrafish. *Nature* **382**: 452–455.
51. Du SJ, Devoto SH, Westerfield M and Moon RT (1997). Positive and negative regulation of muscle cell identity by members of the Hedgehog and TGF-β gene families. *J. Cell Biol.* **139**: 145–156.
52. Coutelle O, Blagden CS, Hampson R, Halai C, Rigby PW and Hughes SM (2001). Hedgehog signaling is required for maintenance of myf5 and myoD expression and timely terminal differentiation in zebrafish adaxial myogenesis. *Dev Biol.* **236**: 136–150.

53. Roy S, Wolff C and Ingham PW (2001). The u-boot mutation identifies a Hedgehog-regulated myogenic switch for fiber-type diversification in the zebrafish embryo. *Genes Dev.* **15**: 1563–1576.
54. Hauschka SD (1994). Development, Anatomy and Cell Biology. In: Engel A and Franzini–Armstrong C (eds.), *Myology*. McGraw Hill Press, New York, pp. 3–73.
55. Hughes SM and Salinas PC (1999) Control of muscle fiber and motoneuron diversification. *Curr. Opin. Neurobiol.* **9**: 54–64.
56. Kelly AM and Rubinstein NA (1994). The diversity of muscle fiber types and its origin during development. In: Engel A and Franzini–Armstrong C (eds), *Myology*. McGraw Hill Press, New York, pp. 119–133.
57. Stockdale FE (1997). Mechanisms of formation of muscle fiber types. *Cell Struct. Funct.* **22**: 37–43.
58. Hatta K, Bremiller R, Westerfield M and Kimmel CB (1991). Diversity of expression of engrailed-like antigens in zebrafish. *Development* **112**: 821–832.
59. Melancon E, Liu DW, Westerfield M and Eisen JS (1997). Pathfinding by identified zebrafish motoneurons in the absence of muscle pioneers. *J. Neurosci.* **17**: 7796–7804.
60. van Raamsdonk W, van der Stelt A, Diegenbach PC, van de Berg W, de Bruyn H, van Dijk J and Mijzen P (1974). Differentiation of the musculature of the teleost Brachydanio rerio. *Z. Anat. Entwickl.-Gesch.* **145**: 321–342.
61. van Raamsdonk W, Pool CW and te Kronnie G (1978). Differentiation of muscle fiber types in the teleost Brachydanio rerio. *Anatomy and Embryology* **153**: 137–155.
62. van Raamsdonk W, Tekronnie G, Pool CW and van de Laarse W (1980). An immune histochemical and enzymic characterization of the muscle fibres in myotomal muscle of the teleost Brachydanio rerio, Hamilton-Buchanan. *Acta Histochem.* **67**: 200–216.
63. Mohler J and Vani K (1992). Molecular organization and embryonic expression of the hedgehog gene involved in cell-cell communication in segmental patterning of Drosophila. *Development* **115**(4):957–971.
64. Lee JJ, von Kessler DP, Parks S and Beachy PA (1992). Secretion and localized transcription suggest a role in positional signaling for products of the segmentation gene hedgehog. *Cell* **71**: 33–50.
65. Tabata T, Eaton S and Kornberg TB (1992). The Drosophila hedgehog gene is expressed specifically in posterior compartment cells and is a target of engrailed regulation. *Genes Dev.* **6**: 2635–2645.

66. Krauss S, Concordet JP and Ingham PW (1993). A functionally conserved homolog of the Drosophila segment polarity gene hh is expressed in tissues with polarizing activity in zebrafish embryos. *Cell* **75**: 1431–44.
67. Ekker SC, Ungar AR, Greenstein P, von Kessler DP, Porter JA, Moon RT and Beachy PA (1995). Patterning activities of vertebrate hedgehog proteins in the developing eye and brain. *Curr. Biol.* **5**: 944–955.
68. Roelink H, Augsburger A, Heemskerk J, Korzh V, Norlin S, Ruiz i Altaba A, Tanabe Y, Placzek M, Edlund T, Jessell TM and Dodd J (1994). Floor plate and motor neuron induction by vhh-1, a vertebrate homolog of *hedgehog* expressed by the notochord. *Cell* **76**: 761–775.
69. Lewis KE, Currie PD, Roy S, Schauerte H, Haffter P and Ingham PW (1999a). Control of muscle cell-type specification in the zebrafish embryo by Hedgehog signaling. *Dev. Biol.* **216**: 469–480.
70. Du SJ and Dienhart M (2001). Gli2 mediation of hedgehog signals in slow muscle induction in zebrafish. *Differentiation* **67**: 84–91.
71. Johnson RL and Scott MP (1998). New players and puzzles in the Hedgehog signaling pathway. *Curr. Opin. Genet. Devl.* **8**: 450–456.
72. Taipale J, Cooper MK, Maiti T and Beachy PA (2002). Patched acts catalytically to suppress the activity of Smoothened. *Nature* **418**: 892–897.
73. Concordet JP, Lewis KE, Moore JW, Goodrich LV, Johnson RL, Scott MP and Ingham PW (1996). Spatial regulation of a zebrafish patched homologue reflects the roles of sonic hedgehog and protein kinase A in neural tube and somite patterning. *Development* **122**: 2835–2846.
74. Lewis KE, Concordet JP and Ingham PW (1999b). Characterisation of a second patched gene in the zebrafish Danio rerio and the differential response of patched genes to Hedgehog signaling. *Dev. Biol.* **208**: 14–29.
75. Varga ZM, Amores A, Lewis KE, Yan YL, Postlethwait JH, Eisen JS and Westerfield M (2001). Zebrafish smoothened functions in ventral neural tube specification and axon tract formation. *Development* **128**: 3497–3509.
76. Karlstrom RO, Talbot WS and Schier AF (1999). Comparative synteny cloning of zebrafish you-too: mutations in the Hedgehog target gli2 affect ventral forebrain patterning. *Genes Dev.* **13**: 388–393
77. Karlstrom RO, Tyurina OV, Kawakami A, Nishioka N, Talbot WS, Sasaki H and Schier AF (2003). Genetic analysis of zebrafish gli1 and gli2 reveals divergent requirements for gli genes in vertebrate Development *Development* **130**: 1549–1564.

78. Barresi MJ, D'Angelo JA, Hernandez LP and Devoto SH (2001). Distinct mechanisms regulate slow-muscle development. *Curr. Biol.* **11**: 1432–1438.
79. Chandrasekhar A, Schauerte HE, Haffter P and Kuwada JY (1999). The zebrafish detour gene is essential for cranial but not spinal motor neuron induction. *Development* **126**: 2727–2737.
80. Karlstrom RO, Trowe T, Klostermann S, Baier H, Brand M, Crawford AD, Grunewald B, Haffter P, Hoffmann H, Meyer SU, Müller BK, Richter S, van Eeden FJ, Nüsslein–Volhard C and Bonhoeffer F (1996). Zebrafish mutations affecting retinotectal axon pathfinding. *Development* **123**: 427–438.
81. Marigo V, Johnson RL, Vortkamp, A and Tabin CJ (1996). Sonic Hedgehog differentially regulates expression of Gli and Gli3 during limb Development. *Dev. Biol. 180*: 273–283.
82. Grindley JC, Bellusci S, Perkins, D and Hogan BL (1997). Evidence for the involvement of the Gli gene family in embryonic mouse lung development. *Dev. Biol.* **188**: 337–348.
83. Hui CC and Joyner AL (1993). A mouse model of Greig cephalopolysyndactyly syndrome: the extra-toes mutation contains an intragenic deletion of the Gli3 gene. *Nature Genet.* **3**: 241–246.
84. Norris W, Neyt C, Ingham PW and Currie PD (2000). Slow muscle induction by Hedgehog signaling in vitro. *J. Cell. Sci.* **113**: 2695–2703.
85. Baxendale S, Davison C, Muxworthy C, Wolff C, Ingham P, and Roy S (2004). The B-cell maturation factor Blimp-1 specifies vertebrate slow-twitch muscle fiber identity in response to Hedgehog signaling. *Nat Genet.* **36**(1): 88–93.
86. Du SJ and Tan X (2003). *mBop*, a new player in regulating skeletal and cardiac muscle differentiation in zebrafish embryos. In: *3rd European Conference on Zebrafish and Medaka Genetics and Development.* Paris, June 11–14. P46.
87. Rescan PY, Collet B, Ralliere C, Cauty C, Delalande JM, Goldspink G and Fauconneau B (2001a). Red and white muscle development in the trout (Oncorhynchus mykiss) as shown by in situ hybridisation of fast and slow myosin heavy chain transcripts. *J. Exp. Biol.* **204**(Pt 12):2097–2101.
88. Rissi M, Wittbrodt J, Delot E, Naegeli M and Rosa FM (1995). Zebrafish Radar: a new member of the TGF-beta superfamily defines dorsal regions of the neural plate and the embryonic retina. *Mech. Dev.* **49**: 223–234.
89. Teillet M-A, Watanabe Y, Jeffs P, Duprez D, Lapointe F and Le Douarin NM (1998). Sonic hedgehog is required for survival of both myogenic and chondrogenic somitic lineages. *Development* **125**: 2019–2030.

90. Cann GM, Lee JW and Stockdale FE (1999). Sonic hedgehog enhances somite cell viability and formation of primary slow muscle fibers in avian segmented mesoderm. *Anat. Embryol.* (*Berl*), **200**: 239–252.
91. Duprez D, Fournier–Thibault C and Le Douarin, N (1998). Sonic Hedgehog induces proliferation of committed skeletal muscle cells in the chick limb. *Development* **125**: 495–505.
92. Roy S and Ingham PW (2002). Hedgehogs tryst with the cell cycle. *J. Cell Sci.* **115**: 4393–4397.
93. Cossu G and Borello U (1999). Wnt signaling and the activation of myogenesis in mammals. *EMBO J.* **18**: 6867–6872.
94. Petropoulos H and Skerjanc IS (2002). Beta-catenin is essential and sufficient for skeletal myogenesis in P19 cells. *J. Biol. Chem.* **277**: 15393–15399.
95. Munsterberg AE, Kitajewski J, Bumcrot DA, McMahon AP and Lassar AB (1995). Combinatorial signaling by Sonic hedgehog and Wnt family members induces myogenic bHLH gene expression in the somite. *Genes Dev.* **9**(23):2911–2922.
96. Ordahl CP and Le Douarin N (1991). Two myogenic lineages within the developing somite. *Development* **114**: 339–353.
97. Cossu G, Kelly R, Tajbakhsh S, Di Donna S, Vivarelli E and Buckingham M (1996). Activation of different myogenic pathways: myf-5 is induced by the neural tube and MyoD by the dorsal ectoderm in mouse paraxial mesoderm. *Development* **122**: 429–437.
98. Tajbakhsh S, Rocancourt D, Cossu G and Buckingham M (1997). Redefining the genetic hierarchies controlling skeletal myogenesis: Pax-3 and Myf-5 act upstream of MyoD. *Cell* **89**: 127–138.
99. Tajbakhsh S, Borello U, Vivarelli E, Kelly R, Papkoff J, Duprez D, Buckingham M and Cossu G (1998) Differential activation of Myf5 and MyoD by different Wnts in explants of mouse paraxial mesoderm and the later activation of myogenesis in the absence of Myf(5). *Development* **125**: 4155–4162.
100. Borycki AG, Mendham L and Emerson CP (1998). Control of somite patterning by Sonic hedgehog and its downstream signal response genes. *Development* **125**: 777–790.
101. Borycki AG, Brunk B, Tajbakhsh S, Buckingham M, Chiang C and Emerson CP (1999). Sonic hedgehog controls epaxial muscle determination through Myf5 activation. *Development* **126**: 4053–4063.

102. Stern HM, Brown AM and Hauschka SD (1995). Myogenesis in paraxial mesoderm: preferential induction by dorsal neural tube and by cells expressing Wnt-(1). *Development* **121**: 3675–3686.
103. Takada S, Stark KL, Shea MJ, Vassileva G, McMahon JA and McMahon AP (1994). Wnt-3a regulates somite and tailbud formation in the mouse embryo. *Genes Dev.* **8**: 174–189.
104. Borello U, Coletta M, Tajbakhsh S, Leyns L, De Robertis EM, Buckingham M and Cossu G (1999). Transplacental delivery of the Wnt antagonist Frzb1 inhibits development of caudal paraxial mesoderm and skeletal myogenesis in mouse embryos. *Development* **126**: 4247–4255.
105. Yamaguchi TP, Takada S, Yoshikawa Y, Wu N and McMahon AP (1999). T (Brachyury) is a direct target of Wnt3a during paraxial mesoderm specification. *Genes Dev.* **13**: 3185–3190.
106. Kelly GM, Lai CJ and Moon RT (1993). Expression of wnt10a in the central nervous system of developing zebrafish. *Dev. Biol.* **158**: 113–121.
107. Kelly GM, Greenstein P, Erezyilmaz DF and Moon RT (1995). Zebrafish wnt8 and wnt8b share a common activity but are involved in distinct developmental pathways. *Development* **121**: 1787–1799.
108. Ungar AR, Kelly GM and Moon RT (1995). Wnt4 affects morphogenesis when mis-expressed in the zebrafish embryo. *Mech. Dev.* **52**(2–3):153–164.
109. Makita R, Mizuno T, Koshida S, Kuroiwa A and Takeda H (1998). Zebrafish wnt**11**: pattern and regulation of the expression by the yolk cell and No tail activity. *Mech. Dev.* **71**: 165–176.
110. Kilian B, Mansukoski H, Barbosa FC, Ulrich F, Tada M and Heisenberg CP (2003). The role of Ppt/Wnt5 in regulating cell shape and movement during zebrafish gastrulation. *Mech. Dev.* **120**(4):467–476.
111. Lekven AC, Buckles GR, Kostakis N and Moon RT (2003). Wnt1 and wnt10b function redundantly at the zebrafish midbrain–hindbrain boundary. *Dev. Biol.* **254**: 172–187.
112. Lekven AC, Thorpe CJ, Waxman JS and Moon RT (2001). Zebrafish wnt8 encodes two wnt8 proteins on a bicistronic transcript and is required for mesoderm and neurectoderm patterning. *Dev. Cell* **1**: 103–14.
113. Thorpe CJ, Lekven AC and Moon RT (2000). The role of Wnt3A in posterior mesoderm development. Abstract 257. In: *The Fourth International Meeting on Zebrafish Development and Genetics.* Cold Spring Harbor, NY., April 26–30.

114. Meng A, Moore B, Tang H, Yuan B and Lin S (1999). A Drosophila doublesex-related gene, terra, is involved in somitogenesis in vertebrates. *Development* **126**: 1259–1268.
115. Sass JB, Martin CC and Krone PH (1999). Restricted expression of the zebrafish hsp90alpha gene in slow and fast muscle fiber lineages. *Int. J. Dev. Biol.* **43**: 835–838.
116. Lele Z, Hartson SD, Martin CC, Whitesell L, Matts RL and Krone PH (1999). Disruption of zebrafish somite development by pharmacologic inhibition of Hsp(90). *Dev. Biol.* **210**: 56–70.
117. Sass JB and Krone PH (1997). HSP90alpha gene expression may be a conserved feature of vertebrate somitogenesis. *Exp. Cell Res.* **233**: 391–394.
118. Yun K and Wold B (1996). Skeletal muscle determination and differentiation: story of a core regulatory network and its context. *Curr. Opin. Cell Biol.* **8**: 877–889.
119. Black BL and Olson EN (1998). Transcriptional control of muscle development by myocyte enhancer factor-2 (MEF2) proteins. *Annu. Rev. Cell Dev Biol.* **14**: 167–96.
120. Weintraub H, Tapscott SJ, Davis RL, Thayer MJ, Adam MA, Lassar AB and Miller AD (1989). Activation of muscle-specific genes in pigment, nerve, fat, liver, and fibroblast cell lines by forced expression of MyoD. *Proc. Natl. Acad. Sci. USA* **86**: 5434–5438.
121. Buskin JN and Hauschka SD (1989). Identification of a myocyte nuclear factor that binds to the muscle-specific enhancer of the mouse muscle creatine kinase gene. *Mol. Cell Biol.* **9(6)**: 2627–2640.
122. Emerson CP. (1990). Myogenesis and developmental control genes. *Curr. Opin. Cell Biol.* **2**: 1065–1075.
123. Olson EN (1990). MyoD family: a paradigm for development? *Genes Dev.* **4**: 1454–1461.
124. Weintraub H (1993). The MyoD family and myogenesis: redundancy, networks and and thresholds *Cell* **75**: 1241–1244.
125. Rudnicki MA, Schnegelsberg PN, Stead RH, Braun T, Arnold HH and Jaenisch R (1993). MyoD or Myf-5 is required for the formation of skeletal muscle. *Cell* **75**: 1351–1359.
126. Hasty P, Bradley A, Morris JH, Edmondson DG, Venuti JM, Olson EN and Klein WH (1993). Muscle deficiency and neonatal death in mice with a targeted mutation in the myogenin gene. *Nature* **364**(6437): 501–506.

127. Nabeshima Y, Hanaoka K, Hayasaka M, Esumi E, Li S, Nonaka I and Nabeshima Y (1993). Myogenin gene disruption results in perinatal lethality because of severe muscle defect. *Nature* **364**(6437):532–535.
128. Rawls A, Valdez MR, Zhang W, Richardson J, Klein WH and Olson EN (1998). Overlapping functions of the myogenic bHLH genes MRF4 and MyoD revealed in double mutant mice. *Development* **125**: 2349–2358.
129. Lin Q, Schwarz J, Bucana C and Olson EN (1997). Control of mouse cardiac morphogenesis and myogenesis by transcription factor MEF2C. *Science* **276**: 1404–7.
130. Weinberg ES, Allende ML, Kelly CS, Abdelhamid A, Murakami T, Andermann P, Doerre OG, Grunwald DJ and Riggleman B (1996). Developmental regulation of zebrafish MyoD in wild-type, no tail and spadetail embryos. *Development* **122**: 271–280.
131. Watabe S (2001). Myogenic regulatory factors. In: Johnston IA (ed.), *Muscle Development and Growth.* Academic Press.
132. Rescan PY. (2001). Regulation and functions of myogenic regulatory factors in lower vertebrates. *Comp. Biochem. Physiol. B. Biochem. Mol. Biol.* **130**: 1–12.
133. Hinits Y, de Barreda CM, Rigby PWJ and Hughes SM (2002). Zebrafish mrf4 expression differs between muscle fibre populations. In: *5th International Conference on Zebrafish Development and Genetics.* June 12–16, (2002). Madison, Wisconsin.
134. Ticho BS, Stainier DY, Fishman MC and Breitbart RE (1996). Three zebrafish MEF2 genes delineate somitic and cardiac muscle development in wild-type and mutant embryos. *Mech Dev.* **59**: 205–218.
135. Rescan PY, Gauvry L, Paboeuf G and Fauconneau B (1994). Identification of a muscle factor related to MyoD in a fish species. *Biochim. Biophys. Acta.* **1218**: 202–204.
136. Rescan PY, Gauvry L and Paboeuf G (1995). A gene with homology to myogenin is expressed in developing myotomal musculature of the rainbow trout and in vitro during the conversion of myosatellite cells to myotubes. *FEBS Lett.* **362**: 89–92.
137. Kobiyama A, Nihei Y, Hirayama Y, Kikuchi K, Suetake H, Johnston IA and Watabe S (1998). Molecular cloning and developmental expression patterns of the MyoD and MEF2 families of muscle transcription factors in the carp. *J. Exp. Biol.* **201**(Pt 20): 2801–2813.

138. Chen Y, Lee W, Cheng C and Tsai H (2000). Muscle regulatory factor gene: zebrafish (Danio rerio) myogenin cDNA. *Comp. Biochem. Physiol. B* **127**: 97–103.
139. Chen YH, Lee WC, Liu CF and Tsai HJ (2001). Molecular structure, dynamic expression, and promoter analysis of zebrafish (Danio rerio) myf-5 gene. *Genesis.* **29**: 22–35.
140. Tan X, Hoang L and Du SJ (2002). Characterization of muscle-regulatory genes, myf5 and myogenin, from striped bass and promoter analysis of muscle-specific expression. *Marine Biotech.* **4**: 537–545.
141. Barresi MJ, Stickney HL and Devoto SH (2000). The zebrafish slow-muscle-omitted gene product is required for Hedgehog signal transduction and the development of slow muscle identity. *Development* **127**: 2189–2199.
142. Hirsinger E and Westerfield M (2003). Uncoupling adaxial cell induction from slow muscle fate. In: *3rd European Conference on Zebrafish and Medaka Genetics and Development.* Paris, June 11–14. L14.
143. Goldhamer DJ, Faerman A, Shani M and Emerson CP Jr. (1992). Regulatory elements that control the lineage-specific expression of myoD. *Science* **256**: 538–542.
144. Goldhamer DJ, Brunk BP, Faerman A, King A, Shani M and Emerson CP Jr. (1995). Embryonic activation of the myoD gene is regulated by a highly conserved distal control element. *Development* **121**: 637–649.
145. Kucharczuk KL, Love CM, Dougherty NM and Goldhamer DJ (1999). Fine-scale transgenic mapping of the MyoD core enhancer: MyoD is regulated by distinct but overlapping mechanisms in myotomal and non-myotomal muscle lineages. *Development* **126**: 1957–1965.
146. Tan X and Du J (2002). Differential expression of two MyoD genes in fast and slow muscles of gilthead seabream (Sparus aurata). *Dev. Genes Evol.* **212**: 207–217.
147. Yang HW, Kanki J and Look T (2002). The upstream enhancer of human MyoD is capable of regulating the expression of zebrafish MyoD transcrtiption during Development Abstract (497). In: *5th International Conference on Zebrafish Development and Genetics.* Madison, WI.
148. Cheng TC, Wallace MC, Merlie JP and Olson EN (1993). Separable regulatory elements governing myogenin transcription in mouse embryogenesis. *Science* **261**: 215–218.

149. Yee SP and Rigby PW (1993). The regulation of myogenin gene expression during the embryonic development of the mouse. *Genes Dev.* 7: 1277–1289.
150. Spitz F, Demignon J, Porteu A, Kahn A, Concordet JP, Daegelen D and Maire P (1998). Expression of myogenin during embryogenesis is controlled by Six/sine oculis homeoproteins through a conserved MEF3 binding site. *Proc. Natl. Acad. Sci. USA* **95**: 14220–14225.
151. Oliver G, Wehr R, Jenkins NA, Copeland NG, Cheyette BN, Hartenstein V, Zipursky SL and Gruss P (1995). Homeobox genes and connective tissue patterning. *Development* **121**: 693–705.
152. Heanue TA, Reshef R, Davis RJ, Mardon G, Oliver G, Tomarev S, Lassar AB and Tabin CJ (1999). Synergistic regulation of vertebrate muscle development by Dach2, Eya2, and Six1, homologs of genes required for Drosophila eye formation. *Genes Dev.* **13**: 3231–3243.
153. Ghanbari H, Seo HC, Fjose A and Brandli AW (2001). Molecular cloning and embryonic expression of Xenopus Six homeobox genes. *Mech. Dev.* **101**: 271–277.
154. Spitz F, Salminen M, Demignon J, Kahn A, Daegelen D and Maire P (1997). A combination of MEF3 and NFI proteins activates transcription in a subset of fast-twitch muscles. *Mol. Cell Biol.* **17**: 656–666.
155. Chen YH and Tsai HJ (2002). Treatment with Myf5-morpholino results in somite patterning and brain formation defects in zebrafish. *Differentiation* **70**: 447–456.
156. Delalande JM and Rescan PY (1999). Differential expression of two nonallelic MyoD genes in developing and adult myotomal musculature of the trout (Oncorhynchus mykiss). *Dev. Genes Evol.* **209**: 432–437.
157. Rescan PY, Delalande JM, Gauvry L and Fauconneau B (1999). Differentail expression of two MyoD genes during early development of the trout:comparison with myogenin. *J. Fish Biol.* **55**: 19–25.
158. Slack JM (1997). Growth control: action mouse. *Curr Biol.* 7: R467–469.
159. Conlon I and Raff M (1999). Size control in animal Development. *Cell* **96**: 235–244
160. Lee SJ and McPherron AC (1999). Myostatin and the control of skeletal muscle mass. *Curr. Opin. Genet.* **9**(5):604–607.
161. McPherron AC, Lawler AM and Lee S-J (1997). Regulation of skeletal muscle mass in mice by a new TGF-fl superfamily member. *Nature* **387**: 83–90.

162. McPherron AC and Lee S-J (1997). Double muscling in cattle due to mutations in the myostatin gene. *Proc. Natl. Acad. Sci. USA* **94**: 12457–12461.
163. Kambadur R, Sharma M, Smith TPL and Bass JJ (1997). Mutations in myostatin (GDF8) in double-muscled Belgian Blue and Piedmontese cattle. *Genome Res.* 7: 910–916.
164. Grobet, L and Martin LJR, Poncelet D, Pirottin D, Brouwers B, Riquet J, Schoeberlein A, Dunner S, Ménissier F, Massabanda J, Fries R, Hanset R and Georges M (1997). A deletion in the bovine myostatin gene causes the double-muscled phenotype in cattle. *Nature Genetics* **17**: 71–74.
165. Grobet L, Poncelet D, Royo LJ, Brouwers B, Pirottin D, Michaux C, Ménissier F, Zanotti M, Dunner S and Georges M (1998). Molecular definition of an allele series of mutations disrupting the myostatin function and causing double-muscling in cattle. *Mammalian Genome* **9**: 210–213.
166. Grobet L, Pirottin D, Farnir F, Poncelet D, Royo LJ, Brouwers B, Christians E, Desmecht D, Coignoul F, Kahn R and Georges M (2003). Modulating skeletal muscle mass by postnatal, muscle-specific inactivation of the myostatin gene. *Genesis* **35**: 227–238.
167. Thomas M, Langley B, Berry C, Sharma M, Kirk S, Bass J and Kambadur R (2000). Myostatin, a negative regulator of muscle growth, functions by inhibiting myoblast proliferation. *J. Biol. Chem.* **275**: 40235–40243.
168. Rios R, Carneiro I, Arce VM and Devesa J (2001). Myostatin regulates cell survival during C2C12 myogenesis. *Biochem. Biophys. Res Commun.* **280**: 561–566.
169. Taylor WE, Bhasin S, Artaza J, Byhower F, Azam M, Willard DH, Kull FC and Gonzalez–Cadavid N (2001). Myostatin inhibits cell proliferation and protein synthesis in C(2)C(12) muscle cells. *Am. J. Physiol. Endocrinol Metab.* **280**: E221–228.
170. Langley B, Thomas M, Bishop A, Sharma M, Gilmour S and Kambadur R (2002). Myostatin inhibits myoblast differentiation by down-regulating MyoD expression. *J. Biol. Chem.* **277**: 49831–49840.
171. Amthor H, Huang R, McKinnell I, Christ B, Kambadur R, Sharma M and Patel K (2002). The regulation and action of myostatin as a negative regulator of muscle development during avian embryogenesis. *Dev. Biol.* **251**: 241–257.
172. Rodgers BD and Weber GM (2001). Sequence conservation among fish myostatin orthologues and the characterization of two additional cDNA

clones from Morone saxatilis and Morone americana. *Comp. Biochem. Physiol. B: Biochem. Mol. Biol.* **129**: 597–603.

173. Rodgers BD, Weber GM, Sullivan CV and Levine MA (2001). Isolation and characterization of myostatin complementary deoxyribonucleic acid clones from two commercially important fish: *Oreochromis mossambicus and Morone chrysops. Endocrinology* **142**: 1412–1418.
174. Maccatrozzo L, Bargelloni L, Radaelli G, Mascarello F and Patarnello T (2001). Characterization of the Myostatin gene in the gilthead seabream (Sparus aurata): sequence, genomic structure, and expression pattern. *Marine Biotech.* **3**: 224–230.
175. Maccatrozzo L, Bargelloni L, Patarnello T, Radaelli G, Mascarello F and Patarnello T (2002). Characterization of the myostatin gene and a linked microsatellite marker in shi drum (umbrina cirrosa, Sciaenidae). *Aquaculture* **205**: 49–60.
176. Kocabas AM, Kucuktas H, Dunham RA and Liu Z (2002). Molecular characterization and differential expression of the myostatin gene in channel catfish (Ictalurus punctatus). *Biochim. Biophys. Acta.* **1575**: 99–107.
177. Roberts SB and Goetz FW (2001). Differential skeletal muscle expression of myostatin across teleost species, and the isolation of multiple myostatin isoforms. *FEBS Lett.* **491**: 212–216.
178. Ostbye TK, Galloway TF, Nielsen C, Gabestad I, Bardal T and Andersen O (2001). The two myostatin genes of Atlantic salmon (Salmo salar) are expressed in a variety of tissues. *Eur. J. Biochem.* **268**: 5249–5257.
179. Rescan PY, Jutel I and Ralliere C (2001b). Two myostatin genes are differentially expressed in myotomal muscles of the trout (Oncorhynchus mykiss). *J. Exp. Biol.* **204**: 3523–3529.
180. Ji S, Losinski RL, Cornelius SG, Frank GR, Willis GM, Gerrard DE, Depreux FF and Spurlock ME (1998). Myostatin expression in porcine tissues: tissue specificity and developmental and postnatal regulation. *Am. J. Physiol.* **275**: R1265–1273.
181. Sharma M, Kambadur R, Matthews KG, Somers WG, Devlin GP, Conaglen JV, Fowke PJ and Bass JJ (1999). myostatin, a transforming growth factor-beta superfamily member, is expressed in heart muscle and is upregulated in cardiomyocytes after infarct. *J. Cell Physiol.* **180**: 1–9.
182. Radaelli G, Rowlerson A, Mascarello F, Patruno M and Funkenstein B (2003). Myostatin precursor is present in several tissues in teleost fish: a comparative immunolocalization study. *Cell Tissue Res.* **311**: 239–250.

183. Xu C, Wu G, Zohar Y and Du SJ (2003). Analysis of myostatin gene structure, expression and function in zebrafish. *J. Exp. Biol.* **206**: 4067–4080.
184. Vianello S, Brazzoduro L, Dalla Valle L, Belvedere P and Colombo L (2003). Myostatin expression during development and chronic stress in zebrafish (Danio rerio). *J. Endocrinol.* **176**: 47–59.
185. McPherron AC and Lee S-J (1996). The transforming growth factor b superfamily. *Growth factors and cytokines in health disease* Volum 1B, 357–(393). JAI press Inc.
186. Thies RS, Chen T, Davies MV, Tomkinson KN, Pearson AA, Shakey QA and Wolfman NM (2001). GDF-8 propeptide binds to GDF-8 and antagonizes biological activity by inhibiting GDF-8 receptor binding. *Growth Factors* **18**: 251–259.
187. Lee SJ and McPherron AC (2001). Regulation of myostatin activity and muscle growth. *Proc. Natl. Acad. Sci. USA.* **98**: 9306–11.
188. Yang J, Ratovitski T, Brady JP, Solomon MB, Wells KD and Wall RJ (2001). Expression of myostatin pro domain results in muscular transgenic mice. *Mol. Reprod. Dev.* **60**: 351–361.
189. Donoghue MJ, Merlie JP, Rosenthal N and Sanes JR (1991). Rostrocaudal gradient of transgene expression in adult skeletal muscle. *Proc Natl Acad Sci* USA **88**: 5847–5851.
190. Morales R, Herrera MT, Arenal A, Cruz O, Hernandez O, Pimentel R, Guillen I, Martinez R and Estrada MP (2001). Tilapia chromosomal growth hormone gene expression accelewrates grwoth in transgenic zebrafish (Danio rerio). *Elec. J. Biotech.* **4**(2) 51–58.
191. Rowlerson A and Vegetti A (2001). Cellular mechanisms of post-embryonic muscle growth in aquaculture species. In: Johnston IA (ed.), *Muscle Development and Growth.* Academic Press.
192. van Raamsdonk W, Mos W, Smit–Onel MJ, van der Laarse WJ and Fehres R (1983). The development of the spinal motor column in relation to the myotomal muscle fibers in the zebrafish (Brachydanio rerio). I. Posthatching development. *Anat. Embryol.* (*Berl*). **167**: 125–139.
193. Weatherley AH and Gill HS (1984). Growth dynamic of white myotomal muscle fibres in the bluntnose minnow, Pimephales notatus Rafinesque, and comparison with rainbow trout, Sdalmo gairdneri Richardson. *J. Fish Biol.* **25**: 13–24.

194. Weatherley AH and Gill HS (1985). Dynamics of increase in muscle fibers in fishes in relation to size and growth. *Experientia* **41**: 353–354.
195. Weatherley AH, Gill HS and Lobo AF (1988). Recruitment and maximal diameter of axial muscle fibres in teleosts and their relationship to somatic growth and ultimate size. *J. Fish Biol.* **33**: 851–859.
196. Bauer H, Meier A, Hild M, Stachel S, Economides A, Hazelett D, Harland RM and Hammerschmidt M (1998). Follistatin and noggin are excluded from the zebrafish organizer. *Dev. Biol.* **204**: 488–507.
197. Spiller MP, Kambadur R, Jeanplong F, Thomas M, Martyn JK, Bass JJ and Sharma M (2002). The myostatin gene is a downstream target gene of basic helix-loop-helix transcription factor MyoD. *Mol. Cell. Biol.* **22**: 7066–7082.
198. Bolanos–Jimenez F, Bordais A, Behra M, Strahle U, Mornet D, Sahel J and Rendon A (2001a). Molecular cloning and characterization of dystrophin and Dp71, two products of the Duchenne Muscular Dystrophy gene and in zebrafish *Gene* **274**: 217–226.
199. Bolanos–Jimenez F, Bordais A, Behra M, Strahle U, Sahel J and Rendon A 2001b. Dystrophin and Dp71, two products of the DMD gene, show a different pattern of expression during embryonic development in zebrafish. *Mech. Dev.* **102**: 239–241.
200. Guyon JR, Mosley AN, Zhou Y, O'Brien KF, Sheng X, Chiang K, Davidson AJ, Volinski JM, Zon LI and Kunkel LM (2003). The dystrophin associated protein complex in zebrafish. *Hum. Mol. Genet.* **12**: 601–615.
201. Parsons MJ, Campos I, Hirst EM and Stemple DL (2002). Removal of dystroglycan causes severe muscular dystrophy in zebrafish embryos. *Development* **129**: 3505–3512.
202. Xu Y, He J, Wang X, Lim TM and Gong Z (2000). Asynchronous activation of 10 muscle-specific protein (MSP) genes during zebrafish somitogenesis. *Dev. Dyn.* **219**: 201–215.
203. Higashijima S, Okamoto H, Ueno N, Hotta Y and Eguchi G (1997). High-frequency generation of transgenic zebrafish which reliably express GFP in whole muscles or the whole body by using promoters of zebrafish origin. *Dev. Biol.* **192**: 289–299.
204. Ju B, Xu Y, He J, Liao J, Yan T, Hew CL, Lam TJ and Gong (1999). Faithful expression of green fluorescent protein (GFP) in transgenic zebrafish embryos under control of zebrafish gene promoters. *Dev. Genet.* **25**: 158–167.

205. Ju B, Chong SW, He J, Wang X, Xu Y, Wan H, Tong Y, Yan T, Korzh V and Gong Z (2003). Recapitulation of fast skeletal muscle development in zebrafish by transgenic expression of GFP under the mylz2 promoter. *Dev. Dyn.* **227**: 14–26.
206. Xu Y, He J, Tian HL, Chan CH, Liao J, Yan T, Lam TJ and Gong Z (1999). Fast skeletal muscle-specific expression of a zebrafish myosin light chain 2 gene and characterization of its promoter by direct injection into skeletal muscle. *DNA Cell Biol.* **18**: 85–95.
207. Nomura–Kitabayashi A, Takahashi Y, Kitajima S, Inoue T, Takeda H and Saga Y (2002). Hypomorphic Mesp allele distinguishes establishment of rostrocaudal polarity and segment border formation in somitogenesis. *Development* **129**: 2473–2481.
208. Wolff C, Roy S, Ingham PW (2003). Multiple muscle cell identities induced by distinct levels and timing of hedgehog activity in the zebrafish embryo. *Curr Biol.* **13**(14): 1169–1181.
209. Nakano Y, Kim HR, Kawakami A, Roy S, Schier AF, Ingham PW (2004). Inactivation of dispatched 1 by the chameleon mutation disrupts Hedgehog signalling in the zebrafish embryo. *Dev Biol.* **269**(2): 381–392.
210. Burke R, Nellen D, Bellotto M, Hafen E, Senti KA, Dickson BJ, Basler K (1999). Dispathced, a novel sterol-sensing domain protein dedicated to the release of cholesterol-modified hedgehog from signaling cells. *Cell.* **99**(7): 803–815.
211. Wolff C, Roy S, Lewis KE, Schauerte H, Joerg–Rauch G, Kirn A, Weiler C, Geisler R, Haffter P, Ingham PW (2004). Iguana encodes a novel zinc-finger protein with coiled-coil domains essential for Hedgehog signal transduction in the zebrafish embryo. *Genes Dev.* Jun 15 [Epub ahead of print] PMID: 15198976.
212. Grimaldi A, Tettamanti G, Martin BL, Gaffield W, Pownall ME, Hughes SM (2004). Hedgehog regulation of superficial slow muscle fibres in Xenopus and the evolution of tetrapod trunk myogenesis. *Development.* **131**(14): 3249–3262.
213. Roberts SB, Goetz FW (2003). Myostatin protein and RNA transcript levels in adult and developing brook trout. *Mol Cell Endocrinol.* Nov 28; **210**(1-2): 9–20.
214. Bassett DI, Bryson–Richardson RJ, Daggett DF, Gautier P, Keenan DG, Currie PD (2003). Dystrophin is required for the formation of stable muscle attachments in the zebrafish embryo. *Development.* **130**(23): 5851–5860.

Chapter 11

Skeletogenesis in Zebrafish *Danio rerio*: Evolutionary and Developmental Aspects

Shannon Fisher[1] and Paula M. Mabee[2]
[1]*McKusick–Nathans Institute of Genetic Medicine*
The Johns Hopkins University School of Medicine
Baltimore, MD 21205
[2]*Department of Biology*
University of South Dakota
Vermillion, SD 57069

The zebrafish, a species within the family Cyprinidae (minnows and carps), has emerged as an important vertebrate model for the study of development, including skeletal development, due to the availability of embryological, molecular, and genetic tools. Ichthyology has a long history, extending from Aristotle through present, fueled by a large number of species relative to most major vertebrate groups and by the considerable fossil record consisting of fish skeletons. The comparative adult osteology of fishes is a traditional area of study within the field, and many of the differences in skeletal anatomy closely track the paths of evolution within various clades. Developmental studies in zebrafish have the potential to clarify historically significant evolutionary questions pertaining to the evolution of the skeleton.

We focus on the study of skeletogenesis in zebrafish, although our emphasis is on the extent to which findings can be generalized to other vertebrates. After a review of descriptive studies, we discuss the availability of molecular markers, and the mutational analysis of skeletal development in zebrafish. Finally, we conclude with a discussion of prospects for additional, targeted mutant screens, and studies to examine lineage relationships among skeletogenic cells in the embryo and the adult.

1. Phylogeny and Evolution

The first fish-like vertebrates are known from the fossil record of the Ordovician Period, nearly 500 million years ago.[1] Today, fishes are the most numerous and diverse of the major vertebrate groups with over 25,000 described species.[2] Bone evolved in most recent common ancestor of vertebrates, and it was present in the earliest members of the major basal groups (lampreys, chondrichthyans, and extinct groups such as the anaspids, placoderms, and acanthodians).[1] In contrast with other metazoan groups, the origin of a fossilizable skeleton occurred early in their radiation, so that most major events in their history are well documented.[3] Because the skeletal system reflects general body form, the mode of locomotion, many aspects of feeding, and the size and details of the external form of the brain, the evolutionary history of vertebrates has been reconstructed in significant detail.[3] Thus the evolutionary history of early vertebrates as recorded in the fossil record is essentially one consisting of hard parts: bones of the exo- and endoskeleton, teeth, scales and spines.

Because of their phylogenetic position at the base of the vertebrates, comparative anatomical data from fishes is critical in correctly determining the directions and patterns of evolution. Careful description and comparisons of skeletal and other anatomical detail has long been a tradition in vertebrate[4,5] and in particular, ichthyological studies.[6–8] Frequently skeletal features were discovered that characterized monophyletic groups. That is, the skeleton provided an excellent record of evolutionary history and consequently, interrelationships.[9–11] Questions concerning the developmental bases of such evolutionary changes emerged, but only recently, with the emergence of zebrafish as a model, have molecular and genetic tools been available to approach skeletal evolution and development ("evo-devo") in fishes.

The zebrafish is a well-established system for genetic studies of development,[12] and because of its basal phylogenetic position within vertebrates, it is a key species for evolutionary and developmental studies. The zebrafish was first described from the Kosi tributary of the Ganges River in India[13] and is also found in Pakistan, Bangladesh, Nepal, Myanmar (Burma), and Sri Lanka.[14] It is a member of the subfamily Leuciscinae, one of the two monophyletic lineages within the Cyprinidae, and it is considered a member of the rasborins, the sister group to all other leuciscines.[15] This clade is sometimes termed the "Rasborinae[16] or the Danioninae."[17] The Cyprinidae is a family in the Cypriniformes, which in turn is a member of the monophyletic Otophysi,[18] which lies within the larger clade Ostariophysi. Otophysan teleosts include the Cypriniformes (zebrafish), and they comprise approximately 75% of all freshwater fishes and 27% of all fishes (~7500 species).[19,20]

The Otophysi are characterized by the Weberian apparatus, a major innovation involving skeletal components. The Weberian apparatus provides a skeletal route for the transmission of sound/vibration from the swim bladder to the inner ear.[21,22] This adaptation is considered key in facilitating the enormous radiation of the Otophysi.[19] Rudiments of the Weberian apparatus are found in sister taxa, and thus it may have evolved gradually.[23–25] The Weberian apparatus is modified but not lost in any otophysan, and the zebrafish retains many of the primitive conditions in this skeletal character complex.

2. Descriptive Embryology and Lineage of the Skeleton

2.1. Methodology

In staging specimens, size is more highly correlated with stage of osteological development than age, and it is thus a better measure of "stage" of skeletal development. This is important in particular for researchers of the skeletal system, as osteogenesis occurs in generally larger (and older) specimens. However, the level of intraspecific variation in relative sequence of ossification in zebrafish is low, regardless of rearing temperature or strain.[26] Most of the variation in developmental timing involves small shifts in sequence, with an average level of variation per bone of about two ranks. Descriptive studies of skeletal morphology in teleosts as well as other species have classically been based on Alcian blue staining to reveal cartilage, and Alizarin red S staining to reveal bone.[27–29] These staining methods give unsurpassed detail, but are labor intensive and only performed on fixed specimens (Fig. 1A).

More recently, methods have been employed to image the skeleton of live fish in considerable detail. The fluorescent calcium chelator calcein has been used as a vital indicator of calcification, to reveal osteogenesis

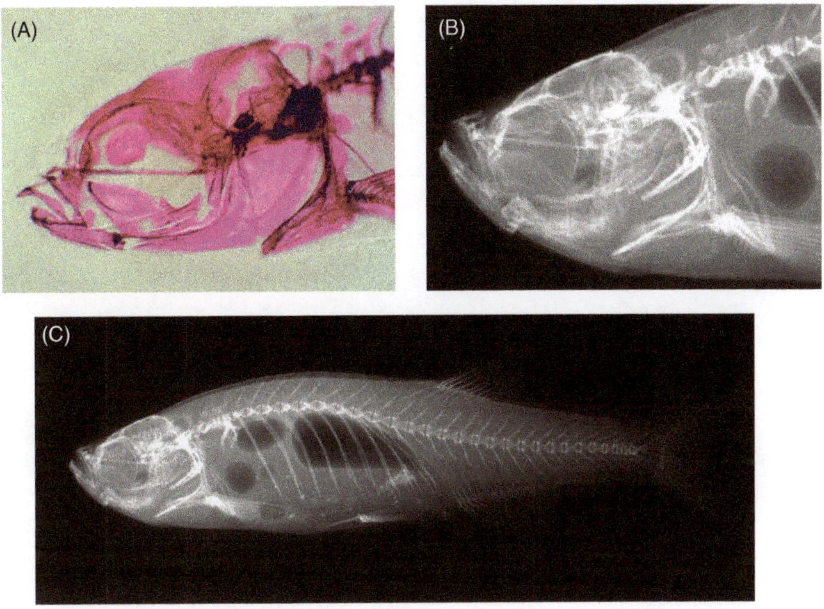

Fig. 1 Visualization of developing and adult skeleton. A comparison is shown of the zebrafish skeleton, visualized in a three-week-old fish by alizarin red staining (A), and in a live adult fish by radiography (B, C).

in zebrafish larvae.[30] It has the advantages of being quick and non-toxic to the embryo, and virtually permanent. The latter property could allow it to be used in histomorphometry to measure bone growth, as is done currently in mice. For most of the skeleton, examination by calcein is limited to young fish when the fluorescence can be seen through the soft tissue.

Radiography of live fish has also been used for the examination of later skeletal morphology (Fig. 1B, C),[31] and to screen for mutations affecting the morphology of the adult skeleton. While radiography is rapid and not harmful to live animals, it does require specialized equipment. Also, it is only useful on older fish with well calcified skeletons. Neither of these methods will replace traditional histological staining of fixed specimens, but they are more easily adaptable to the examination of large numbers of fish and to a mutational screening strategy.

2.2. Craniofacial Development

The skull of fishes, including that of zebrafish, is far more complex in terms of number of bones,[32] diversity of skeletal types,[33] and articulations, than the skulls of other model organisms including amphibians, "reptiles," birds, and mammals. Zebrafish has 74 cranial ossifications, compared to, for example, the frog *Spea bombifrons* which has 19[34] or the marsupial *Monodelphis domestica* which has 27.[35]

The development of the chondrocranium and pharyngeal cartilages has been described in detail for selected ostariophysans including the cyprinids *Barbus barbus*,[36] *Rasbora daniconius*,[37] and *Cyprinus carpio*,[38] and the characid *Hepsetus odoé*.[39] The developing chondrocranium of zebrafish appears to be similar in most respects to these fishes. In zebrafish, chondrocranial and pharyngeal arch development has been described,[40,41] and a brief description of cartilage development has been made.[42] Six different types of cartilage are present in the cranium of zebrafish, the predominant type being cell-rich hyaline cartilage.[33] As is true for other vertebrates, some pharyngeal arch cartilages have been shown to derive from neural crest,[43] but the embryological origin of the osteoblasts that invade and ossify these cartilages, transforming them into the functional bones of the gill arches in adult zebrafish, remains unknown.

Two developmental types of cellular bone — acellular bone is present in higher teleosts[44] — are present in zebrafish, dermal bone and cartilage bone. Bones that develop directly within a connective tissue membrane and are homologous with similarly formed bones in primitive vertebrates are termed "dermal bones." Twenty-nine of the 74 bones present in the skull of zebrafish are dermal bones. Bones that develop directly within a connective tissue membrane and are homologous with cartilage bones in primitive vertebrates are termed "membrane bones" following Patterson.[45] The intercalar is the only membrane bone in *D. rerio*. Bones that ossify around (perichondral) or within (endochondral) a cartilage model are termed "cartilage bones." Forty-three of the 74 bones are cartilage bones. The anterior end of the notochord ossifies, and it is classified as a perichordal bone.

Often in the development of a cartilage bone, membrane bone extensions from the perichondral ossification form off of the bone, e.g. the palatine, pterosphenoid, and sphenotic.[45]

Dermal bones are generally reported to develop before cartilage bones in fishes[28,36,46] and in other vertebrates. Dermal bones do appear before cartilage bones in *Betta, Barbus*, and *Oryzias latipes*, but dermal and cartilage bones do not differ in timing of ossification in zebrafish.[47] This is because a small group of dermal bones (anguloarticular, preopercle, premaxilla, exoccipital, and quadrate) develop at a very different time in zebrafish than in other species.

Bones of the lateral line system develop in close proximity to neuromasts, the mechanoreceptors on the skin of fishes. During development, neuromasts become enclosed in a bony canal system that transverses the cranium of fishes in stereotypic and highly conserved lines. Bones associated with lateral line canals have been hypothesized to develop earlier than (and evolve before) bones not so associated.[48] However, in *Danio, Betta*, and *Oryzias*, lateral line bones ossify significantly later than other dermal bones. The trend observed in *Barbus* followed this pattern as well. Additionally, in *Danio* and *Betta*, ossification of lateral line-associated dermal bones was later than all other bones.

2.3. Vertebral Column

In fishes, the axial skeleton includes the vertebral column and associated median (unpaired) fins. In amniotes, the vertebral column develops from sclerotome, a mesenchymal cell population derived from the ventral somite, which migrates to surround axial midline structures and differentiate as cartilage and bone. Lineage of single cells in the ventral somite of zebrafish embryos has confirmed that they give rise to sclerotome, although some single cells can give rise to both myotome and sclerotome cells.[49] Further work has characterized the contribution of specific somites to the vertebrae; as has been shown for other vertebrates, cells from individual somites contribute to two adjacent vertebrae.[50] However, more recent work has revealed discrepancies between this accepted model of vertebral formation and the actual mechanisms in zebrafish.[51] Unlike in amniotes, most of the vertebral bone in zebrafish does not form via an intermediate cartilage stage. More surprisingly, the bone matrix forming the centra is secreted directly by the notochord in the early larva, and the bone is devoid of osteoblasts. These observations raise important questions, such as the actual contribution of the sclerotome cells in zebrafish and whether this mechanism of bone formation is shared by other vertebrates.

The vertebral column is regionalized into precaudal and caudal vertebrae. Precaudal vertebrae include the centra, neural arches and spines, parapophyses, and ribs, and in ostariophysans, the precaudal vertebrae are anteriorly regionalized as the Weberian vertebrae. The Weberian vertebrae in otophysans are so named for the Weberian apparatus,[22] a novel adaptation of otophysans[18] that provides a route for the transmission of sound/vibration from the swim bladder to the inner ear.[22] Development and anatomy of the inner ear[52–54] and the specific role of the Weberian apparatus in sound transmission[21] are well understood. Caudal vertebrae consist of centra, dorsal neural arches and neural spines, and ventral hemal arches and hemal spines of vertebra 15 through 28. Caudal fin vertebrae, the most posterior three caudal vertebrae, are modified to support the caudal fin. The last precaudal and/or the first caudal vertebra is frequently categorized as a "transitional" vertebra, exhibiting elongated unfused hemal arches or parapophyses, drastically shortened ribs, and absence of a hemal spine. Associated with many vertebrae, particularly the most anterior and most posterior, are basidorsals and basiventrals, paired cartilaginous structures which give rise to neural arches or hemal arches, respectively.[4] Basiventrals also develop into parapophyses.

Molecular genetic studies of *Hox* genes in fishes and other vertebrates[50,55,56] have revealed the correlation of the rib-bearing vertebrae (the "thoracic" vertebrae in amniotes) with the anterior expression boundary of *Hoxc6*. In the zebrafish this boundary lies between vertebra 2 and vertebra 3.[50] This is of significance in interpretation of the homologies of the Weberian apparatus. Specifically, whether the ribs of v3 and v4 contribute to the tripus and os suspensorium respectively, has been debated.[27] The expression of *Hoxc6a* and *Hoxc6b* in the myomeres from which vertebrae 3 and 4 (as well as v5–13) are derived, lends support to a rib homology for these elements.

An overall pattern of development within the axial skeleton emerges from the analysis of the relative timing of ossification events.[27] Development within the axial skeleton begins simultaneously within the Weberian region (third and fourth centra) and the caudal fin region (hypural 1 and parhypural). This is notable in that the two regions are at the anterior and posterior limits of the axial skeleton. Two centers of ossification within the vertebral column are common among ostariophyans.[57] This may be related to a possible difference in molecular patterning mechanisms between these two regions.[50] Development proceeds posteriorly and anteriorly from the third and fourth (Weberian) vertebrae, such that hemal vertebrae are last to form. Development of associated vertebral structures, such as neural arches and spines, mirrors this pattern. In contrast, the vertebral column of amniotes develops from anterior to posterior.[58,59] The supraneurals develop from anterior to posterior as in other fishes.[60]

Homologies of the Weberian ossicles (scaphium, claustrum, intercalarium, tripus, os suspensorium), as well as Weberian support structures (roofing cartilage and supraneurals), are issues of intense debate in the ichthyological literature.[61,62] The developmental complexity of Weberian elements, which involves variability and modification of the centra, neural arches and spines, parapophyses, ribs, and supraneurals of the first four vertebrae, has made interpretation of their homologies quite difficult. In addition, individual bones may have a compound origin and may develop differently among species. New molecular genetic data, such as that from Morin–Kensicki and colleagues,[50] have provided a new level of understanding for the evolutionary origin of such structures.[27]

2.4. Median Fins

Most fishes have two sets of fins: a group of midline fins along the anterior to posterior body axis termed the unpaired or median fins, as well as two pairs of fins, the homolog of paired limbs in vertebrates, in the abdominal region. The median fins, however, evolved before paired fins in early craniates, approximately 400 million years ago.[63] Thus the mechanisms patterning median fins may be ancestral to those used by the paired fins and limbs, and their mechanisms may in fact have been co-opted by the paired fins. Despite the evolutionary significance, little is known about the mechanisms underlying median fin development. Conserved patterns of development within the median fins, though, have contributed to our understanding of their development and evolution.

The median fins include the dorsal, anal, and caudal fins, their bony internal support elements (radials) and external supports (fin rays). Potential trunk neural crest contribution to the caudal fin has been examined in zebrafish.[64] The caudal fin is first to form in most fishes,[65] and in zebrafish the caudal fin develops first, followed by the anal and finally the dorsal fin radials and fin rays. Of the dorsal and anal fins, the more posterior fin develops before the more anterior fin. Thus, in *D. rerio*, the anal fin, which is slightly more posterior than the dorsal fin, develops first. The exoskeleton and endoskeleton of both the dorsal and anal fins develop bi-directionally, a co-patterning similarity first noted by Mabee and colleagues.[65] Bi-directional development of these fins appears to be the basal condition for teleost fishes.[65]

2.5. Paired Fins

There is a high level of interest and wealth of data pertaining to the genetic regulation underlying paired fin and limb development,[66–68] and many inductive interactions and signaling molecules have been conserved between fishes and limbed vertebrates.[69,70] The evolution and details of morphology of the limb endoskeleton of gnathostomes are well synthesized.[63,71,72] Basal actinopterygians possess a pectoral fin with propterygial and metapterygial elements; the homology of the mesopterygium is less well understood, but middle mesopterygial radials are present.[63] This tribasal condition is primitive at the level of gnathostomes. Teleost fishes have lost the metapterygium.[63,71,72] The propterygium, present in all teleosts, is a retained primitive feature. It is sometimes misidentified as the distal radial one.[73,74] The four proximal radials that are primitive for teleosts[75] are considered homologous to the mesopterygium. Six or seven distal radials are present in the pectoral fin, and the lepidotrichia articulate with these.

The pectoral girdle is composed of three dermal bones and 12–13 cartilage bones. The dermal bones are the paired cleithra, postcleithra, and supracleithra. The cartilage bones are the paired coracoids, mesocoracoids, scapulae. The pelvic fin is composed of three radials; the girdle of four cartilage bones.[29]

2.6. Phylogenetic Comparisons

Most aspects of the morphology of the axial skeleton of *D. rerio* reflect the basal (primitive) conditions for various hierarchical phylogenetic levels: the slender-bodied clade within *Danio, Danio*, Leuciscinae, Cyprinidae, and Otophysi.[76,77] For example, *D. rerio* retains the shared derived features for the slender-bodied clade of *Danio*,[77] such as a broad fourth neural spine, a reduced first lateral process and a reduced supraneural three. It also retains the shared derived features for *Danio*, such as a reduced ascending process of the intercalarium,[77] and *D. rerio* retains the primitive rib condition (thin and straight), among other characters, for cyprinids.[76] *Danio rerio* shares 9 of 13 derived characters associated with the axial skeleton of basal leuciscins.[76] It differs in that the fourth rib is short and stout, the fifth parapophysis is cup-like and the associated rib head is rounded and reduced, the third uroneural is absent, and the double neural arch of preural 2 is fused, but the neural spines remain unfused.[27] Relative to the basal condition for otophysans,[78] the Weberian apparatus of *D. rerio* has been modified in several aspects.[27] Correlating the recency of evolutionary origin of these skeletal features with the type of genetic network that underlies each, may elucidate the nature of evo-devo changes in genetics and morphology.

3. Molecular Aspects of Skeletal Development

Many markers for tissue formation, differentiation, and specialization of the skeleton have been characterized in mammals. The identification of homologous genes in teleosts has confirmed the similarities at the molecular level between mammals and teleosts, and revealed details about the early events of skeletogenesis that are not morphologically obvious.

3.1. Early Inductive Signals

The *Bmp* (bone morphogenetic protein) genes are members of the *Tgf-β* superfamily, a large group of genes encoding secreted signaling molecules that regulate a diverse range of biological processes during normal development and growth.[79–81] Many of the *Bmp* genes were originally isolated on the basis of their remarkable ability to induce the entire process of endochondral bone formation.[82] Specific BMP signals play roles in both bone and joint formation in tetrapods. Zebrafish BMP signals are critical in embryonic dorsal-ventral patterning.[83–85] However, their later expression has not been previously examined in the developing skeleton.

In zebrafish there are two ortholog of *Bmp2, bmp2a* and *bmp2b*. *Bmp2b* is expressed in the region of segmentation and in the hypertrophic chondrocytes in the developing radials of the median fins in 10–20 dph larvae (Crotwell P and Mabee P, unpubl. obs.). The highly localized expression of *bmp2b* in the region of segmentation of zebrafish mirrors its expression in the developing joints of tetrapods, and is indicative of a potentially conserved function. In contrast, *bmp2a* is expressed in developing dermal bone including fin rays and neural and hemal spines, and is not expected to contribute to radial segmentation (Crotwell P and Mabee P, unpubl. obs.). It is likely that other roles for BMPs in skeletal induction are conserved in zebrafish as well, but detailed expression analysis of the individual zebrafish genes will be required to define their roles.

3.2. Cartilage Formation

In mammals, a controlling step in cartilage differentiation is the expression of the transcription factor *Sox9*. In tetrapods, *Bmp2* expression is activated by Hedgehog signaling, and BMP2 itself regulates the expression of *Sox9*.[86] *Sox9* induces chondrogenesis via *Noggin* provided BMP signals are present.[87] Two *sox9* orthologues have been identified in zebrafish.[88,89] Both are expressed in cartilage, with some overlap of expression in the craniofacial skeleton.

Mutational analysis in the mouse has confirmed that *Sox9* is a direct transcriptional regulator of *Col2a1*, the gene encoding type II collagen, the major collagen component of cartilage. The *col2aI* gene has also been identified in zebrafish.[90] It is expressed in midline cells of all germ layers in the early embryo, and has been extensively used as a marker for axial development. However, it also shows later strong expression in skeletal elements and has proven a useful marker for cartilage differentiation in the zebrafish. For instance, the developing dorsal, anal, and caudal fins, cartilage cells, perichondrial cells, and inter-radial mesenchyme all strongly express type II collagen (Crotwell P and Maybee P, unpubl. obs.). Zebrafish may possess two *col2a1* splice variants,[91] but splice variant-specific probes have not been developed for zebrafish (Postlethwait J pers. comm.).

3.3. Bone Formation

The transcription factor *Runx2* (formerly called *Cbfa1*) is required for bone formation in mouse;[92–94] in humans, heterozygous mutations in *RUNX2* cause cleidocranial dysplasia.[95] However, *Runx2* is also expressed in hypertrophic cartilage and required for normal development of a subset of cartilage elements,[96–98] suggesting that it is a more general marker for pre-skeletogenic mesenchymal cells. A medaka gene homologous to *Runx2* has been described.[99] It is expressed in immature osteoblasts as well as a subset of chondrocytes. Two orthologues of *Runx2* have been identified in zebrafish.[100] Expression of *runx2a* is observed early in scattered head mesoderm cells during gastrulation,

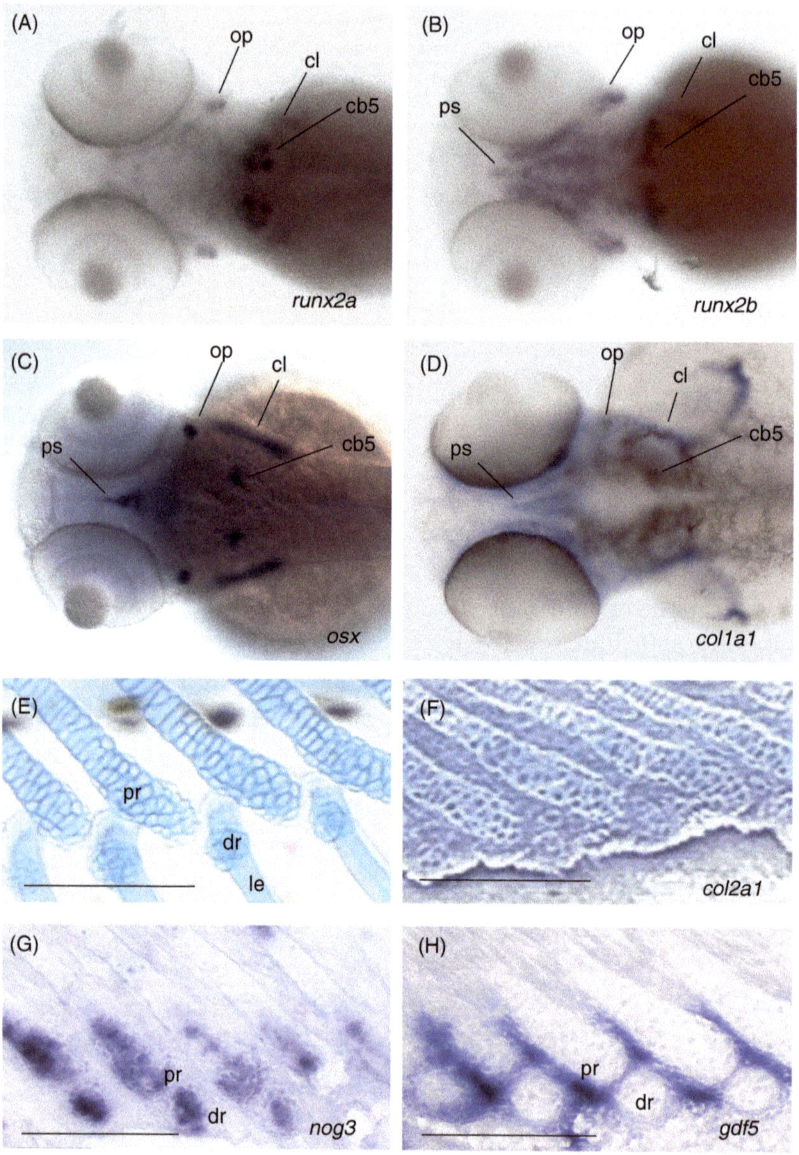

Fig. 2 Conservation of gene pathways in skeletal development. (A–D) The successive expression of markers defines a pathway of differentiation in the craniofacial skeleton, shown here in 3 dpf larvae. (A, B) At this stage, the two *runx2* orthologues are expressed in numerous elements, including cartilage and bone. Expression is seen in early forming

and later in the pharyngeal arches, in both the mesodermal cores and the neural crest component. In the larva, its expression marks some early sites of osteogenesis, such as the opercle, but like the medaka gene it is also expressed in cartilage elements (Fig. 2A). *runx2b* expression is more widespread in the larva, and is more similar to that reported for the medaka gene (Fig. 2B). All three fish genes are expressed widely in developing cartilage elements. In contrast, expression of the mouse gene has been described only in hypertrophic cartilage. This apparent discrepancy may be resolved with more detailed expression analysis of the fish and mammalian genes, or it may reflect an important regulatory difference between teleost and mammalian genes.

Another transcription factor, *Osterix* (*Osx*), has been identified in mouse that is also required for bone formation.[101] *Osx* appears to be downstream of *Runx2*, since its expression is absent in *Runx2* mutants. We have identified a zebrafish homolog of *osx* (Chung J and Fisher S, unpubl. obs.); its expression in osteoblasts appears to be well conserved, and both later and more specific to that cell type than the *runx2* genes (Fig. 2C). In general, its expression precedes histological evidence of bone formation by one or more days, making it an important marker for examining the early process of bone formation.

---

membranous bones, such as the opercle (*op*) and cleithrum (*cl*), and in cartilage bones such as the fifth ceratobranchial (*cb5*) and the ossifications around the trabeculae cranii which give rise to the parasphenoid (*ps*). (C) Expression of *osx*, which is thought to be downstream of *runx2*, is apparently confined to osteoblasts. Because it is expressed later than the *runx2* genes in specific bones, it is expressed in only a subset at this stage. (D) Strong *col1a1* expression is a marker of differentiated osteoblasts, and its expression at this stage is still weak, particularly in the *op* and *cb5*. (E–H) Gene expression during median fin radial segmentation is consistent with that observed in tetrapod joint development. (E) Segmentation of the anal fin radials is apparent morphologically in an Alcian blue stained specimen; the proximal radial (*pr*), distal radial (*dr*), and lepidotrichia (*le*) can be easily distinguished. (F) Expression of *col2a1* is seen throughout the forming radial cartilages, and in the mesenchymal tissue surrounding them. (G) *nog3* expression is observed in cartilage outside the region of segmentation. In contrast, *gdf5* expression shows a reciprocal pattern of expression in the mesenchymal tissue surrounding the proximal and distal radial cartilages (H). The scale bars in E–H are 1.0 mm.

The major protein component of adult bone is type I collagen, which in mammals is a heterotrimer whose chains are encoded by the *Col1a1* and *Col1a2* genes. Orthologues of both genes have been identified in zebrafish,[31,102] and other ESTs may represent additional orthologues. The expression of *col1a1* is a good marker for early bone formation, although it is not specific to bone and is expressed also in fin folds and skin (Fig. 2D). Together with the *runx2* genes and *osx*, expression of these markers defines a molecular sequence of osteoblast differentiation in zebrafish that parallels that seen in mammals, and precedes morphological evidence of osteogenesis by hours to days in the larva.

3.4. Joint Formation

Some of the molecular events in joint formation have been defined in mammals, and these events as well seem to be conserved in zebrafish (Fig. 2E–H). Mutations in the BMP inhibitor *Noggin* are associated with failures in joint formation in both mouse and humans.[103,104] There are three zebrafish *noggin* genes, two of which are only expressed in the early embryo.[105] However, *nog3* shows a pattern of expression in craniofacial and median fin cartilage suggesting that it may play a role in zebrafish joint formation.[105,106] Mouse *Gdf5*, a member of the BMP family, is also expressed in developing joints.[107] Similarly, zebrafish *gdf5* is expressed during segmentation of the fin radial bones, suggesting a conserved function for this gene as well.[108]

4. Mutational Analysis of Zebrafish Skeletal Development

Mutational analysis of craniofacial cartilage has been particularly productive in zebrafish, largely because development of those cartilages during the early larval period is morphologically apparent. Two large-scale screens were carried out to detect craniofacial abnormalities during larval development.[190–111] Although the classification schemes varied between the screens, the mutants could generally be put in three categories:

mutations that affected general cartilage formation and differentiation; mutations that affected cartilages in the posterior (branchial) arches primarily; and mutants that showed defects primarily in the anterior (mandibular and hyoid) arch cartilages. In addition, mutants with abnormalities in midline development (many subsequently shown to act in the hedgehog signaling pathway) showed craniofacial defects affecting ventral midline and anterior cartilages. Subsequent identification of some of these mutant loci has revealed the conservation of signaling pathways in vertebrate craniofacial development; specific examples are discussed below.

Craniofacial cartilage formation is an early aspect of skeletal development, amenable to standard screening strategies. However, many important events in skeletogenesis take place after the larval period. Mutants affecting these later processes have sometimes been saved because of other, earlier phenotypes. One example is the collection of mutants in somitogenesis.[112,113] In some cases these mutants are viable, and as adults, they show striking defects in organization of the vertebral spinous processes and arches. Curiously, the vertebral bodies themselves are properly segmented, despite the absence of morphological and molecular somite boundaries. This implies a separate mechanism for segmentation of the vertebral column, and there is evidence that the signal originates from the notochord.[51,114] These observations have raised intriguing questions about the role of sclerotome and somites in the segmentation of the vertebral column, and how well the mechanisms of segmentation are conserved in evolution.

A similar vertebral phenotype has been described in adult fish homozygous for mutations in the BMP inhibitor *chordin*.[115] However, in *chordin* (*din*) mutant embryos, somite segmentation appears normal, so BMP signaling may be important in aligning the spinous processes with the boundaries between the myotomes. Despite the relatively late appearance of the skeletal phenotype, vertebral patterning in *din* mutants can be fully rescued by the replacement of *chordin* RNA during gastrulation. This suggests that the skeletogenic cells are induced and patterned during very early embryonic development.

Mutants with early defects in pectoral fin formation generally do not shed light on processes of skeletogenesis; they are either entirely missing the fins,[116,117] or display transient defects in the embryonic fin folds but have normal adult fins.[116] Perhaps of greater interest are mutants in later fin growth and regeneration.[118–120] Growth and segmentation of the lepidotrichia are prominent features of fin growth; mutations affecting these processes may reveal important aspects of general bone growth.

Relatively late aspects of skeletal development, such as formation of the post-cranial skeleton and general bone maturation, are not accessible through embryonic mutational screens. In an effort to identify mutants specifically affecting these later processes, we are carrying out a screen based on radiography of adult fish.[31] Although radiographs do not afford the same detail as fixed, histological specimens, the fact that they can be performed on live animals makes them more practical for large-scale screens.

5. Molecular Identification of Zebrafish Skeletal Mutants

The molecular identification of zebrafish mutants affecting skeletogenesis has revealed both genes with similar functions in mammals, and novel gene products. Several specific examples are discussed below; Table 1 lists some additional zebrafish orthologues of mammalian genes important in skeletal development, including their mutant or morphant phenotypes where known. One mutant classified as having a general defect in cartilage formation is *jellyfish* (*jef*). Mutants are characterized by an almost complete lack of cartilages, with small remnants of the ceratohyal and in some embryos Meckel's and palatoquadrate cartilages. In two *jef* alleles a lesion was identified in *sox9a*, confirming a conserved role for the zebrafish gene in cartilage development.[121] In humans, mutations in the transcription factor *SOX9* are responsible for the severe skeletal deformity campomelic dysplasia. Human patients are heterozygous rather than homozygous for *SOX9* mutations, and *Sox9*$^{+/-}$ mutations in mice causes perinatal lethality,[122] making it impossible to examine the homozygous phenotype in the whole organism.[123] Direct comparison of the zebrafish

Table 1 Zebrafish orthologues of mammalian skeletal genes.

Mammalian gene	1) Associated human disease 2) Mouse null phenotype	Zebrafish gene(s)	Mutant or morphant phenotype
Ap2	2) Abnormalities in cranial neural crest development	*lockjaw/ap2*[111,132]	Severe reduction of hyoid arch derivatives, partial loss of branchial arches
Col1a1	1) Osteogenesis imperfecta	*chihuahua/col1a1*[31]	Abnormal bone growth and fragile bones
Col2a1	1) Stickler syndrome, hypo- and achondrogenesis 2) Lack of endochondral ossifications	*col2a1*[133]	
Endothelin	2) Defects in first and second pharyngeal arch cartilages	*sucker/et-1*[129]	Defects in ventral pharyngeal arch cartilages
Nog	1) Proximal symphalangism 2) Early patterning defects; failure of limb joints to form	*nog3*[134,135]	
Osx	2) Lack of bone	*osx* (Chung J and Fisher S, unpubl. obs.)	
Runx2	1) Cleidocranial dysplasia 2) Lack of bone, abnormalities in hypertrophic cartilage	*runx2a*[101] *runx2b*[101]	
Sox9	1) Campomelic dysplasia 2) Failure of cartilage formation	*jellyfish/sox9a*[90,136] *sox9b*[90]	Loss of most cartilage elements
Tbx1	1) DiGeorge syndrome (defects in cartilage and other derivatives of cranial neural crest)	*van gogh*[110,137]	Pharyngeal arch defects

and mammalian phenotypes is also complicated by the existence of two zebrafish genes; however, redundant function of the two fish genes may be the reason heterozygous mutants are viable, making it possible to study the phenotype in *jef* homozygous mutants.

Several mutants display a common phenotype of cartilage formation, pointing to an important interaction between Wnt signals and the extracellular matrix in that process. One of these is *jekyll* (*jek*),[124] which was isolated based on a defect in heart valve formation; however, the craniofacial cartilage elements also are smaller than wild-type and fail to stain with Alcian blue. Through positional cloning the *jek* gene was identified as a UDP-glucose dehydrogenase, homologous to the *Drosophila* gene *sugarless*.[125] Another mutant with a similar cartilage phenotype, the gastrulation mutant *knypek*, has been shown to encode glypican.[126] Finally, the *pipetail* (*ppt*) mutant, which also shows a similar cartilage defect, has been shown to encode *wnt5*.[127] It has been suggested that the *jek* and *kny* gene products are required for the transport or internalization of the *ppt* signal, which is required for correct cartilage morphogenesis.[125] Thus these three mutants appear to define a pathway for the processing of Wnt signals that had not previously been identified in mammals, and which is crucial for cartilage formation.

Among the class of mutants primarily affecting anterior arch cartilages was the mutant *sucker* (*suc*).[109] In a striking parallel with mammalian development, *suc* has been identified as the zebrafish *endothelin-1* (*et-1*) gene.[128] Endothelin signaling had been previously shown to operate in mouse craniofacial development through analysis of knock-outs in the *Et-1* gene and in its receptor, encoded by the *Ednr-A* gene. Further analysis in zebrafish, through the use of morpholino antisense gene targeting, has shown that two downstream targets of *et-1*, the *bapx1* and *hand2* genes, mediate its effects on development of the jaw joint and ventral pharyngeal cartilages respectively.[129]

In a screen for dominant, adult skeletal phenotypes, the *chi* mutant was identified.[31] Although their cartilaginous larval skeletons appear normal, *chi* heterozygous mutants display a generalized defect in bone growth and as adults show evidence of frequent fractures. The phenotype is similar to human osteogenesis imperfecta (OI), which is usually caused by a missense mutation in one of the two genes encoding type I collagen. A similar mutation was identified in the *col1a1* coding sequence in *chi* mutants, confirming a conserved role for type I collagen in bone maturation in zebrafish.

The identification of the mutants described above has confirmed the similarity of the molecular pathways in teleost skeletogenesis to that in other vertebrates. In addition, novel genes are also being identified, revealing pathways not previously described in other organisms. In zebrafish, detailed molecular and embryological studies can be performed in these mutants that are often not possible in mouse embryos. This combination of genetic and embryological tools holds out great promise for the use of zebrafish in the study of skeletal development.

6. Conclusions and Future Directions

We have summarized a growing body of work on the formation of the skeleton in the zebrafish *Danio rerio*, including molecular and genetic studies, as well as detailed descriptions of morphogenesis. It is reassuring that results in zebrafish have been largely consistent with what is known about other vertebrates. However, if studies in zebrafish served only to confirm results from other systems, there would be little excitement in the field. The promise of zebrafish lies in the types of experiments that are not easily performed in other organisms, such as targeted forward genetic screens and single cell lineage experiments.

Understanding morphogenesis was identified by developmental biologists as the most important unanswered question in developmental biology, and also as the area in which rapid progress is expected in the next five years.[130] Understanding the evolution of developmental mechanisms was next on this list of most important questions. Many important evolutionary questions regarding the morphogenesis and evolution of the vertebrate skeleton remain unanswered, but someday will be addressed using the zebrafish, as more recent molecular and genetic studies are rapidly added to the large body of descriptive literature. Among these are the absence of teeth on the maxilla and premaxilla, absences that characterize not only zebrafish, but all Ostariophysi (approximately 6000 species). Another example discussed above is the specialization of anterior vertebrae into the Weberian apparatus in the Otophysi. By comparing the genetic basis and embryological formation of teeth in non-ostariophysan fishes with zebrafish, or of the Weberian specialization in zebrafish with its absence in non-Otophysi, the developmental basis for major evolutionary changes might be addressed.

The power of mutant screens to dissect zebrafish skeletal development has not yet been fully exploited. In the early embryo, bone formation is not as morphologically apparent as cartilage formation, and few mutants have been identified that specifically affect osteogenesis. Enhancement of future embryonic screens, by the use of a transgenic reporter line or a specific stain or marker for bone formation, may reveal additional mutants in this category. In addition, continued screens for adult phenotypes promise to yield mutants affecting later patterning and general bone morphogenesis, many of which may serve as models of human diseases affecting the skeleton.

Zebrafish also offers an ideal system in which to carry out single cell lineage experiments. Lineage studies in the skeleton have been confined to craniofacial cartilages and sclerotome, and have been consistent with findings in other vertebrates. However, the origins of other cartilage elements, and of osteoblasts, have not been examined. In addition, questions of cell commitment, e.g. whether single cells give rise to both cartilage and bone in the embryo as is thought to occur in adult tissues, and when cells are determined to give rise to skeletal lineages, are difficult to address in other model organisms. The zebrafish, with its accessible and optically clear embryos, lends itself to these detailed studies.

References

1. Janvier P (1996). *Early Vertebrates.* Claredon Press, Oxford.
2. Nelson JS (1994). *Fishes of the World.* Wiley, New York.
3. Carroll RL (1997). Development and Evolution. In: *Patterns and Processes of Vertebrate Evolution.* Cambridge University Press, New York.
4. Goodrich ES (1930). *Studies on the Structure and Development of*

Vertebrates. Macmillan and Company, London.
5. de Beer GR (1937). *The Development of the Vertebrate Skull.* Oxford University Press, Oxford.
6. Daniel JF (1922). *The Elasmobranch Fishes.* University of California Press, Berkeley, California.
7. Gregory WK (1959). *Fish Skulls: A Study of the Evolution of Natural Mechanisms.* Noble Offset Printers, Inc., New York.
8. Stiassny MLJ, Parenti LR and Johnson GD (1996). *Interrelationships of Fishes.* Academic Press, San Diego.
9. Nelson GJ (1969). Gill arches and the phylogeny of fishes, with notes on the classification of vertebrates. *Bull. Am. Mus. Nat. Hist.* **141**: 475–552.
10. Nelson GJ (1969). Infraorbital bones and their bearing on the phylogeny and geography of Osteoglossomorph fishes. *Am. Mus. Nov.* **2394**: 1–37.
11. Rosen DE (1973). Interrelationships of higher euteleostean fishes. In: *Interrelationships of Fishes.* Academic Press, New York.
12. Nüsslein–Volhard C and Dahm R (2002). *Zebrafish.* Oxford University Press, Oxford.
13. Hamilton F (1822). *An Account of the Fishes Found in the River Ganges and its Branches.* Archibald Constable and Co., Edinburgh and London.
14. Barman RP (1991). A taxonomic revision of the Indo-Burmese species of *Danio* Hamilton–Buchanan (Pisces: Cyprinidae). *Rec. Zool. Sur. India, Occ. Paper* **1371**: 1–91.
15. Cavender TM and Coburn MM (1992). Phylogenetic relationships of North American Cyprinidae. In: *Systematics, Historical Ecology, and North American Freshwater Fishes.* Stanford University Press, Stanford.
16. Meyer A, Biermann C and Orti G (1993). The phylogenetic position of zebrafish (*Danio rerio*), a model system in developmental biology: an invitation to the comparative method. *Proc. R. Soc. Lond. B* **252**: 231–236.
17. Meyer A, Ritchie PA and Witte KE (1995). Predicting developmental processes from evolutionary patterns: a molecular phylogeny of the zebrafish (*Danio rerio*) and its relatives. *Phil. Trans. R. Soc. Lond. B* **349**: 103–111.
18. Fink SV and Fink WL (1996). Interrelationships of ostariophysan fishes (Teleostei). In: *Interrelationships of Fishes.* Academic Press, San Diego.

19. Berra TM (2001). *Freshwater Fish Distribution.* Academic Press, San Diego.
20. Helfman GS, Collette BB and Facey DE (1997). *The Diversity of Fishes.* Blackwell Science, Malden, Massachusetts.
21. Popper AN and Fay RR (1999). The auditory periphery in fishes. In: *Comparative Hearing: Fish and Amphibians.* Springer–Verlag, New York.
22. Weber EH (1820). *De Aure et Auditu Hominis et Animalium, Pars I. De Aure Animalium Aquatilium.* Leipzig.
23. Rosen DE and Greenwood PH (1970). Origin of the Weberian apparatus and the relationships of the Ostariophysan and Gonorynchiform fishes. *Am. Mus. Nov.* **2428**:
24. Grande T and Braun CB (2002). Evolution of the Weberian apparatus. *Bioacoustics* **12**: 120–122.
25. Fink SV and Fink WL (1981). Interrelationships of the ostariophysan fishes (Teleostei). *Zool. J. Linnean Soc.* **72**: 297–353.
26. Mabee PM, Olmstead KL and Cubbage CC (2000). An experimental study of intraspecific variation, developmental timing and heterochrony in fishes. *Evolution* **54**: 2091–2106.
27. Bird NC and Mabee PM (2003). Developmental morphology of the axial skeleton of the zebrafish, *Danio rerio* (Ostariophysi: Cyprinidae). *Dev. Dyn.* **228**: 337–357.
28. Mabee PM and Trendler TA (1996). Development of the cranium and paired fins in *Betta splendens* (Teleostei: Percomorpha): Intraspecific variation and interspecific comparisons. *J. Morphol.* **227**: 249–287.
29. Cubbage CC and Mabee PM (1996). Development of the cranium and paired fins in the zebrafish *Danio rerio* (Ostariophysi, Cyprinidae). *J. Morphol.* **229**: 121–160.
30. Du SJ, Frenkel V, Kindschi G and Zohar Y (2001). Visualizing normal and defective bone development in zebrafish embryos using the fluorescent chromophore calcein. *Dev. Biol.* **238**: 239–246.
31. Fisher S, Jagadeeswaran P and Halpern ME (2003). Radiographic analysis of zebrafish skeletal defects. *Dev. Biol.* **264**: 64–76.
32. Hyman LH (1979). *Hyman's Comparative Vertebrate Anatomy.* University of Chicago Press, Chicago.
33. Benjamin M (1990). The cranial cartilages of teleosts and their classification. *J. Anat.* **169**: 153–172.
34. Wiens J (1989). Ontogeny of the skeleton of *Spea bombifrons* (Anura:

Pelobatidae). *J. Morphol.* **202**: 29–51.

35. Clark CT and Smith KK (1993). Cranial osteogenesis in *Monodelphis domestica* (Didelphidae) and *Macropus eugenii* (Macropodidae). *J. Morphol.* **215**: 119–149.
36. Vandewalle P, Focant B, Huriaux F and Chardon M (1992). Early development of the cepahlic skeleton of *Barbus barbus* (Teleostei, Cyprinidae). *J. Fish Biol.* **41**: 43–62.
37. Tewari SK (1971). The development of the chondrocranium of *Rasbora daniconius* (Ham. Buch). *Gegenbaurs morph. Jahrb.* **116**: 491–502.
38. Pashine RG and Marathe VB (1977). The development of the chondrocranium of *Cyprinus carpio* Linn. *Proc. Ind. Acad. Sci. B* **85**: 351–363.
39. Bertmar G (1959). On the ontogeny of the chondral skull in Characidae, with a discussion on the chondrocranial base and the visceral chondrocranium in fishes. *Acta Zool.* **40**: 203–364.
40. Kimmel CB, Miller CT, Kruze G, Ullmann B, BreMiller RA, Larison KD and Snyder HC (1998). The shaping of pharyngeal cartilages during early development of the zebrafish. *Dev. Biol.* **203**: 245–263.
41. Kimmel CB, Miller CT and Moens CB (2001). Specification and morphogenesis of the zebrafish larval head skeleton. *Dev. Biol.* **233**: 239–257.
42. Miyake T and Hall BK (1994). Development of *in vitro* organ culture techniques for differentiation and growth of cartilages and bones from teleost fish and comparisons with *in vivo* skeletal development. *J. Exp. Zool.* **268**: 22–43.
43. Schilling TF and Kimmel CB (1994). Segment and cell type lineage restrictions during pharyngeal arch development in the zebrafish embryo. *Dev.* **120**: 483–494.
44. Parenti LR (1986). The phylogenetic significance of bone types in euteleost fishes. *Zool. J. Linn. Soc.* **87**: 37–51.
45. Patterson C (1977). Cartilage bones, dermal bones and membrane bones, or the exoskeleton versus the endoskeleton. In: *Problems in Vertebrate Evolution.* Academic Press, London.
46. Arratia G (1990). Development and diversity of the suspensorium of Trichomycterids and comparison with Loricarioids (Teleostei: Siluriformes). *J. Morphol.* **205**: 193–218.
47. Mabee PM (2000). The usefulness of ontogeny in interpreting morphological characters. In: *Phylogenetic Analysis of Morphological Data.*

Smithsonian Institution Press, Washington.

48. Parrington FR (1949). A theory of the relations of lateral lines to dermal bones. *Proc. Zool. Soc. London* **119**: 65–78.
49. Morin–Kensicki EM and Eisen JS (1997). Sclerotome development and peripheral nervous system segmentation in embryonic zebrafish. *Dev.* **124**: 159–167.
50. Morin–Kensicki EM, Melancon E and Eisen JS (2002). Segmental relationship between somites and vertebral column in zebrafish. *Dev.* **129**: 3851–3860.
51. Fleming A, Keynes R and Tannahill D (2004). A central role for the notochord in vertebral patterning. *Dev.* **131**: 873–880.
52. Haddon C and Lewis J (1996). Early ear development in the embryo of the zebrafish, *Danio rerio*. *J. Comp. Neurol.* **365**: 113–128.
53. Bang PI, Sewell WF and Malicki JJ (2001). Morphology and cell type heterogeneities of the inner ear epithelia in adult and juvenile zebrafish (*Danio rerio*). *J. Comp. Neurol.* **438**: 173–190.
54. Bever MM and Fekete DM (2002). Atlas of the developing inner ear in zebrafish. *Dev. Dyn.* **223**: 536–543.
55. Burke AC, Nelson CE, Morgan BA and Tabin C (1995). *Hox* genes and the evolution of vertebrate axial morphology. *Dev.* **121**: 333–346.
56. Prince VE, Price AL and Ho RK (1998). *Hox* gene expression reveals regionalization along the anteroposterior axis of the zebrafish notochord. *Dev. Genes Evol.* **208**: 517–522.
57. Emelianov SV (1939). Sequence of the ontogenetic appearance of vertebral arches in teleosts and the omission of chondral stages in their development. *Academie de Leningrad. Comptes Rendus de L'Academie des Sciences. Nouvelle Serie.* **23**: 978–981.
58. Arey LB (1940). *Developmental Anatomy: A Textbook and Laboratory Manual of Embryology.* W. B. Saunders Company, Philadelphia.
59. Rieppel O (1993). Studies on skeleton formation in reptiles. v. Patterns of ossification in the skeleton of *Alligator mississippiensis* DAUDIN (Reptilia, Crocodylia). *Zool. J. Linn. Soc.* **109**: 301–325.
60. Mabee PM (1988). Supraneural and predorsal bones in fishes: development and homologies. *Copeia* **1988**: 827–838.
61. Fink SV, Greenwood PH and Fink WL (1984). A critique of recent work

on fossil ostariophysan fishes. *Copeia* **1984**: 1033–1041.

62. Gayet M (1986). About ostariophysan fishes: a reply to S. V. Fink, P. H. Greenwood and W. L. Fink's criticisms. *Bull. Mus. Natn. Hist. nat., Paris* **4**: 393–409.
63. Coates MI (1994). The origin of vertebrate limbs. *Dev. Suppl.* 169–180.
64. Smith M, Hickman A, Amanze D, Lumsden A and Thorogood P (1994). Trunk neural crest origin of caudal fin mesenchyme in the zebrafish *Brachydanio rerio. Proc. R. Soc. Lond. B* **256**: 137–145.
65. Mabee PM, Crotwell PL, Burke AC and Bird NC (2002). Evolution of median fin modules in the axial skeleton of fishes. *J. Exp. Biol.: Mol. Dev. Evol.* **294**: 77–90.
66. Nelson CE, Morgan BA, Burke AC, Laufer E, DiMambro E, Murtaugh LC, Gonzales E, Tessarollo L, Parada LF and Tabin C (1996). Analysis of Hox gene expression in the chick limb bud. *Dev.* **122**: 1449–1466.
67. Shubin N, Tabin C and Carroll S (1997). Fossils, genes and the evolution of animal limbs. *Nature* **388**: 638–648.
68. Arthur W, Jowett T and Panchen A (1999). Segments, limbs, homology, and co-option. *Evol. Dev.* **1**: 74–76.
69. Ekker M, Akimenko MA, Bremiller R and Westerfield M (1992). Regional expression of three homeobox transcripts in the inner ear of zebrafish embryos. *Neuron* **9**: 27–35.
70. Logan C, Hornbruch A, Campbell I and Lumsden A (1997). The role of Engrailed in establishing the dorsoventral axis of the chick limb. *Dev.* **124**: 2317–2324.
71. Coates MI and Cohn MJ (1998). Fins, limbs, and tails: outgrowths and axial patterning in vertebrate evolution. *Bioessays* **20**: 371–381.
72. Mabee PM (2000). Developmental data and phylogenetic systematics: evolution of the vertebrate limb. *Am. Zool.* **40**: 789–800.
73. Sordino P, van der Hoeven F and Duboule D (1995). Hox gene expression in teleost fins and the origin of vertebrate digits. *Nature* **375**: 678–681.
74. Grandel H and Schulte–Merker S (1998). The development of the paired fins in the zebrafish (*Danio rerio*). *Mech. Dev.* **79**: 99–120.
75. Johnson GD and Patterson C (1996). Relationships of lower euteleostean

fishes. In: *Interrelationships of Fishes.* Academic Press, San Diego.

76. Coburn MM and Cavender TM (1992). Interrelationships of North American cyprinid fishes. In: *Systematics, Historical Ecology, and North American Freshwater Fishes.* Stanford University Press, Stanford, California.
77. Sanger TJ and McCune AR (2002). Comparative osteology of the *Danio* (Cyprinidae: Ostariophysi) axial skeleton with comments on *Danio* relationships based on molecules and morphology. *Zool. J. Linn. Soc.* **135**: 529–546.
78. Chardon M and Vandewalle P (1997). Evolutionary trends and possible origin of the Weberian apparatus. *Neth. J. Zool.* **47**: 383–403.
79. Balemans W and Van Hul W (2002). Extracellular regulation of BMP signaling in vertebrates: a cocktail of modulators. *Dev. Biol.* **250**: 231–250.
80. Hogan BL (1996). Bone morphogenetic proteins in development. *Curr. Opin. Genet. Dev.* **6**: 432–438.
81. Kingsley DM (1994). The TGF-beta superfamily: new members, new receptors, and new genetic tests of function in different organisms. *Genes Dev.* **8**: 133–146.
82. Clark RM, Marker PC, Roessler E, Dutra A, Schimenti JC, Muenke M and Kingsley DM (2001). Reciprocal mouse and human limb phenotypes caused by gain- and loss-of-function mutations affecting Lmbr1. *Genetics* **159**: 715–726.
83. Dick A, Hild M, Bauer H, Imai Y, Maifeld H, Schier AF, Talbot WS, Bouwmeester T and Hammerschmidt M (2000). Essential role of Bmp7 (snailhouse) and its prodomain in dorsoventral patterning of the zebrafish embryo. *Dev.* **127**: 343–354.
84. Schmid B, Furthauer M, Connors SA, Trout J, Thisse B, Thisse C and Mullins MC (2000). Equivalent genetic roles for bmp7/snailhouse and bmp2b/swirl in dorsoventral pattern formation. *Dev.* **127**: 957–967.
85. Willot V, Mathieu J, Lu Y, Schmid B, Sidi S, Yan YL, Postlethwait JH, Mullins M, Rosa F and Peyrieras N (2002). Cooperative action of ADMP- and BMP-mediated pathways in regulating cell fates in the zebrafish gastrula. *Dev. Biol.* **241**: 59–78.
86. Zehentner BK, Dony C and Burtscher H (1999). The transcription factor Sox9 is involved in BMP-2 signaling. *J. Bone Miner. Res.* **14**: 1734–

1741.
87. Zeng L, Kempf H, Murtaugh LC, Sato ME and Lassar AB (2002). Shh establishes an Nkx3.2/Sox9 autoregulatory loop that is maintained by BMP signals to induce somitic chondrogenesis. *Genes Dev.* **16**: 1990–2005.
88. Li M, Zhao C, Wang Y, Zhao Z and Meng A (2002). Zebrafish sox9b is an early neural crest marker. *Dev. Genes Evol.* **212**: 203–206.
89. Chiang EF, Pai CI, Wyatt M, Yan YL, Postlethwait J and Chung B (2001). Two sox9 genes on duplicated zebrafish chromosomes: expression of similar transcription activators in distinct sites. *Dev. Biol.* **231**: 149–163.
90. Yan YL, Hatta K, Riggleman B and Postlethwait JH (1995). Expression of a type II collagen gene in the zebrafish embryonic axis. *Dev. Dyn.* **203**: 363–376.
91. Lele Z and Krone PH (1997). Expression of genes encoding the collagen-binding heat shock protein (Hsp47) and type II collagen in developing zebrafish embryos. *Mech. Dev.* **61**: 89–98.
92. Otto F, Thornell AP, Crompton T, Denzel A, Gilmour KC, Rosewell IR, Stamp GW, Beddington RS, Mundlos S, Olsen BR, Selby PB and Owen MJ (1997). Cbfa1, a candidate gene for cleidocranial dysplasia syndrome, is essential for osteoblast differentiation and bone development. *Cell* **89**: 765–771.
93. Komori T, Yagi H, Nomura S, Yamaguchi A, Sasaki K, Deguchi K, Shimizu Y, Bronson RT, Gao YH, Inada M, Sato M, Okamoto R, Kitamura Y, Yoshiki S and Kishimoto T (1997). Targeted disruption of Cbfa1 results in a complete lack of bone formation owing to maturational arrest of osteoblasts. *Cell* **89**: 755–764.
94. Ducy P, Zhang R, Geoffroy V, Ridall AL and Karsenty G (1997). Osf2/Cbfa1: a transcriptional activator of osteoblast differentiation. *Cell* **89**: 747–754.
95. Mundlos S, Otto F, Mundlos C, Mulliken JB, Aylsworth AS, Albright S, Lindhout D, Cole WG, Henn W, Knoll JH, Owen MJ, Mertelsmann R, Zabel BU and Olsen BR (1997). Mutations involving the transcription factor CBFA1 cause cleidocranial dysplasia. *Cell* **89**: 773–779.
96. Kim IS, Otto F, Zabel B and Mundlos S (1999). Regulation of chondrocyte differentiation by Cbfa1. *Mech. Dev.* **80**: 159–170.
97. Inada M, Yasui T, Nomura S, Miyake S, Deguchi K, Himeno M, Sato

M, Yamagiwa H, Kimura T, Yasui N, Ochi T, Endo N, Kitamura Y, Kishimoto T and Komori T (1999). Maturational disturbance of chondrocytes in Cbfa1-deficient mice. *Dev. Dyn.* **214**: 279–290.

98. Enomoto H, Enomoto–Iwamoto M, Iwamoto M, Nomura S, Himeno M, Kitamura Y, Kishimoto T and Komori T (2000). Cbfa1 is a positive regulatory factor in chondrocyte maturation. *J. Biol. Chem.* **275**: 8695–8702.
99. Inohaya K and Kudo A (2000). Temporal and spatial patterns of cbfa1 expression during embryonic development in the teleost, *Oryzias latipes. Dev. Genes Evol.* **210**: 570–574.
100. Flores MV, Tsang VWK, Hu W, Kalev–Zylinska ML, Postlethwait J, Crosier PS, Crosier KE and Fisher S (2004). Duplicate zebrafish *runx2* orthologues are expressed in developing skeletal elements. *Gene Expr. Patt.* in press.
101. Nakashima K, Zhou X, Kunkel G, Zhang Z, Deng JM, Behringer RR and de Crombrugghe B (2002). The novel zinc finger-containing transcription factor osterix is required for osteoblast differentiation and bone formation. *Cell* **108**: 17–29.
102. Dubois GM, Haftek Z, Crozet C, Garrone R and Le Guellec D (2002). Structure and spatio temporal expression of the full length DNA complementary to RNA coding for alpha2 type I collagen of zebrafish. *Gene* **294**: 55–65.
103. Brunet LJ, McMahon JA, McMahon AP and Harland RM (1998). Noggin, cartilage morphogenesis, and joint formation in the mammalian skeleton. *Science* **280**: 1455–1457.
104. Gong Y, Krakow D, Marcelino J, Wilkin D, Chitayat D, Babul–Hirji R, Hudgins L, Cremers CW, Cremers FP, Brunner HG, Reinker K, Rimoin DL, Cohn DH, Goodman FR, Reardon W, Patton M, Francomano CA and Warman ML (1999). Heterozygous mutations in the gene encoding noggin affect human joint morphogenesis. *Nat. Genet.* **21**: 302–304.
105. Fürthauer M, Thisse B and Thisse C (1999). Three different *noggin* genes antagonize the activity of bone morphogenetic proteins in the zebrafish embryo. *Dev. Biol.* **214**: 181–196.
106. Bauer H, Meier A, Hild M, Stachel S, Economides A, Hazelett D, Harland RM and Hammerschmidt M (1998). Follistatin and noggin are

excluded from the zebrafish organizer. *Dev. Biol.* **204**: 488–507.
107. Storm EE and Kingsley DM (1999). GDF5 coordinates bone and joint formation during digit development. *Dev. Biol.* **209**: 11–27.
108. Crotwell PL, Clark TG and Mabee PM (2001). *Gdf5* is expressed in the developing skeleton of median fins of late-stage zebrafish, *Danio rerio. Dev. Genes Evol.* **211**: 555–558.
109. Piotrowski T, Schilling TF, Brand M, Jiang YJ, Heisenberg CP, Beuchle D, Grandel H, van Eeden FJ, Furutani–Seiki M, Granato M, Haffter P, Hammerschmidt M, Kane DA, Kelsh RN, Mullins MC, Odenthal J, Warga RM and Nüsslein–Volhard C (1996). Jaw and branchial arch mutants in zebrafish II: anterior arches and cartilage differentiation. *Dev.* **123**: 345–356.
110. Schilling TF, Piotrowski T, Grandel H, Brand M, Heisenberg CP, Jiang YJ, Beuchle D, Hammerschmidt M, Kane DA, Mullins MC, van Eeden FJ, Kelsh RN, Furutani–Seiki M, Granato M, Haffter P, Odenthal J, Warga RM, Trowe T and Nüsslein–Volhard C (1996). Jaw and branchial arch mutants in zebrafish I: branchial arches. *Dev.* **123**: 329–344.
111. Neuhauss SC, Solnica–Krezel L, Schier AF, Zwartkruis F, Stemple DL, Malicki J, Abdelilah S, Stainier DY and Driever W (1996). Mutations affecting craniofacial development in zebrafish. *Dev.* **123**: 357–367.
112. van Eeden FJ, Holley SA, Haffter P and Nüsslein–Volhard C (1998). Zebrafish segmentation and pair-rule patterning. *Dev. Genet.* **23**: 65–76.
113. van Eeden FJ, Granato M, Schach U, Brand M, Furutani–Seiki M, Haffter P, Hammerschmidt M, Heisenberg CP, Jiang YJ, Kane DA, Kelsh RN, Mullins MC, Odenthal J, Warga RM, Allende ML, Weinberg ES and Nüsslein–Volhard C (1996). Mutations affecting somite formation and patterning in the zebrafish, *Danio rerio. Dev.* **123**: 153–164.
114. Fleming A, Keynes RJ and Tannahill D (2001). The role of the notochord in vertebral column formation. *J. Anat.* **199**: 177–180.
115. Fisher S and Halpern ME (1999). Patterning the zebrafish axial skeleton requires early *chordin* function. *Nat. Genet.* **23**: 442–446.
116. van Eeden FJ, Granato M, Schach U, Brand M, Furutani–Seiki M, Haffter P, Hammerschmidt M, Heisenberg CP, Jiang YJ, Kane DA, Kelsh RN, Mullins MC, Odenthal J, Warga RM and Nusslein–Volhard C (1996). Genetic analysis of fin formation in the zebrafish, *Danio rerio. Dev.* **123**: 255–262.
117. Yelon D, Ticho B, Halpern ME, Ruvinsky I, Ho RK, Silver LM and

Stainier DY (2000). The bHLH transcription factor hand2 plays parallel roles in zebrafish heart and pectoral fin development. *Dev.* **127**: 2573–2582.

118. Johnson SL and Weston JA (1995). Temperature-sensitive mutations that cause stage-specific defects in Zebrafish fin regeneration. *Genetics* **141**: 1583–1595.
119. Goldsmith MI, Fisher S, Waterman R and Johnson SL (2003). Saltatory control of isometric growth in the zebrafish caudal fin is disrupted in *long fin* and *rapunzel* mutants. *Dev. Biol.* **259**: 303–317.
120. Johnson SL and Bennett P (1999). Growth control in the ontogenetic and regenerating zebrafish fin. *Meth. Cell Biol.* **59**: 301–311.
121. Yan YL, Miller CT, Nissen RM, Singer A, Liu D, Kirn A, Draper B, Willoughby J, Morcos PA, Amsterdam A, Chung BC, Westerfield M, Haffter P, Hopkins N, Kimmel C, Postlethwait JH and Nissen R (2002). A zebrafish sox9 gene required for cartilage morphogenesis. *Dev.* **129**: 5065–5079.
122. Bi W, Huang W, Whitworth DJ, Deng JM, Zhang Z, Behringer RR and de Crombrugghe B (2001). Haploinsufficiency of Sox9 results in defective cartilage primordia and premature skeletal mineralization. *Proc. Natl. Acad. Sci. U.S.A.* **98**: 6698–6703.
123. Bi W, Deng JM, Zhang Z, Behringer RR and de Crombrugghe B (1999). Sox9 is required for cartilage formation. *Nat. Genet.* **22**: 85–89.
124. Stainier DY, Lee RK and Fishman MC (1993). Cardiovascular development in the zebrafish. I. Myocardial fate map and heart tube formation. *Dev.* **119**: 31–40.
125. Walsh EC and Stainier DY (2001). UDP-glucose dehydrogenase required for cardiac valve formation in zebrafish. *Science* **293**: 1670–1673.
126. Topczewski J, Sepich DS, Myers DC, Walker C, Amores A, Lele Z, Hammerschmidt M, Postlethwait J and Solnica–Krezel L (2001). The zebrafish glypican knypek controls cell polarity during gastrulation movements of convergent extension. *Dev. Cell* **1**: 251–264.
127. Rauch GJ, Hammerschmidt M, Blader P, Schauerte HE, Strahle U, Ingham PW, McMahon AP and Haffter P (1997). Wnt5 is required for tail formation in the zebrafish embryo. *Cold Spring Harb. Symp. Quant. Biol.* **62**: 227–234.
128. Miller CT, Schilling TF, Lee K, Parker J and Kimmel CB (2000). *sucker* encodes a zebrafish Endothelin-1 required for ventral pharyngeal arch

development. *Dev.* **127**: 3815–3828.

129. Miller CT, Yelon D, Stainier DY and Kimmel CB (2003). Two *endothelin 1* effectors, *hand2* and *bapx1*, pattern ventral pharyngeal cartilage and the jaw joint. *Dev.* **130**: 1353–1365.
130. Baringa M (1994). Looking to development's future. *Science* **266**: 561–564.
131. Knight RD, Nair S, Nelson S, Javidan Y, Afshar A, Geisler R, Rauch GJ and Schilling TF (2003). The *lockjaw* mutation reveals requirements for zebrafish *tfap2a* in early neural crest development. *Dev.* **130**: 5755–5768.
132. Piotrowski T, Ahn DG, Schilling TF, Nair S, Ruvinsky I, Geisler R, Rauch GJ, Haffter P, Zon LI, Zhou Y, Foott H, Dawid IB and Ho RK (2003). The zebrafish van gogh mutation disrupts tbx1, which is involved in the DiGeorge deletion syndrome in humans. *Dev.* **130**: 5043–5052.

Chapter 12

Endoderm Formation in Zebrafish

Nicolas B. David, Philippe Mourrain and Frédéric M. Rosa*
Groupe Danio, U 368 INSERM, Ecole Normale Supérieure
46, rue d'Ulm F-75230 Paris Cedex 05, France

Endoderm, the innermost embryonic germ layer, gives rise to the epithelium of the digestive tract and of the respiratory system. Contrary to ectoderm and mesoderm, endoderm formation had been poorly analyzed. These three germ layers are formed during gastrulation via specific cell movements. An in-depth understanding of endoderm formation requires the defination of the molecular basis leading to the establishment of endodermal identity but also the defination of the mechanisms driving cell migration during gastrulation. Studies in Xenopus, mouse and zebrafish have converged on one conserved signaling and transcription pathway responsible for endoderm formation: the TGF-β/Nodal pathway. These past years, zebrafish mutants analyses have allowed the isolation of several loci necessary for endoderm formation. They encode elements of the Nodal ligand/receptor complex or transcription factors that have been shown to act downstream of Nodal ligands, in the Nodal signaling pathway. Using the genetic and embryologic advantages of the zebrafish, recent studies have started linking molecular data to cellular behavior during gastrulation. The different mutant backgrounds and the manipulation of Nodal signaling have allowed researchers to start deciphering key embryological events such as fate choice decision between mesoderm and endoderm, endodermal specification and commitment as well as the mechanism triggering and controlling cell movements during gastrulation.

During gastrulation, the few thousands cells resulting from the cleavage of the fertilized egg become organized into three distinct germ layers: the ectoderm, the mesoderm and the endoderm. Endoderm, the innermost layer, will give rise to the vast majority of the digestive tract. Endoderm derivatives populate the entire epithelium of the gastrointestinal tract and form the associated organs: liver, pancreas and gall bladder. In addition, endoderm derivatives also participate in the build-up of the respiratory system by forming, respectively, the epithelium of the gills in lower vertebrates and the epithelium of the lungs in tetrapod. Last, they contribute to the thymus, the thyroid gland and the swim bladder. During early development, additional roles for endoderm include the induction, the patterning and proper morphogenesis

*Corresponding author Tel: 33.(0)1.44.32.39.78; Fax: 33.(0)1.44.32.39.88; E-mail: rosa@wotan.ens.fr

of neighboring structures, including the heart and the head.[1–3] Therefore, endoderm development represents an attractive system to define the rules governing early patterning and differentiation of the embryo, as well as late morphogenesis.

In contrast to the ectoderm and mesoderm, the formation of which has now been studied for many years, endoderm development had been poorly addressed, probably because its deep position inside the embryo prevents easy observation. Within the past few years though, endoderm formation has become a field of intense scrutiny, and zebrafish has proven to be a fruitful model system in these studies. Zebrafish embryos are optically clear, permitting direct observation of endoderm development. Combined with techniques for the labeling of living cells, this optical clarity allows high-resolution imaging of morphogenetic movements and the construction of detailed fate maps. Fish also offer the opportunity to carry out genetic approaches and a number of mutants affecting endoderm development have been identified. Many of them have now been cloned and a first molecular pathway controlling endoderm formation can be assembled.

1. Fate Map and Morphogenetic Movements

After fertilization, fish development begins with a segmentation period that progressively divides the initial cell into a mass of a few thousand cells (the blastoderm) lying on top of a large, unsegmented yolk. Between the 9th and 10th cleavage, marginal blastomeres collapse into the yolk cell, thus forming a syncytial layer (the Yolk Syncytial Layer, YSL). During blastula stage, the blastoderm and YSL start expanding vegetally over the yolk cell, in a movement called epiboly.[4] Gastrulation starts when the blastoderm has covered half of the yolk cell (Fig. 1). At this stage, the embryo is composed of a large yolk ball covered by 'an inverted cup' of cells. While gastrulation movements take place and form the three germ layers, epiboly proceeds so that, at the end of gastrulation, the blastoderm has completely covered the yolk.

1.1. Endoderm Progenitors are Located at the Margin of the Zebrafish Embryo at the Onset of Gastrulation

Fate mapping studies have revealed that, at the mid-blastula stage, animal pole cells will give rise to the ectoderm, whereas marginal cells will form both mesoderm and endoderm.[5,6] At this stage, a single marginal

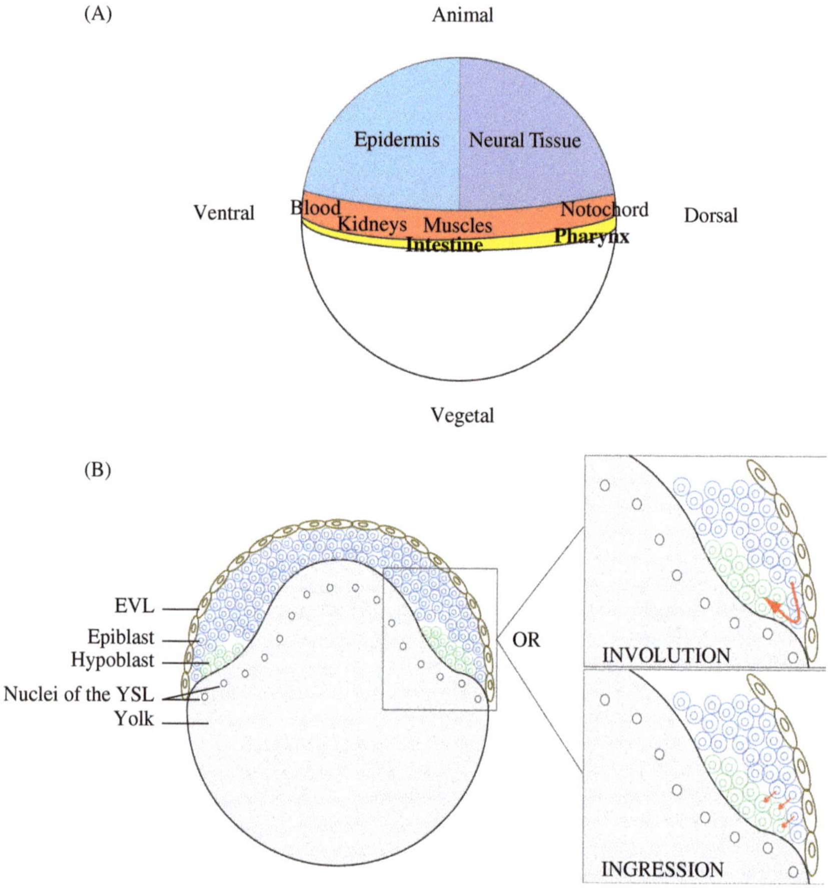

Fig. 1 A. Fate map of the zebrafish embryo at the onset of gastrulation. Endoderm progenitors are located at the very margin of the embryo, within the first four tiers of cells. Future pharyngeal cells are located dorsally, whereas future intestinal cells are located more laterally and ventrally. B. During gastrulation, marginal cells dive into the embryo creating two layers of cells, an outside layer, the epiblast, and an inside layer, the hypoblast. The epiblast corresponds to the ectoderm, whereas the hypoblast contains both endoderm and mesoderm. The exact nature of the cell movement remains controversial (involution or ingression). (After Solnica–Krezel *et al.*, 1995)

cell frequently gives rise to both endodermal and mesodermal derivatives, demonstrating that these two lineages have not yet separated. This separation occurs during the late blastula period since late blastula — early gastrula marginal cells generally contribute to one germ layer only.[5]

At the onset of gastrulation, endodermal precursors are found within the four tiers of blastomeres located closest to the margin, with higher density in the two lowest tiers (Fig. 1A). The progeny of these blastomeres will give rise to all the endodermal derivatives. According to this fate map, the position of endodermal precursors resembles the topographic arrangement of the future digestive tract. Although each organ derives from a large field of precursor cells that overlaps with that of other organs, the position of cells along the dorsoventral axis before gastrulation prefigures the location of their derivatives in the digestive tract along the anteroposterior axis. The dorsal-most cells will result in the anterior-most structures, such as the pharynx; the lateral ones will form the oesophagus and the stomach; and the ventral-most cells will give rise to intestinal cells (Fig. 1A).[6,7]

1.2. Involution/Ingression and Convergent Extension form Endoderm

The onset of gastrulation is marked by the beginning of the morphogenetic movements giving rise to the different germ layers. One of these movements, called involution or ingression, brings marginal cells (both endodermal and mesodermal precursors) to underlie the prospective ectoderm. Marginal cells leave the outside of the embryo to reach the surface of the YSL and start migrating towards the animal pole. This movement creates two multicellular layers: the outer epiblast corresponding to the ectoderm, and the inner hypoblast, including both endoderm and mesoderm (Fig. 1B).[4] In fish, the exact nature of this movement remains controversial. Some reports describe an involution movement[8] in which cells at the very margin of the embryo move inward and start migrating toward the animal pole at the surface of the YSL (Fig. 1B). Such an involution movement is very similar to the cell movements occurring at the blastopore lip of amphibians, except that, in fish, involution is observed as coordinated movement of individual cells rather than a movement of a layer of cells. However, other studies have proposed that the hypoblast is formed by ingression.[9] Ingression is a delamination process in which cells of the epiblast translocate from their superficial position to the surface of the YSL to form the hypoblast. The main difference between involution and ingression is in the position of moving cells with regards to the margin. Involution involves translocation of cells from epiblast to hypoblast only at the very margin, whereas ingression can occur at a distance from the margin (Fig. 1B). Recent transplantation experiments have demonstrated that cells can indeed ingress to form the hypoblast, confirming previous time lapse analysis of marginal labeled cells.[1,10,11] These transplantation experiments also demonstrated that individual cells can ingress even though the neighboring cells are not gastrulating. Together, these results suggest that the hypoblast can be formed by ingression of individual cells, but that ingression is restricted to the close proximity of the margin, possibly by cell-cell interactions.[10]

Once they have reached the surface of the YSL, endodermal cells start migrating towards the animal pole so that, by the end of gastrulation, they have colonized the entire extent of the anteroposterior axis of the embryo. At that time, they form a sparse but uniform monolayer of cells. During these movements on the yolk syncytium, endodermal cells adopt a characteristic morphology: between 70–80% epiboly, they progressively flatten and develop filopodia. Thus, although endodermal cells express specific markers from the onset of gastrulation (see below), it is only at late gastrulation that they become morphologically distinguishable from mesodermal cells, which remain spherical and lack visible filopodia.[6]

The anteroposterior distribution of endodermal cells is also due to another important gastrulation movement: dorsal convergence. During gastrulation, hypoblastic cells move towards the dorsal side of the embryo. Medio-lateral intercalations between cells accumulating dorsally lead to anteroposterior extension of the converging endoderm. This convergence/extension process results in the gathering of endodermal cells on the dorsal side, so that at the end of gastrulation, the endoderm is a sparse monolayer on the ventral-most side of the developing embryo. Dorsal convergence continues during somitogenesis, leading to the formation of a rod of endoderm (except in the pharyngeal region which remains monolayered).[12] Polarization and rearrangements of endodermal cells will then transform the rod into a tube.[13]

Beyond the description of gastrulation movements, intense work has been done in the past few years to unravel the mechanisms controlling endoderm fate and differentiation. As we shall see in the first part of this chapter, taking advantage of previous results in amphibians and the endoderm mutants available in fish (Table 1), a molecular pathway leading to endoderm formation can now be assembled (Fig. 2). In the

Table 1 Components of the zebrafish Nodal pathway. Expression patterns observed by *in toto in situ* hybridization are described in a simplified manner to point out the reminiscent presence of transcripts at the margin i.e. the presumptive territory of mesendoderm. Mutant (when isolated) and other phenotypes are also merely described to underline the mesendodermal defects. Abbreviations: B blastula, EB early blastula, MB mid-blastula, LB late blastula, G gastrula, EG early gastrula.

Mutant/Gene	Protein	Nature	Expression pattern	Phenotype
squint (sqt)	nodal	TGF-β ligand	EB: dorsal margin MB: margin G: off	*sqt*: no prechordal plate and dorsal mesoderm defects.
cyclops (cyc)	nodal	TGF-β ligand	B: margin EG: dorsal margin G: axial	*cyc*: reduction of prechordal plate. Double mutant *sqt*; *cyc*: no prechordal plate, no endoderm and dorsal mesoderm.
one-eyed pinhead (oep)	Oep	EGF-CFC co-receptor	B: ubiquitous EG: margin G: axial	Z*oep*: no prechordal plate and endoderm. MZ*oep*: no prechordal plate, no endoderm and dorsal mesoderm.
	Lefty/ Antivin	TGF-β antagonist	MB: dorsal margin LB: margin EG: dorsal margin G: prechordal plate (+ectoderm)	Overexpression leads to absence of endoderm and dorasal mesoderm.

Table 1 (*Continued*)

Mutant/Gene	Protein	Nature	Expression pattern	Phenotype
	Taram-A	Type I TGF-β receptor	B: ubiquitous EG: margin G: axial	Dominant negative phenotype (TarMR): no endoderm and *cardia bifida*. The constitutive activated form (Tar*) is able to commit cells to an endodermal fate.
schmalspur (sur)	Fast1/ FoxH1	Winged helix Transcription Factor	B: ubiquitous G: ubiquitous	Z*sur*: reduction of prechordal plate. MZ*sur*: no prechordal plate.
bonnie and clyde (bon)	Mixer	Homeodomain Transcription Factor	B: margin EG: margin G: off	*bon*: 70% reduction of endodermal cells number, reduction of prechordal plate and *cardia bifida*.
faust (fau)	Gata5	Zinc finger Transcription Factor	B: margin EG: margin G: hypoblastic cells	*fau*: 10% reduction of endodermal cells number and *cardia bifida*.
	Mezzo	Homeodomain Transcription Factor	B: margin EG: margin G: off	MO *mezzo* increases the *bon* phenotype: no prechordal plate and endoderm.
casanova (cas)	Casanova	Sox Transcription Factor	B: marginal-most cells EG: marginal-most cells G: endodermal cells	*cas*: no endoderm and *cardia bifida*.
	Sox17	Sox Transcription Factor	B: dorsal marginal-most cells EG: dorsal marginal-most cells G: endodermal cells	
	Axial/ foxA2/ HNF3β	Winged helix Transcription Factor	EG: dorsal margin G: endodermal cells and axial mesoderm	

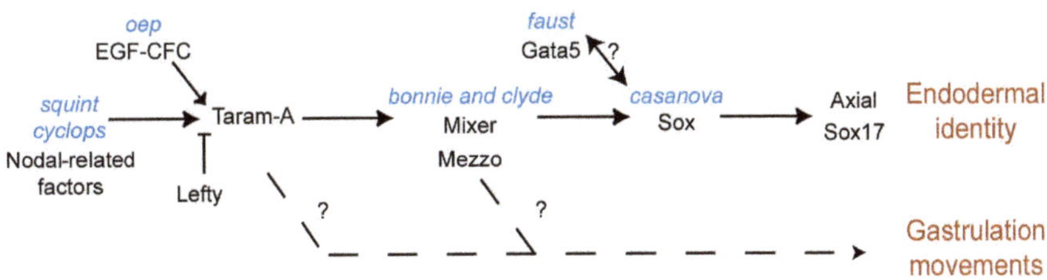

Fig. 2 Zebrafish Nodal signaling pathway.

second part, we will show that identifying the molecules involved in endoderm formation is only part of the job, and we now need to integrate both molecular and cellular studies to get a more comprehensive and thorough understanding of endoderm development.

2. Molecular Basis of Endoderm Formation: The TGF-β/Nodal signaling pathway

2.1. The Xenopus Contribution

The molecular mechanisms, underlying endodermal fate and behavior acquisition, are only partially understood. The first data have been obtained in frogs (*Xenopus laevis*) and have allowed the establishment of a two-step model.[14] The first step involves maternal factors (β-catenin, VegT, Vg1) present in vegetal blastomeres of the Nieuwkoop center that activate the zygotic expression of extracellular factors belonging to the Nodal/TGF-β subfamily (Xnr1, Xnr2 and Derrière).[15–17] Nodal, the founding member of this subfamily, was isolated in mice and is the only member identified in mammals.[18] Mice lacking Nodal are not able to form the node, the mouse organizer, and die shortly after gastrulation.[18,19] The *Xenopus* Nodal-related factors are also localized in the Spemann organizer and control axis formation. Moreover, they are involved in the second step of the endoderm formation model because once induced by maternal factors, they in turn activate endodermal genes (Mix/Bix, Gata5, Sox17, described below).

Overexpression of the Nodal-related factors 1, 2 and 4 or the mouse Nodal in ectodermal explants induces mesendodermal tissue, and inhibition of Nodal signaling in *Xenopus* embryos prevents mesoderm induction at the margin and endoderm formation in the vegetal region.[17,20–23] Nodal signaling controls gene expression, cell differentiation and the specific migratory behavior of marginal cells. Another TGF-β member, Activin, is also able to influence mesendoderm formation but its endogenous role is still controversial.[24–26]

2.2. The Zebrafish Contribution

2.2.1. *The Nodal-related ligands*

Despite its remarkable embryological qualities, *Xenopus laevis* is not adapted to genetic analysis because of its pseudo-tetraploidy and long generation time. However, vertebrates share common signaling pathways such as Bmp, Wnt, FGF, Hedgehog, Delta/Notch and retinoic acid, and this is also true for the TGF-β/Nodal pathway. Several laboratories took advantage of the zebrafish as a genetic model to dissect this latter pathway. Analyses of zebrafish mutants and morphants (embryos injected with antisense morpholino oligonucleotides[27,28]) allow the unraveling of genetic networks and have led to a more detailed model that shares several steps with the *Xenopus* model (Fig. 2). As in mice, genetic experiments in zebrafish have shown that Nodal-related factors are required for mesendoderm formation.[29–32] Mice and zebrafish carrying mutations in Nodal-related genes fail to form mesendoderm and the Spemann organizer. The zebrafish genome harbors two Nodal-related genes named *squint* (*sqt*) and *cyclops* (*cyc*)[29,31] (Table 1). Both are expressed in marginal cells of the zebrafish blastula. *Squint* is also expressed in the dorsal YSL which underlies the prospective zebrafish organizer (shield region).[29] Thus, as described above in the fate map, *cyc* and *sqt* are expressed in domains where mesoderm and endoderm precursors are localized: the blastoderm marginal region. *sqt* and *cyc* appear to be essential for endoderm (and dorsal mesoderm) formation since *sqt;cyc* double mutant completely lack these structures.[29–33]

Interestingly, single *sqt* or *cyc* mutants still form endoderm showing a functional redundancy between these two factors. Although factors related to the maternal *Xenopus* molecules like VegT and Vg1 have been described, their requirement for mesendoderm formation has never been clearly established. The maternal expression of *sqt* is the earliest event involved in endoderm induction. So, *sqt* and *cyc* are the earliest known regulators of endoderm formation in zebrafish.

2.2.2. *The type I receptors Taram-A/ALK4 and the EGF-CFC co-factors Oep/Cripto*

Similar to other TGF-β extracellular ligands, Nodals are thought to bind and activate serine/threonine kinase receptors that in turn activate a signaling cascade leading to endoderm formation.[34] These serine/threonine kinase receptors are tetrameric complexes consisting of two type I and two type II receptors. Both type I and type II receptors cooperate for ligand binding but the specificity is mainly determined by the type I receptor. Several type I receptors have been known for some time, including ALK4, an activin receptor,[35] but receptors transducing Nodal signals remained elusive. Recently, biochemical and genetic analyses have provided evidence that Nodals bind to and signal through ALK4 or its zebrafish most related counterpart Taram-A (Table 1), and through the orphan receptor ALK7.[36,37] First, in mice, ALK4 and Nodal loss-of-function lead to very similar phenotypes including absence of mesoderm and endoderm.[38] Second, in vertebrates, Nodals and Taram-A or ALK4 are expressed in overlapping domains. Third, blocking Taram-A or ALK4 function via dominant negative isoforms inhibits Nodal signaling and endoderm formation.[39] Conversely, co-injection of *taram-A* with *cyclops* or *squint* mRNAs have a synergistic activating effect on mesendodermal markers expression.[39] Last, similar to Nodals, injection of an activated form of Taram-A (Tar*) can also induce ectopic mesendodermal markers expression.[1,3] This receptor variant (Tar*) rendered constitutively active by a point mutation in the GS-box, is a highly effective, cell-autonomous inducer of Nodal-responsive mesendodermal genes.

To bind and transduce Nodal signals, ALK4 and Taram-A require a transmembrane EGF-CFC co-receptor, mouse Cripto or zebrafish Oep (Table 1), respectively.[40–48] Zebrafish *oep* mutants (devoid of the maternal and zygotic contributions, MZ*oep*) have a phenotype highly similar to the double mutant *sqt;cyc* i.e. they are devoid of endoderm and of dorsal mesoderm, suggesting a role for Cripto/Oep in the Nodal pathway.[40] In total absence of *oep*, no mesendodermal Nodal targets are expressed even when Nodal ligands and receptors are overexpressed. However, the constitutively active Tar* or ALK4* variants are able to rescue the MZ*oep* mutant phenotype, suggesting a role of the Oep factor in parallel to or upstream of the Nodal ligands/receptor complex.[3,39] In fact, *oep* gene is likely necessary as a co-receptor for Nodal signaling transduction via the serine/threonine kinase receptor complex. Recently, Cripto (the murine Oep homolog) has been shown to interact directly with the ALK4 type I receptor but not with the type II ActRIIb, and is strictly necessary for the responsiveness of ALK4 to Xnr1.[36] Likewise, Cripto can bind directly to Nodals. The ability of Cripto to interact with ALK4 but not ActRIIb, suggests that EGF-CFC proteins target type I and not type II receptors in the complex and acts as an adaptor between ALK4/Tar and Nodals.

Unlike Taram-A and ALK4, ALK7, a type I receptor recently discovered in mice, does not appear to require Cripto to directly interact with Nodal ligands and to transduce the Nodal signal, although Cripto levels have not been measured. However, Cripto is able to interact with ALK7 and increase the responsiveness of the ALK7/actRIIB receptor complex to Xnr1 and Nodal, showing again an EGF-CFC potentiating influence.[36] ALK7 is similar in its serine/threonine kinase domain to ALK4 and Taram-A, but very divergent in its extracellular domain from all Alks. No zebrafish orthologues and mice ALK7 mutants have been yet described and ALK7 function in mesendoderm formation is not determined.

2.2.3. *The TGF-β/leftys inhibitors*

Nodals signaling can also be controlled by the extracellular factors, Lefty1 and Lefty2.[49] These factors also belong to the TGF-β superfamily but, in contrast, behave as Nodal antagonists (Table 1). Overexpression of Leftys inhibits formation of endoderm and dorsal mesoderm, and phenocopies the *sqt;cyc* or MZ*oep* mutants.[50,51] *lefty* genes expression is induced by Nodal signaling, suggesting a negative feedback mechanism quantitatively controlling endoderm and mesoderm specification.[52] Indeed, overexpression of low doses of Lefty suppresses endoderm formation while higher doses also affect the mesoderm.[53] Hence Leftys would act, like EGF-CFC factors described above, outside the cell to modulate the Nodal signal but in a negative way.

2.2.4. *Smads and Fast1 transducers*

Experiments in cell cultures and in *Xenopus* embryos have shown that signal transduction from Nodal/Activin receptors requires the maternally and ubiquitously expressed Smad2 transcription factor. After phosphorylation by the activated receptor complex, Smad2 homo- and hetero-dimers with other Smads and translocates to the nucleus. Once in the nucleus, Smad factors bind to co-factors such as the forkhead winged-helix transcription factor FoxH1/Fast1 through a Smad Interacting Motif (SIM), and activate downstream target genes.[54–56] For example, FoxH1/Fast1 forms a complex with activated Smad2/Smad4 dimers to bind to a responsive element of the *mix2* promoter.[57] Zebrafish *uncle freddy/schmalspur*, carry mutations in the *foxH1/fast1* (Table 1). Embryos lacking maternal and zygotic transcripts of *schmalspur* (MZ*sur*) have axial mesoderm deficiency but still develop endoderm.[58,59] Compared to *cyc;sqt* double mutant and MZ*oep* mutant, the phenotype of MZ*sur* mutants is clearly less severe. However injection of FoxH1/Fast1 fused to the strong transcriptional repressor domain of *engrailed* (Fast1-EnR) in zebrafish embryos induces a dramatic phenotype strikingly similar to that of *sqt;cyc* and MZ*oep* mutants. Since the *schmalspur* allele encodes a completely dysfunctional protein but does not lead to a complete *sqt;cyc*/MZ*oep* phenocopy, similar factors with overlapping function may be required to mediate the full response to Nodals. These factors could be yet undescribed Fast1-related proteins or other Smads interactors like the Mixer transcription factor (see below).

2.2.5. *An epistatic pathway leading to endoderm formation*

In addition to Nodal ligands, their receptors and the Smads transducers, several transcription factors, induced by Nodal signaling have been identified as major regulators of zebrafish mesendoderm and endoderm development (Fig. 2). Loss of function in the corresponding genes lead to embryos exhibiting different levels of endodermal defects (Table 1). These defects can be recognized by the altered expression patterns of the endoderm-specific markers *Sox17* and *axial/foxA2/HNF3β*,[60] which encodes respectively transcription factors of the SRY-box and winged helix type factors. Genetic analyses of these loci, involved in mesendoderm formation, have allowed to determine their epistatic position within the Nodal pathway, downstream of receptors, Smads and Fast1. Involved genes are *mixer* and *mezzo*, encoding related homeobox proteins;[61,62] *Gata5*, encoding a zinc finger factor;[63] and *casanova,* encoding a SRY-box factor related to *Sox17*.[64–66]

cyc;sqt double mutants and MZ*oep* mutants display a very severe phenotype.[40] Completely lacking the endoderm and dorsal mesoderm, they fail to form a prechordal plate (the anterior mesendoderm), and the gut tube, possessing just a small amount of caudal tissue showing that Nodal signaling is required for the formation of most of the mesendoderm. During gastrulation they do not express *mixer*, *mezzo*, *gata5*, *casanova*, *sox17* nor *foxA2* genes, confirming the upstream role of Nodal ligands and their EGF-CFC co-factor in Nodal pathway and mesendoderm formation (Fig. 2).

One of the transcriptional targets of Nodal signals is the *mixer* gene (Table 1), which, like Nodals, is also expressed in a large marginal region overlapping endodermal and mesodermal precursors before gastrulation. Then *mixer* expression is down-regulated at the onset of gastrulation. The *mixer* mutant, *bonie and clyde* (*bon*),[61] possesses a prechordal plate and somites but develops a very reduced gut tube. *bon* gastrulae display a strong reduction of the number of cells expressing the endodermal markers *foxA2* and *sox17*. Thus, *mixer* is a transcriptional target of Nodal signals but would be only necessary for transduction of part of the signal involved in endoderm formation. The homeobox protein

Mixer is a good candidate to interact, like Fast1, with Smads interactors. Zebrafish and *Xenopus* Mixers are members of the Mix/Bix family. Studies of Mix/Bix homeoprotein function in *Xenopus* have shown that they are downstream transcriptional targets in the TGF-β superfamily pathway that regulates mesendodermal patterning.[67,68] Moreover, the C-terminal region of Mixers contain a SIM domain required for interaction with phosphorylated Smads, suggesting that Mixers like FoxH1/Fast1 can cooperate with Smads to transduce Nodal signaling.[55] Interestingly *bon* mutants fail to develop most of the posterior axial mesendoderm (very primitive gut tube) whereas MZ*sur* (Fast1) mutant embryos display deficiencies in anterior axial mesendoderm (prechordal plate).[58,59,61] This complementarity of defects may reflect a division of Nodal signaling via the different Smads interactors.

Like *bon* mutants, *casanova* (*cas*) mutants are affected in endoderm formation but their defects are more severe: they do not develop a gut tube at all and completely lack endodermal *sox17* and *foxA2* expression during gastrulation (Table 1).[69] However *casanova* gastrulae have normal *Mixer* expression while *bon* gastrulae display a very strong reduction of the number of *Casanova*-expressing cells showing that *casanova* acts downstream of *mixer*.[70] In contrast to *mixer*, which is expressed in mesoderm and endoderm precursors, *cas* and *sox17* expression patterns are strikingly similar suggesting that *cas* and *sox17* may label the very same endodermal population (Fig. 3).[65,66] However, in contrast to *Sox17* which is only expressed when endodermal cells have involuted, *cas* is also expressed at an earlier stage, i.e. before involution when endodermal precursors are still located at the margin (Fig. 3). Moreover *cas* appears as a major regulator of endoderm formation since its overexpression in embryos completely devoid of Nodal signaling (MZ*oep*) is able to induce *sox17* expression. Thus, *Cas* acts clearly upstream of *sox17* and would be the major inducer of *sox17* expression.[65,66,70,71]

These first genetic analyses demonstrate that Nodals (Cyc and Sqt) and their co-receptor Oep are required for *bon/mixer* expression which is necessary to control *casanova* that in turn regulate *sox17* expression (Fig. 2).[71] Genetically *bon* clearly acts upstream of *cas*. However *bon*

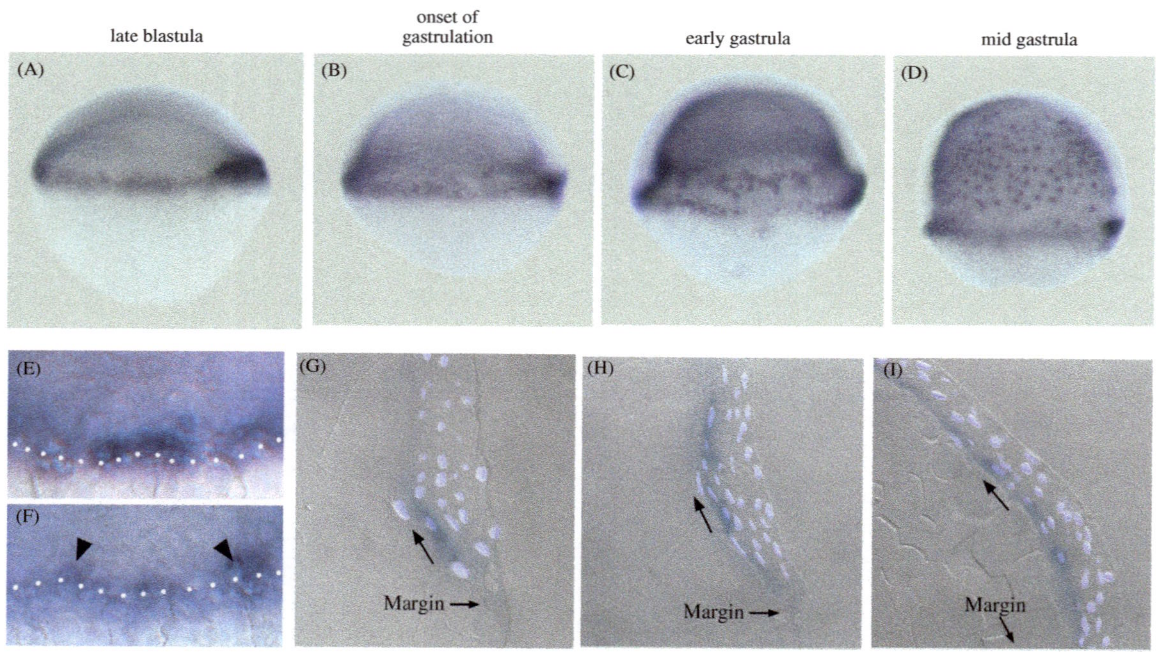

Fig. 3 *cas* expression pattern, between late blastula and mid gastrula. (A) At late blastula, *cas* is first expressed in the YSL and in scattered marginal cells, mostly on the dorsal side. (E,F) Close up of the dorsal (E) and (F) lateral margin. Notice the isolated blastodermal *cas*-expressing cells (arrowheads). The doted line marks the YSL-blastoderm frontier. (B) At the onset of gastrulation, the *cas* pattern is still mosaic but is found throughout the margin. (C) In the young gastrula, *cas*-expressing cells have begun to involute and abut the YSL. (D) During gastrulation, *cas* is expressed in the scattered endodermal cells. (G–I) Cross-sections of embryos showing that during gastrulation, *cas*-expressing cells involute at the margin, abut the YSL and spread over the whole embryo with a scattered pattern.

is less affected in endoderm formation than *casanova*, suggesting the existence of other factors with overlapping functions acting in parallel to Mixer to regulate expression of *casanova*. Two candidates for such a redundant role have been identified (Table 1): the zinc finger-containing factor Gata5 encoded by the *faust* gene[63] and another Mixer-homeobox-related factor (no mutant identified yet) Mezzo.[62] Like *mixer*, they are expressed in presumptive mesoderm and endoderm and can induce ectopic expression of *sox17* in wild-type embryos. In *Xenopus*, Gata5 and Mixer co-injection has a synergistic effect on *sox17* expression suggesting parallel roles upstream of endoderm-specific genes. In zebrafish on the contrary, the position of *gata5* in this pathway is not clearly established (Fig. 2). Prior to gastrulation, *Gata5* expression appears normal in *casanova* mutants, but the endodermal expression of *gata5* is then lost during gastrulation, suggesting that *Casanova* is required to maintain *gata5* expression. However, the *gata5/faust* mutants have a much weaker phenotype than *cas*, presenting a slightly abnormal gut tube. During gastrulation they have a very slight reduction in the number of *cas*-expressing cells suggesting a limited effect of *gata5* on *casanova*. Moreover, overexpression experiments in MZ*oep* embryos have shown that *casanova* is able to induce *gata5* in the absence of Nodal signaling. However in the same conditions *Gata5* is not able to induce *casanova*. So, in contrast to the *Xenopus* model, it is not clear that *gata5* can mainly act upstream of *casanova*. Moreover, while *casanova* can autonomously induce *sox17* in MZ*oep*, *mixer* and *gata5* alone or co-injected are able to induce neither *sox17* nor *cas*.[70] So the simultaneous presence of both *mixer* and *gata5* is not sufficient to restore the Nodal signal(s), showing that other Nodal transducing factors are needed, upstream of *Cas*, to allow endoderm development.

The recently identified Mezzo could be one of these factors. Indeed, Mezzo is regulated by Nodals and is able to induce *casanova* and *sox17* expression in wild-type embryos.[62] Moreover, whereas Mezzo is devoid of SIM box, its function appears redundant with that of *bon*: *mezzo* mRNA can partially rescue *bon* mutants and injection of anti-*mezzo* morpholinos into *bon* embryos abolishes the residual *sox17* expression.[62]

However, unfortunately it is not known whether co-injection of *mixer* and *mezzo* mRNAs or *mixer*, *mezzo* and *gata5* mRNAs can restore *casanova* and *sox17* expression in MZ*oep* mutants and are thus sufficient to transduce completely the Nodal signal required for endoderm formation.

3. Cellular Analysis of Endoderm Formation

Mutant analyses have allowed the identification of a number of genes involved in endoderm formation, and although some important genes are probably still to be discovered, epistatic studies have led to a quite coherent model from Nodal ligands to the expression of endoderm specific markers (Table 1 and Fig. 2). However, in spite of these great genetical advances, our understanding of endoderm formation is still quite limited in terms of cellular biology. So far, the involvement of a given gene in endoderm formation has usually been tested considering its effect on the expression of a few endodermal markers. A loss-of-function should lead to a reduction of endodermal markers, whereas a-gain-of function should induce these markers. But for an in-depth understanding of endoderm formation we need to define this phenomenon in cellular terms. Critical issues need to be addressed: when and where are cells specified to the endodermal fate, when and where is their fate restricted to endoderm versus mesoderm, and when and where are they committed to an endodermal fate? And for each of these questions, which signals or which effectors, expressed in the right position and at the right time, might mediate these cellular changes.

3.1. Endoderm Specification

Fate mapping studies have shown that, at the blastula stage, endoderm progenitors are located at the margin of the embryo, animal pole cells being precursors of ectoderm.[5,6] However grafting experiments show that an animal pole cell, transplanted to the margin at this time, can take part to endodermal derivatives.[1,69] This demonstrates that the margin provides, at the blastula stage or later during development, extracellular signals that are sufficient to drive cells to an endodermal fate. This excludes the strict requirement for cell-autonomous maternal or early factors that would specify endodermal cells from the first cleavages. Nodal ligands are of course among the extracellular signals involved in this specification. As previously described, they are indeed expressed at the margin of the embryo, from blastula to mid-gastrula stages, and genetic studies have demonstrated their absolute requirement for endoderm formation (see above). When does endoderm specification begin? By injecting Oep protein into MZ*oep* mutants, one can control the timing of activation of the Nodal pathway. Using this system, Aoki and colleagues demonstrated that Nodal signaling is not required early on but has to be activated during the mid-late blastula period to allow endoderm formation.[70] Later activation of the Nodal pathway allows formation of mesoderm but progressively leads to the loss of endodermal structures. These results suggest that endoderm specification by Nodal signals starts soon after the mid-blastula, at the margin of the embryo. Interestingly, fate mapping experiments have shown that at this stage one marginal cell can still give rise to both endodermal and mesodermal derivatives, potentially because specification by Nodals is progressive or because other mechanisms are at stake.[6]

3.2. Endodermal Commitment

Commitment or determination of a cell marks the time when the cell fate is irreversibly established. At this point, the cell fate is no longer sensitive to external cues. The classical way for testing the state of commitment of a cell is to transplant it into an ectopic position and determine whether its fate is affected by its new environment.[72] Such an experiment demonstrates that in fish, cells are committed to an endodermal fate at the onset of gastrulation, around the time they involute.[1] Before gastrulation endodermal precursors transplanted to the animal pole give rise to classical animal pole derivatives, i.e. epidermis and neural tissue (Fig. 4). On the contrary, once they have reached the surface of the YSL, endodermal cells transplanted to the animal pole of a host embryo will keep their endodermal identity.

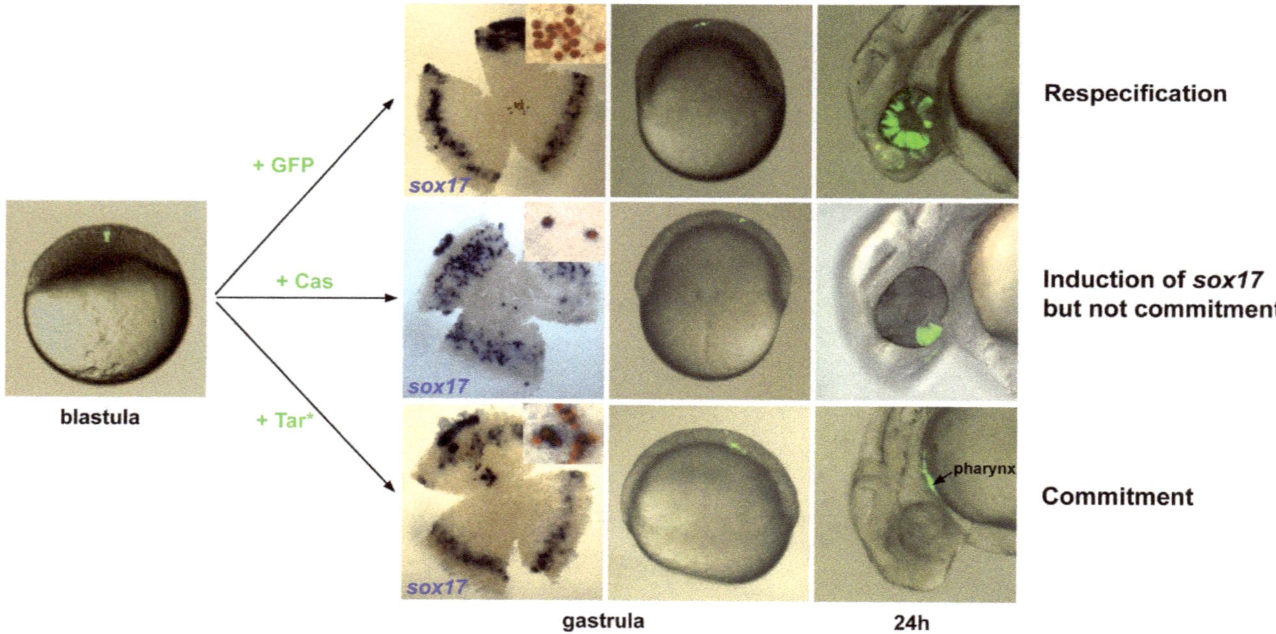

Fig. 4 At blastula stage, cells are transplanted to the animal pole of a host embryo. When these cells express only a lineage tracer (GFP) they adopt a classical animal pole fate. They stay in the epiblast during gastrulation, and take part to the brain and to the eyes at 24 hours of development. When *cas*-expressing cells are transplanted to the same position, they express the endodermal marker *sox17* at the beginning of gastrulation. But they stay epiblastic and are finally respecified towards animal pole fates. On the contrary, *tar**-expressing cells do not only express *sox17* during gastrulation, they also translocate to reach the surface of the YSL and at 24h, they take part to endodermal derivatives (e.g. pharynx), demonstrating that full activation of the Nodal pathway is sufficient to commit cells to an endodermal fate.

What are the signals and molecules leading to endodermal commitment? The most obvious candidate is the Nodal signaling pathway and its downstream effectors. As a matter of fact, they are absolutely required for endoderm formation (mutants are deprived of endoderm), and appear involved in endoderm specification. But is this pathway sufficient on its own to commit cells to an endodermal fate? Transplants experiments were performed with cells in which the Nodal pathway has been activated by expression of a constitutively active version of the receptor Taram-A (Tar*).[1] Strikingly, activation of the Nodal pathway by this means led to the full commitment to an endodermal fate: cells kept their endodermal identity when transplanted to the animal pole and differentiated into endodermal derivatives (Fig. 4). The conclusion of this experiment is that full activation the Nodal pathway is sufficient to commit cells to an endodermal fate. Clearly, Nodal signaling appears as a major player in endoderm determination. It is unlikely though to be the only factor involved since in other species other mechanisms are at stake (see Stainier[73] and Shivdasani[74] for reviews). However, one can postulate that modulation of Nodal signaling by other pathways or by time might be a key element in the induction of the proper amount of endodermal tissue.

Genetic and epistatic studies have identified some of the downstream components of the Nodal pathway, including *faust/gata5, bon/mixer, mezzo, casanova, sox17*. Are these effectors sufficient for commitment? Unfortunately, the effect of overexpression of either *gata5, mixer* or *mezzo* alone, or in combination, has not yet been tested on the state of commitment of cells. Such an experiment has however been carried out for *casanova*. Interestingly, cells overexpressing *Casanova*, when transplanted to the animal pole, express the endodermal marker *sox17*, but contrary to Tar*-expressing cells, they lose *sox17* expression during gastrulation and are respecified towards classical animal pole fates (Fig. 4) (David, unpublished data). This reveals that *casanova* mediates only part of the Nodal signaling, and strongly suggests that there are still a number of components of the Nodal pathway to be identified.

3.3. Endoderm Gastrulation Movements

During gastrulation, endodermal cells do not only acquire their endodermal identity, they also display a very specific behavior: they involute to reach the surface of the YSL and stay in close contact to the YSL during gastrulation. What the exact nature of these cell movements is (active migration, cell sorting ...) and how they are controlled remain two largely unresolved questions.

One of the most striking results of the commitment experiments was that committed cells (endodermal cells that have involuted as well as cells expressing Tar*) not only keep their identity during gastrulation, but also display a typical endodermal behavior during gastrulation. When transplanted to the animal pole, they stay epiblastic till the beginning of gastrulation, but during gastrulation they segregate from neighboring cells and translocate to adopt a classical endodermal location within the embryo.[1] They leave the epiblast to reach the surface of the YSL. There they join the other endodermal cells and finally differentiate into endodermal derivatives according to their anteroposterior location (Fig. 4). Interestingly, since Tar*-expressing cells systematically display this endodermal behavior, Nodal signaling appear sufficient to trigger these gastrulation movements, and genes controlling these movements are thus probably to be found within the downstream targets of Nodals. However, among the targets identified so far, *casanova* is not able to commit cells to an endodermal fate and in particular, *cas*-expressing cells do not display any particular movement when transplanted to the animal pole (Fig. 4). This suggests that the downstream targets of Nodal-controlling cell behaviors remain to be identified. Cell adhesion molecules such as cadherins or proteins interacting with the cytoskeleton could be good candidates.

The exact nature of gastrulation movement also remains unclear. Transplants of wild-type cells in MZ*oep* mutant[10] demonstrate that a single cell can gastrulate, suggesting that involution is a largely cell-autonomous process. These experiments also demonstrate that endodermal cells can gastrulate even in the absence of mesodermal cells that normally surround them at the margin. But how this involution is achieved is still unknown. Endodermal cells can be expelled from the epiblast, in a sorting-out process. Or, they might migrate actively to reach the surface of the YSL. In such a hypothesis, YSL would provide guidance cues to direct the endodermal cell movements. Further investigations will be required to discriminate these possibilities and to identify the genes triggering these movements.

3.4. Endoderm Versus Mesoderm Fate Choice

Both endoderm and mesoderm derive from the margin of the late blastula. Fate maps have defined the position of endodermal progenitors within the most marginal tiers whereas mesoderm derives from both marginal-most tiers and tiers of cells located further away. Both endoderm and dorsal mesoderm are specified by Nodals. What then leads to the differential specification of endoderm versus dorsal mesoderm? Several arguments suggest that fate choice may rely on differential levels of Nodal signaling, as has been proposed by Thisse and colleagues[51] and Dougan and colleagues.[75]

First, reduction of Nodal signaling by injection of the Nodal antagonist Lefty leads to embryos devoid of endoderm but still developing dorsal mesodermal tissues.[53] Second, reduction of Nodal signaling by providing only early expression of *oep* (*Zoep* mutants) also leads to embryos mostly devoid of endoderm but developing at least some dorsal mesodermal structures (notochord).[40] Third, overexpression of Nodals at low doses lead to the induction of dorsal mesodermal tissues and to endoderm at higher doses.[76] Such a differential level of signaling may happen in different ways. According to the so-called "morphogen" model cells could exhibit different responses to different concentrations of ligands. Ligands would have to be expressed in a gradient concentration. The differential sensitivity of cells to Nodals, the capacity of Squint to diffuse and probably establish a gradient[77] are fully consistent with the idea that endoderm and dorsal mesoderm are differentially induced by gradient of Nodals. However, it is also equally possible that other factors are involved, for instance FGF which has inducing activity and is expressed at the margin of the blastula.[76] In addition, time might be an important component to control Nodal signaling. First, in other species (*Xenopus*), early cells have a stronger capacity to respond to Nodal-like signals than late embryonic cells. Furthermore, the duration of the signals might also be important. These hypotheses need to be rigorously tested. However, some hints that time is an important factor come from the fact that late activation of the Nodal pathway by injecting Cripto protein into MZ*oep* embryos only rescues dorsal mesodermal structures but not endodermal structures.[70]

An additional problem is coming from the fact that very near the margin, endodermal and mesodermal progenitors are intermingled. It thus appears unlikely that a single concentration gradient of diffusible morphogen, even complemented by the diffusion of an antagonist, can achieve such a salt-and-pepper pattern. On the contrary, this is suggestive of the existence of additional signals that can modulate Nodal signaling. In particular, a system modulating locally the amount of available ligand (via ligand-matrix interaction) or a process of lateral inhibition, such as those mediated by the Notch pathway, could be envisioned.[78,79] In this respect, it should be borne in mind that Notch signaling has been implicated in endoderm development in sea urchin.[80]

Among the components of the Nodal pathway, *casanova* is the first whose expression is strictly restricted to endodermal cells. It is thus a good candidate for being directly involved in the endoderm versus mesoderm fate choice. Furthermore, in the *casanova* mutant, fate mapping experiments demonstrate that cells normally fated to endoderm are respecified to a mesodermal fate.[66] Conversely, overexpression of *casanova* in marginal cells can transfate mesodermal cells to endoderm[65,70] although the level of transfating seems to be controversial. Taken together these results demonstrate that *Casanova* plays a key instructive role in the endoderm versus mesoderm fate choice. However, as previously mentioned, *casanova* alone is not sufficient to commit cells to an endodermal fate, and it cannot transfate ectodermal cells to endoderm.[65] This means that some other signals, present at the margin, act with *casanova* to direct cells to an endodermal fate. Since activating the whole Nodal pathway (Tar*) is sufficient to commit cells to an endodermal fate, these other signals are likely to be regulated by Nodals. This is in agreement with the fact that *casanova* is unable to induce strong expression of endodermal markers such as *fkd7* and *nkx2.3* in MZ*oep* mutants.[65,70,81]

3.5. Early Endoderm Patterning

At the late blastula stage, fate mapping experiments have shown that endodermal precursors are located at the dorsal margin and that their number decreases ventrally. A correlation exists between the location of endodermal progenitors along the dorsoventral axis in blastula and the future anteroposterior (AP) position of their derivatives in the larva: ventral cells will populate the posterior part of the intestine whereas dorsal cells will take part to the anterior region of the digestive tract such as the pharynx.[6] Does it mean that some stable patterning events have already occurred before gastrulation? This seems unlikely since blastomeres collected from any position of Tar*-expressing embryos and transplanted, before gastrulation, into host untreated embryos correctly differentiate according to their new position along the AP axis, likely under the influence of external cues.[1] Then does AP patterning of endoderm occur during gastrulation? Two reports have analyzed this issue in zebrafish. First, similar to mesoderm and ectoderm, endoderm appears to be sensitive to the ventralizing activity of BMP signaling occurring during gastrulation. Hyperdorsalized mutants resulting from a null mutation in the *bmp2* gene *swirl* exhibit an expansion of dorsal endoderm at gastrulation and thus of future anterior endoderm derivatives[82] (Bally–Cuif *et al.*, unpublished). Second, Bally–Cuif *et al.* have analyzed a more discrete patterning event, with the help of a marker specific for a subregion of dorsal endoderm in the early gastrula, *her5*.[7] As expected from a hairy-related gene, *her5* expression is regulated by the Notch pathway. However this regulation is negative and *her5* expression in dorsal endoderm results from the local inhibition of Notch signaling. In addition, *her5* exhibit a patterning activity with regard to dorsal endoderm. Dorsal endoderm in fish has the peculiarity to generate both the most extreme part of endoderm along the AP axis, e.g. the mesendodermal most anterior prechordal plate (PP), the intermediate anterior pharyngeal endoderm and the most posterior forerunner (FR) cells, eventually located behind the posterior tip of the gut. Overexpression of dominant gain and loss-of-function Her5 isoforms have shown that *her5* represses the most extreme anterior PP and posterior FR endodermal fates to allow the development of intermediate pharyngeal endoderm. These two reports clearly show that some endoderm patterning events occur during gastrulation. Nodals have also been implicated in controlling AP patterning,[53,83] with high levels of Nodal signaling inducing anterior structures and low levels inducing posterior structures. However, varying the dose of Nodal signaling does not appear to have a clear influence on the degree of endoderm AP patterning.[70]

Thus, patterning of endodermal cells into respiratory or digestives tracts tissues appears to initiate at the onset of gastrulation, but later patterning events are expected to occur, as suggested by the relative scarcity of regional endodermal markers during gastrulation and experiments carried out in other vertebrate species.

3.6. Convergence and Migratory Behavior of Endodermal Cells

At the onset of gastrulation, endodermal and mesodermal precursors concomitantly involute in a coordinate fashion. When they have reached the YSL, endodermal cells flatten, develop filopodia and migrate onto the YSL.[6] Even though detailed fate mapping experiments would be required, the behavior of the ventro-lateral endodermal cells can be inferred from the expression patterns of *cas* and *sox17* during gastrulation. These cells are scattered and keep isolated from each other in gastrulae, but at the end of gastrulation, they all migrate in a coordinate way to reach the dorsal axis. Although they are clearly initiated by Nodal signals, almost nothing is known about the mechanisms controlling these movements. What triggers the transition between early resting endodermal progenitors and the development of motility during gastrulation? What are the cues controlling their change in morphology, their stereotyped behavior during gastrulation (involution, convergence)? Do cells rely on chemoattraction or chemorepulsion? Is their movement dependent on the movements of neighboring mesodermal or ectodermal cells, or on cues provided by them? Do they follow the same rules as mesoderm and ectoderm to converge? All these intriguing questions have to be addressed to unravel the migratory behavior of the endodermal cells during gastrulation and to understand how gastrulation can be achieved in such a highly coordinated fashion.

Conclusion

In the past few years, results in fish and in other species have widely improved our understanding of endoderm formation. In particular, genetic analyses have confirmed *Xenopus* results identifying the Nodal pathway as a key regulator of endoderm formation, and have identified many of its downstream components. However, some more investigations are needed to better understand how Nodal signaling controls endoderm formation. In particular, all the Nodal targets identified so far are transcription factors. But which effector genes are activated by these regulators remains to be established. Other transcription factors leading to a progressive commitment of cells to an endodermal fate are likely to be part of them. But these targets most probably also include effectors genes directly controlling cell behavior and cell movements. Identifying these targets will be of particular interest and due to its optical clarity, zebrafish should prove a particularly convenient tool to tackle these issues. Another gap in our knowledge concerns the later development of endoderm: how endodermal organs such as pancreas or liver form. This question has been so far poorly studied in fish but will certainly be addressed in the coming years since fish appears again as a very attractive model system to answer these questions. The optical clarity of the embryo allows direct observation of the forming organs. The establishment of stable transgenic lines expressing GFP in endodermal cells will even facilitate these observations, thus allowing an easy screening for mutants affecting organ formation. A better understanding of these late differentiation steps, together with our knowledge of early endoderm specification will probably allow exciting attempts to drive stem cells to form endodermal tissues, offering the possibility to form endodermal organs *de novo.*

References

1. David NB and Rosa FM (2001). *Development* **128**: 3937.
2. David NB, Saint-Etienne L, Tsang M, Schilling TF and Rosa FM (2002). *Development* **129**: 4457.
3. Peyrieras N, Strahle U and Rosa F (1998). *Curr. Biol.* **8**: 783.
4. Solnica–Krezel L, Stemple DL and Driever W (1995). *Bioessays* **17**: 931.
5. Kimmel CB, Warga RM and Schilling TF (1990). *Development* **108**: 581.
6. Warga RM and Nusslein–Volhard C (1999). *Development* **126**: 827.
7. Bally–Cuif L, Goutel C, Wassef M, Wurst W and Rosa F (2000). *Genes Dev.* **14**: 1664.
8. Warga RM and Kimmel CB (1990). *Development* **108**: 569.
9. Trinkaus JP (1996). *Dev. Biol.* **177**: 356.
10. Carmany–Rampey A and Schier AF (2001). *Curr. Biol.* **11**: 1261.
11. Shih J and Fraser SE (1995). *Development* **121**: 2755.
12. Ober EA, Field HA and Stainier DY (2003). *Mech. Dev.* **120**: 5.
13. Horne–Badovinac S, Lin D, Waldron S, Schwarz M, Mbamalu G, Pawson T, Jan Y, Stainier DY and Abdelilah–Seyfried S (2001). *Curr. Biol.* **11**: 1492.
14. Yasuo H and Lemaire P (1999). *Curr. Biol.* **9**: 869.
15. Weeks DL and Melton DA (1987). *Cell* **51**: 861.
16. Thomsen G, Woolf T, Whitman M, Sokol S, Vaughan J, Vale W and Melton DA (1990). *Cell* **63**: 485.
17. Jones CM, Kuehn MR, Hogan BL, Smith JC and Wright CV (1995). *Development* **121**: 3651.
18. Zhou X, Sasaki H, Lowe L, Hogan BL and Kuehn MR (1993). *Nature* **361**: 543.
19. Conlon FL, Lyons KM, Takaesu N, Barth KS, Kispert A, Herrmann B and Robertson EJ (1994). *Development* **120**: 1919.
20. Joseph EM and Melton DA (1997). *Dev. Biol.* **184**: 367.

21. Piccolo S, Agius E, Leyns L, Bhattacharyya S, Grunz H, Bouwmeester T and De Robertis EM (1999). *Nature* **397**: 707.
22. Agius E, Oelgeschlager M, Wessely O, Kemp C and De Robertis EM (2000). *Development* **127**: 1173.
23. Osada SI and Wright CV (1999). *Development* **126**: 3229.
24. Schulte-Merker S, Smith JC and Dale L (1994). *Embo J* **13**: 3533.
25. Matzuk MM, Kumar TR, Vassalli A, Bickenbach JR, Roop DR, Jaenisch R and Bradley A (1995). *Nature* **374**: 354.
26. Dyson S and Gurdon JB (1997). *Curr. Biol.* 7: 81.
27. Summerton J and Weller D (1997). *Antisense Nucleic Acid Drug Dev.* 7: 187.
28. Nasevicius A and Ekker SC (2000). *Nat. Genet.* **26**: 216.
29. Erter CE, Solnica-Krezel L and Wright CV (1998). *Dev. Biol.* **204**: 361.
30. Feldman B, Gates MA, Egan ES, Dougan ST, Rennebeck G, Sirotkin HI, Schier AF and Talbot WS (1998). *Nature* **395**: 181.
31. Rebagliati MR, Toyama R, Haffter P and Dawid IB (1998). *Proc. Nat. Acad. Sci. USA* **95**: 9932.
32. Sampath K, Rubinstein AL, Cheng AM, Liang JO, Fekany K, Solnica-Krezel L, Korzh V, Halpern ME and Wright CV (1998). *Nature* **395**: 185.
33. Hatta K, Kimmel CB, Ho RK and Walker C (1991). *Nature* **350**: 339.
34. Massague J and Chen YG (2000). *Genes Dev.* **14**: 627.
35. Harrison CA, Gray PC, Koerber SC, Fischer W and Vale W (2003). *J. Biol. Chem.*
36. Reissmann E, Jornvall H, Blokzijl A, Andersson O, Chang C, Minchiotti G, Persico MG, Ibanez CF and Brivanlou AH (2001). *Genes Dev.* **15**: 2010.
37. Renucci A, Lemarchandel V and Rosa F (1996). *Development* **122**: 3735.
38. Gu Z, Nomura M, Simpson BB, Lei H, Feijen A, van den Eijnden-van Raaij J, Donahoe PK and Li E (1998). *Genes Dev.* **12**: 844.
39. Aoki TO, Mathieu J, Saint-Etienne L, Rebagliati MR, Peyrieras N and Rosa FM (2002). *Dev. Biol.* **241**: 273.
40. Gritsman K, Zhang J, Cheng S, Heckscher E, Talbot WS and Schier AF (1999). *Cell* **97**: 121.
41. Yeo C and Whitman M (2001). *Mol. Cell* 7: 949.
42. Minchiotti G, Manco G, Parisi S, Lago CT, Rosa F and Persico MG (2001). *Development* **128**: 4501.

43. Minchiotti G, Parisi S, Liguori G, Signore M, Lania G, Adamson ED, Lago CT and Persico MG (2000). *Mech. Dev.* **90**: 133.
44. Minchiotti G, Parisi S, Liguori GL, D'Andrea D and Persico MG (2002). *Gene* **287**: 33.
45. Yan YT, Liu JJ, Luo Y, E C, Haltiwanger RS, Abate–Shen C and Shen MM (2002). *Mol. Cell Biol.* **22**: 4439.
46. Yabe SI, Tanegashima K, Haramoto Y, Takahashi S, Fujii T, Kozuma S, Taketani Y and Asashima M (2003). *Development* **130**: 2071.
47. Zhang J, Talbot WS and Schier AF (1998). *Cell* **92**: 241.
48. Rosa FM (2002). *Sci STKE* **2002**: PE47.
49. Meno C, Shimono A, Saijoh Y, Yashiro K, Mochida K, Ohishi S, Noji S, Kondoh H and Hamada H (1998). *Cell* **94**: 287.
50. Meno C, Gritsman K, Ohishi S, Ohfuji Y, Heckscher E, Mochida K, Shimono A, Kondoh H, Talbot WS, Robertson EJ, Schier AF and Hamada H (1999). *Mol. Cell* **4**: 287.
51. Thisse B, Wright CV and Thisse C (2000). *Nature* **403**: 425.
52. Meno C, Takeuchi J, Sakuma R, Koshiba–Takeuchi K, Ohishi S, Saijoh Y, Miyazaki J, ten Dijke P, Ogura T and Hamada H (2001). *Dev. Cell* **1**: 127.
53. Thisse C and Thisse B (1999). *Development* **126**: 229.
54. Randall RA, Germain S, Inman GJ, Bates PA and Hill CS (2002). *EMBO J.* **21**: 145.
55. Germain S, Howell M, Esslemont GM and Hill CS (2000). *Genes Dev.* **14**: 435.
56. Yamamoto M, Meno C, Sakai Y, Shiratori H, Mochida K, Ikawa Y, Saijoh Y and Hamada H (2001). *Genes Dev.* **15**: 1242.
57. Chen X, Weisberg E, Fridmacher V, Watanabe M, Naco G and Whitman M (1997). *Nature* **389**: 85.
58. Pogoda HM, Solnica–Krezel L, Driever W and Meyer D (2000). *Curr. Biol.* **10**: 1041.
59. Sirotkin HI, Gates MA, Kelly PD, Schier AF and Talbot WS (2000). *Curr. Biol.* **10**: 1051.
60. Strahle U, Blader P, Henrique D and Ingham PW (1993). *Genes Dev.* 7: 1436.
61. Kikuchi Y, Trinh LA, Reiter JF, Alexander J, Yelon D and Stainier DY (2000). *Genes Dev.* **14**: 1279.

62. Poulain M and Lepage T (2002). *Development* **129**: 4901.
63. Reiter JF, Alexander J, Rodaway A, Yelon D, Patient R, Holder N and Stainier DY (1999). *Genes Dev.* **13**: 2983.
64. Sakaguchi T, Kuroiwa A and Takeda H (2001). *Mech. Dev.* **107**: 25.
65. Kikuchi Y, Agathon A, Alexander J, Thisse C, Waldron S, Yelon D, Thisse B and Stainier DY (2001). *Genes Dev.* **15**: 1493.
66. Dickmeis T, Mourrain P, Saint–Etienne L, Fischer N, Aanstad P, Clark M, Strahle U and Rosa F (2001). *Genes Dev.* **15**: 1487.
67. Lemaire P, Darras S, Caillol D and Kodjabachian L (1998). *Development* **125**: 2371.
68. Rosa FM (1989). *Cell* **57**: 965.
69. Alexander J, Rothenberg M, Henry GL and Stainier DY (1999). *Dev. Biol.* **215**: 343.
70. Aoki TO, David NB, Minchiotti G, Saint–Etienne L, Dickmeis T, Persico GM, Strahle U, Mourrain P and Rosa FM (2002). *Development* **129**: 275.
71. Alexander J and Stainier DY (1999). *Curr. Biol.* **9**: 1147.
72. Spemann H (1938). *(Reprinted in 1967 by Hafner, New York).*
73. Stainier DY (2002). *Genes Dev.* **16**: 893.
74. Shivdasani RA (2002). *Dev. Biol.* **249**: 191.
75. Dougan ST, Warga RM, Kane DA, Schier AF and Talbot WS (2003). *Development* **130**: 1837.
76. Rodaway A, Takeda H, Koshida S, Broadbent J, Price B, Smith JC, Patient R and Holder N (1999). *Development* **126**: 3067.
77. Chen Y and Schier AF (2001). *Nature* **411**: 607.
78. Muskavitch MA (1994). *Dev. Biol.* **166**: 415.
79. Chitnis AB (1995). *Mol. Cell Neurosci.* **6**: 311.
80. Sherwood DR and McClay DR (2001). *Development* **128**: 2221.
81. Odenthal J and Nusslein–Volhard C (1998). *Dev. Genes Evol.* **208**: 245.
82. Tiso N, Filippi A, Pauls S, Bortolussi M and Argenton F (2002). *Mech. Dev.* **118**: 29.
83. Gritsman K, Talbot WS and Schier AF (2000). *Development* **127**: 921.

Chapter 13

Gene 'Knockdown' Approaches Using Unconventional Antisense Oligonucleotides

Eleanor Chen,[1,2] Perry B. Hackett[1,3] and Stephen C. Ekker[1]

[1]*Arnold and Mabel Beckman Center for Transposon Research Department of Genetics, Cell Biology and Development, University of Minnesota, 6-160 Jackson Hall, 321 Church Street SE, Minneapolis, MN 55455, USA*

[2]*Combined MD/PhD and Molecular, Cellular, Developmental Biology and Genetics Graduate Program, University of Minnesota, 6-160 Jackson Hall, 321 Church Street SE, Minneapolis, MN 55455, USA*

[3]*Discovery Genomics, Inc., 614 McKinley Place, NE, Minneapolis, MN 55413*

Derived from therapeutic tools developed for the clinic, unconventional antisense technology has emerged as a new and broadly applicable RNA-based gene inhibition approach. Its targeting mechanisms of action are both RNaseH-independent and distinct from other sequence-based tools such as small inhibitory RNAs. Among the novel classes of antisense oligonucleotides, morpholino phosphorodiamidate oligonucleotides (morpholinos) have emerged as the preferred effector molecules of targeted gene "knockdown" strategies. Morpholinos have been shown to be extremely effective, specific and convenient for elucidating gene functions in a variety of model systems. This chapter provides an overview on technical aspects of morpholino usage and some examples of the many potential biological applications of this technology ranging from human disease modeling to functional genomics.

Corresponding author: Dr. Stephen C. Ekker
Address: Department of Genetics, Cell Biology and Development
University of Minnesota, 6-160 Jackson Hall, 321 Church Street SE
Minneapolis, MN 55455, USA
Tel: (612) 626-4509
Fax: (612) 626-7031
E-mail: ekker001@mail.med.umn.edu

1. Introduction

RNase-H-mediated antisense gene inhibition (conventional antisense) utilizes first generation antisense oligonucleotides such as phosphorothioate-linked DNA for a wide variety of applications, including use as a clinical therapeutic agent.[1] Broad adoption of these compounds has been hampered by the relatively low targeting specificity, modest efficacy, and significant sequence-independent effects which reduce their utility to very specific applications.[2] Phosphorothioate-linked antisense oligos are therefore not suitable for large-scale study of gene function. In contrast, morpholino phosphorodiamidate (MO)-based antisense oligonucleotides, artificial DNA analogs first designed with the clinic in mind, have been shown to exhibit higher gene targeting efficacy compared to phosphorothioate-based DNA oligos in cultured cells.[3] Subsequent studies in many model organisms have demonstrated MOs as an effective antisense approach *in vivo*.[4–6] The high targeting efficacy and specificity of morpholinos allow for their potential use from clinical to genomic applications. This chapter surveys technical aspects of morpholino usage in different model systems and applications in many areas of scientific endeavors.

2. Mechanisms of Unconventional Antisense Targeting

A large number of DNA and RNA analogs have been developed over the past two decades. The major goal of these studies has been to generate a readily-manufactured agent for antisense gene inhibition studies. Conventional antisense works through an RNase-H-mediated RNA/DNA duplex, and many groups have focused their chemistry to the subset of DNA analogs that can be recognized by this enzyme.[1] The approach was to develop a polymer with long-term and specific biological efficacy. Unfortunately, *in vivo* stability was usually inversely proportional to activity, with the longest-lasting agents completely unrecognized by most cellular machinery (see examples below).

Phosphorothioate-based oligonucleotides, one of the lead first generation compounds, are nuclease resistant, but at the cost of lower targeting specificity and efficacy due to their reduced RNA-binding ability; indeed, these agents bind RNA more poorly than DNA-based polymers, and they bind a large number of proteins resulting in significant sequence-independent effects *in vivo*.[7] Despite these challenges, effective gene inhibition in specific applications has been achieved through the identification of sequences with especially high specific activity that allows the use of these compounds at relatively low cellular concentrations.[1]

The development of DNA analogs with very long cellular half-life and no RNase-H activation potential lead to the exploration for effective novel gene inhibition approaches. These "unconventional antisense" studies have demonstrated extremely high efficacy, good specificity, and low toxicity. Each polymer has a higher affinity for RNA than RNA itself, allowing these agents to invade RNA:RNA secondary structure and providing a high reliability in effective, oligonucleotide sequence design. Examples include the polyamide nucleic acids (PNAs),[8] their charged derivative the trans-4-hydroxy-L-proline/phosphonate polyamides (HypNA-pPNA),[9] and morpholino phosphorodiamidate oligonucleotides (MOs).[10] Each of these compounds, the PNAs, the HypHA-pPNAs[11] and MOs,[12] have all been shown to be effective for use in gene-targeting in the model teleost, *Danio rerio* (the zebrafish), an ideal system for performing *in vivo* gene inhibition studies.[13] The richest literature is found using MOs, with multiple gene targeting approaches and detailed efficacy and specificity studies now available. These approaches would in principle apply to any new generation antisense compound. This review is focused on MO development and use, exemplifying the rich array of approaches that have been taken with this class of "unconventional antisense" molecules.

2.1. Morpholino Phosphorodiamidate Oligonucleotides

Morpholinos (MOs) are synthetic antisense oligonucleotides with a 6-membered morpholine ring in place of the deoxyribose moiety (Fig. 1).[10] MOs contain a neutral charge backbone due to its phosphorodiamidate

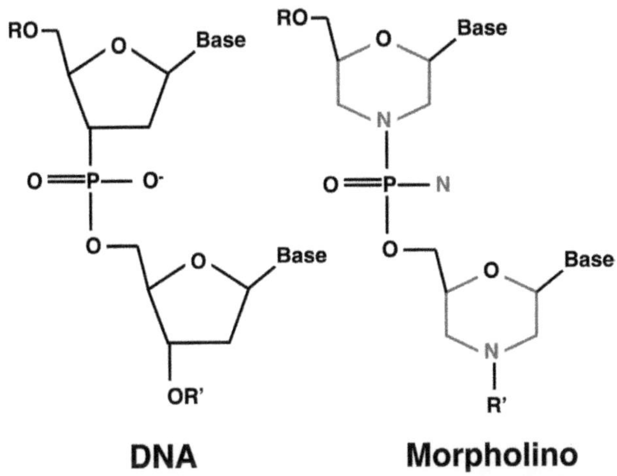

Fig. 1 Structures of DNA and morpholino oligonucleotides. Unlike DNA, morpholino has a 6-membered morpholine ring and a neutral phosphorodiamidate linkage. R, 5' end; R', 3'end.

linkage and have high affinity for RNA.[14] In contrast to phosphorothioate-linked DNA analogs, MOs appear resistant to nuclease cleavage.[14]

Morpholinos were first used for clinical application as inhibitors of the *c-myc* gene.[15] In this study, *c-myc* antisense MOs were extremely effective at reducing protein levels. Subsequent analysis demonstrated that this pilot therapeutic agent achieved its potency through two independent mechanisms: the blocking of translation and the alteration of pre-mRNA splicing.[16] To date, these are still the primary mechanisms of action for unconventional antisense strategies.

2.2. Translational Inhibition

Morpholinos are effective at blocking translation when targeted against sequences in the 5′ untranslated region and about 25 bases past the translational initation site of mRNA (Fig. 2). MOs targeted to sites beginning more than 20 bases 3′ to the AUG translational start site show no significant translation inhibition using a cell-free assay system.[3]

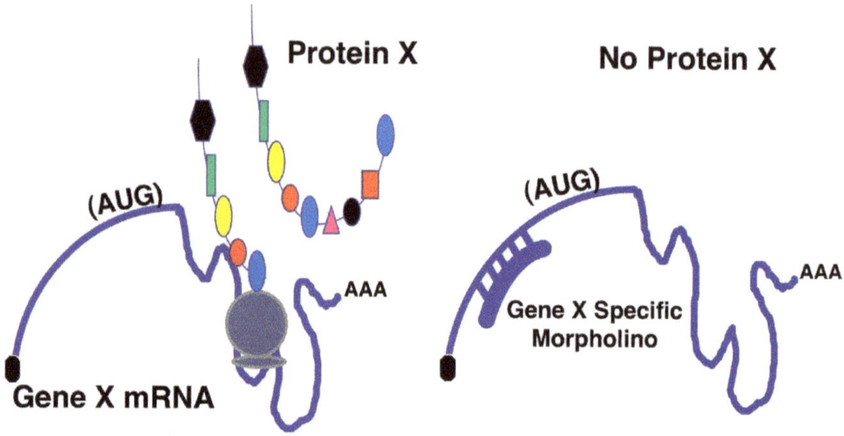

Fig. 2 Morpholinos can function as translational blockers. Morpholinos targeted against the 5′-untranslated region of mRNA and about 25 bases past the translational start site block the access of the ribosome to the mRNA and consequently the translation of the gene product.

Multiple oligonucleotide target sequences in the same gene have been shown to be effective for translational inhibition approaches,[4] and dual oligonucleotide targeting is further used as a specificity test (see below).

2.3. Alteration of Pre-mRNA Splicing

Disruption of gene function by MOs can also be achieved by targeting against exon-intron junctions (Fig. 3). The most commonly employed tactic is to design the MO to bind the splice donor site to induce aberrant pre-mRNA splicing, resulting in either exon skipping or the use of a cryptic splice donor site (Fig. 3).[17] Aberrantly spliced transcripts often encode a non-functional gene product, generating phenotypes, a subset of which are similar to those generated by MOs acting as translational blockers (Fig. 4). Alternatively, splice acceptor targeting can result in mRNA splice forms which contain sequences normally intronic in wild-type messages (Fig. 4D). One major difference between splice-site and translational inhibition targeting approaches is especially

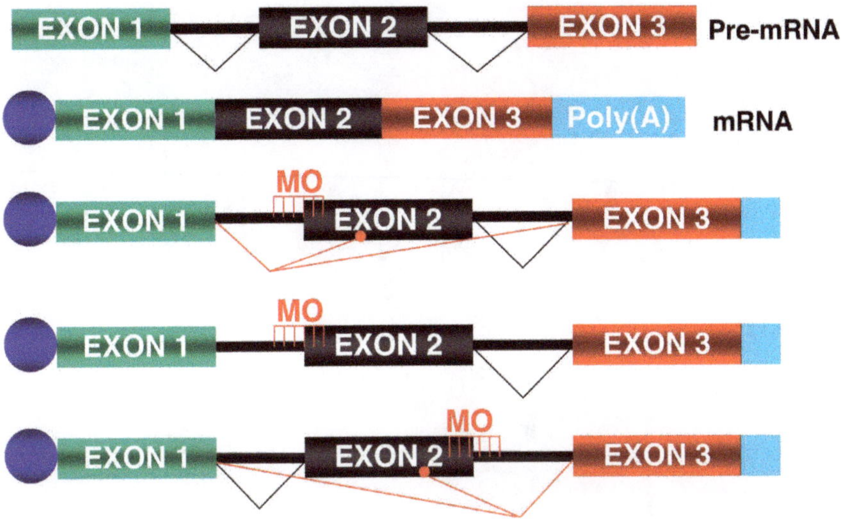

Fig. 3 Morpholinos can alter pre-mRNA splicing. Morpholinos targeted against the intron/exon or exon/intron boundaries can cause abnormal splicing of the targeted transcript, as the splicing machinery either might select a cryptic splice site in the adjacent exon or skip the exon completely (lines 3 and 5) or skip the splice site completely resulting in a message with an incorporated intron (line 4; see also Fig. 4).

noteworthy for organisms where a large pool of maternal RNAs contributes to early processes in embryogenesis. Translational-blocking oligonucleotides are capable of inhibiting maternal messages as well as later transcripts[2,12] whereas splice-site targeting approaches only affect messages generated after transcription from the zygotic genome has initiated.[17]

2.3. Morpholino Delivery and Distribution

In model systems such as frog, sea urchin and zebrafish, morpholinos are commonly introduced by microinjection.[2,12,18] As demonstrated by the use of fluorescein isothiocyanate (FITC)-labeled morpholinos in zebrafish and frog embryos, morpholinos can spread from the injection site and be segregated from cell to cell during mitosis, resulting in uniform distribution

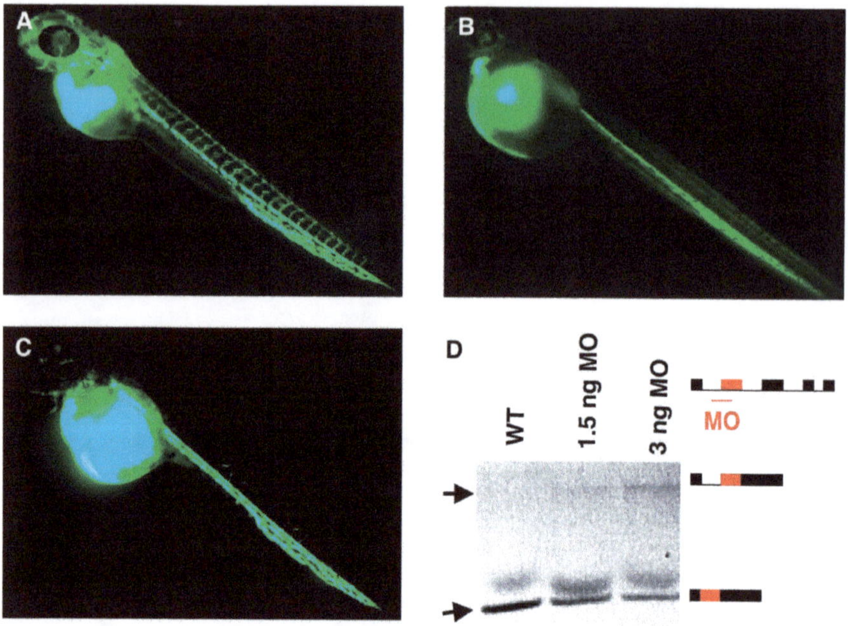

Fig. 4 Splice-site morpholino targeting. Morpholinos targeted against splice junctions often generate phenotypes similar to those generated by morpholinos targeted against the region around the translational start site of the same gene. An MO targeted against the initiation codon of zebrafish *syndecan-2* specifically generated embryos with vascular defects in the trunk, a defect that can be readily visualized by microangiography analysis (compare B to the wild-type embryo in A; Chen *et al.*[49]). An MO directed MO against a *syndecan-2* splice site also generated embryos with vascular defects similar to those of a weak loss-of-function phenotype due to the use of a translational blocking *syndecan-2* MO (C). (D) In these embryos, the splice-site-directed morpholino resulted in transcripts which included an intron. An RT-PCR analysis of embryos injected with the indicated MO is shown next to a cartoon of the genomic structure of the *syndecan-2* gene. Note the reduction in wild-type RNA (indicated by the bottom arrow) and the concomitant appearance of a novel, intron-containing transcript (marked by the top arrow).

throughout the embryo.[12,19] In chick embryos, efficient MO delivery has been observed using the method of electroporation.[20] Other physical methods such as scrape-loading were first used to enhance the uptake of morpholinos by cultured cells.[3] Due to their neutral backbone, morpholinos are poorly delivered into cultured mammalian cells via traditional lipid-based delivery systems. Subsequently, delivery agents such as ethoxylated polyethylenimine (EPEI) have been modified using DNA as a carrier bridge for morpholino loading and effective delivery in tissue culture cells.[21] Delivery agents such as lipofectin and lysolecithin have also been used for delivery into cultured pre-implantation mouse embryos,[22] suggesting that in some applications the direct coupling of the morpholino to the carrier agent is not always required to achieve effective delivery.

3. Duration of Morpholino Effects

Antisense oligonucleotide gene inhibition strategies are inherently transient in nature. Once delivered, the *in vivo* concentration will diminish with each cell division, and upon degradation, filtration or sequestration of the active agent. The duration of these effects will be a function of all these factors and is a major constraint on the general utility of antisense for specific applications. This function has been extensively measured for morpholino gene targeting studies in the zebrafish.

3.1. Translational Blockers

To assess the persistence of morpholino effects, phenotypes resulting from morpholino targeting can be observed over time until the severity of the same phenotypes begin to diminish. An early study in zebrafish has shown that the effects of morpholino can persist at least throughout the first two days of development, during which early patterning events and fundamental organogenesis occur.[12] The same study showed that the effects of MOs directed against late acting genes essential for pigmentation such as *nacre* and *sparse* persist through the first 50 hours and 10 days of development, respectively.[12] No known limitations on MO effectiveness have thus been noted within the period of embryogenesis with the zebrafish model. For some genes, this estimate of perdurance is complicated by the potential that a particular visible phenotypic consequence may only appear until long after a critical gene inhibition time point. Work targeting genes with measurable enzymatic activity strongly suggests that morpholinos can inhibit gene function through day four and beyond (Essner and Clark, personal communication). The design of longer-lasting oligonucleotides appears to be dependent upon many factors including optimizing for efficacy and minimizing non-specific side-effects, which can reduce long-term viability of the embryos.

3.2. Splice-Site Targeting

The persistence of effects generated by splice-directed MOs has also been assessed and shows a similar range of duration *in vivo*. A splice-directed MO against zebrafish *endothelin* mimicked the phenotypes resulting from loss-of-function mutations at 5–6 days of development.[23] However, another study showed that the inhibitory effect of the splice-directed MO on *sox9a* mRNA splicing is diminished by 4 days of development, as evidenced by an increase in the amount of normal-sized transcript and a concomitant decrease in the amount of aberrantly spliced transcript.[24] Although less well characterized, the duration of splice-site targeting appears to be comparable to noted for translational blockers.

4. Widespread Use and Limitations of Morpholino Targeting

4.1. Model Systems

Morpholinos have been shown to inhibit translation of mRNA *in vitro*.[3] Effective gene targeting by morpholinos has also been shown in a wide range of model organisms including two species of frog *X. laevis*[2] and *X. tropicalis*,[19] chicken *G. gallus*,[25] zebrafish *D. rerio*,[12] sea urchin *S. purpuratus*,[18] ascidian *Ciona savignyi*,[26] mouse oocytes,[27] leech *Helobdella robusta*,[28] fly *D. melanogaster* and nematode *Oscheius sp.*1.[29] Indeed, work has now progressed to human trials for clinical studies.[30] The major summary of this data suggests that when delivery was achieved, effective gene targeting was noted. The general utility of morpholinos will increase as delivery technology progresses in each model system.

4.2. Mistargeting

Antisense molecules long been known to be limited by unpredictable and undesirable effects on cellular processes above and beyond the targeting of the selected gene product.[7] Indeed, non-specific gene inhibition has been demonstrated with RNA interference approaches.[31] Characterizing these undesirable effects is critical for the utility of sequence-based gene inhibition technologies.

Mistargeting by morpholinos has been best documented when MOs fail to phenocopy genes with known loss-of-function phenotypes, e.g. MOs against *dharma* and *pax-2.1* in zebrafish.[4] Commonly observed mistargeting phenotypes in zebrafish include localized and transient cell death in areas surrounding developing eyes and ventricles in the developing brain in the less severe case. In more extreme cases, a variety of abnormalities including severely reduced body axis and cell death at each somite boundary can be observed.[4] A study in *Xenopus* showed that MOs designed to inhibit translation of GFP in transgenic lines harboring the GFP reporter effectively blocked GFP fluorescence at lower doses, but generated non-specific phenotypes such as anterior-posterior truncations and microcephaly at higher doses.[19] Thus, the effects of mistargeting are both sequence- and dose-dependent, strongly suggesting they are due to "off-target" inhibition of genes of related sequence. Rules for antisense oligonucleotide design for maximizing efficacy while minimizing mistargeting effects are still being refined but are likely to include strategies for gene-specific target sequence identification that is unique within the genome and transcriptome of that model organism.

5. Efficacy and Specificity Tests for Morpholino Targeting

5.1. Efficacy Measurements

In zebrafish, the efficacy of gene targeting by morpholinos has been demonstrated by phenocopies of known mutations such as *chordin*, *swirl* and *cyclops*,[12,32,33] and potentiation of specific phenotypes in mutants, e.g. *gridlock*[35] upon introduction of morpholinos. Introduction of MOs at different doses often result in phenotypes of graded severity, mimicking both hypomorphic and null mutations as evidenced by morphant phenocopies of weak and strong alleles of *chordin* and *cyclops*.[12,33]

5.1.1. *Translational inhibition*

To test the efficacy of morpholino targeting, several quantitative assays can be performed. If antibodies specific for target proteins are available, Western-blot or related protein detection analyses can be performed. Dose-dependent decrease in the target protein level is observed if the morpholino knockdown is effective.[2,12]

Alternatively, a surrogate reporter assay can also be designed to assess the efficacy of morpholino targeting (Fig. 5). Messenger RNA is first synthesized *in vitro* from a DNA expression construct in which the 5′ untranslated region containing the MO targeting sequence is fused in frame to a reporter gene, e.g. green fluorescent protein. The chimeric mRNA and the morpholino are then introduced into the model system. If the morpholino blocks translation effectively, the reporter gene expression level should decrease in a morpholino-dose-dependent manner. To increase the stringency of the efficacy test, one can show that the reporter gene expression is retained in the presence of a control morpholino targeted against another sequence not found in the fusion construct. A potential disadvantage of this test is that the accessibility of the chimeric reporter mRNA may not mimic that of the endogenous mRNA.

5.1.2. *Transcriptional processing inhibition*

To assess the efficacy of targeting by splice site-directed MOs, RT-PCR analysis can be performed to assess the presence of aberrantly spliced transcripts in MO-injected embryos.[17] In addition to assessment of final RNA processing levels, one study recently showed that the introduction of a splice-inhibiting MO resulted in the accumulation of the modified transcript in the nucleus while the wild-type transcript

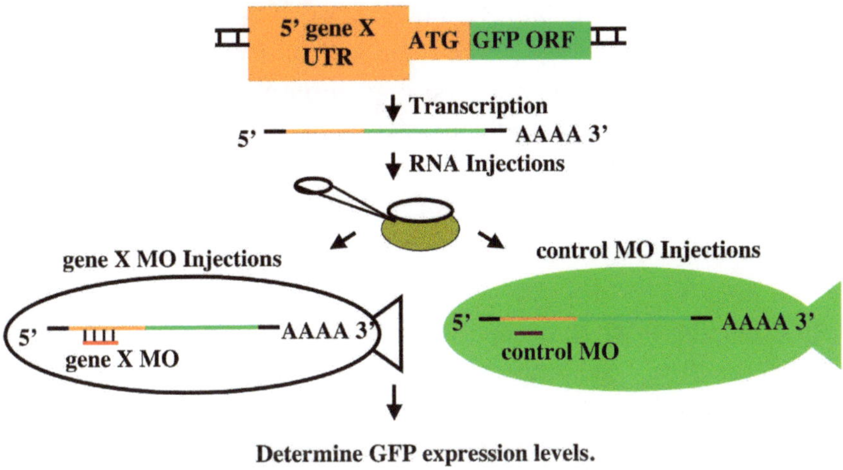

Fig. 5 Morpholino efficacy test. To assess the efficacy of morpholino targeting, chimeric mRNA is synthesized *in vitro* from an expression construct in which the 5' untranslated region including morpholino targeting sequence is fused in frame with a reporter gene (green fluorescent protein in this example). This mRNA along with either the test MO or a control, is introduced into the model system (such as a zebrafish embryo) and the reporter expression subsequently determined as a measure of the inhibition potential of the experimental oligonucleotide for a selected gene sequence.

was found in the cytoplasm.[24] The detection of the accumulation of a transcript in the nucleus presents as a novel assay for assessing the targeting efficacy of splice-directed MOs.

5.2. Specificity Assessments

5.2.1. *Mutant phenocopy*

One potential method now possible in zebrafish for confirming any identified phenotype from morpholino knockdown analyses would be to compare loss-of-function phenotypes with that of a mutant. One potential source of mutations for a given gene, a series of insertional alleles using a retrovirus, has been generated recently with annotation primarily at the gene sequence level.[34] Another strategy would be to generate an allele in the gene of interest through the analysis of a collection of chemically-induced mutations.[36] These two approaches are limited by their coverage of the genome and cost, respectively, and thus currently represent only rare opportunities for direct phenocopy comparisons of morphant phenotypes.

5.2.2. *Multi-oligonucleotide targeting*

While the efficacy tests described in section 5.1 reveal effective targeting of the gene of interest by a MO, they do not exclude non-specific knockdown of other genes by the same MO. Additional specificity tests can be performed to ensure that phenotypes generated by MOs are specific to the loss of endogenous gene functions. One simple test is to assess whether MOs targeted against independent sequence of the same gene generate similar phenotypes. Synergistic interaction between two gene-specific MOs of non-overlapping sequence in generating knockdown phenotypes has been observed (Fig. 6).[4] The synergy is generally be abolished if a 4-base mismatch is introduced into one of the MOs. Alternatively, a control MO targeted against another gene should not potentiate the MO effects observed.

5.2.3. *"Rescue" with exogenous gene product*

If MO targeting results in specific loss of endogenous gene functions, then the phenotypes observed can be ameliorated with the addition of exogenous target gene products. The reintroduction of specific gene products can sometimes be difficult to achieve with the same temporal and/or spatial restrictions normally placed on a specific gene product due to the limitations inherent in RNA and DNA injection, the most common methods of delivering gene activity.

In addition, many MOs are designed against the region overlapping the translational start site of the target gene. To avoid binding of the morpholino to the same sequence in the mRNA generated from the rescue expression construct, one can take advantage of the degeneracy of the genetic code by introducing silent mutations behind the translational start site. This modified open reading frame is thus suitable for "rescue" studies as it will not be an effective target of morpholino-based inhibition.

6. Applications of Morpholinos

6.1. Therapeutic Agents

Several studies have also demonstrated MOs as promising therapeutic agents for human diseases. One study showed that the growth rate of Lewis lung carinoma established in mice decreased upon treatment combing chemotherapeutic drugs and MO targeted against *c-myc*.[37] In addition, administration of MO against *c-myc* in four different animal models resulted in a reduction of vessel wall thickening, showing promise as a therapeutic option for restenosis, a phenomenon occasionally observed following coronary interventions.[38] Splice-blocking MOs also have potential therapeutic uses. For example, an MO targeted against an aberrant 5′ splice site can repair a splicing defect in erythroid cells from patients with β-thalassemia in a dose-dependent manner.[39]

6.2. Modeling of Biological Processes and Human Diseases

Morpholinos are highly effective at "knocking down" the expression of specific genes in zebrafish.[12] Zebrafish eggs are fertilized externally and embryos are optically clear, allowing easy visualization of a variety of developmental events during embryogenesis. Moreover, essential biological processes in vertebrates such as cardiovascular development and hematopoiesis are highly conserved in zebrafish.[40–42] Efficient gene inhibition by morpholinos in zebrafish facilitates the study of conserved vertebrate processes. For example, vascular endothelial growth factor or VEGF is essential for proper vascular formation.[43,44] Zebrafish VEGF morphants show similar vascular phenotypes as observed in VEGF-deficient mice,[13] demonstrating the feasibility of zebrafish as a model system for vascular development.

Phenotypes generated due to morpholino-based gene targeting can also be used to model human diseases. For example, morpholinos targeted against *sonic hedgehog* and *tiggy-winkle hedgehog* in zebrafish resulted in partial cyclopia and other midline abnormalities, mimicking holoprosencephaly in humans.[12] A splice-donor blocking morpholino

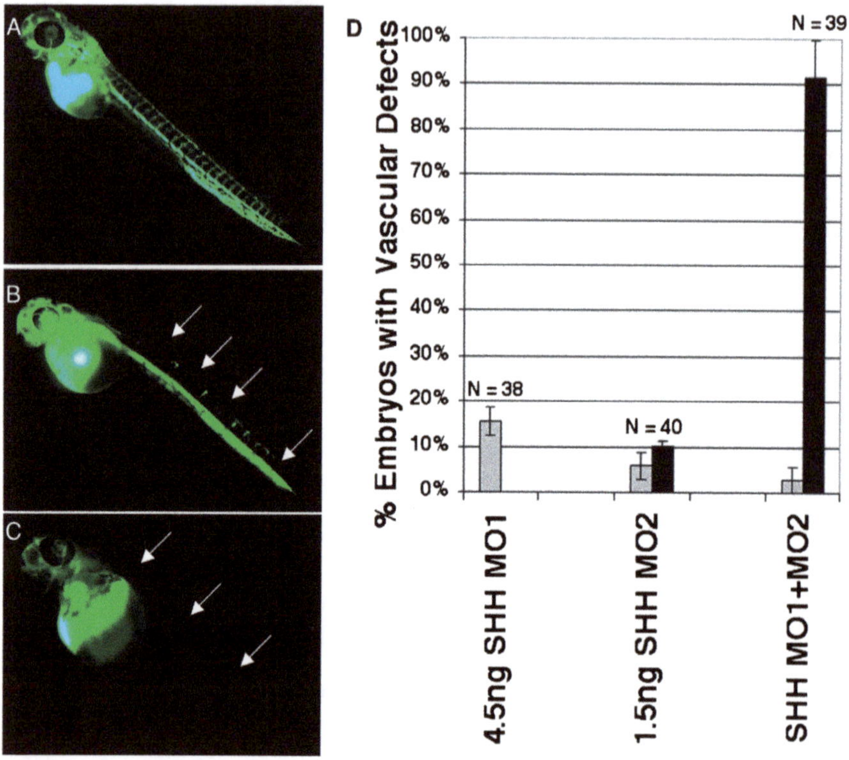

Fig. 6 Synergistic inhibition by two MOs targeted to the same gene results in both an increase in penetrance and an increase in severity of gene inhibition. Two independent morpholinos against the *sonic hedgehog* gene[4] were injected into zebrafish embryos and assayed for the loss of *hedgehog* gene function in vascular development (panels A–C). Introduction of a small dose (1.5–4.5 ng) or either MO alone resulted in defects restricted to effects on vascular sprouting in these embryos and at low penetrance (panel B and grey bars, panel D). When both are injected, however, the embryos display a much higher penetrance of the vascular phenotypes (from ~15% to > 90%) as well as a much greater increase in the severity of the phenotype (panel C, black bars in panel D).

targeted against *Nek8* in zebrafish generated a pronephric cyst phenotype. Mutations in *Nek8* were also found in mice with autosomal recessive juvenile cystic kidney (*jck*) mutation, a model of human polycystic kidney disease.[45]

6.3. Gene Function Discovery and Functional Genomics

The vertebrate genome consists of 30,000+ genes. Traditional methods of positional cloning in forward genetics do not allow rapid identification of genes required for a variety of biological processes. The effectiveness of the morpholino knockdown technology in zebrafish as the vertebrate model system facilitates high-throughput assignment of gene functions on a genomic scale. Sequencing of the zebrafish genome, begun by the Sanger Center, is projected to be completed by the end of 2005. Currently, >250,000 expressed sequence tags (ESTs) comprise >25,000 zebrafish tentative consensus transcripts, as compiled by the Institute for Genomic Research (http://tigr.org/tdb/tgi/zgi/). A summary of using morpholinos for both random and specific subsets of the genome for functional genomics applications in zebrafish has recently been reviewed.[6]

Some vertebrate gene products perform redundant functions. Moreover, the zebrafish genome has undergone partial duplication during early evolution of teleosts. Hence, a subset of genes found in humans have two orthologs in zebrafish, each of which usually have distinct expression patterns.[46,47] To circumvent the limitation of functional redundancy using morpholinos in elucidating gene functions, several studies have demonstrated the success of simultaneously targeting two genes by morpholinos, e.g. double knockdowns of *shh* and *twhh*,[12] and those of *hoxb1a* and *hoxb1b*.[48] A systematic targeting of gene pairs selected based on their primary sequence similarity is a viable option using morpholinos for assessing functional requirements during vertebrate development.

7. Future Directions

In less than three years, morpholinos have become a critical tool for illuminating the function of genes using a number of different model systems. In zebrafish, their use in a comprehensive screen for function on a genome scale is just beginning. Screening methods using sensitized genetic or morpholino-based partial loss-of-function backgrounds have also yet to be explored. By analogy with temperature-sensitive genetic screening approaches, chemistries that regulate the activity of these antisense compounds would potentially uncover late functions that are currently masked by early developmental requirements. Finally, tomorrow's delivery tools for selective oligonucleotide introduction into specific cell-types will open the door to an array of cell biological studies that are technically-challenging to achieve today. These tools, and their creative applications in the future, promise to continue to help the modern biologist tackle the complex biological challenges that lie ahead.

Acknowledgments

This work was supported by NIH grants to EC and SCE. (GM55877 and GM63904).

References

1. Yacyshyn BR and Crooke ST (2001). The concept and application of antisense oligonucleotides. *Dis. Colon Rectum* **44:** 1241–1243.
2. Heasman J, Kofron M and Wylie C (2000). Beta-catenin signaling activity dissected in the early Xenopus embryo: a novel antisense approach. *Dev. Biol.* **222:** 124–134.
3. Summerton J (1999). Morpholino antisense oligomers: the case for an RNase H-independent structural type. *Biochim. Biophys. Acta.* **1489:** 141–158.
4. Ekker SC and Larson JD (2001). Morphant technology in model developmental systems. *Genesis* **30:** 89–93.
5. Heasman J (2002) Morpholino oligos: making sense of antisense. *Dev. Biol.* **243**: 209–214.
6. Sumanas S and Larson JD (2002). Morpholino phosphorodiamidate oligonucleotides in zebrafish: a recipe for functional genomics? *Briefings Functional Genomics Proteomics* **1**: 239–256.
7. Agrawal S (1999). Importance of nucleotide sequence and chemical modifications of antisense oligonucleotides. *Biochim. Biophys. Acta.* **1489:** 53–68.

8. Nielsen PE, Egholm M, Berg RH and Buchardt O (1991). Sequence-selective recognition of DNA by strand displacement with a thymine-substituted polyamide. *Science* **254:** 1497–1500.
9. Efimov VA, Buriakova AA, Chub MB and Chakhmakhcheva OG (1998). Peptide nucleic acids and their phosphonate analogues: synthesis and hybridization characteristics. *Bioorg. Khim.* **24:** 696–709.
10. Summerton J and Weller D (1997). Morpholino antisense oligomers: design, preparation, and properties. *Antisense Nucleic Acid Drug Dev.* **7:** 187–195.
11. Urtishak KA, Choob M, Tian X, Sternheim N, Talbot WS, Wickstrom E and Farber SA (2003). Targeted gene knockdown in zebrafish using negatively charged peptide nucleic acid mimics. *Dev. Dyn.*, in press.
12. Nasevicius A and Ekker SC (2000). Effective targeted gene 'knockdown' in zebrafish. *Nat. Genet.* **26:** 216–220.
13. Nasevicius A and Ekker SC (2001). The zebrafish as a novel system for functional genomics and therapeutic development applications. *Curr. Opin. Mol. Ther.* **3:** 224–228.
14. Hudziak RM, Barofsky E, Barofsky DF, Weller DL, Huang SB and Weller DD (1996). Resistance of morpholino phosphorodiamidate oligomers to enzymatic degradation. *Antisense Nucleic Acid Drug Dev.* **6:** 267–272.
15. Giles RV, Spiller DG, Clark RE and Tidd DM (1999). Antisense morpholino oligonucleotide analog induces missplicing of C-myc mRNA. *Antisense Nucleic Acid Drug Dev.* **9:** 213–220.
16. Arora V, Knapp DC, Smith BL, Statdfield ML, Stein DA, Reddy MT, Weller DD and Iversen PL (2000). c-Myc antisense limits rat liver regeneration and indicates role for c-Myc in regulating cytochrome P-450 3A activity. *J. Pharmacol. Exp. Ther.* **292:** 921–928.
17. Draper BW, Morcos PA and Kimmel CB (2001). Inhibition of zebrafish fgf8 pre-mRNA splicing with morpholino oligos: a quantifiable method for gene knockdown. *Genesis* **30:** 154–156.
18. Howard EW, Newman LA, Oleksyn DW, Angerer RC and Angerer LM (2001). SpKrl: a direct target of beta-catenin regulation required for endoderm differentiation in sea urchin embryos. *Development* **128:** 365–375.
19. Nutt SL, Bronchain OJ, Hartley KO and Amaya E (2001). Comparison of morpholino based translational inhibition during the development of Xenopus laevis and Xenopus tropicalis. *Genesis* **30:** 110–113.

20. Kos R, Tucker RP, Hall R, Duong TD and Erickson CA (2003). Methods for introducing morpholinos into the chicken embryo. *Dev. Dyn.* **226:** 470–477.
21. Morcos PA (2001). Achieving efficient delivery of morpholino oligos in cultured cells. *Genesis* **30:** 94–102.
22. Siddall LS, Barcroft LC and Watson AJ (2002). Targeting gene expression in the preimplantation mouse embryo using morpholino antisense oligonucleotides. *Mol. Reprod. Dev.* **63:** 413–421.
23. Kimmel CB, Ullmann B, Walker M, Miller CT and Crump JG (2003). Endothelin 1-mediated regulation of pharyngeal bone development in zebrafish. *Development* **130:** 1339–1351.
24. Yan YL, Miller CT, Nissen RM, Singer A, Liu D, Kirn A, Draper B, Willoughby J, Morcos PA, Amsterdam A, Chung BC, Westerfield M, Haffter P, Hopkins N, Kimmel C, Postlethwait JH and Nissen R (2002). A zebrafish sox9 gene required for cartilage morphogenesis. *Development* **129:** 5065–5079.
25. Kos R, Reedy MV, Johnson RL and Erickson CA (2001). The winged-helix transcription factor FoxD3 is important for establishing the neural crest lineage and repressing melanogenesis in avian embryos. *Development* **128:** 1467–1479.
26. Satou Y, Imai KS and Satoh N (2001). Action of morpholinos in Ciona embryos. *Genesis* **30:** 103–106.
27. Coonrod SA, Bolling LC, Wright PW, Visconti PE and Herr JC (2001). A morpholino phenocopy of the mouse mos mutation. *Genesis* **30:** 198–200.
28. Song MH, Huang FZ, Chang GY and Weisblat DA (2002). Expression and function of an even-skipped homolog in the leech Helobdella robusta. *Development* **129:** 3681–3692.
29. Louvet–Vallee S, Kolotuev I, Podbilewicz B and Felix MA (2003). Control of Vulval Competence and Centering in the Nematode Oscheius sp. 1 CEW1. *Genetics* **163:** 133–146.
30. Iversen PL, Arora V, Acker AJ, Mason DH and Devi GR (2003). Efficacy of antisense morpholino oligomer targeted to c-myc in prostate cancer xenograft murine model and a Phase I safety study in humans. *Clin. Cancer Res.* **9:** 2510–2519.

31. Jackson AL, Bartz SR, Schelter J, Kobayashi SV, Burchard J, Mao M, Li B, Cavet G and Linsley PS (2003). Expression profiling reveals off-target gene regulation by RNAi. *Nat. Biotechnol.* **21:** 635–637.
32. Lele Z, Bakkers J and Hammerschmidt M (2001). Morpholino phenocopies of the swirl, snailhouse, somitabun, minifin, silberblick, and pipetail mutations. *Genesis* **30:** 190–194.
33. Karlen S and Rebagliati M (2001). A morpholino phenocopy of the cyclops mutation. *Genesis* **30:** 126–128.
34. Golling G, Amsterdam A, Sun Z, Antonelli M, Maldonado E, Chen W, Burgess S, Haldi M, Artzt K, Farrington S, Lin SY, Nissen RM and Hopkins N (2002). Insertional mutagenesis in zebrafish rapidly identifies genes essential for early vertebrate development. *Nat. Genet.* **31:** 135–140.
35. Zhong TP, Childs S, Leu JP and Fishman MC (2001). Gridlock signalling pathway fashions the first embryonic artery. *Nature* **414:** 216–220.
36. Wienholds E, Schulte–Merker S, Walderich B and Plasterk RH (2002). Target-selected inactivation of the zebrafish ragl gene. *Science* **297:** 99–102.
37. Knapp DC, Mata JE, Reddy MT, Devi GR and Iversen PL (2003). Resistance to chemotherapeutic drugs overcome by c-Myc inhibition in a Lewis lung carcinoma murine model. *Anticancer Drugs* **14:** 39–47.
38. Kipshidze N, Moses J, Shankar LR and Leon M (2001). Perspectives on antisense therapy for the prevention of restenosis. *Curr. Opin. Mol. Ther.* **3:** 265–277.
39. Suwanmanee T, Sierakowska H, Fucharoen S and Kole R (2002). Repair of a Splicing Defect in Erythroid Cells from Patients with beta-Thalassemia/HbE Disorder. *Mol. Ther.* **6:** 718–726.
40. Roman BL and Weinstein BM (2000). Building the vertebrate vasculature: research is going swimmingly. *Bioessays* **22:** 882–893.
42. Yelon D (2001). Cardiac patterning and morphogenesis in zebrafish. *Dev. Dyn.* **222:** 552–563.
42. Thisse C and Zon LI (2002). Organogenesis — heart and blood formation from the zebrafish point of view. *Science* **295:** 457–462.
43. Carmeliet P, Ferreira V, Breier G, Pollefeyt S, Kieckens L, Gertsenstein M, Fahrig M, Vandenhoeck A, Harpal K, Eberhardt C, Declercq C, Pawling J, Moons L, Collen D, Risau W and Nagy A (1996). Abnormal blood vessel development and lethality in embryos lacking a single VEGF allele. *Nature* **380:** 435–439.

44. Ferrara N, Carver–Moore K, Chen H, Dowd M, Lu L, O'Shea KS, Powell–Braxton L, Hillan KJ and Moore MW (1996). Heterozygous embryonic lethality induced by targeted inactivation of the VEGF gene. *Nature* **380:** 439–442.
45. Liu S, Lu W, Obara T, Kuida S, Lehoczky J, Dewar K, Drummond IA and Beier DR (2002). A defect in a novel Nek-family kinase causes cystic kidney disease in the mouse and in zebrafish. *Development* **129:** 5839–5846.
46. Force A, Lynch M, Pickett FB, Amores A, Yan YL and Postlethwait J (1999). Preservation of duplicate genes by complementary, degenerative mutations. *Genetics* **151:** 1531–1545.
47. Postlethwait J, Amores A, Force A and Yan YL (1999). The zebrafish genome. *Methods Cell Biol.* **60:** 149–163.
48. McClintock JM, Kheirbek MA and Prince VE (2002). Knockdown of duplicated zebrafish hoxb1 genes reveals distinct roles in hindbrain patterning and a novel mechanism of duplicate gene retention. *Development* **129:** 2339–2354.
49. Chen E, Hermanson S and Ekker SC (2004). Syndecan-2 is essential for angiogenic sprouting during zebrafish development. *Blood* **103**: 1710–1719.

Further Reading

Ekker SC (2000). Morphants: a new systematic vertebrate functional genomics approach. *Yeast* **17:** 302–306.

Garrity DM, Childs S and Fishman MC (2002). The heartstrings mutation in zebrafish causes heart/fin Tbx5 deficiency syndrome. *Development* **129:** 4635–4645.

Klee EW, Ekker SC and Ellis LB (2001). Target selection for Danio rerio functional genomics. *Genesis* **30:** 123–125.

Lo J, Lee S, Xu M, Liu F, Ruan H, Eun A, He Y, Ma W, Wang W, Wen Z and Peng J (2003). 15000 unique zebrafish EST clusters and their future use in microarray for profiling gene expression patterns during embryogenesis. *Genome Res.* **13:** 455–466.

Miller CT and Kimmel CB (2001). Morpholino phenocopies of endothelin 1 (sucker) and other anterior arch class mutations. *Genesis* **30:** 186–187.

Penberthy WT, Shafizadeh E and Lin S (2002). The zebrafish as a model for human disease. *Front Biosci.* 7: d1439–1453.

Talbot WS and Hopkins N (2000). Zebrafish mutations and functional analysis of the vertebrate genome. *Genes Dev.* **14:** 755–762.

Yee NS, Yusuff S and Pack M (2001). Zebrafish pdx1 morphant displays defects in pancreas development and digestive organ chirality, and potentially identifies a multipotent pancreas progenitor cell. *Genesis* **30:** 137–140.

Chapter 14

Transgenic Fish for Developmental Biology Studies

Zhiyuan Gong, Yi Lian Wu and Sudha Puttur Mudumana
Department of Biological Sciences
National University of Singapore, Singapore

Shuo Lin
Department of Molecular, Cellular, and Developmental Biology
University of California Los Angeles, Los Angeles, CA 90095-1606, USA

Transgenic technology is the introduction of foreign DNA into a host organism so that the function and regulation of the inserted foreign DNA can be studied in the transgenic organism. Over the past decade, most of the key transgenic techniques have been developed in various fish models. In the present review, we focus on the transgenic studies in two experimental fish models, the zebrafish (*Danio rerio*) and medaka (*Oryzias latipes*). These include transient *transgenic* assays, stable *transgenic* lines focusing on *GFP* transgenic fish and their applications, conditional activation of transgene expression using inducible *promoters* and binary transgenic systems (GAL4-UAS and Cre-loxP), *cell* lineage ablation, *insertional* mutagenesis, *gene* traps and the potential of *gene* targeting approaches in fish. Future prospects of transgenic fish studies are also discussed.

1. Introduction

Transgenic technology in fish has come a long way since 1985 when the first successful transgenic fish was reported by Zhu *et al.*[1] Since then, many advances have been made (see Table 1) and transgenic fish have been produced in over 30 fish species. Currently, numerous laboratories around the world are working on transgenic fish, resulting in more than 3000 publications related to transgenic fish. In the 1980s and early 1990s, the impetus to develop transgenic fish was largely driven by the interest of generating superior fish stocks for aquaculture. This has led to the development of fish with useful traits in aquaculture,

Table 1 Milestones in development of transgenic fish technology.

Year	Milestone	Reference
1985	First report of transgenic fish by microinjection of a hybrid gene containing mouse metallothionein-1 promoter fused to human growth hormone into goldfish	Zhu *et al.*[1]
1986	First transient transgenic medaka study using chicken *?-crystalline* gene	Ozato *et al.*[3]
1988	Generation of stable transgenic zebrafish lines containing SV40-CAT chimeric gene	Stuart *et al.*[7]
1992	Production of rapidly growing salmon for aquaculture using an "all fish" GH gene construct	Du *et al.*[99]
1992	Demonstration that developmental regulation of mammalian promoters is conserved in zebrafish	Westerfield *et al.*[5]
1995	GFP first used as a reporter in zebrafish	Amsterdam *et al.*[35]
1994	Development of murine leukemia virus/vesicular stomatitis pseudotyped *retroviral* vectors for transgenic insertion in zebrafish	Lin *et al.*[79]
1996	Insertional mutagenesis screen	Gaiano *et al.*[80,81]
1997	Generation of *GFP* transgenic zebrafish under control of promoters of zebrafish origin	Long *et al.*[14] Higashijima *et al.*[12]
1998	Production of medaka fish chimeras from a stable *embryonic* stem cell line	Hong *et al.*[93]
1999	Demonstration of GAL4-UAS binary transgenic systems in zebrafish	Scheer and Campos–Ortega[69]
2000	Morpholino as effective gene knockdown approach in zebrafish	Nasevicius and Ekker[45]
2001	Production of zebrafish germline chimeras from embryo cell cultures	Ma *et al.*[94]
2001	Successful *nuclear* transplantation in medaka using embryonic cells	Wakamatsu *et al.*[40]
2002	Successful nuclear transplantation in zebrafish using long-term-cultured donor cells	Lee *et al.*[41]
2002	First targeted gene mutant generated in zebrafish	Wienholds *et al.*[96]
2003	First transgenic oncofish generated in zebrafish using a *c-myc* oncogene	Langenau *et al.*[97]

e.g. increased growth rates, freezing resistance and improved disease resistance, especially in species of commercial interest such as salmonids, tilapia and carp (for a recent review, see Maclean and Laight[2]). Till 1995, most transgenic fish research was focused on the methodology and its application in aquaculture, with few reports on the establishment of a transgenic fish model for developmental analyses.

The value of transgenic organisms in investigating developmental processes had been well recognized and demonstrated in mammalian and invertebrate models such as the mouse, *Drosophila* and *Caenorhabditis elegans* since the development of the transgenic technology in early 1980s. While the potential of transgenic fish for analyzing developmental mechanisms was also recognized as early as 1986,[3] it has only been in the recent few years that the transgenic fish has been gaining prominence. The increasing popularity for the transgenic fish model can in part be attributed to the successful application of the *green fluorescent protein* (*GFP*) gene from the jellyfish (*Aequorea victoria*) as a reporter gene.[4] Aptly termed "living color" fluorescent protein by Clontech, GFP is intrinsically fluorescent, allowing direct visualization without the need of exogenous substrates. Two freshwater fish, the zebrafish (*Danio rerio*) and medaka (*Oryzias latipes*), have proven amenable for such transgenic analyses owning to their short generation time, easy maintenance, rapid development and transparency of embryos. In the past few years, tremendous progresses have been made in transgenic analysis of gene expression and function in development. In addition, the transgenic technology has also been used for insertional mutagenesis in zebrafish. In this chapter, we will review these progresses in these two fish models. Current research and future prospects will also be discussed.

2. Transient Transgenic Expression

The transgenic technology has been increasingly used for analysis of gene expression and function in development. Transgenic organisms provide an excellent *in vivo* system for such analysis and generally produce more accurate and reliable information than an *in vitro* cell culture system. There are two transgenic analysis systems: transient transgenic expression and stable transgenic lines. The transient transgenic system is the analysis of gene expression immediately after the introduction of the foreign gene into embryos while the introduced gene remains largely in extrachromosomal form. This system provides a rapid and convenient assay. However, due to mosaic segregation of injected DNA during the cleavage stage,[5] differential and mosaic gene expression from the same transgenic construct are frequently observed among injected embryos. To overcome this problem, the expression profile of the transgene is generally derived from a large pool of injected embryos (e.g. Westerfield *et al.*[5]; Muller *et al.*[6]).

The pioneer transgenic works by Stuart *et al.*[7] and Chong and Vielkind[8] in zebrafish and medaka respectively demonstrated that exogenous DNA underwent extensive replication, likely in extrachromosomal form, when introduced into early embryos. The amplified exogenous DNA peaked at blastula stage, followed by a gradual decrease during late embryogenesis, and eventually became undetectable in most injected embryos. Expression of reporter gene activity could be detected in zebrafish as early as 8 hours after injection and most of the expression was likely from the extrachromosomal DNA before integration.

2.1. Analysis of Gene Promoters/Enhancer Elements

The transgenic expression observed in embryos is largely dependent on the promoter used. A promoter generally contains many distinct *cis*-elements that control where, when and how strongly the gene is expressed in the organism. By dissecting different regions of a promoter and splicing them to a reporter gene, assays of the temporal and spatial expression pattern of the reporter gene allow a thorough investigation into the activity of a promoter. Much of the earlier works in this area took advantage of the zebrafish and medaka model to analyze gene promoters of heterologous origin. For example, Inoue *et al.*[9] introduced a chicken *δ-crystallin* gene into medaka embryos to study the regulation of gene expression of vertebrate crystallin genes. Although the gene was expressed in lens tissues at an early stage, expression in other tissues was also observed in the later stages. Gong *et al.*[10] injected into medaka embryos several fish antifreeze protein gene promoters fused to the *CAT* (*chloramphenicol acetyltransferase*) reporter gene, and identified several positive and negative regulatory regions, consistent with transfection data in cultured fish cells. Using the transient transgenic zebrafish embryos, Westerfield *et al.*[5] examined the activation of two mammalian *Hox* genes in zebrafish and found that the same *cis*-elements that specified the spatial expression in mice also functioned in zebrafish.

In early transgenic fish studies, the problem of mosaic expression was compounded by the fact that many of the reporter genes used, such as *CAT* and *β-galactosidase*, required the sacrifice of embryos for analysis. In addition, many of the promoters used in these studies were of heterologous origin, making it difficult to predict the expression of the transgene. Thus, the transgenic approach was not a favorable tool for developmental analysis. These problems have largely been overcome by use of the GFP reporter gene and homologous gene promoters. Thus, a turning point was the successful reports of faithful expression of GFP driven by tissue-specific, homologous promoters in transgenic zebrafish.[11–14] The advantage of the GFP system is the live observation of transgene expression, implying that the same embryos can be used for observation at multiple stages for a dynamic pattern of gene expression. The work by Meng *et al.*[12] represented the first attempt using the *GFP* reporter gene to dissect the regulatory region of a zebrafish promoter, *GATA-2*. By microinjecting zebrafish embryos with a deletion series of the zebrafish *GATA-2* promoter ligated to the *GFP* reporter gene, Meng *et al.*[12] was able to identify three distinct tissue-specific elements that enhance gene expression specifically in blood cell precursors, the enveloping layer, or the central nervous system. Since then, this approach has been successfully employed to analyze the regulatory elements from several other zebrafish genes, e.g. the intronic enhancers in *sonic hedgehog*,[6] several conserved intergenic domains in *Dlx* gene clusters,[15] muscle-specific elements from *myogenin*[16] etc. Similarly, several medaka gene promoters have also been analyzed in the transient transgenic medaka systems (e.g. Kusakabe *et al.*[17], Kusakabe and Suzuki[18]).

In our own laboratory, we have also used the GFP reporter to demonstrate the tissue-specificity of several zebrafish gene promoters, including ones from muscle-specific *creatine kinase* (*MCK*), fast muscle-specific *myosin light polypeptide 2* (*mylz2*), skin-specific *keratin8* (*krt8*), exocrine pancrease-specific *elastase A*, and a ubiquitous *ARP* (*acidic ribosomal phosphoprotein P0*).[14,19–21] In all cases, the promoters are capable of driving *GFP* to express in patterns as expected based on the expression of their endogenous gene. By generating 5' promoter deletions of zebrafish *mylz2* promoter linked to *GFP*, we were able to correlate the strength of promoter activity with the number of muscle *cis*-elements present in the promoter. We also found that a minimal −77-bp region was sufficient for relatively strong promoter activity in muscle cells.[19]

It is worth to mention another powerful approach to rapidly scan and identify *cis*-acting activator elements of genes in transient transgenic embryos. This method is based on rapid concatamerization of injected DNA molecules. Instead of constructing a series of plasmid DNAs with different potential *cis-elements*, embryos are simply co-injected with a minimal promoter linked to a reporter gene together with putative enhancer fragments. Muller *et al.*[6] demonstrated the feasibility of this method by injecting different heterologous enhancers together with a carp *myosin heavy chain* (*MyHC*) promoter/*lacZ* reporter construct, and showed that β-galactosidase activity increased similarly whether the enhancer was co-injected separately or ligated to the promoter/reporter construct. This method has been used successfully to identify intronic enhancers for *sonic hedgehog* expression in the floor plate and notochord[22] and for *netrin* expression in the floor plate and hypochord of transgenic embryos.[23] This method is especially useful for some genes, particularly those important for developmental regulation, where their promoters are highly complicated with numerous *cis*-regulatory modules spread over a large genomic region.

2.2. Analysis of Gene Function in Development and Mutants

Transient transgenic assay also provides a rapid and effective way to test the function of a gene. One way to achieve this is to inject sense-strand

RNA into embryos. However, the injected RNA is generally ubiquitously expressed. A more-specific method of expressing a gene is injection of a DNA construct with a tissue-specific promoter. Du *et al.*[24] have employed this approach to inject a *dorsalin-1* DNA construct with a notochord-specific promoter from *twhh* (*tiggy-winkle hedgehog*) and found that the expression of Dorsalin-1 in the notochord inhibited development of muscle pioneers in adjacent somites.

Another application of the transient expression assay is the mutant rescue experiment. In many cases, a native promoter is required and thus an unmodified and complete genomic clone can be conveniently injected. For example, Yan *et al.*[25] have injected a zebrafish BAC clone containing *floating head* (*flh*) gene into the *flh* mutant embryos, which lack a differentiated notochord and have a reduced, discontinuous floor plate; they found the sign of restoration of the notochord and floor plate. This approach could be used for a large-scale functional screen of genes responsible for a mutant. Rescue of mutant studies have also been carried out in medaka using both medaka and mouse tyrosinase genes.[26–28] In all these reports, pigmentation was rescued not only in transgenic founders but also their offspring.

3. Stable Transgenic Lines

In contrast to the transient transgenic expression system, stable transgenic lines refer to germline transmitted transgenic organisms. Offspring from the same transgenic founder usually display an identical pattern of transgene expression as the transgene is already stably integrated into the host genome. The advantage of stable transgenic lines is the availability of a large number of transgenic individuals for repeated analyses. The disadvantages are a longer time required for generation and extra labors for maintenance of transgenic lines. Another set of potential disadvantages is the chromosomal effect and variegated transgene expression within the same transgenic line, both of which have been frequently reported in transgenic mice but rarely in transgenic fish. The former is due to the integration site in the chromosome where some neighboring gene regulatory elements may affect the transgene expression. The latter may be due to the difference of DNA methylation patterns among individuals. Sometimes, the chromosomal effect could be beneficial in research. For example, Field *et al.*[29] reported a *GFP* transgenic zebrafish line under the ubiquitous *Xenopus EF-1α* (*elongation factor 1α*) promoter that displayed GFP expression throughout the digestive system. Thus this transgenic line, termed gutGFP, becomes an excellent model to study development of endoderm and digestive organs.

Stable transgenic lines of zebrafish were first generated over a decade ago by Stuart *et al.*,[7,30] who demonstrated that plasmid DNA injected into the cytoplasm of fertilized eggs could integrate into the genome and be stably transmitted to subsequent generations.[7] The same transgene expression pattern was observed from all transgenic offspring within the same line of transgenic fish containing *CAT* reporter gene under an SV40 viral promoter.[31] Similar successes were also reported by other groups.[31–33] Despite these early attempts, there was a paucity of early transgenic research for developmental analyses. It has only been in the recent years that there has been a boom in transgenic fish studies in developmental biology. This can be attributed to the use of homologous promoters and *GFP* reporter gene. The following is a review of various applications of *GFP* transgenic fish.

3.1. Labeling of Cells with GFP and Recapitulation of Gene Expression Programs

GFP as a new reporter gene for transgenic fish study was first used in zebrafish by Amsterdam *et al.*,[34] who injected a *GFP* construct under a *Xenopus EF-1α* promoter. The optical clarity of zebrafish embryos makes it possible to observe GFP expression throughout embryogenesis. Moreover, the germline transmitted *GFP* transgenic embryos in F1 and F2 also expressed detectable GFP fluorescence and it appeared that a single copy of *GFP* gene under the *Xenopus* promoter was sufficient to express detectable amount of GFP fluorescence.

The first stable line of *GFP* transgenic zebrafish under a tissue-specific zebrafish promoter was reported by Long *et al.*[13] In this study, a 5.6 kb erythroid-specific *GATA-1* promoter was isolated and linked to the *GFP* reporter gene. Analyses of expression in the transgenic progeny showed that GFP was specifically expressed in erythroid cells, similar to the endogenous *GATA-1* expression. Since then, there has been a rapid increase of reports on *GFP* transgenic zebrafish under tissue-specific promoters. Now over 40 *GFP* transgenic zebrafish lines with different promoter-GFP chimeric genes have been reported and these transgenic lines encompass a wide range of tissue-specificity from neurons to neural crest, retina, notochord, floor plate, epithelia, lymphoid cells, blood cells, blood vessels, pineal gland, liver, endocrine and exocrine pancreatic cells, thymus, pituitary, muscle, germ cells, etc (Table 2). To a less extent, *GFP* transgenic medaka under tissue-specific promoters have also been produced. Essentially in all cases, faithful GFP expression patterns have been observed. Examples of GFP expression in selected stable lines *GFP* transgenic zebrafish are shown in Fig. 1.

Because GFP expression can be observed in living embryos, *GFP* transgenic fish under a tissue-specific promoter have become a powerful tool for recapitulating gene expression programs. For example, in the *GATA-1:GFP* line, the erythroid cell lineage can be traced based on visible green fluorescence from the earliest progenitor cells to adult circulating blood cells.[13] In the *mylz2:GFP* transgenic lines we created using a fast muscle-specific *mylz2* promoter, GFP expression can be detected in both the trunk and head skeletal muscles. Particularly in the head muscle region, the sequence of development of over 20 pairs of eye, jaw and gill muscles can be easily tracked.[19] This reflects a true expression pattern of the endogenous *mylz2* and the muscle development in the cranial region, consistent with the previous report by morphological criteria and other molecular markers for cranial muscle development.[35]

In addition, the ease of generation and analysis of *GFP* transgenic zebrafish also provide an excellent model to study the genes from other organisms. For example, *GFP* transgenic zebrafish have been generated using the promoters from mouse *Tie2*[36] and rat *GAP-43* (*growth associated protein*).[37] In the *Tie2:GFP* transgenic lines, GFP expression is observed globally in endothelial cells and thus these transgenic lines provide an excellent model for investigation of vascular development.

Table 2 Summary of stable lines of *GFP* transgenic zebrafish and medaka.

Promoters	GFP expression	Reference
Zebrafish		
Homologous promoters		
GATA-1	erythroid cell lineage	Long *et al.*[13]
α-actin	muscle	Higashijima *et al.*;[11] Hsiao *et al.*[120]
β-actin	ubiquitous	Higashijima *et al.*[11]
rag1	thymus, olfactory neurons	Jessen *et al.*[64]
rag2	thymus, olfactory neurons	Jessen *et al.*[64]
islet-1	cranial motor *neurons*	Higashijima *et al.*[52]
Hsp70	heat-shock and laser inducible	Halloran *et al.*[53]
HuC	neurons	Park *et al.*[59]
shh	neural retina and others	Neumann and Nusslein–Volhard[99]
twhh	notochord, floor plate, branchial arches, pectoral fin buds, retina, epithelial lining cells of Kupffer's vesicle	Du and Dienhart[100]
α-tubulin	neurons	Goldman *et al.*[101]
insulin	endocrine pancreas	Huang *et al.*[102]
Pdx-1	pancreas	Huang *et al.*[102]
Rhodopsin	rod photoreceptors in retina	Kennedy *et al.*;[103] Hamaoka *et al.*[104]
histone variant H2A.F	ubiquitous	Pauls *et al.*[47]
keratin8	skin and intestinal epithelia	Gong *et al.*[20]
keratin8(GFP)/ mylz2(RFP)	two-color transgenic zebrafish with GFP expression in skin epithelia and RFP in muscle	Wan *et al.*[21]
exorh (exo-rhodopsin)	pineal gland	Asaoka *et al.*[105]
zFoxD3/fkd6	migrating neural crest lineages	Gilmour *et al.*[106]
serotonin-N-acetyl-transferase-2	pineal gland (epiphysis)	Gothilf *et al.*[107]
vasa	germ cells	Knaut *et al.*;[50] Krovel and Olsen[51]
fli1	blood vessels	Lawson and Weinstein[108]
pax 2.1	MHB, hindbrain, spinal cord, ear and pronephros	Picker *et al.*[109]

Table 2 (*Continued*)

Promoters	GFP expression	Reference
netrin	floor plate and hypochord	Rastegar *et al.*[23]
deltaD	mesodermal and neuroectodermal tissues	Hans and Campos–Ortega[110]
nicotinic acetylcholine receptor β3	retinal ganglion cell	Tokuoka *et al.*[111]
olfactory marker protein	olfactory neurons	Yoshida *et al.*[112]
neurogenin	neural plate	Blader *et al.*[113]
L-FABP	liver	Her *et al.*[114]
mylz2	fast skeletal *muscle*	Ju *et al.*[19]
pro-opiomelanocortin	pituitary corticotrophs	Liu *et al.*[115]
VEGFR2 (flk-1)	blood vessels	Cross *et al.*[61]
elastaseA	exocrine pancreas	Wan and Gong, unpublished[116]
Heterologous promoters		
Xenopus *elongation factor 1α*	ubiquitous	Amsterdam *et al.*[34] Linney *et al.*[117]
Carp β-actin	ubiquitous	Gibbs and Schmale[118]
Mouse *Tie2*	endothelial cells	Motoike *et al.*[119]
Rat *GAP-43*	neurons	Udvadia *et al.*[37]
Medaka β-actin	ubiquitous and strong expression in female gonad	Hsiao *et al.*[120]
RARE+basal promoter	domains of retinoic acid responsiveness	Perz–Edwards *et al.*[56]
4 Lef binding sites-minimal mouse *cFos*	domains of Wnt/β-catenin signaling	Dorsky *et al.*[57]
Xenopus opsin	rod photoreceptor	Perkins *et al.*[121]
Xenopus ef1α	digestive tract presumably due to chromosome effect	Field *et al.*[29]
Medaka		
Medaka β*-actin*	ubiquitous	Hamada *et al.*;[122] Yamauchi *et al.*;[123] Chou *et al.*[124]
Medaka elongation *factor α*	ubiquitous	Kinoshita *et al.*[125]
Medaka *vasa*	germ cells	Tanaka *et al.*[49]
zebrafish *mylz2*	muscle	Zeng and Gong[126]
Medaka vitellogenin	estrogen-inducible expression in liver	Zeng and Gong[127]

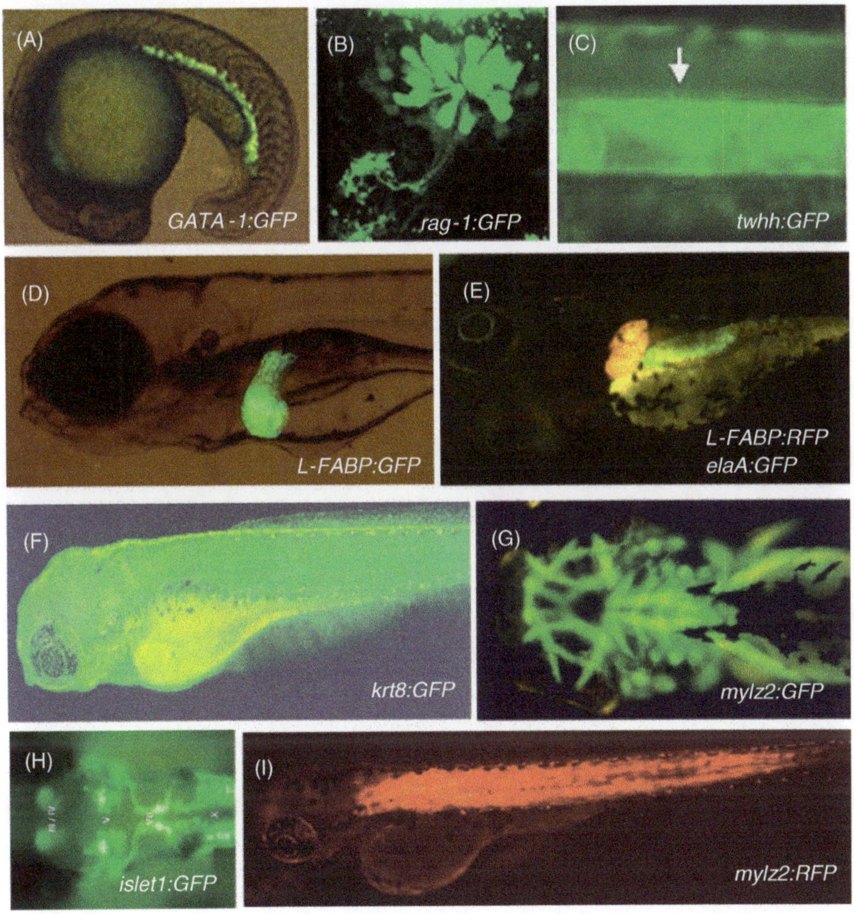

Fig. 1 Examples of *GFP* and *RFP* transgenic zebrafish lines. (A) Expression of GFP in erythroid progenitor cells in a *GATA-1:GFP* transgenic embryo (22 hpf, hour post-fertilization). (B) Expression of GFP in olfactory sensory *neurons* in a *rag1:GFP* transgenic zebrafish embryo. (C) Notochord and floorplate expression of GFP in a *twhh:GFP* transgenic embryo. (D) Liver expression of GFP in a *L-FABP:GFP* transgenic fry (5 dpf, day post-fertilization). (E) Exocrine pancreatic expression of GFP and liver expression of RFP in an *L-FABP:RFP* and *ElaA:GFP* double transgenic fry (3 dpf). Photo was taken under an Ex BP 450-490 filter (Zeiss). (F) Skin epithelial expression of GFP in a *krt8:GFP* transgenic fry (~4 dpf). (G) Cranial skeletal *muscle* expression of GFP in a *mylz2: GFP* transgenic fry (ventral view, 4 dpf). (H) Neuronal expression of GFP in an *islet1:GFP* transgenic embryo (42 hpf). (I) Skeletal muscle expression of RFP in the trunk region of a *mylz2:RFP* transgenic fry (~3 dpf). Panel C by contributed by Dr. S.J. Du, panel D by Dr. G.M. Her and panel H by Dr. H. Okamoto.

In the *GAP-43:GFP* lines, GFP expression occurs in differentiating neurons that extend long axons. In both cases, the expression of transgenic GFP is similar to the expression of the corresponding genes in the mouse and rat; thus, many of the tissue-specific regulatory elements are evolutionarily conserved.

Targeted GFP expression in transgenic fish also aids many other experimental operations such as cell and nuclear transplantation and microsurgery. Transgenic GFP labeled cells serve as an excellent donor or recipient for cell or nuclear transplantation, as contribution of the GFP expressing cells to the hosts (or a wild-type cell to a *GFP* transgenic host) can be easily identified in the developing embryos. *GFP* transgenic fish lines had been used to characterize the *eyeless* (*el*) mutation in medaka, in which either the wild-type transgenic fish donated the cells to the mutant host or vice versa.[38] Study of the interaction between donor and host cells led to the conclusion that the *el* gene functions in a cell-autonomous manner. *GFP* transgenic fish have also successfully served as donors for nuclear transplantation in medaka[39] and zebrafish.[40] Recently, by aiming a laser beam at the GFP labeled retinal fiber using the *Shh:GFP* transgenic zebrafish line, Roeser and Baier[41] selectively ablated the tectum to study visuomotor behaviors in larval zebrafish.

GFP transgenic fish also play an important role in development of experimental tools in developmental biology. One application is to use GFP expressing transgenic lines to evaluate several gene knockdown approaches such as injection of double-stranded RNA (RNA interference or RNAi)[42,43] and morpholino antisense RNA.[44] In these studies, *GFP* transgenic embryos, either from transient expression or from stable lines, have been used for injection of double-stranded or morpholino antisense RNA. Although the results from injection of double-stranded RNA are controversial, the morpholino approach appears to be a very promising gene knockdown approach.

3.2. Expression of GFP Fusion Protein

A target gene may be expressed with the GFP reporter as a GFP-fusion protein. GFP serves as a visible marker to monitor the target protein and the function of the target protein is generally not affected by the GFP fusion.[45] This approach has also been used in transgenic zebrafish. Wang *et al.*[46] have introduced into zebrafish a fusion construct in which the *uroporphyrinogen decarboxylase* (*UROD*) gene is linked in-frame with the *GFP* reporter gene and they found that the *UROD-GFP* fusion gene can rescue the *urod* null mutant, which is regarded as a model for human hepatoerythropoietic porphyria.

GFP fusion proteins have also been widely used in determination of cellular localization of target proteins. Previously, Long *et al.*[47] introduced into zebrafish blood cells a GFP fusion protein with a newly isolated hematopoietic death receptor, ZH-DR, and found that the fusion protein was preferentially located in the membrane. More importantly, this study established the role of negative regulation of erythropoiesis by death receptor *in vivo.* Pauls *et al.*[48] showed a nuclear localization of a histone-GFP fusion protein. We also demonstrated the incorporation of a cytokeratin-GFP fusion protein into intermediate filaments in zebrafish epithelia.[20]

3.3. RNA Localization

GFP transgenic lines are not only useful for analyzing *cis*-regulatory DNA elements for transcription, but also have a potential in the analysis of regulatory elements for RNA localization. One excellent example is GFP expression of *vasa:GFP* transgenic lines. *Vasa* is expressed specifically in germ cells, but the localization of *vasa* mRNA in germ cells requires the 3' UTR (untranslated) region. This has been demonstrated in three species of fish: rainbow trout,[49] medaka[50] and zebrafish.[51,52] The important region in the 3' UTR for germ cell localization has been mapped in transgenic zebrafish.[51]

3.4. Cell Lineage and Cell Migration

The *GFP* transgenic fish also have the potential to be used to trace cell lineage and cell migration. Because of the transparency of the zebrafish and medaka embryos, the internal structures and certain cell-types in developing embryos can be readily visualized. This feature, aided by GFP expression under a tissue-specific promoter, offers a further advantage to trace the developmental process of essentially all tissues throughout the embryogenesis and even in adult fish. In the *islet1:GFP* transgenic line, GFP expression was strong enough to allow a dynamic analysis of the migration of differentiating motor neurons and pathfinding of their extending growth cones in the developing embryos, demonstrating the feasibility of using the *GFP* transgenic fish for dissection of the neuronal development pathways.[53] In another report, Halloran *et al.*[54] have generated *GFP* transgenic lines under a heat-shock inducible promoter from *hsp70.* In these transgenic lines, GFP expression occurs only after heat shock induction. The activation of *GFP* gene can also be induced by a sublethal laser microbeam; thus, this system provides an unique opportunity to activate different target cells at different stages to study detailed cell lineage, migration, axon outgrowth, etc.

3.5. Analysis of Upstream Regulatory Genes

Tissue differentiation and organogenesis are complex processes involving the interaction of different genes and their products. Stable *GFP* transgenic fish lines showing tissue-specific GFP expression could be used to study the genes regulating the development of these tissues/ organs. For example, in our *mylz2:GFP* transgenic lines, GFP expression is specifically in fast skeletal *muscles* but not in slow *muscles.*[19] Previously, Hammerschmidt *et al.*[55] demonstrated that the development of slow muscle requires the Sonic hedgehog signal and overexpression of a dominant negative form of protein kinase A (dnPKA), a key component in the Sonic hedgehog pathway, can induce slow muscle at the expense of fast muscle. We injected dnPKA RNA into *mylz2:GFP* transgenic embryos and indeed we observed a great reduction of GFP expression because of the overwhelming development of slow muscles.[20] Another example is the knockdown of Pdx-1, a transcription factor important for pancreas development, severely reduced GFP expression driven by insulin promoter in living transgenic zebrafish embryos.[56] Thus, the direct visualization of GFP labeled cells/tissues greatly facilitates the analysis of molecular events involved in the development of these labeled cells/tissues and the screening of upstream molecules that affect the development of these tissues/organs.

3.6. Monitoring of Signaling Molecules

Another area of application of *GFP* transgenic fish is to use multiple responsive elements linked to a basal promoter to drive GFP expression to monitor signaling molecules during embryonic development. Thus the regions with such signaling pathway will be displayed by visible GFP fluorescence. Perz–Edwards *et al.*[57] have developed several stable lines of transgenic zebrafish that use retinoic acid responsive elements (RARE) to drive GFP or YFP (yellow fluorescent protein) expression. The expression was located in the regions of conceivably high concentrations of retinoic acid (RA), such as the neural tube, retina, notochord, somites, heart, pronephric ducts, branchial arches, jaw *muscles*, etc. Furthermore, by treatment with exogenous RA, other regions of RA inducibility were identified in additional regions of the neural tube and retina as well as in immature notochord, hatching gland, enveloping layer and fin. Similarly, Dorsky *et al.*[58] have examined the targets of Wnt/β-catenin signaling by creating *GFP* transgenic zebrafish under a β-catenin responsive promoter containing four consensus Lef/Tcf binding sites. Early zygotic expression of GFP in this transgenic line mimics known regions of Wnt signaling. Such a system has also been proposed for the development of transgenic fish for environmental monitoring. Examples of responsive elements proposed include aromatic hydrocarbon response elements (AHREs), electrophilic response elements (EPREs), metal response elements (MREs), estrogen response elements (EREs) and retinoic acid response elements (RARE, RXRE).[59] These transgenic fish, if successfully developed, could hopefully be used in the detection of environmental pollutants based on the activated reporter gene expression. Recently, we developed a *GFP* transgenic medaka line using an estrogen-inducible *vitellogenin* promoter and observed estrogen-inducible GFP expression in male medaka; thus, this transgenic line has promises to be used in biomonitoring of estrogenic endocrine disrupters (Zeng and Gong[127].)

3.7. Mutant Generation and Characterization

There are two areas that *GFP* transgenic fish can serve as a good tool in fish mutant studies. First, transgenic fish lines showing GFP expression in specific tissues/organs can be used for mutant screens as the GFP labeled tissues/organs are readily recognizable. Both genetic and chemical screens can benefit from *GFP* transgenic fish. In the second area, *GFP* transgenic fish may help to characterize existing mutants after introduction of the *GFP* transgene into relevant mutants through standard breeding. For example, Park *et al.*[60] have crossed their neuron-specific *HuC:GFP* transgenic line with *mind bomb* (*mib*) mutant zebrafish and produced GFP-labeled *mib* mutant, enabling visualization of the neurogenesis phenotype in living $mib^{-/-}$ mutant embryos. Because of the ability to observe a dynamic GFP expression in living embryos, a high resolution of characterization of mutant can be achieved. Field *et al.*[29] transferred the *GFP* transgene from the gutGFP line into *ntl* (*no tail*) mutant and discovered the presence of the liver rudiment in $ntl^{-/-}$ embryos, though it was previously regarded to be absent from the $ntl^{-/-}$ embryos. Similarly, we also transferred muscle-specifically expressed *mylz2:GFP* transgene into several zebrafish mutants including *spadetail, chordino, mind bomb, floating head*, etc. and we have observed some new phenotypes of these mutants.[19]

3.7. Cell Sorting for *in vitro* Culture and Cell Type Specific cDNA Library Construction

With the initiation of the zebrafish genome program, we would expect a shift of efforts to understanding of the complete set of gene transcripts or transcriptome, an area in which *GFP* transgenic fish may also play an important role. Using tissue or cell type-specific gene promoters, many cell types can be tagged by GFP expression. These specific cells can then be purified by fluorescence activated cell sorting (FACS) and used for *in vitro* cell culture and for construction of cell type-specific cDNA libraries, which will be an important source for isolation of transcripts only expressed in a single cell type. This is especially desirable if the cells of interest are limited in number, such as the primordial germ cells and pancreatic cells in zebrafish. A successful application of this technique in *GFP* transgenic fish has been reported by Long *et al.*[13,16] who constructed a cDNA library using purified *GATA-1*-positive cells from *GATA:GFP* transgenic zebrafish embryos and isolated a novel hematopoietic death receptor gene, *ZH-DR*, from this cDNA library.

3.8. Screening and Analyzing Compounds Using Transgenic Zebrafish

Cross *et al.*[61] demonstrated that angiogenesis drugs can be rapidly analyzed by a live fluorescent zebrafish assay using transgenic embryos expressing GFP under the control of *flk* promoter. Another study by Liu *et al.*[62] showed that only a specific sub-domain of pituitary expressing GFP driven by proopiomelanocortin regulatory sequences was selectively suppressed by the treatment of dexamethasone as a feedback response to glucocorticoid. Therefore, transgenic zebrafish embryos expressing GFP in specific cell lineages provide a unique opportunity allowing rapid analysis of small molecules or drug candidates on the whole animal system. This has the potential to be scaled up because embryos can be directly incubated with a large number of substances of interest in a very small volume of liquid, and examined under a fluorescence microscope or fluorescence reader.

3.9. Chi-Recombination and Artificial Chromosome Transgenesis

So far, most transgenic lines were produced by using promoters from terminally differentiation marker genes such as *α-actin*, *β-actin*, *krt8*, *mylz2*, etc. These promoters are relatively simple and a short 5' flanking region (< 2 kb) is generally sufficient for full tissue-specificity. However, in many other genes, particularly those important for developmental regulation, their promoters are highly complex with numerous *cis*-regulatory modules. For example, the transcription factor *islet1* gene is expressed broadly in many different types of neurons as well as in some endocrine cells. A 4.1 kb promoter upstream of *islet1* was insufficient to drive GFP expression in any neurons and the enhancer elements for cranial motor *neurons* were located ~62 kb from the transcriptional start site.[53] Thus, a full range of analysis of gene promoters from this type of promoters is unlikely to be accomplished by conventional cloning using a plasmid vector. In view of this, an alternative and effective homologous recombination system has been developed for inserting a reporter gene directly into a BAC (bacterial artificial chromosome) clone.[63] In this method, a reporter gene is inserted into a target region flanked with two sets of Chi sites in a plasmid vector and homologous recombination with a corresponding BAC clone will be stimulated by the Chi sites in recombination competent *Escherichia coli* cells. Using this method, over 10 kb of flanking sequences of zebrafish *GATA-2* genes were linked to *GFP* in BAC clones. The whole BAC DNA was then introduced into zebrafish embryos and improved tissue-specificity of GFP expression was observed in transient transgenic assay.[63] By the same method, they have generated stable transgenic zebrafish line with a *rag1:GFP* PAC (P1-derived artificial chromosome) clone containing 80 kb of 5' flanking region and demonstrated a higher tissue-specificity using a longer 5' flanking region compared to shorter constructs.[64] An alternative method, based on RecA recombination system in *E. coli*, was also used to insert *GFP* into a PAC clone containing the *netrin1* locus.[23]

4. Conditional Gene Activation in Transgenic Systems

In most of the transgenic systems, expression of a transgene is controlled by a constitutively functional promoter. Expression of a reporter gene under such a promoter is usually not a problem as reporter proteins/ enzymes generally have no adverse effect on the normal physiology of the expressing cells. However, when a functional gene is expressed, whose product affects early development and survival, the host cells may not be able to survive embryogenesis and thus a stable line of transgenic organism may never be established. Similarly, if a gene affecting reproduction is used for transgenic studies, it may also be difficult to establish a stable transgenic line. To overcome these problems, conditional gene activation systems are required. The simplest idea is to use an inducible promoter such as a heat-shock inducible promoter. Transgenes under this type of promoter will be activated only when the heat-shock condition applies. More sophisticated conditional gene activation systems are two binary transgenic systems: GAL4-UAS and Cre-*loxP*. In both systems, the transgene in test remains silent in transgenic organism and is activated only in double transgenic offspring after crossing the two transgenic lines.

4.1. Inducible Systems

Manipulation of gene expression in a directed fashion is a useful means of analyzing its role in development. One way to achieve this is to use an inducible promoter and the gene under the inducible promoter can be turned on at a specific developmental stage or in a specific cell by an inducer. In the *hsp70:GFP* transgenic line generated by Halloran *et al.*,[54] GFP was generally not expressed in transgenic embryos at normal temperature, but was robustly expressed following heat-shock treatment. Moreover, GFP expression in individual cells can be activated by focusing a sublethal laser microbeam on them and these cells labeled by GFP expression can be followed for studies of cell migration, cell lineage and axon projection, etc. By using the heat-shock inducible transgenic system, they also transferred a semaphorin (guidance molecule) gene under the *hsp70* promoter. They found that motor axon outgrowth was retarded by the expression of semaphorin in laser-induced individual muscle fibers.[54] In another study, Slit2, a regulatory factor of mesodermal cell movement during gastrulation, was expressed under the same *hsp70* promoter; when it was induced to express by heat-shock, the convergent extension movement of the mesoderm as well as the rostral migration of the cells in the dorsal diencephalon was impaired.[65] Recently, using the *hsp70:GFP* transgenic zebrafish, Blechinger *et al.*[66]

established an *in vivo* system using *hsp70* gene activation as a measure of cadmium toxicity in living larvae of transgenic zebrafish, thus widening the scope of inducible transgenic systems to toxicological analysis and environmental monitoring.

There are several other inducible systems by different inducers, such as heavy metals, interferon, hormones, antibiotics, etc. To our knowledge, none of these inducible systems have been seriously tested in the transgenic fish system. The popular Tet-on and Tet-off systems have been used in transgenic mice,[67] but they remain untested in the transgenic fish system. The most evident advantage of the inducible transgenic system is the controllable expression at correct timing and sometimes in correct cells. There are also several disadvantages, including ubiquitously ectopic expression, sometimes basal levels of expression, and potential side-effects of inducers. However, if the induction condition is well-controlled, these effects may be minimized and negligible. For example, by applying the heat-shock condition in the *hsp70:GFP* transgenic zebrafish lines, Halloran *et al.*[54] reported normal embryonic development under their heat-shock condition.

4.2. GAL4-UAS System

The Gal4-UAS binary system consists of two different transgenic lines: the activator and the effector. In the activator line, the gene for the yeast transcriptional activator Gal4 is placed under a specific promoter and in the effector line, the gene of interest is fused to multiple copies of GAL4 DNA-binding motif called UAS (upstream activating sequence). The transgene in the effector line is silent until it is crossed to the activator line expressing the transcriptional activator. The expression of effector gene depends on the presence of GAL4 product in the progeny, which in turn is controlled by the promoter used for *Gal4.* Scheer and Campos–Ortega[68] first used this method in zebrafish and demonstrated the efficacy of transgenic Gal4 in activating transcription of effector genes. This method allowed them to study the role of Notch receptor in promoting gliogenesis in the developing retina. The two activator lines, *deltaD:Gal4* and *hsp70:Gal4,* could drive strong

Gal4 expression in the retina. When the activator lines were crossed with the *UAS:notch1a-intra* effector line, Notch receptor was activated in the retina and blocked neuronal differentiation, causing cells either to enter gliogenesis or remain undifferentiated and/or die.[69]

An important feature of the GAL4-UAS system is that we can make a library of different GAL4 transgenic lines with different tissue-specificities. On one hand, a tissue-specific *GAL4* transgenic line can be used to activate different effector genes from different *UAS:effector* transgenic lines in a particular tissue; on the other hand, the effector gene in the same *UAS:effector* transgenic line can be activated in different tissues by crossing with different tissue-specific *GAL4* transgenic lines. A similar binary expression system has been described in transient assays in zebrafish in which the activator line expresses the bacteriophage T7 RNA polymerase and the effector line expresses the gene of interest under the regulation of the T7 promoter.[70]

4.3. Cre-*loxP* System

The Cre-*loxP* system mediates site-specific DNA recombination and was originally described in bacteriophage P1.[71] Two components are involved: a 34-bp DNA sequence, termed *loxP*, containing 13-bp inverted repeats and an asymmetric 8-bp spacer region; and a 343-amino acid monomeric recombinase, Cre, which recognizes the *loxP* sites. Any DNA sequence flanked by two *loxP* sites in the same orientation will be excised by Cre. The Cre-*loxP* transgenic system has been used successfully in transgenic mice for targeted gene activation and conditional gene knock-out.[67] In general, Cre is expressed under a tissue-specific promoter in one transgenic line, while in another transgenic line, expression of an effector gene is blocked by insertion of a *loxP*-flanked DNA fragment and the blocking DNA fragment can be removed by Cre. Similar to the GAL4-UAS binary system, a *Cre* transgenic line can be crossed with different *loxP* transgenic lines of different effector transgenes and a *loxP* transgenic line may be crossed with different tissue-specifically expressed *CRE* transgenic lines depending on the experimental objectives.

To test whether the CRE-*loxP* system functions in a fish system, we designed a double-reporter gene construct: a *loxP*-flanked *GFP* gene followed by a *RFP* (*red fluorescent protein*) gene. When this double-reporter gene construct was injected into zebrafish embryos, the injected embryos expressed GFP only. When the same construct was co-injected with a *Cre* construct, the co-injected embryos expressed GFP first and later also RFP, indicating that the *loxP*-flanked *GFP* gene has been excised as confirmed by polymerase chain reaction assay. Recently, we have developed a stable transgenic line with the *loxP*-flanked double reporter gene construct. By injection of a *Cre* construct into transgenic embryos collected from the stable transgenic line, we also demonstrated that *loxP*-flanked GFP gene can be excised from integrated chromosomal DNA.[72]

5. Cell Lineage Ablation and Genetic Ablation

Cell lineage ablation, or genetic ablation, is to eliminate certain cells by tissue-specific expression of a toxin gene. This approach is useful in the analysis of cell lineage and cell-cell interaction during embryonic development. The most commonly used toxin is the bacterial diphtheria toxin A chain (DTA). DTA is an ADP ribosyltransferase, which catalyzes ADP-ribosylation of elongation factor 2, leading to inhibition of protein synthesis and subsequent cell death. DTA can penetrate into cells only when diphtheria toxin B chain coexists.[73] Thus the expression of the DTA fragment affects only the cells expressing DTA, and does not penetrate neighboring cells. Tissue-specific expression of DTA has been widely used in transgenic mice to study cell lineage and cell-cell interactions. For example, Palmiter *et al.*[74] have generated transgenic mice using DTA fused to *elastase I* promoter/enhancer and shown the ablation of pancreatic acinar cells by the toxin. The cell lineage ablation technique is just emerging in fish studies. Recently, Kurita *et al.*[75] introduced the DTA gene under a lens-specific *αA-crystallin* promoter and found that not only the lens development was abrogated but also the neural retina, which did not express DTA, was disorganized, suggesting the importance of lens cells to the development of neural retina.

6. Transgenic Insertional Mutagenesis

An important and powerful tool in genetic analyses is mutation. The consequence or phenotype of the mutation provides important clue to the function of the mutated gene. In the classical genetic models, *Drosophila melanogaster* and *Caenorhabditis elegans,* large-scale or saturated mutagenesis on early developmentally important genes have been carried out using chemical mutagens. In the vertebrate mouse model, such a large-scale of mutagenesis may be too costly to carry out; thus, the zebrafish becomes a vertebrate species of choice. Two large-scale screens of ethylnitrosourea (ENU) induced mutants have been carried out in zebrafish and thousands of mutations in genes that are essential for embryonic development have been isolated.[76,77] However, the cloning of these mutated genes is difficult as it requires arduous positional cloning or candidate gene approach. So far, only a handful of genes from these mutants have been cloned.

An alternative mutagenesis strategy is transgenic insertion in which exogenous DNA is used as a mutagen to randomly insert into the genome. Although DNA is a less efficient mutagen than ENU, it serves as a molecular tag to aid in the isolation of the mutated gene. DNA can be inserted by microinjection, retroviral integrases or transposable elements. So far, the most efficient way is to use a pseudotyped retrovirus that, upon injection into embryos, infects many cells and results in a mosaic host with insertions into different chromosomal sites.[78] These cells harboring insertional mutation may be incorporated into germ cells, producing mutant progeny. In a pilot insertional mutagenesis screen in zebrafish, 16% of the injected embryos transmitted proviral insertions to their offspring and each founder transmitted an average of 11 different insertions through their germline.[79] Out of 217 proviral insertions, three insertional mutants with embryo lethal phenotypes were identified.[80] In a large-scale mutagenesis screen, Golling *et al.*,[81] using a high titer of the retrovirus, could obtain mutations in 450 to 500 different genes, representing roughly 20% of the genes that can be mutated to produce a visible embryonic phenotypes. Recently, Chen *et al.*[80] used the retroviral system to develop a gene-trap vector containing the *lacZ* reporter gene. The pattern of reporter gene expression may facilitate the identification and analysis of mutated genes in mutant screen.

Transposons have also been evaluated for their efficacy and use in insertional mutagenesis in zebrafish.[83] While still in its infancy, several transposon systems show great potential as a tool to develop insertional mutagenesis, e.g. the synthetic *Sleeping Beauty* (SB) transposon systems[84] and the *Tol2* element from medaka.[85] Sleeping Beauty (SB), a member of the Tc1/mariner superfamily of transposable elements, is the only active DNA-based transposon system of vertebrate origin and has recently been demonstrated to be functional in transgenic mice.[86] Furthermore, the *C. elegans* Tc3 transposon carrying *GFP* has been integrated into zebrafish genome and transferred to the germline, showing potential in transposon mediated transgenesis and mutagenesis.[87] For more information on fish transponsons and their applications in transgenic fish, readers may refer to Chapter 16 of this book.

Another application of gene disruption in transgenic organisms is the identification of endogenous genomic regulatory elements. The rationale of the approach is to insert a reporter gene without a promoter (promoter trap) or with a truncated weak promoter (enhancer trap) randomly into the host genome; the expression of the reporter gene depends on the site of integration in the host chromosome or whether a promoter or enhancer is available from the neighboring region of the integration. Such a technique is useful in identification and isolation of tissue-specific promoters and enhancers. In an early study, Bayer and Campos–Ortega[32] used this strategy in zebrafish and successfully isolated an enhancer trap line after injection of a *lacZ* construct under a truncated mouse heat-shock promoter. The enhancer line showed lacZ expression exclusively in primary sensory neurons, including Rohon-beard neurons of the spinal cord and trigeminal ganglion neurons in the head.

Other than the production of insertional mutants and gene trap, transgenic fish may also be used for detection of mutagens, which is especially valuable for environmental monitoring. Winn *et al.*[88] introduced a λ bacteriophage vector into medaka and used the λ *cII* gene as a mutational target. After treatment of the transgenic medaka with mutagen ENU, *cII* mutants were recovered from medaka genomic DNA, packaged *in vitro* and re-infected into an appropriate bacterial host. Under selective conditions for *cII* mutants, they observed a dosage-dependent increase of mutation rates.

7. Gene Targeting

There are two main principles for determination of the function of a gene. One is to overexpress or ectopically express the gene and to observe the effect caused; this is called gain-of-function analyses. A more powerful approach is loss-of-function analyses, i.e. suppression of gene expression. Suppression of gene expression can be achieved by mutating the gene from the genome (knock-out) or by inhibition of gene expression pathway (knockdown).

Over the past 10 years, a powerful technology has been developed and used in transgenic mice and this is the well-known gene targeting (or gene knock-out) technology as a reverse genetic tool. This technology allows researchers to mutate any gene of interest from the mouse genome by homologous recombination. To achieve this, the totipotent embryonic stem (ES) cells derived from the inner cell mass of mouse embryos at the blastocyst stage are cultured *in vitro* and modified by DNA transfection. The ES cell colonies whose genomes have been replaced with a mutated gene through homologous recombination are then selected to construct chimeric mouse embryos. The genetically modified ES cells with the mutated gene have the potential to incorporate into the germline and thus mutant mice can be created. So far, hundreds of mouse genes have been mutated or knocked out in this way. However, the ES cell-based gene-knock technology is so far only reliably available in certain strains of mice but not in other animals. Therefore, to knock-out a gene in other animals, a different approach has been used, i.e. nuclear transplantation using *in vitro* modified cultured cells as nuclear donors. This technology has been successfully used in several large farm animals.[89,90]

Currently, the loss-of-function analyses in fish have been performed largely by inhibition of gene expression through a variety of gene knockdown approaches. One is to inject antisense RNA into embryos and this seems to be not very effective, presumably due to the short half-life of injected RNA and that a large excess of injected RNA is required to significantly inhibit endogenous RNA. A more popular approach is to use the dominant negative form of RNA, which generally encodes a truncated form of protein that competes with the endogenous native protein for targets; in this way, the function of the endogenous protein is inhibited. Another approach is RNA interference or RNAi, which has been demonstrated successfully in several model organisms in developmental biology, including *C. elegans*, *Drosophila* and mouse. However, in zebrafish, the effectiveness of this technology remains controversial and only non-specific inhibitory roles have been observed in a few recent reports.[43,91] In recent years, another powerful gene knockdown technology using morpholino antisense oligonucleotides has been developed in zebrafish.[44] Morpholino is a nucleotide analog that can be incorporated into DNA but cannot be degraded by RNAse. This feature enables it to survive longer *in vivo* and thus a better inhibition of gene expression is achieved (see Chapter 13).

Gene knock-out technology will allow researchers to mutate a specific gene and has long been a dream of many developmental biologists. The key technology for developing the gene knock-out technology is the establishment of ES cell culture. The ES culture has been tried in both zebrafish and medaka. Hong *et al.*[92] reported the successful production of medaka chimeras from a stable embryonic stem cell line and a high frequency (90%) of chimera formation that was not compromised by cryostorage or DNA transfection. Recently, Ma *et al.*[93] reported successful production of germline zebrafish chimeras from short term embryonic cell culture and the chimeric zebrafish were capable of producing offspring from both genetic backgrounds. An alternative approach to the ES cell method is the use of primordial germ cells (PGCs) that have been used successfully in production of transgenic poultry by introduction of genetically modified PGCs to chick embryos.[94] The availability of *GFP* transgenic fish under the *vasa* promoter that drives the GFP expression specifically in germ cells greatly facilitates the purification of PGCs through fluorescence activated cell sorting (FACS) technique.[49,50,52] All these progresses propel gene targeting technology in fish closer to success (see Chapter 19).

Another potential way to develop gene targeting technology is through nuclear transplantation. The nuclear transplantation technique has been recently reported in both medaka[39] and zebrafish.[40] The detail of the progress in nuclear transplantation has been reviewed in Chapters 15 and 19. Recently, a target-selected mutagenesis approach has been reported in zebrafish and the first targeted mutant for *rag1* was successfully generated.[95] In this method, male fish were first mutagenized by ENU and crossed with wild-type females. Sperm was then collected from individual F1 fish and genomic DNA was isolated from sperm aliquot to screen for mutation of a targeted gene. Sperm from the same F1 individual with the positive target gene mutation was then used to fertilize eggs *in vitro* to recover the mutant fish. Although tedious, this work represents the first successful report of generation of a fish mutant by reverse genetics.

8. Concluding Remarks

The transgenic technology was first developed in mice at the beginning of 1980s. Although the first transgenic fish was reported in 1985,[1] the real attempt to use the transgenic technology in the zebrafish model only took place in 1988,[7] coinciding with the emerging popularity of the zebrafish model in developmental analyses. In early years, due to the scarcity of cloned fish gene resources, the application of transgenic fish technology in developmental biology was largely limited to technology development. Only with the emergence of the living color *GFP* reporter gene and use of homologous fish gene promoters, did transgenic fish become an increasingly popular tool in developmental biology. However, most of the transgenic fish studies are still limited to technology development and characterization of gene promoters and regulatory elements. So far, transgenic expression of functional genes for developmental analyses remains rare.

In the near future, we expect the transgenic approach to be increasingly important in several areas. First, there will be an increase in the use of tissue-specific promoters to drive a functional gene to be expressed for analysis of its role in specific tissues. Second, tissue-specifically expressed *GFP* transgenic fish may be used for mutagenesis screening for genes affecting development of particular organs/tissues; similarly, chemical and drug screens can also be performed in *GFP* transgenic fish for their roles in specific tissues/organs.[61] Third, the transgenic approach can also be used to rescue mutants, particularly those related to human diseases.[46] Fourth, by tissue-specific expression of an oncogene, more and more transgenic oncofish will be developed for medical studies, as reported recently in zebrafish by Langenau *et al.*;[96] such transgenic oncofish may be used in drug screens for prevention and treatment of tumors as well as in genetic screens for identifying mutations that suppress or enhance tumorigenesis. Fifth, it has also been proposed to use transgenic zebrafish to scan for gene regulatory elements from the compact *Fugu* genome.[97] Similarly, heterologous promoters from other species such as the mouse and human may also find the transgenic fish system a convenient and economical system for pre-tests. Finally, a gene targeting technique will be developed and used in developmental analyses; so far, several different approaches are being used to develop the gene targeting technology in fish and successful mutagenesis of a target gene by a target-selected method has already been reported.[95] While the whole zebrafish genome sequence is expected to complete soon, a full fledge of applications of transgenic fish technology is also coming of age. (The sequencing for the zebrafish genome is expected to be completed soon, and a diverse spread of applications for transgenic fish technology will come of age.)

Acknowledgments

Our research is supported by National University of Singapore and Biomedical Research Council, A*STAR (Agency for Sciences, Technology and Research), Singapore. We thank Drs. S.J. Du, G.M. Her and H. Okamoto for providing *GFP* transgenic zebrafish pictures.

References

1. Zhu Z, Li G, He L and Chen S (1985). Novel gene transfer into the fertilized eggs of goldfish (*Carassius auratus* L. 1758). *Z. Angew. Ichthyol.* **1**: 31–34.
2. Maclean N and Laight RJ (2000). Transgenic fish: an evaluation of benefits and risks. *Fish Fisher* **1**: 146–172.
3. Ozato K, Kondoh H, Inohara H, Iwamatsu T, Wakamatsu T and Okada TS (1986). Production of transgenic fish: Introduction and expression of chicken delta-crystallin gene in medaka embryos. *Cell Diff.* **19**: 237–244.
4. Chalfie M, Tu Y, Euskirchen G, Ward WW and Prasher DC (1994). Green fluorescent protein as a marker for gene expression. *Science* **263**: 802–805.
5. Westerfield M, Wegner J, Jegalian BG, DeRobertis EM and Puschel AW (1992). Specific activation of mammalian Hox promoters in mosaic transgenic zebrafish. *Genes Dev.* **6**: 591–598.
6. Muller F, Williams DW, Kobolak J, Gauvry L, Goldspink G, Orban L and Maclean N (1997). Activator effect of coinjected enhancers on the muscle-specific expression of promoters in zebrafish embryos. *Mol. Reprod. Dev.* **47**: 404–412.
7. Stuart GW, McMurray JV and Westerfield M (1988). Replication, integration and stable germ-line transmission of foreign sequences injected into early zebrafish embryos. *Development* **103**: 403–412.
8. Chong SSC and Vielkind JR (1989). Expression and fate of CAT reporter gene microinjected into fertilized medaka *Oryzias latipes* eggs in the form of plasmid DNA, recombinant phage particles and its DNA. *Theoret. Appl. Genet.* **78**: 369–380.
9. Inoue K, Ozato K, Kondoh H, Iwamatsu T, Wakamatsu Y, Fijita T and Okada TS (1989). Stage-dependent expression of the chicken δ-crystallin gene in transgenic fish embryos. *Cell Diff. Dev.* **27**: 57–68.
10. Gong Z, Hew CL and Vielkind JR (1991). Functional analysis and temporal expression of promoter regions from fish antifreeze protein genes in transgenic Japanese medaka embryos. *Mol. Mar. Biol. Biotechnol.* **1**: 64–72.
11. Higashijima S, Okamoto H, Ueno N, Hotta Y and Eguchi G (1997). High-frequency generation of transgenic zebrafish which reliably express

GFP in whole muscles or the whole body by using promoters of zebrafish origin. *Dev. Biol.* **192**: 289–299.

12. Meng A, Tang H, Ong BA, Farrell MJ and Lin S (1997). Promoter analysis in living zebrafish embryos identifies a cis-acting motif required for neuronal expression of GATA-2. *Proc. Natl. Acad. Sci. USA* **94**: 6267–6272.
13. Long Q, Meng A, Wang H, Jessen JR, Farrell MJ and Lin S (1997). GATA-1 expression pattern can be recapitulated in living transgenic zebrafish using GFP reporter gene. *Development* **124**: 4105–4111.
14. Ju B, Xu Y, He J, Liao J, Yan T, Hew CL, Lam TJ and Gong Z (1999). Faithful expression of Green Fluorescent Protein (GFP) in transgenic zebrafish embryos under control of zebrafish gene promoters. *Dev. Genet.* **25**: 158–167.
15. Ghanem N, Jarinova O, Amores A, Long Q, Hatch G, Park BK, Rubenstein JL and Ekker M (2003). Regulatory roles of conserved intergenic domains in vertebrate dlx bigene clusters. *Genome Res.* **13**: 533–543.
16. Du SJ, Gao J and Anyangwe V (2003). Muscle-specific expression of *myogenin* in zebrafish embryos is controlled by multiple regulatory elements in the promoter. *Comp. Biochem. Physiol. B Biochem. Mol. Biol.* **134**: 123–134.
17. Kusakabe R, Kusakabe T and Suzuki N (1999). *In vivo* analysis of two striated muscle actin promoters reveals combinations of multiple regulatory modules required for skeletal and cardiac muscle-specific gene expression. *Int. J. Dev. Biol.* **43**: 541–554.
18. Kusakabe T and Suzuki N (2001). A cis-regulatory element essential for photoreceptor cell-specific expression of a medaka retinal guanylyl cyclase gene. *Dev. Genes. Evol.* **211**: 145–149.
19. Ju B, Chong SW, He J, Wang X, Xu Y, Wan H, Tong Y, Yan T, Korzh V and Gong Z (2003). Recapitulation of fast skeletal muscle development in zebrafish by transgenic expression of GFP under the *mylz2* promoter. *Dev. Dyn.* **227**: 14–26.
20. Gong Z, Ju B, Wang X, He J, Wan H, Sudha PM and Yan T (2002). Green fluorescent protein expression in germ-line transmitted transgenic zebrafish under a stratified epithelial promoter from keratin8. *Dev. Dyn.* **223**: 204–215.

21. Wan H, He J, Ju B, Yan T, Lam TJ and Gong Z (2002). Generation of two-color transgenic zebrafish using the green and red fluorescent protein reporter genes *gfp* and *rfp*. *Mar. Biotechnol.* **4**: 146–154.
22. Muller F, Chang B, Albert S, Fischer N, Tora L and Strahle U (1999). Intronic enhancers control expression of zebrafish sonic hedgehog in floor plate and notochord. *Development* **126**: 2103–2116.
23. Rastegar S, Albert S, Roux IL, Fischer N, Bl'ader P, Müller F and Strähle U (2002). A floor plate enhancer of the zebrafish netrin1 gene requires Cyclops (nodal) signaling and the winged helix transcription factor FoxA2. *Dev. Biol.* **252**: 1–14.
24. Du SJ, Devoto SH, Westerfield M and Moon RT (1997). Positive and negative regulation of muscle cell identity by members of the hedgehog and TGF-beta gene families. *J. Cell Biol.* **139**: 145–156.
25. Yan YL, Talbot WS, Egan ES and Postlethwait JH (1998). Mutant rescue by BAC clone injection in zebrafish. *Genomics* **50**: 287–289.
26. Hyodo–Taguchi Y, Winkler C, Kurihara Y, Schartl A and Schartl M (1997). Phenotypic rescue of the albino mutation in the medakafish (*Oryzias latipes*) by a mouse tyrosinase transgene. *Mech. Dev.* **68**: 27–35.
27. Inagaki H, Koga A, Bessho Y and Hori H (1998). The tyrosinase gene from medakafish: transgenic expression rescues albino mutation. *Pigment Cell Res.* **11**: 283–290.
28. Fu L, Mambrini M, Perrot E and Chourrout D (2000). Stable and full rescue of the pigmentation in a medaka albino mutant by transfer of a 17 kb genomic clone containing the medaka tyrosinase gene. *Gene 2000* **241**: 205–211.
29. Field HA, Ober EA, Roeser T and Stainier DYR (2003). Formation of the digestive system. I. Liver morphogenesis. *Dev. Biol.* **253**: 279–290.
30. Stuart GW, Vielkind JR, McMurray JV and Westerfield M (1990). Stable lines of transgenic zebrafish exhibit reproducible patterns of transgene expression. *Development* **103**: 403–412.
31. Culp P, Nusslein-Volhard C and Hopkins N (1991). High-frequency germ-line transmission of plasmid DNA sequences injected into fertilized zebrafish eggs. *Proc. Natl. Acad. Sci. USA* **88**: 7953–7957.
32. Bayer TA and Campos-Ortega JA (1992). A transgene containing *lacZ* is expressed in primary sensory neurons in zebrafish. *Development* **115**: 421–426.

33. Gibbs PD, Gray A and Thorgaard G (1994). Inheritance of P element and reporter gene sequences in zebrafish. *Mol. Mar. Biol. Biotechnol.* **3**: 317–326.
34. Amsterdam A, Lin S and Hopkins N (1995). The *Aequorea victoria* green fluorescent protein can be used as a reporter in live zebrafish embryos. *Dev. Biol.* **171**: 123–129.
35. Schilling TF and Kimmel CB (1997). Musculoskeletal patterning in the pharyngeal segments of the zebrafish embryo. *Development* **124**: 2945–2960.
36. Motoike T, Loughna S, Perens E, Roman BL, Liao W, Chau TC, Richardson CD, Kawate T, Kuno J, Weinstein BM, Stainier DY and Sato TN (2000). Universal GFP reporter for the study of vascular development. *Genesis* **28**: 75–81.
37. Udvadia AJ, Koster RW and Skene JH (2001). GAP-43 promoter elements in transgenic zebrafish reveal a difference in signals for axon growth during CNS development and regeneration. *Development* **128**: 1175–1182.
38. Winkler S, Loosli F, Henrich T, Wakamatsu Y and Wittbrodt J (2000). The conditional medaka mutation *eyeless* uncouples patterning and morphogenesis of the eye. *Development* **127**: 1911–1919.
39. Wakamatsu Y, Ju B, Pristyaznhyuk I, Niwa K, Ladygina T, Kinoshita M, Araki K and Ozato K (2001). Fertile and diploid nuclear transplants derived from embryonic cells of a small laboratory fish, medaka (*Oryzias latipes*). *Proc. Natl. Acad. Sci. USA* **98**: 1071–1076.
40. Lee KY, Huang H, Ju B, Yang Z and Lin S (2002). Cloned zebrafish by nuclear transfer from long-term-cultured cells. *Nat. Biotechnol.* **20**: 795–799.
41. Roeser T and Baier H (2003). Visuomotor behaviors in larval zebrafish after GFP-guided laser ablation of the optic tectum. *J. Neurosci.* **23**: 3726–3734.
42. Li YX, Farrell MJ, Liu R, Mohanty N and Kirby ML (2000). Double-stranded RNA injection produces null phenotypes in zebrafish. *Dev. Biol.* **217**: 394–405.
43. Zhao Z, Cao Y, Li M and Meng A (2001). Double-stranded RNA injection produces nonspecific defects in zebrafish. *Dev. Biol.* **229**: 215–223.
44. Nasevicius A and Ekker SC (2000). Effective targeted gene 'knockdown' in zebrafish. *Nat. Genet.* **26**: 216–220.

45. Peters KG, Rao PS, Bell BS and Kindman LA (1995). Green fluorescent fusion proteins: powerful tools for monitoring protein expression in live zebrafish embryos. *Dev. Biol.* **171**: 252–257.
46. Wang H, Long Q, Marty SD, Sassa S and Lin S (1998). A zebrafish model for hepatoerythropoietic porphyria. *Nat. Genet.* **20**: 239–243.
47. Long Q, Huang H, Shafizadeh E, Liu N and Lin S (2000). Stimulation of erythropoiesis by inhibiting a new hematopoietic death receptor in transgenic zebrafish. *Nat. Cell Biol.* **2**: 549–552.
48. Pauls S, Geldmacher–Voss B and Campos–Ortega JA (2001). A zebrafish histone variant H2A.F/Z and a transgenic H2A.F/Z:GFP fusion protein for *in vivo* studies of embryonic development. *Dev. Genes Evol.* **211**: 603–610.
49. Yoshizaki G, Takeuchi Y, Sakatani S and Takeuchi T (2000). Germ cell-specific expression of green fluorescent protein in transgenic rainbow trout under control of the rainbow trout vasa-like gene promoter. *Int. J. Dev. Biol.* **44**: 323–326.
50. Tanaka M, Kinoshita M, Kobayashi D and Nagahama Y (2001). Establishment of medaka (*Oryzias latipes*) transgenic lines with the expression of green fluorescent protein fluorescence exclusively in germ cells: a useful model to monitor germ cells in a live vertebrate. *Proc. Natl. Acad. Sci. USA* **98**: 2544–2549.
51. Knaut H, Steinbeisser H, Schwarz H and Nusslein–Volhard C (2002). An evolutionary conserved region in the vasa 3′UTR targets RNA translation to the germ cells in the zebrafish. *Curr. Biol.* **12**: 454–466.
52. Krovel AV and Olsen LC (2002). Expression of a vas::EGFP transgene in primordial germ cells of the zebrafish. *Mech. Dev.* **116**: 141–150.
53. Higashijima S, Hotta Y and Okamoto H (2000). Visualization of cranial motor neurons in live transgenic zebrafish expressing green fluorescent protein under the control of the islet-1 promoter/enhancer. *J. Neurosci.* **20**: 206–218.
54. Halloran MC, Sato-Maeda M, Warren JT, Su F, Lele Z, Krone PH, Kuwada JY and Shoji W (2000). Laser-induced gene expression in specific cells of transgenic zebrafish. *Development* **127**: 1953–1960.
55. Hammerschmidt M, Pelegri F, Mullins MC, Kane DA, Brand M, van Eeden FJ, Furutani–Seiki M, Granato M, Haffter P, Heisenberg CP, Jiang YJ, Kelsh RN, Odenthal J, Warga RM, Nusslein–Volhard C (1996). Mutations affecting morphogenesis during gastrulation and tail formation in the zebrafish, *Danio rerio*. *Development* **123**: 143–151.

56. Huang H, Liu N and Lin S (2001). Pdx-1 knockdown reduces insulin promoter activity in zebrafish. *Genesis* **30**: 134–136.
57. Perz–Edwards A, Hardison NL and Linney E (2001). Retinoic acid-mediated gene expression in transgenic reporter zebrafish. *Dev. Biol.* **229**: 89–101.
58. Dorsky RI, Sheldahl LC and Moon RT (2002). A transgenic Lefl/beta catenin-dependent reporter is expressed in spatially restricted domains throughout zebrafish development. *Dev. Biol.* **241**: 229–237.
59. Carvan MJ 3rd, Dalton TP, Stuart GW and Nebert DW (2000). Transgenic zebrafish as sentinels for aquatic pollution. *Ann. NY Acad. Sci.* **919**: 133–147.
60. Park HC, Kim CH, Bae TK, Yeo SY, Kim SH, Hong SK, Shin J, Yoo KW, Hibi M, Hirano T, Miki N, Chitnis AB and Huh TL (2000). Analysis of upstream elements in the HuC promoter leads to the establishment of transgenic zebrafish with fluorescent neurons. *Dev. Biol.* **227**: 279–293.
61. Cross LM, Cook MA, Lin S, Chen JN and Rubinstein (2003). Rapid analysis of angiogenesis drugs in a live fluorescent zebrafish assay. *Arterioscler. Thromb. Vasc. Biol.* **23**: 911–912.
62. Liu NA, Huang H, Yang Z, Herzog W, Hammerschmidt M, Lin S and Melmed S (2003). Pituitary corticotroph ontogeny and regulation in transgenic zebrafish. *Mol. Endocrinol.* **17**: 959–966.
63. Jessen JR, Meng A, McFarlane RJ, Paw BH, Zon LI, Smith GR and Lin S (1998). Modification of bacterial artificial chromosomes through chi-stimulated homologous recombination and its application in zebrafish transgenesis. *Proc. Natl. Acad. Sci. USA* **95**: 5121–5126.
64. Jessen JR, Willett CE and Lin S (1999). Artificial chromosome transgenesis reveals long-distance negative regulation of rag1 in zebrafish. *Nat. Genet.* **23**: 15–16.
65. Yeo SY, Little MH, Yamada T, Miyashita T, Halloran MC, Kuwada JY, Huh TL and Okamoto H (2001). Overexpression of a slit homologue impairs convergent extension of the mesoderm and causes cyclopia in embryonic zebrafish. *Dev. Biol.* **230**: 1–17.
66. Blechinger SR, Warren JT Jr, Kuwada JY and Krone PH (2002). Developmental toxicology of cadmium in living embryos of a stable transgenic zebrafish line. *Environ. Health Perspect.* **110**: 1041–1046.
67. Lewandoski M (2001). Conditional control of gene expression in the mouse. *Nat. Rev. Genet.* **2**: 743–755.

68. Scheer N and Campos–Ortega JA (1999). Use of the GAL4-UAS technique for targeted gene expression in the zebrafish. *Mech. Dev.* **80**: 153–158.
69. Scheer N, Groth A, Hans S and Campos–Ortega JA (2001). An instructive function for Notch in promoting gliogenesis in the zebrafish retina. *Development* **128**: 1099–1107.
70. Verri T, Argenton F, Tomanin R, Scarpa M, Storelli C, Costa R, Colombo L and Bortolussi M (1997). The bacteriophage T7 binary system activates transient transgene expression in zebrafish (*Danio rerio*) embryos. *Biochem. Biophys. Res. Commun.* **237**: 492–495.
71. Sternberg N and Hamilton D (1981). Bacteriophage P1 site-specific recombination. I. Recombination between loxP sites. *J. Mol. Biol.* **150**: 467–486.
72. Pan X, Wan H, Chia W, Tong Y and Gong Z (2004). Demonstration of site-directed recombination in transgenic zebrafish using the Cre/loxP system. Submitted.
73. Pappenheimer AM Jr (1977). Diphtheria toxin. *Annu. Rev. Biochem.* **46**: 69–94.
74. Palmiter RD, Behringer RR, Quaife CJ, Maxwell F, Maxwell IH and Brinster RL (1987). Cell lineage ablation in transgenic mice by cell-specific expression of a toxin gene. *Cell* **50**: 435–443.
75. Kurita R, Sagara H, Aoki Y, Link BA, Arai K and Watanabe S (2003). Suppression of lens growth by alphaA-crystallin promoter-driven expression of diphtheria toxin results in disruption of retinal cell organization in zebrafish. *Dev. Biol.* **255**: 113–127.
76. Driever W, Solnica–Krezel L, Schier AF, Neuhauss SC, Malicki J, Stemple DL, Stainier DY, Zwartkruis F, Abdelilah S, Rangini Z, Belak J and Boggs C (1996). A genetic screen for mutations affecting embryogenesis in zebrafish. *Development* **123**: 37–46.
77. Haffter P, Granato M, Brand M, Mullins MC, Hammerschmidt M, Kane DA, Odenthal J, van Eeden FJ, Jiang YJ, Heisenberg CP, Kelsh RN, Furutani–Seiki M, Vogelsang E, Beuchle D, Schach U, Fabian C and Nusslein–Volhard C (1996). The identification of genes with unique and essential functions in the development of the zebrafish, *Danio rerio*. *Development* **123**: 1–36.
78. Lin S, Gaiano N, Culp P, Burns JC, Friedmann T, Yee JK and Hopkins N (1994). Integration and germ-line transmission of a pseudotyped retroviral vector in zebrafish. *Science* **265**: 666–669.

79. Gaiano N, Allende M, Amsterdam A, Kawakami K and Hopkins N (1996a). Highly efficient germ-line transmission of proviral insertions in zebrafish. *Proc. Natl. Acad. Sci. USA* **93**: 7777–7782.
80. Gaiano N, Amsterdam A, Kawakami K, Allende M, Becker T and Hopkins N (1996b). Insertional mutagenesis and rapid cloning of essential genes in zebrafish. *Nature* **383**: 829–832.
81. Golling G, Amsterdam A, Sun Z, Antonelli M, Maldonado E, Chen W, Burgess S, Haldi M, Artzt K, Farrington S, Lin SY, Nissen RM and Hopkins N (2002). Insertional mutagenesis in zebrafish rapidly identifies genes essential for early vertebrate development. *Nat. Genet.* **31**: 135–140.
82. Chen W, Burgess S, Golling G, Amsterdam A and Hopkins N (2002). High-throughput selection of retrovirus producer cell lines leads to markedly improved efficiency of germ line-transmissible insertions in zebra fish. *J. Virol.* **76**: 2192–2198.
83. Ivics Z, Izsvak Z and Hackett PB (1999). Genetic applications of transposons and other repetitive elements in zebrafish. *Methods Cell Biol.* **60**: 99–131.
84. Ivics Z, Hackett PB, Plasterk RH and Izsvak Z (1997). Molecular reconstruction of Sleeping Beauty, a Tc1-like transposon from fish, and its transposition in human cells. *Cell* **91**: 501–510.
85. Kawakami K, Shima A and Kawakami N (2000). Identification of a functional transposase of the Tol2 element, an Ac-like element from the Japanese medaka fish, and its transposition in the zebrafish germ lineage. *Proc. Natl. Acad. Sci. USA* **97**: 11403–11408.
86. Dupuy AJ, Clark K, Carlson CM, Fritz S, Davidson AE, Markley KM, Finley K, Fletcher CF, Ekker SC, Hackett PB, Horn S and Largaespada DA (2002). Mammalian germ-line transgenesis by transposition. *Proc. Natl. Acad. Sci. USA* **99**: 4495–4499.
87. Raz E, van Luenen HG, Schaerringer B, Plasterk RH and Driever W (1998). Transposition of the nematode *Caenorhabditis elegans* Tc3 element in the zebrafish *Danio rerio*. *Curr. Biol.* **8**: 82–88.
88. Winn RN, Norris MB, Brayer KJ, Torres C and Muller SL (2000). Detection of mutations in transgenic fish carrying a bacteriophage lambda cII transgene target. *Proc. Natl. Acad. Sci. USA* **97**: 12655–12660.

89. McCreath KJ, Howcroft J, Campbell KH, Colman A, Schnieke AE and Kind AJ (2000). Production of gene-targeted sheep by nuclear transfer from cultured somatic cells. *Nature* **405**: 1066–1069.
90. Lai L, Kolber–Simonds D, Park KW, Cheong HT, Greenstein JL, Im GS, Samuel M, Bonk A, Rieke A, Day BN, Murphy CN, Carter DB, Hawley RJ and Prather RS (2002). Production of alpha-1,3-galactosyltransferase knockout pigs by nuclear transfer cloning. *Science* **295**: 1089–1092.
91. Oates AC, Bruce AE and Ho RK (2000). Too much interference: injection of double-stranded RNA has nonspecific effects in the zebrafish embryo. *Dev. Biol.* **224**: 20–28.
92. Hong Y, Winkler C and Schartl M (1998). Production of medakafish chimeras from a stable embryonic stem cell line. *Proc. Natl. Acad. Sci. USA* **95**: 3679–3684.
93. Ma C, Fan L, Ganassin R, Bols N and Collodi P (2001). Production of zebrafish germ-line chimeras from embryo cell cultures. *Proc. Natl. Acad. Sci. USA* **98**: 2461–2466.
94. Wong EA, Wentworth AI and Wentworth BG (1999). Generation of transgenic poultry by transfection of primordial germ cells. In: Murray JD, Anderson GB, Oberbauer AM and McGloughlin (eds.), *Embryology: Transgenic Animal in Agriculture*. CABI Publishing, Wallingford, UK, pp. 117–129.
95. Wienholds E, Schulte–Merker S, Walderich B and Plasterk RH (2002). Target-selected inactivation of the zebrafish rag1 gene. *Science* **297**: 99–102.
96. Langenau DM, Traver D, Ferrando AA, Kutok JL, Aster JC, Kanki JP, Lin S, Prochownik E, Trede NS, Zon LI and Look AT (2002). Myc-induced T cell leukemia in transgenic zebrafish. *Science* **299**: 887–890.
97. Rothenberg EV (2001). Mapping of complex regulatory elements by pufferfish/zebrafish transgenesis. *Proc. Natl. Acad. Sci. USA* **98**: 6540–6542.
98. Du SJ, Gong Z, Fletcher GL, Shears MA, King MJ, Idler DR and Hew CL (1992). Growth enhancement in transgenic Atlantic salmon by use of fish antifreeze/growth hormone chimeric gene constructs. *Bio./Tech.* **10**: 176–181.

99. Neumann CJ and Nusslein–Volhard C (2000). Patterning of the zebrafish retina by a wave of sonic hedgehog activity. *Science* **289**: 2137–2139.
100. Du SJ and Dienhart M (2001). Zebrafish tiggy-winkle hedgehog promoter directs notochord and floor plate green fluorescence protein expression in transgenic zebrafish embryos. *Dev. Dyn.* **222**: 655–666.
101. Goldman D, Hankin M, Li Z, Dai X and Ding J (2001). Transgenic zebrafish for studying nervous system development and regeneration. *Transgenic Res.* **10**: 21–33.
102. Huang H, Vogel SS, Liu N, Melton DA and Lin S (2001). Analysis of pancreatic development in living transgenic zebrafish embryos. *Mol. Cell Endocrinol.* **177**: 117–124.
103. Kennedy BN, Vihtelic TS, Checkley L, Vaughan KT and Hyde DR (2001). Isolation of a zebrafish rod opsin promoter to generate a transgenic zebrafish line expressing enhanced green fluorescent protein in rod photoreceptors. *J. Biol. Chem.* **276**: 14037–14043.
104. Hamaoka T, Takechi M, Chinen A, Nishiwaki Y and Kawamura S (2002). Visualization of rod photoreceptor development using GFP transgenic zebrafish. *Genesis* **34**: 215–220.
105. Asaoka Y, Mano H, Kojima D and Fukada Y (2002). Pineal expression-promoting element (PIPE), a cis-acting element, directs pineal-specific gene expression in zebrafish. *Proc. Natl. Acad. Sci. USA* **99**: 15456–15461.
106. Gilmour DT, Maischein HM and Nusslein–Volhard S (2002). Migration and function of a glial subtype in the vertebrate peripheral nervous system. *Neuron* **34**: 577–588.
107. Gothilf Y, Toyoma R, Coon SL, Du SJ, Dawid IB and Klein DC (2002). Pineal-specific expression of green fluorescent protein under the control of the serotonin-N-acetyltransferase gene regulatory regions in transgenic zebrafish. *Dev. Dyn.* **225**: 241–249.
108. Lawson ND and Weinstein BM (2002). *In vivo* imaging of embryonic vascular development using transgenic zebrafish. *Dev. Biol.* **248**: 307–318.
109. Picker A, Scholpp S, Bohli H, Takeda H and Brand M (2002). A novel transcriptional feedback loop in midbrain-hindbrain boundary development is revealed through analysis of the zebrafish pax2.1 promoter in transgenic lines. *Development* **129**: 3227–3239.

110. Hans S and Campos-Ortega JA (2002). On the organisation of the regulatory region of the zebrafish deltaD gene. *Development* **129**: 4773–4784.
111. Tokuoka H, Yoshida T, Matsuda N and Mishina M (2002). Regulation by glycogen synthase kinase-3β of the arborization field and maturation of retinotectal projection in zebrafish. *J. Neurosci.* **22**: 10324–10332.
112. Yoshida T, Ito A, Matsuda N and Mishina M (2002). Regulation by protein kinase A switching of axonal pathfinding of zebrafish olfactory sensory neurons through the olfactory placod-olfactory bulb boundary. *J. Neurosci.* **22**: 4964–4972.
113. Blader P, Plessy C and Strähle U (2003). Multiple regulatory elements with spatially and temporally distinct activities control neurogenin 1 expression in primary neurons of the zebrafish embryo. *Mech. Dev.* **120**: 211–218.
114. Her GM, Chiang C, Chen W and Wu J (2003). *In vivo* studies of liver-type fatty acid binding protein (L-FABP) gene expression in liver of transgenic zebrafish (*Danio rerio*). *FEBS Lett.* **538**: 125–133.
115. Liu NA, Huang H, Yang Z, Herzog W, Hammerschmidt M, Lin S and Melmed S (2003). Pituitary corticotroph ontogeny and regulation in transgenic zebrafish. *Mol. Endocrinol.* **17**: 959–966.
116. Wan H, Mudumana SP, Korzh V and Gong Z (2004). Generation of exocrine pancreas specific GFP transgenic zebrafish and analyses of pancreas development. Submitted.
117. Linney E, Hardison NL, Lonze BE, Lyons S and DiNapoli L (1999). Transgene expression in zebrafish: a comparison of retroviral-vector and DNA-injection approaches. *Dev. Biol.* **213**: 207–216.
118. Gibbs PD and Schmale MC (2000). GFP as a genetic marker scorable throughout the life cycle of transgenic zebra fish. *Mar. Biotechnol.* **2**: 107–125.
119. Motoike T, Loughna S, Perens E, Roman BL, Liao W, Chan TC, Richardson CD, Kawate T, Kuno J, Weinstein BM, Stainier DY and Sato TN (2000). Universal GFP reporter for the study of vascular development. *Genesis* **28**: 75–81.
120. Hsiao CD, Hsieh FJ and Tsai HJ (2001). Enhanced expression and stable transmission of transgenes flanked by inverted terminal repeats from adeno-associated virus in zebrafish. *Dev. Dyn.* **220**: 323–336.

121. Perkins BD, Kainz PM, O'Malley DM and Dowling JE (2002). Transgenic expression of a GFP-rhodopsin COOH-terminal fusion protein in zebrafish rod photoreceptors. *Vis. Neurosci.* **19**: 257–264.
122. Hamada K, Tamaki K, Sasado T, Watai Y, Kani S, Wakamatsu Y, Ozato K, Kinoshita M, Kohno R, Takagi S and Kimura M (1998). Usefulness of the medaka beta-actin promoter investigated using a mutant GFP reporter gene in transgenic medaka (*Oryzias latipes*). *Mol. Mar. Biol. Biotechnol.* 7: 73–80.
123. Yamauchi M, Kinoshita M, Sasanuma M, Tsuji S, Terada M, Morimyo M and Ishikawa Y (2000). Introduction of a foreign gene into medakafish using the particle gun method. *J. Exp. Zool.* **287**: 285–293.
124. Chou CY, Horng LS and Tsai HJ (2001). Uniform GFP expression in transgenic medaka (*oryzias latipes*) at the Fo generation. *Transgenic Res.* **10**: 303–315.
125. Kinoshita M, Kani S, Ozato K and Wakamatsu Y (2000). Activity of the medaka translation elongation factor 1alpha-A promoter examined using the GFP gene as a reporter. *Dev. Growth Differ.* **42**: 469–478.
126. Zeng Z and Gong Z (2004). Faithful expression of flurescent reporter genes under two tissue-specific zebrafish promoters in transgenic medaka. In preparation.
127. Zeng Z, Shan T and Gong Z (2004). Development of estrogen-responsive transgenic medaka for environmental monitoring of endocrine disrupters. In preparation.

Chapter 15

Cloning the Zebrafish

Bensheng Ju, Haigen Huang, Ki-Young Lee and Shuo Lin
Department of Molecular, Cellular, and Developmental Biology
University of California Los Angeles
Los Angeles, CA 90095-1606

Although the efforts started 40 years ago, resources dedicated to fish cloning and achievements made so far lag far behind those in other vertebrates, especially mammals. In this review, we summarize the fish cloning work carried out by a number of laboratories and our data of generating cloned zebrafish from long-term cultured embryonic fibroblast cells. This success lays the foundation for developing cloning-based gene-trapping and homologous recombination technologies for gene function studies in zebrafish. In addition, fish cloning should contribute to studies addressing basic issues such as development and aging of cloned animals.

1. Introduction

In 1938, Hans Spemann[1] first conceived the idea of nuclear transfer in his book *Embryonic Development and Introduction*. His idea only became reality in 1952 when Robert Briggs and Thomas King[2] obtained *Rana pipiens* tadpoles by transferring blastula cells into enucleated eggs. Ever since, the frogs of *Rana pipens* and *Xenopus* became the primary targets for cloning, culminating in 1962 when John Gurdon announced cloning frogs using nuclei of fully differentiated adult intestinal cells.[3,4] The works in frogs inspired researchers worldwide to try their luck in their own favorite model animals. Tong Dizhou (T. C. Tung) in China initiated nuclear transplantation in fish in early 1960s.[5] Works on mammals started in early 1980s, with the first major breakthrough achieved in 1997 by the birth of Dolly the sheep from an adult somatic cell nucleus.[6]

Since Dolly the sheep came into this world, animal cloning has become a buzzword. The number of animals that were successfully cloned just keeps growing, from sheep to cow,[7,8] mouse,[9] pig,[10] rabbit,[11] goat,[12] and domestic cat.[13] The spectacular successes in mammals had been forecasted by Di Bernardino in her 1997 book *Genomic Potential of Differentiated Cells*, in which she stated "my intuition is that some of the cell cycle problems, and hopefully all, can be solved in the experiments on mammalian species, because mammalian cell cycles are longer than those in insects, fish and amphibians." And she was right.

The question that needs to be addressed is whether same successes could be replicated in other vertebrates such as amphibians and fish, which have longer history for nuclear transplantation.[14,15] Unlike mammals whose embryos generally develop rather slowly and initiate transcription early in embryonic development, amphibians and fish generally show fast embryonic development and late transcription at the mid-blastula stage (mid-blastula transition). Therefore, transplanted donor nuclei need to switch quickly from transcription to DNA replication to keep pace with the extremely rapid cleavage cycle imposed by the recipient cytoplasm, a scenario that is thought to be not favorable for correct reprogramming.[14]

Fish represent an important system for nuclear transplantation because of their value as an important food source and also as important model organisms for physiology, genetics and developmental studies. Several fish species have been used for nuclear transplantation studies, such as the loach *Misgurnus fossilis*,[16] cyprinids (reviewed by Yan[15]) and the small freshwater fish, medaka *Oryzias latipes*. Here we summarize the achievements made on those few fish species and describe the process leading to the cloning of zebrafish from long-term cultured cells in our laboratory.

2. Fish Cloning

2.1. Cyprinids

The late professor T. C. Tung initiated nuclear transplantation in fish in China.[5] He and his colleagues mainly focused on studying the effect of the nucleus and the cytoplasm on development and producing fish clones of commercial importance for agriculture. They used fish of different genus (mainly in cyprinid fish) and even different subfamilies as nuclei donors or recipients to produce so-called nucleocytoplasmic hybrids, in which the nuclei of one species were introduced into cytoplasm of another species from which the pronuclei were manually removed. In one of the combinations involving two species of different genus, blastula nuclei from the common carp *Cyprinus carpio* L. were transplanted into enucleated eggs of the crucian carp *C. auratus* L and they were able to obtain adult and fertile hybrid fish.[5,15] Their work demonstrated that teleost blastula nuclei are totipotent, and phenotypic traits of the nucleocytoplasmic hybrids, though mostly controlled by the nucleus, sometimes were controlled by cytoplasm or both nucleus and cytoplasm. Professor Shaoyi Yan superbly summarized these works in his book published by IUBS Educational and Cultural Press Ltd., Hong Kong.[15]

2.2. Loach

Gasaryan *et al.*'s work in 1979[16] on loach *Misgurnus fossilis* L. is considered to be an important contribution since he was the first one to introduce genetic markers into fish nuclear transplantation. The loach species they used was known to have two alleles of carboxylesterase genes coding for two distinct forms of the enzyme E2, electrophoretically fast and slow. They introduced nuclei of blastulae homozygous for the fast E2 into either enucleated (by X-ray irradiation) or non-enucleated eggs homozygous for the slow E2. When enucleated eggs were used as nuclei recipient, only larvae carrying the fast E2 were obtained, demonstrating that the donor blastulae nuclei contributed to the embryonic development; when non-enucleated eggs were used as recipient, five out of six larvae obtained carried only the fast E2, again showing perhaps only the donor nuclei contributed to embryonic development, and the recipient pronuclei were eliminated during embryonic development, an intriguing observation that could not easily explained. Gasaryan's nuclear transplants survived only to the active feeding larvae stage; therefore it did not establish the totipotency of the loach blastula cells.

2.3. Medaka

Medaka is a small (3–4 cm) egg-laying freshwater fish favored by the Japanese researchers for various studies.[17,18] Laboratory of Kenjiro Ozato and Yoko Wakamatsu at Nagoya University of Japan has systematically worked on this species and established the procedure for nuclear transplantation in medaka.[19] Taking advantage of various strains and transgenic lines available to them, they introduced blastulae nuclei with markers into non-enucleated eggs of an orange-red strain that has non-melanized melanophores in the skin. An inbred strain with the wild-type body color (HNI-I), which is genetically dominant to the orange-red color, was used as nuclei donors. In this combination of donor and recipient, successful nuclear transplants were identified by the appearance of densely pigmented melanophores in the skin. Out of 845 eggs operated, 52 hatched, representing 6.1% of the eggs operated. Over 80% hatchlings had the pigmented melanophores in their skin, showing contribution from the nucleus donors and they were all triploid.[19] They also obtained fish without the pigmented melanophores. These fish grew to adult stage to become fertile females, presumably originated from parthenogenesis. In another series of experiments, when the donor strain was an outbred wild-type in the body color, except for obtaining triploid fish with the wild-type body color and diploid fish without the wild-type body color (parthenogenetic individuals), diploid nuclear transplants carrying the donor marker could also be obtained.[20] These nuclear transplants grew normally to adult stage and reproduced like the donor fish. Although diploid larvae carrying only donor markers were observed in loach *Misgurnus fossilis*[16] nuclear transplantation, the results from medaka is the first to demonstrate that these nuclear transplants could grow normally to adult stage and become fertile individuals. In summary, three phenomena were observed in nuclear transfer involving transplantation of blastula nuclei into the non-enucleated eggs: (1) triploid fish could be produced by fusion of donor and recipient nuclei; (2) only recipient genome contributed to the embryonic development, producing fish of parthenogenetic origin; and (3) donor nuclei solely contributed to embryonic development. The last two scenarios resulted in diploid and fertile nuclear transplants. The mechanism leading to the diploid nuclear transplants of seemingly sole donor origin is unknown. What is known is that fish are generally more permissive to unconventional developmental fates, as evidenced by the presence of polyploids and reproduction by gynogenesis in some species, such as loach *Misgurnus anguillicaudatus*[21] and Crucian carp *Carassius auratus*.[22] This permissiveness may have allowed the loach and medaka eggs to develop normally from the donor cell nuclei, without obvious contribution from the recipient genome.

Wakamatsu *et al.*[23] further introduced blastula nuclei carrying different genetic markers into enucleated eggs of the orange-red strain. The first donor strain was a transgenic HNI-I strain (an inbred strain with the wild-type body color). Out of 588 eggs operated, only 2 hatched; one survived to adult stage and was fertile, representing 0.2% of total transplanted eggs. This fish carried the endogenous PGM allozyme and the transgenic GFP markers of the donor fish. When the same orange-red strain carrying the GFP transgene was used as nuclei donors, out of 291 eggs operated, seven larvae were obtained; five survived to adult stage and were fertile, representing 1.7% of total eggs transplanted. These experiments unequivocally demonstrated the totipotency of the medaka blastula nuclei.

An important lesson from loach and medaka nuclear transplantation experience is that we need to be careful in both planning the experiment and interpreting the results from fish nuclear transplantations. So far, most of nuclear transplantation experiment in fish involves manual enucleation. Since the pronucleus of a fish egg is not visible under most optic microscopes, there is no guarantee that every enucleation operation will be successful. Therefore, it is essential that stable genetic markers are used to determine cloned animals. Additionally, it should be noted that genetic background of the donors and recipients may affect the efficiency of the nuclear transplantation.

2.4. Cloning the Zebrafish

Although fish cloning efforts started 40 years ago,[5] it is still in its infancy stage. Most of the work done in model systems, such as medaka, is still on establishing the basic procedure. Very few studies have been carried out to study the potency of the differentiated or cultured fish cells. Chen *et al.*[24] first attempted to clone the crucian carp using short-term cultured kidney cells. Kidney cells from a triploid fish were transplanted into enucleated eggs of a diploid fish. After two rounds of nuclear transplantation (serial transplantation), one triploid adult fish was obtained. This could represent the first cloned animal from an adult somatic cell. Unfortunately, the study did not receive any international attention and was not followed by additional studies.

Zebrafish has become an important model organism and is now used to study almost every aspect of biology, such as genetics, developmental biology, behavior and human diseases.[25,26] A genome-sequencing project will be completed soon to understand the genetic makeup of this model system and the challenge has now shifted to understanding the functions of more than 30,000 predicated genes present in a typical vertebrate genome. Our rationale to take on the challenge to clone this fish is to seek alternative approaches for homologous recombination-based gene targeting mutagenesis, which has been so powerful in analyzing mouse genetics and development.

The nuclear transplantation procedure is summarized in Fig. 1. The first step towards cloning the zebrafish is to efficiently remove the pronuclei of the recipient eggs. Our enucleation procedure is essentially same as that described by Yan.[15] Matured eggs of fish are arrested at the metaphase of the second meiosis. Upon contact with water, the eggs are activated and begin to release their second polar bodies. In zebrafish the second polar bodies are visible under inverted microscope after egg activation. The relative size and location of the pronucleus in an unfertilized egg can be revealed using Hoechst 33342 staining.[27] The size of the polar body is approximately 1 μm, which is situated immediately above the underlying nucleus. Using the polar body as a reference, we attempted to remove the pronucleus by aspirating a small amount of egg cytoplasm just underneath the polar body. The unfertilized eggs were treated with protease treatment to remove the egg chorions. The eggs were then held by a holding needle with an inner diameter of ~260 μm, and enucleation and nuclear transfer were

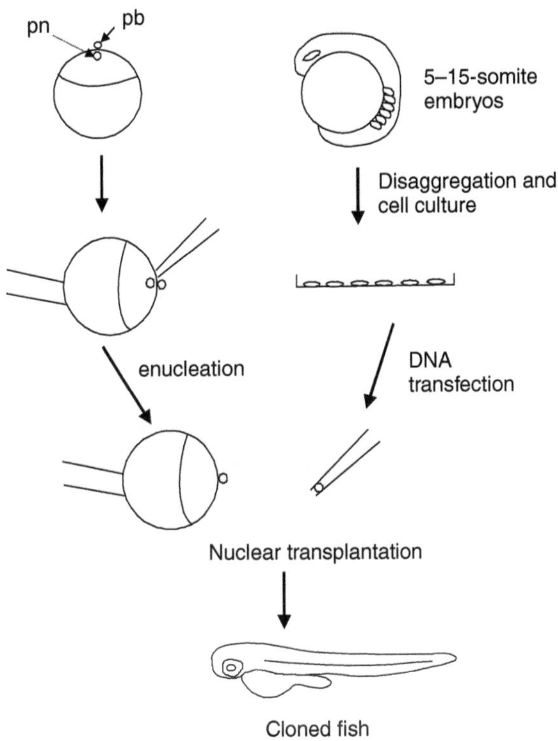

Fig. 1 Zebrafish cloning procedure. An egg was held by a holding needle, its pronucleus (pn) was sucked out by a transfer needle using the polar body (pb) as the reference point. Embryos at 5–15 somites were disaggregated and cultured for 8 weeks, then were transfected by proviral infection and selected for another 4 weeks. The nuclei of these donor cells were transplanted into the enucleated eggs, again using the polar body as the reference point, to create cloned fish. Please note that though the egg's polar body is visible under the microscope, the egg's pronucleus is not.

achieved by a transfer needle with a diameter of ~10 μm. To avoid compromising the egg's developmental potential, we tried to remove the nucleus in a volume that is as small as possible.

Since the totipotency of blastula nuclei has been proven previously,[15,23] we used primary cells from embryos at more advanced developmental stages to test the feasibility of the nuclear transplantation. Embryos at 5–15 somite stages were chosen because it is easier to manipulate the embryos at these stages and also because cell differentiation has already begun. The embryos used for cell culture were homozygous for a transgene expressing the GFP, so the GFP transgene served as a donor marker to help us determine if the donor nuclei indeed contributed to developing the embryos. The individual donor cell was ruptured by repeatedly aspirating with the injection needle and then transplanted into the enucleated egg at the position of enucleation. We normally perform nuclear transfers in the morning and counted each day's operation as one experimental group. Overall, approximately 80% of experiments never yielded any developing embryos, most likely due to poor egg quality. For those experiments that produced developing nuclear transplants, the embryos exhibited various degrees of abnormity, but normal individuals were also obtained. In a series of 8 successful nuclear transfer experimental groups involving 67 transplanted eggs, we obtained a total of 20 embryos (30%) that reached blastula stage, 12 embryos (18%) that hatched, and 11 (16%) of them that survived to adulthood (9 females and 2 males). All the nuclear transplants expressed GFP, indicating the donor cell nuclei contributed to development of the nuclear transplants (Table 1).

Table 1 Nuclear transplants generated using primary embryonic cells.

		Number of individuals		
Experimental groups	Eggs operated	Blastula	Hatched	Adult
N1	9	3	2	2
N2	9	4	2	2
N3	8	2	2	1
N4	9	2	1	1
N5	7	3	1	1
N6	8	2	2	2
N7	8	2	1	1
N8	9	2	1	1
Total	67	20(30)	12(18)	11(16)

*Numbers in parentheses represent the percentage of the total number of transplants from 8 experiments that yield developing embryos. Such experiments represent approximately 30% of total operations.

Encouraged by the transplantation results using primary cells as nuclei donors, we continued to determine the cloning competence of long-term cultured somatic cells. We again used 5–15-somite stage embryos. Embryos were disaggregated and cultured initially for 8 weeks, then a concentrated stock of pseudotyped retroviral vector containing GFP reporter gene driven by the *Xenopus* elongation 1 alpha (XeX) promoter[28] was used to infect these cells. The infection process usually lasted for 4 weeks. The titer of the virus was approximately 2×10^7 colony-forming units per milliliter on these cultured cells. We obtained approximately 20% cells expressing GFP, as determined by FACS analysis and visual inspection under a fluorescent microscope. The infection rate was rather low — we chose this low rate because we wished that each cell would carry a single viral integration.

To ensure that the cells used for nuclear transplantation still possessed the euploid status, a prerequisite for successful nuclear transfer, we performed chromosomal analysis and FACS analyses to determine ploidy of the 12-week-old virus-infected cells. We found that around 77% cultured cells contained a normal chromosomal complement (50 chromosomes, 2N) and FACS results further confirmed this ratio, suggesting that majority of the long-term cultured cells still had the desirable karyotype to support normal zebrafish development after nuclear transplantation. It is generally believed that cells at the G0 and G1 stages are better candidates for nuclear transplantation. In line with the common practices in mammals, we also serum-starved our cultured cell. FACS analyses indicated that none-starved cells had around 59% cells at G0 + G1 stage whereas serum starvation for 4 days increased G0 + G1 stage cells to around 80%. Based on this observation, a 4-day serum-starvation was applied to all donor cells used for nuclear transfer.

By the time the cells were used for nuclear transplantation, they had been in culture for a minimum of 12 weeks. As experiments proceeded, the oldest cells in the experiments were approximately 26 weeks old. As experienced in our initial study using primary cells as donors, more than 80% of experiments failed to produce developing embryos. From the 10 experimental groups that produced embryos that went through cell cleavages, a total of 34 embryos (36%) reached

Table 2 Nuclear transplants generated using long-term cultured fibroblast cells.

Experimental groups	Eggs operated	Number of individuals		
		Blastula	Hatched	Adult
N1	15	5(33)	4(29)	2(13)
N2	14	7(50)	3(21)	2(14)
N3	7	5(71)	2(29)	1(14)
N4	9	3(33)	2(22)	2(22)
N5	9	3(33)	2(22)	2(22)
N6	8	2(25)	0(0)	0(0)
N7	9	1(11)	0(0)	0(0)
N8	6	2(33)	0(0)	0(0)
N9	9	3(33)	0(0)	0(0)
N10	8	3(38)	2(25)	2(25)
Total	94	34(36)	15(15)	11(12)

*Numbers in parentheses represent the percentage of the total number of transplants. Such experiments represent approximately 30% of total operations.

the blastula stage, 15 embryos (16%) in 6 groups hatched, and 11 (12%) of these embryos reached adulthood (Table 2). The 11 adults represent approximately 2% of total embryos operated.

All of the transplanted embryos we obtained expressed GFP, confirming that donor cells contributed to their development. The transgenic embryo produced by cloning has a ubiquitous expression pattern reminiscent of that of retroviral XeX-GFP transgenic zebrafish. Adult zebrafish produced by nuclear transfer appear indistinguishable from normal wild-type zebrafish.

Of those normal individuals, 9 adults produced offspring after mating with wild-type fish. Around 50% of the offspring also expressed GFP and the expression patterns were the same as those of the nuclear transplants throughout the embryonic development. The GFP donor marker gene was transmitted to the subsequent generations following a Mendelian fashion. We also performed southern blot analyses on 8 cloned fish. Each fish had a junction fragment that was different from the others, confirming that they were derived from different donor cells. These data suggest that the viral insertions are stably integrated into the zebrafish genome. We have also examined the karyotype of F1 embryos at 24 hours post-fertilization and no detectable chromosomal abnormalities were observed. Together, these data indicate that the nuclear transplants produced zebrafish that are diploid and fertile.

Cloning lower vertebrates such as fish and amphibians from long-term cultured cells has been a major challenge due to the dramatic difference between cell cycles of early embryos and cultured cells. During zebrafish embryonic development it takes approximately 20 minutes to complete each cell cycle for the first 10 cell divisions. However, the same process requires more than 24 hours in cultured cells, so it is amazing to know that the cell cycles of the developing embryos and the cultured cells are able to synchronize and normal zebrafish could be produced. To the best of our knowledge, this study is the first case in which a lower vertebrate animal has been successfully cloned using long-term cultured cells. Even though the cells had been maintained for an extended period of time, we could not exclude the possibility that there were still some undifferentiated stem cells present in the cultures and it was these undifferentiated cells that contributed to the development of normal cloned zebrafish.

3. Applications of Zebrafish Cloning

We have established the procedure for nuclear transplantation in zebrafish. The cells could be cultured for up to 26 weeks, frozen and thawed, and their capacity for producing viable nuclear transplants remains. This long window period provides ample opportunity for various genetic manipulations to the cells, such as proviral infection and DNA transfection. The availability of such a cloning technology may have many potential applications.

3.1. Production of Transgenic Fish Through Cloning

Although transgenic fish can be produced by direct injection of DNA into fertilized eggs, it requires screening a large number of founder fish to identify germline transmission. Whereas by nuclear transplantation, donor cells containing stable transgene integrations can be selected in cell culture prior to fish cloning. We have achieved this by obtaining heterozygous transgenic fish at the first generation.

3.2. Development of Techniques for Generating Zebrafish from Cultured Cells Carrying Gene-Traps

This involves identifying and cloning trapped genes from cultured cells carrying the gene-trap constructs and generating cloned zebrafish using these cells for functional study. We have already carried out initial gene-trapping studies in the cultured cells and successfully produced cloned zebrafish carrying gene-trap events.

3.3. Development of Techniques for Targeted Mutagenesis in Zebrafish

This involves designing targeting constructs, selecting cells carrying homologous recombination events and cloning zebrafish using these cells. We are now testing whether DNA homologous recombination could be achieved in our cultured cells and whether cells carrying a homologous recombination event can be used to produce normal fertile zebrafish. If successful, zebrafish will have all the genetic tools available to the mouse system, and fully realize its potential as a model system to study vertebrate gene function and human diseases.

3.4. Study of Effects of Cloning on Animal Development

Cloned zebrafish could be excellent models for studying effects of the cloning has on animal development. Developmental abnormalities in cloned zebrafish can be easily found due to *in vitro* and transparent embryogenesis. The short generation time and easy access to large number of progenies mean that effects of cloning can be monitored thoroughly in multiple generations in a relatively short time.

Acknowledgments

We thank members of our laboratory for technical assistance and discussion. This work was supported by a grant from the National Institutes of Health (R01 RR13227) to S.L.

References

1. Spemann H (1938). *Embryonic Development and Induction.* New Haven, CT, Yale University Press.
2. Briggs R and King TJ (1952). Transplantation of living nuclei from blastula cells into enucleated frog's eggs. *Proc. Natl. Acad. Sci. USA* **38**: 455–463.
3. Gurdon JB (1962). The developmental capacity of nuclei taken from intestinal epithelium cells of feeding tadpoles. *J. Embryol. Exp. Morph.* **10**: 622–640.
4. Gurdon JB and Uehlinger V (1966). "Fertile" intestine nuclei. *Nature* **210**: 1240–1241.
5. Tung TC, Wu SC, Tung YYF, Yan SY, Tu M and Lu TY (1963). Nuclear transplantation in fish. *Science Bulletin, Acad. Sin.* (In Chinese) 7: 60–61.
6. Wilmut I, Schnieke AE, McWhir J, Kind AJ and Campbell KH (1997). Viable offspring derived from fetal and adult mammalian cells. *Nature* **385**: 810–813.
7. Cibelli JB, Stice SL, Golueke PJ, Kane JJ, Jerry J, Blackwell C, Ponce de Leon FA and Robl JM (1998). Cloned transgenic calves produced from nonquiescent fetal fibroblasts. *Science* **280**: 1256–1258.
8. Kasinathan P, Knott JG, Wang Z, Jerry DJ and Robl JM (2001). Production of calves from G1 fibroblasts. *Nat. Biotechnol.* **19**: 1176–1178.
9. Wakayama T, Perry AC, Zuccotti M, Johnson KR and Yanagimachi R (1998). Full-term development of mice from enucleated oocytes injected with cumulus cell nuclei. *Nature* **394**: 369–374.
10. Polejaeva IA, Chen SH, Vaught TD, Page RL, Mullins J, Ball S, Dai Y, Boone J, Walker S, Ayares DL, Colman A and Campbell KH (2000). Cloned pigs produced by nuclear transfer from adult somatic cells. *Nature* **407**: 86–90.

11. Chesne P, Adenot PG, Viglietta C, Baratte M, Boulanger L and Renard JP (2002). Cloned rabbits produced by nuclear transfer from adult somatic cells. *Nat. Biotechnol.* **20**: 366–369.
12. Baguisi A, Behboodi E, Melican DT, Pollock JS, Destrempes MM, Cammuso C, Williams JL, Nims SD, Porter CA, Midura P, Palacios MJ, Ayres SL, Denniston RS, Hayes ML, Ziomek CA, Meade HM, Godke RA, Gavin WG, Overstrom EW and Echelard Y (1999). Production of goats by somatic cell nuclear transfer. *Nat. Biotechnol.* **17**: 456–461.
13. Shin T, Kraemer D, Pryor J, Liu L, Rugila J, Howe L, Buck S, Murphy K, Lyons L and Westhusin M (2002). A cat cloned by nuclear transplantation. *Nature* **415**: 859.
14. Di Berardino MA (1997). *Genomic Potential of Differentiated Cells.* Columbia University Press, New York.
15. Yan SY (1998). *Cloning in Fish: Nucleocytoplasmic Hybrids.* Educational and Cultural Press, Hong Kong.
16. Gasaryan KG, Hung NM, Neyfakh AA and Ivanenkov VV (1979). Nuclear transplantation in teleost *Misgurnus fossilis* L. *Nature* **280**: 585–587.
17. Ozato K and Wakamatsu Y (1994). Developmental genetics of medaka. *Dev. Growth. Differ.* **36**: 437–443.
18. Wittbrodt J, Shima A and Schartl M (2002). Medaka — a model organism from the far East. *Nat. Rev. Genet.* **3**: 53–64.
19. Niwa K, Ladygina T, Kinoshita M, Ozato K and Wakamatsu Y (1999). Transplantation of blastula nuclei to non-enucleated eggs in the medaka, *Oryzias latipes. Dev. Growth. Differ.* **41**: 163–172.
20. Wakamatsu Y, Ju B, Pristyaznhyuk I, Niwa K and Ozato K. Fertile and diploid medaka (Oryzias latipes) generated by transplantation of blastula nuclei into non-enucleated eggs. Unpublished.
21. Arai K, Matsubara K and Suzuki R (1991). Karyotype and erythrocyte size of spontaneous tetraploid and triploid in the loach *Misgurnus anguillicaudatus. Nippon Suisan Gakkaishi* **57**: 2167–2172.
22. Onozato H, Torisawa M and Kusama M (1983). Distribution of the gynogenetic polyploid crucian carp *Carassius auratus* in Hokkaido, Japan. *Japan J. Ichthyol.* **30**: 184–190.
23. Wakamatsu Y, Ju B, Pristyaznhyuk I, Niwa K, Ladygina T, Kinoshita M, Araki K and Ozato K (2001). Fertile and diploid nuclear transplants derived

from embryonic cells of a small laboratory fish, medaka (*Oryzias latipes*). *Proc. Natl. Acad. Sci. USA* **98**: 1071–1076.
24. Chen H, Yi Y, Chen M and Yang X (1986). Studies on the developmental potentiality of cultured cell nuclei of fish. *Acta. Hydrobiol. Sin.* (In Chinese) **10**: 1–7.
25. Dooley K and Zon LI (2000). Zebrafish: a model system for the study of human disease. *Curr. Opin. Genet. Dev.* **10**: 252–256.
26. Zon LI (1999). Zebrafish: a new model for human disease. *Genome Res.* **9**: 99–100.
27. Lee KY, Huang H, Ju B, Yang Z and Lin S (2002). Cloned zebrafish by nuclear transfer from long-term-cultured cells. *Nat. Biotechnol.* **20**: 795–799.
28. Linney E, Hardison NL, Lonze BE, Lyons S and DiNapoli L (1999). Transgene expression in zebrafish: A comparison of retroviral-vector and DNA-injection approaches. *Dev. Biol.* **213**: 207–216.

Chapter 16

Applications of Transposable Elements in Fish for Transgenesis and Functional Genomics

Perry B. Hackett*†, Karl J. Clark†, Stephen C. Ekker* and Jeffrey J. Essner†

**Department of Genetics*
Cell Biology and Development
Arnold and Mabel Beckman Center for Transposon Research
University of Minnesota
6-106 Jackson Hall
Minneapolis, MN 55455, USA

†*Discovery Genomics, Inc.*
614 McKinley Place, NE
Minneapolis, MN 55413, USA

Transgenic fish were first made more than 30 years ago. Since then a variety of methods and constructs have been tested for introducing genetic sequences into fish for scientific investigations as well as commercial purposes. Here we review transposable elements and their applications in fish. Transposons can be used to deliver genes to chromosomes to confer new traits or as insertional agents and traps to uncover the functions and expression patterns of natural genes in chromosomes. Two DNA transposons have been characterized for transposon-based gene transfer and insertional mutagenesis. The first is the *Sleeping Beauty* transposon system that was reconstituted from a *Tc1/mariner*-like relic in salmonid genomes after more than a 10 million year evolutionary sleep. The second is a naturally occurring transposon from medaka, the *Tol2* transposon that belongs to the *hAT* family of mobile elements. In comparison with random integration of plasmid sequences and pseudotyped retroviral genomes, transposons have several advantages for genetic studies in fish. These include introduction of a single, defined DNA sequence into a cellular chromosome, stable expression from the integrant for multiple generations, no absolute

Tel: 612-656-4485
Email: PerryH@DiscoveryGenomics.net

size restrictions on the transferred gene, ease in construction and use, and safety. Early experiments have validated the versatility of the *Sleeping Beauty* transposon for all of these purposes. The applications of transposon systems surpass use just in fish; the *Sleeping Beauty* transposon system is being used in mice to discover functions of genes and is being developed for gene therapy in humans.

1. Introduction

Research involving small model fish such as zebrafish and medaka are being increasingly used for a two broad applications as the 21st century dawns. The first application is for basic research. Here the goal is to discover the functions of vertebrate genes in order to uncover their coordinated expression patterns that take a single cell through a myriad of developmental programs to produce a multicellular adult capable of producing more such single cells. A second broad set of applications is using fish as model vertebrates for testing a variety of biomedical treatments such as drugs and other biopharmaceuticals. Here we review development of transgenic procedures that led to the construction and discovery of active transposable elements that could be used to deliver precise units of genetic information to vertebrate cells.

1.1. Transgenesis in Fish

The advantages of small fish in general, and zebrafish in particular, as model vertebrates are well documented.[1–12] Essentially they are: (1) developmental rates of small fish are rapid;[13] (2) the embryos of many species develop outside the mother so that non-invasive analysis of living embryos can be performed; (3) fish share approximately 90% of their genes with other vertebrates — this leads to common physiological pathways and responses to drugs;[14,15] (4) large numbers of embryos and fish can be efficiently generated at low cost so that large sample sizes can be examined during genetic screening of mutants;[4,16–26,230] (5) transgenic procedures have been developed for delivering genes to many species of fish (reviewed by Hackett;[27] Gong and Hew;[28] Maclean;[29] Hyatt and Ekker;[30] Meng *et al.*;[31] Hackett and Alvarez;[6] and

Zbikowska[32]); (6) a wide range of mutations in medaka[33] and zebrafish[1,2,34–36] have been characterized in hundreds of genes so that a myriad of genetic interactions can be examined, many of which have medical relevance; and (7) the optical clarity of the zebrafish, medaka, and other fish is especially suitable for visualizing activities in developing embryos.[37,38] Furthermore, the optical clarity allows efficient visualization of the expression of fluorescent proteins[39] under the control of specific promoters,[8,31,40–52,228,231,232] fluorescent proteins fused to other polypeptide sequences,[53–55] and other fluorescent marker macromolecules.[9,56,57]

The powerful advantages listed above place fish in a prime position to address problems in vertebrate development, physiology that often center on the following questions: (1) What are the functions of the approximate 35,000 genes in vertebrate genomes? (2) How are these genes coordinately regulated — that is, what are the regulatory elements for these genes? To answer these questions, geneticists have developed a number of transgenic tools to investigate several aspects of gene expression and its control in fish.[218]

1.2. Retroviruses and Transposons for Gene Delivery in Fish

Gene delivery to fish chromosomes is not new, and in fact predated the recognition of zebrafish as a model system. The first study of gene transfer was reported by Vielkind *et al.*,[58] who introduced heterologous fragments of *E. coli* DNA. Later large fragments of chromosomal DNA containing a locus that caused formation of tumors in Xiphorphorine fish were transferred by microinjection.[59,60] The first report of integration of a specific gene into a fish genome was by Zhu *et al.*[61] wherein a single gene was microinjected into fish that appeared able to transmit the transgene through the germline. Since then there have been several hundred papers on fish transgenesis that have reported wide variations in efficacy of gene delivery.[6] Nearly all of these studies used purified DNA that entered genomes by random integration. The process worked, but was highly inefficient. As genome projects were initiated, the emphasis in the 1990s shifted from production of transgenics for single gene analysis to high frequency insertion of DNA for random insertional mutagenesis screens. For this two techniques were developed. The first was the use of pseudotyped retroviruses that have a broad host range.[62] This form of retrovirus is encapsulated in an envelope glycoprotein that permits the virus to infect most cell types rather than the normal species/tissue-specific manner that most retroviruses employ. The second was the development of DNA-based transposons.

Viruses are attractive for introducing DNA sequences into vertebrate chromosomes; indeed, they are the primary delivery tools for human gene therapy.[63] Retroviruses are able to circumvent three barriers, cellular and nuclear membranes and chromosomal integrity; these steps are necessary before DNA sequences can be integrated into a genome. In general, viruses are very picky about the cells they infect. The specificity resides in particular glycoproteins on their exterior membranes, which they pick up as they bud from their host cell. There are only a few known fish retroviruses[64–66] so a common mouse retrovirus was developed that had a replacement of its normal species/tissue-specific *env* protein gene by the G-protein of vesicular stomatitis virus.[62] These retroviruses can infect any cell (fish or human) and once inside the infected cell, the viral integrase protein carried inside the virus catalyzes the integration of a DNA copy of the retroviral genome (called a provirus) into a chromosome of the host cell. However, the pseudotyped virus cannot penetrate fish chorions so they must be injected. In zebrafish, customarily about $10–20 \times 10^4$ viral particles are injected into blastula-stage embryos (512–2000 cells), leading to several insertions into zebrafish genomes.[67] This leads to embryos that are highly mosaic with from 5 to 22 proviral insertions transmitted to F_1 fish. For screening of mutations that affect growth and development, this is a useful strategy because it allows a number of mutations to be screened simultaneously per fish.[68–70] A large-scale retroviral screen has isolated approximately 500 insertional mutants[20,71] with an estimated mutagenic frequency of 1 mutation in 70 insertions.[19]

There are several drawbacks of using pseudotype retroviruses as vectors for gene delivery. First, pseudotyped retroviruses are difficult to prepare at the high titers required for efficient integration. Second, the high-titer retroviruses apparently do not express transgenes.[72] As a result, to date they have only been effective as insertional mutagens for tagging genes, *after* molecular analysis has detected their presence in chromosomes. Third, they do not integrate randomly; rather they have a preference for transcriptional control motifs close to the 5' ends of expressed genes.[73] Fourth, for some purposes, such as genetically engineering food fish, retroviruses had distinct disadvantages because these elements involve retroviral-like integration and therefore had the potential of being misunderstood by the public due to their association with cancer- and leukemia-causing viruses. Lastly, there are questions about safety of using these retroviruses because they can infect the human conducting the experiments (e.g., Smith *et al.*[74]). Consequently, we developed an alternative vector system, the *Sleeping Beauty* transposon system.[75]

In this review, we will examine the application of DNA-based transposons for transgenesis and functional genomics in fish. We will concentrate on three advantages of DNA transposons for genetic investigations in fish. They are: (1) ease of delivery and detection; (2) integration of precise sequences; (3) a relatively high efficiency of

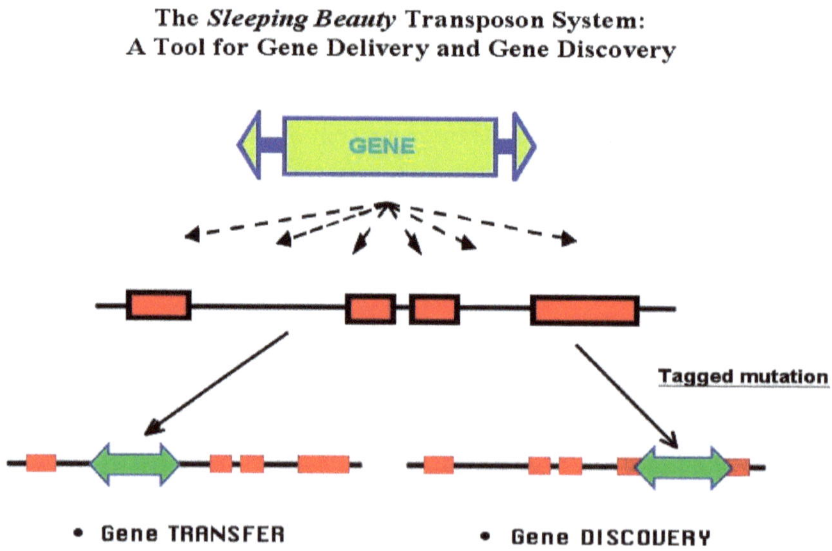

Fig. 1 The transposon containing a gene, with an appropriate transcriptional regulator, is shown flanked between the inverted terminal repeats of the transposon, which are indicated by the inverted arrowheads.

gene transfer into chromosomes; and (4) random integration. Fig. 1 summarizes the activities of the transposons discussed in this review.

2. Transposons

This review concentrates on DNA-based transposons. However, these mobile elements are just one type of mobile element, in fact a minority group in most vertebrate (and plant) genomes. About 90%[76] of mobile elements in fish genomes are retro-elements, so called because they operate as intracellular retrovirus in which an mRNA copy of the element is copied by cellular RNA-dependent DNA polymerase into a double-stranded DNA that invades the chromosome using a cellular integrase activity. In humans, the LINE and SINE families comprise the largest number of retro-elements, approximately 33% of the genome.[77] Furthermore, these elements are still active and cause random mutations in humans.[78] In zebrafish, the same family of SINEs has been given the names DANA[79] and *mermaid*.[80] Although they are distributed throughout the genomes of vertebrates, LINE and SINE elements often appear in nests that have likely resulted from repeated integration into or close to each other over the course of millions of years.[81] This clustering may also be a consequence of selection against deleterious integrations.[82]

2.1. Structures and Mechanisms of Transposition of DNA-Type Transposons

Attractive candidates for introducing DNA into fish chromosomes are DNA transposases that were responsible for the widespread distribution of transposons in fish and other animal genomes.[76,83] DNA transposons move in a simple, cut-and-paste manner (Fig. 2) in which a precise DNA segment is excised from one DNA molecule and moved to another site in the same or different DNA molecule.[84] The protein that catalyzes this reaction, the transposase, is encoded within the transposon for an autonomous element or can be supplied in *trans* by another source for a non-autonomous element. *Tc1/mariner*-type transposases require a

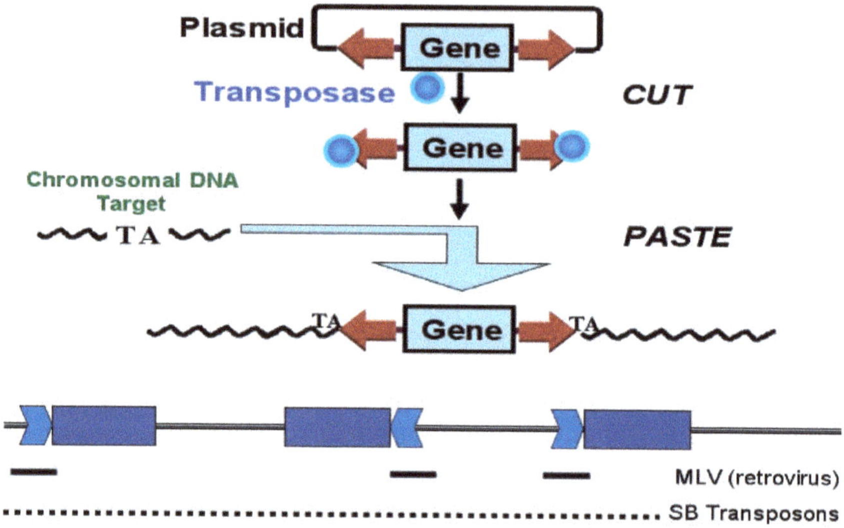

Fig. 2 The cut-and-paste mechanism of transposition. The transposon contains a gene with an appropriate transcriptional regulator. Two transposase molecules bind to each inverted terminal repeat. In experiments using transposons as a delivery vehicle, the transposase is supplied from another source, either from a plasmid carrying the transposase gene or from an mRNA co-delivered with the transposon. The schematic at the bottom illustrates the integration site preferences for retrovirus (promoter-proximal regions) compared to *Tc1/ mariner* transposons (nearly random). Promoter regions are shown as arrowheads and genes are shown as blocks. The long horizontal lines show the integration pattern of retroviruses and the dotted line shows the more random pattern of integration by transposons.

TA dinucleotide basepair for an integration site, which is duplicated during the integration process (described further in Fig. 3). The excised DNA is flanked by inverted terminal repeats to which the appropriate transposase molecule binds. Emphasizing this simplicity of action, *Tc1/mariner*-type transposable elements are ubiquitous in animal genomes, suggesting that they require few, if any, species-specific host factors. Indeed, both the *Tc1* transposon from *Caenorhabditis elegans*[85] and the *Mos* transposon from *Drosophila mauritiana*[86] were shown to transpose in cell-free systems in the presence of their respective transposase enzymes made in *E. coli*. Consequently, transposons carrying a gene-of-interest could be mobilized by transposase provided in *trans*.

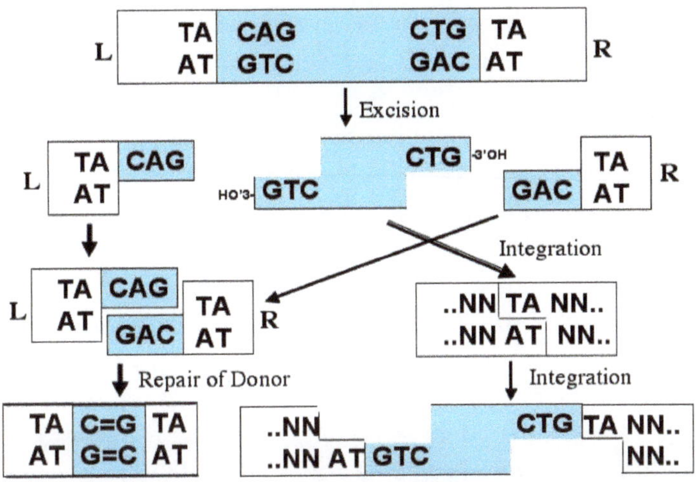

Fig. 3 *Tc1/mariner*-type transposition. The transposon is shown as a shaded sequence. The excision step is shown on the second line. Integration follows to the right and repair of the transposon donor plasmid on the left. The last step of the integration process is repair of the gaps (lower right corner) by DNA repair enzymes. In the lower left block, the stacked = signs represent an A–T or T–A basepair. The excision and integration steps are coordinated, with the 3' hydroxyl groups of the excised transposon attacking the TA dinucleotide basepairs in the target DNA sequence.

Unlike retroviruses, *Tc1/mariner*-type transposable elements integrate into TA sequences nearly at random (more on this later).

A more detailed schematic of the excision-integration process, with concomitant repair of the "donor" DNA molecule whence the transposon was excised is shown in Fig. 3. TA basepairs flank *Tc1/mariner*-type transposons. In the excision step transposase cleaves the transposon in a staggered manner such that three bases, CTG, overhang at each 3' end of the transposon (top two lines in the Fig. 3). The 3' ends of the excised transposon invade the target DNA molecule (indicated by the .. NNTANN .. sequence) at a TA sequence by cutting it in a staggered manner to expose two 5' ends with overhanging TA nucleotides (third line in Fig. 3). Integration is completed by repair of the 5-base gaps on both strands. Note that the original TA basepair target sequence is duplicated on both flanks of the transposon following integration. This is referred to as target site duplication. The donor DNA's left and right ends (indicated by L and R) that have extending CAG overhangs on their 3' ends, are brought together with a single A–A mispairing in the center. Repair enzymes then correct one of the two bases to produce an A–T basepair. The result is that upon remobilization, a "footprint" is left in the original site such that the initial sequence of TA is changed to TAC(A or T)GTA; that is, a five-basepair canonical sequence is inserted in a site into which a *Tc1/mariner*-type transposon entered and then left. Imprecise repair can alter the canonical footprint. Overall, the excision-integration-repair process is a series of breaking and joining of phosphodiester bonds that is highly coordinated by transposase. Thus, transposases are not like restriction enzymes that release a product that is free to wander to the next reaction site. Rather the excision and integration reactions are coordinated events.[84,87]

The theory looks simple in Figs. 2 and 3. However, when we tested the activity of *C. elegans Tc1* transposons in zebrafish and cultures of human HeLa cells, we found no indication of transposition; the levels of integration were roughly equal whether the transposase was present or not.[88] These results were similar to those obtained for another DNA transposon, the Drosophila *P*-element that has specific requirements for host cofactors.[88,89] In a search to find endogenous elements that could circumvent these difficulties, the Emmons lab showed evidence of transposable elements belonging to the *Tc1/mariner* family in salmonids and zebrafish.[90] The report led us to search for *Tc1/mariner*-like elements in many species of fish, but all of the sequences we isolated had many mutations and gaps in the transposase genes and inverted repeats of the transposon vector.[91,92] The search uncovered two families of *Tc1/mariner*-like elements in zebrafish, *Tdr1* and *Tdr2* that evolved from a common ancestor.[92] Others have found similar incomplete elements in various species of fish.[91,93–95] All of the reported elements are defective, with gaps, stop codons, and frame-shift mutations in the putative transposase-coding sequences.

Despite the frustrating failure to find an active transposase gene in a large number of inactive transposase sequences, it was possible to derive a theoretical sequence for an active transposase enzyme from phylogenetic principles. Based on the theoretical sequence, an active transposase was assembled in a 10-step process of site-specific mutagenesis of a salmonid transposase gene that entered an evolutionary sleep more than 10 million years ago. The awakened transposase was named *Sleeping Beauty* (SB).[75] We refer to the transposon and transposase as the SB transposon system. The SB transposase was able to improve integration from 20- to 40-fold in mammalian cells[75] and about 20-fold in zebrafish embryos.[76] This represents about an order of magnitude higher rate of integration than that from two heterologous transposon systems, *Tc3* from the nematode *C. elegans*[96] and *mariner* from insects.[97] However, it should be noted that *mariner* transposons can also transpose at low rates into chicken germline cells[98] and human cells.[99] In a head-to-head competition, Fischer *et al.*[100] compared the rates of transposition by the SB transposon system and a variety of transposons from nematodes and flies and found that the SB system was about 10-fold more effective in delivering a neomycin phosphotransferase (*neo*) gene to chromosomes in cultured human HeLa cells than the others. In these experiments as with all others reviewed here, *non-autonomous* transposons were used, i.e. transposons in which the transposase gene was replaced with alternative genetic cargo. The transposase is generally supplied by another plasmid carrying the transposase gene or an mRNA encoding the transposase.

A non-*Tc1/mariner*-type, active mobile element, *Tol2*, was identified in an albino medaka after it transposed into and inactivated a pigment gene.[101–103] The *Tol2* transposon belongs to the Ac/Ds family of transposable elements with sequence similarities to *hobo* in Drosophila and the *Tam3* transposon found in snapdragons.[104,105] It encodes a transposase with excision[105,106] and integration[107,108] activity. The transposon is active in mammalian cells as well as fish.[109,174] Unlike the transposase genes of *Tc1/mariner*-like elements, the *Tol2* transposase gene is divided into three exons.[110] The structure of the *Tol2* element is compared to that of various *Tc1/mariner* transposons in Fig. 4. Note that there are two sub-categories of *Tc1/mariner*-like transposons, those with two repeats within each inverted terminal repeat (called IR-DRs for

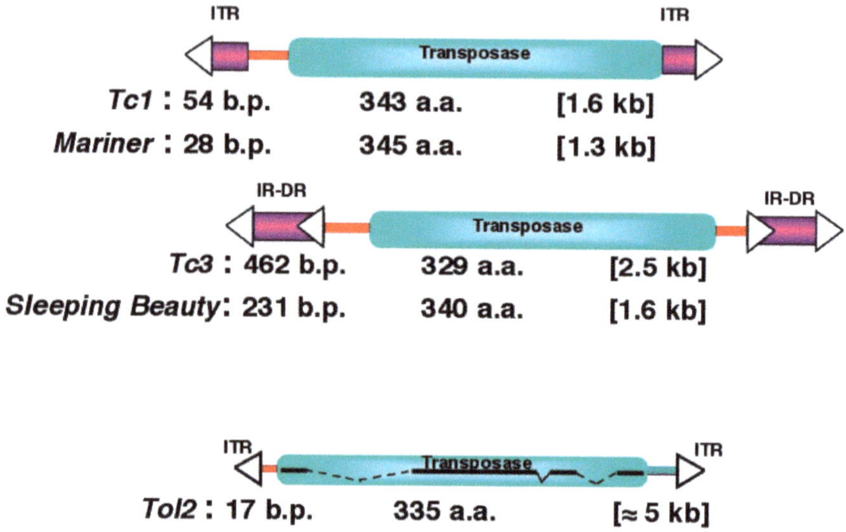

Fig. 4 Comparative structures of *Tc1/mariner*-like transposons and the *Ac*-like *Tol2* transposon. Each of the DR sequences in the SB transposon is 31 bp. The schematic of the SB transposon is hypothetical. An active, autonomous element has not been produced since it could have uncontrolled transposition activity in vertebrate cells. ITR, inverted terminal repeat sequence; IR-DR, inverted repeat containing direct repeated sequences.

inverted repeats containing direct repeats) and those with a single inverted repeats. The two inner DRs vary from the outer DRs.[111]

2.2. Advantages of Transposon-Mediated Gene Transfer

Transgenesis by transposition has several subtle benefits. First, the transposition reaction delivers a defined DNA sequence to a chromosome. Cutting and pasting is precise to the basepair on both ends. Prokaryotic sequences that might contain CpG-rich, methylation-sensitive sequences that could lead to silencing of the transgenes do not accompany the integrated transposon. Second, only single copies of transposons integrate at a given transgenic locus. In contrast, concatemers of transgenes often enter chromosomes by gene transfer via random integration of plasmids.[112,113] Concatemerization has been associated with gene silencing.[114–116] For these reasons, and maybe others, expression of genes delivered in transposons is stable. Third, studies from gene transfer into cells of non-dividing tissues of mammals, e.g. liver and lung, suggests that the SB transposon system can pass through nuclear membrane to the DNA in the nucleus.[117,118,156] This may be the result of the binding of four transposase proteins, each of which has a nuclear localization motif, to the inverted terminal repeats of the transposon. These proteins may be sufficient to facilitate transport of the transposon, presumably still in the donor plasmid, through a nuclear membrane.[119] Fourth, compared to viruses, transposons are easy to construct and prepare. Moreover, the transposon system has two components, an active transposase and the DNA transposon that is mobilized. This binary requirement renders the SB system relatively safe when the source of transposase is either a short-lived mRNA or an unintegrated gene. Lastly, compared with viruses that have limitations on the sizes of the genomes that can be transduced, transposons do not have an *a priori* upper limit beyond which transposition does not occur.

While size limitations are not expected for transposition, studies have shown that the efficiency of integration of transposons decreases with size. Three studies of this effect have been done in tissue culture using transposons that carried an antibiotic resistance gene so that cells carrying an integrated transposon could be quantified. Two groups[120,121] compared the sizes of transposons in which sequences from lambda virus were inserted into the transposons and found that enhanced colony formation was essentially lost when transposons exceeded 6–8 kilobase pairs (kbp). In the third study, the "stuffer" DNA was from the salmon beta-actin gene and different size transposons containing different antibiotic resistance genes were co-transfected into cells. In this case, elevated gene transfer to chromosomes was evident even in transposons larger than 10 kbp.[122] Figure 5 shows a comparison of two of the sets of results. The difference in outcomes for the larger transposons could be due to the nature of the assay. If the prokaryotic lambda sequences induced silencing of the selective marker gene, then those transposition events would not have been recorded. If so, then the probability of silencing would increase as the length of the lambda stuffer sequence increased. Alternatively, plasmids of increased length might have greater difficulty traversing either or both the plasma and nuclear membranes of the transfected cells, and prokaryotic sequences might exacerbate difficulties

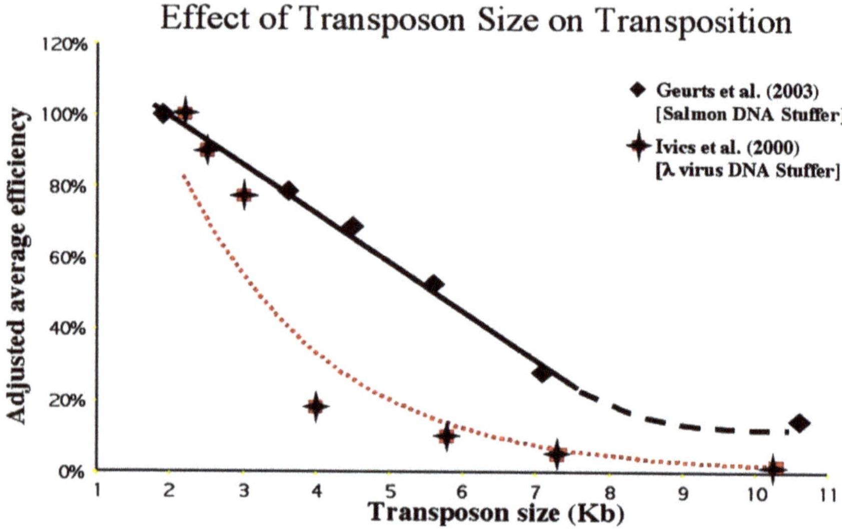

Fig. 5 Gene transfer in SB transposons as a function of transposon length in HeLa cells. [Data from Ivics *et al.*[120] and Geurts *et al.*[122]]

in membrane passage. Regardless of the reason, it appears that transposons up to 5–6 kbp, which would accommodate an estimated 80% of vertebrate cDNAs,[123,124] can efficiently deliver genes to chromosomes.

3. Applications of Transposons in Fish

Transposon systems are powerful vectors for integrating genes into chromosomes. But, first they must be introduced into cells. Active transposons require a transposase, whose activity can be delivered by the transposase gene, an mRNA copy, or presumably the encoded enzyme. For most applications in fish, the method of delivery of a transposon system is similar to that of any DNA. The next steps include analyzing whether integration occurred, whether it was by transposition, and, in cases where the sequence of the insertion site is of interest. Determining the integration status, defined as the exact sequence inserted and the locus into which it integrated, is essential for most genetic analyses. There are several methods for gene transfer and analysis of integration.

3.1. Delivery of Transposon Systems into Fish Embryos

Several methods of DNA delivery to fish embryos have been reported. They are reviewed in the following sections. Only microinjection has been reported for delivery of transposon systems in fish. While methods for DNA delivery other than microinjection have not been efficient for integration and expression of plasmid DNA, these alternative methods may be more useful for effective delivery of transposons. For example, it may be that earlier introduction of DNA and transposase into the cell will lead to transposition at the one- or two-cell stage of embryogenesis. This consideration is discussed further at the end of the review in Section 4.4.

3.1.1. Microinjection

Microinjection is the most popular form of gene transfer into fish embryos because of its reliability.[6] One reason is that fish eggs are about 1000–30,000 times larger than mammalian eggs, which makes injection into the embryo or yolk fairly easy — an experienced person can inject more than 1000 embryos per hour. However, injection into fish pronuclei is very difficult because they make up only about 0.001% of the one-cell embryo compared with about 5% in a mammalian embryo. As a result, pronuclear injections are often impractical, although in some species such as medaka this technique can be used. Moreover, soon after fertilization the chorion hardens so penetration by the injection needles becomes harder with time. Injection of DNA into oocytes has been examined[125,126] but the technique has gained little interest owing to the difficulties in harvesting both the oocytes and sperm for fertilization afterinjection. For certain types of fish embryos the chorion is removed prior to gene delivery (e.g., Culp *et al.*;[127] Müller *et al.*[128]). However, the procedure is labor-intensive, slow, and embryo survival and transgene integration rates do not appear to be better than those obtained by microinjection though the chorion.

In most cases, about 10^6 to 10^7 molecules of DNA in an aqueous, buffered volume of 1 to 2 nl, about 20 to 100 times the average nuclear volume of fish nuclei, are injected into embryos or fertilized eggs.[6,31] It appears that most of the microinjected DNA remains as a distinct aggregate in zebrafish embryos as visualized by labeling with ethidium bromide.[129] Of this large number of molecules, in general less than 100 will integrate; an efficiency of about 0.01%. This is nearly 100-fold less than that achieved using the pseudotyped viruses. This aggregation explains in part the relatively poor integration results. The conclusions from hundreds of microinjection experiments in fish, using a wide variety of genes, promoters, and methods of assay, are that (1) embryonic survival decreases to less than about 10% when more than 100 pg of DNA were injected; (2) only a portion of the injected embryos express the transgenic DNA in a mosaic fashion; and (3) only a proportion of fish that develop from injected embryos are able to pass on the transgenic construct as an integrant in a fish chromosome.[6] However, this is generally sufficient because hundreds of embryos can be injected in a single day. For this reason, microinjection is the most popular method of gene transfer and for standard gene transfer in fish. Nevertheless, microinjection is perceived to be tedious and inefficient. Consequently, other methods for gene transfer have been developed and reported, mainly in zebrafish and medaka.

3.1.2. *Electroporation*

Electroporation is used to depolarize cellular membranes to allow entry of DNA through the plasma membrane.[130] Electroporation of DNA has been tried in several species of fish embryos following removal of their chorions.[125,126,128,131–137] However, as noted above, this is a time-consuming process that defeats the objective of mass transfer of DNA without treatment of each embryo. Many variations of conditions have been reported, but the bottom line is that electroporation has not caught on with most labs and few labs use it routinely.

3.1.3. *Sperm-mediated DNA delivery*

Sperm-coated with DNA was first reported to work in mice with relatively high frequency and no damage to the embryo,[138] but Brinster *et al.*[139]

reported their inability to reproduce the results. DNA does bind to sperm[140–142] and can be taken up into fish embryos,[134,143–145] but the genes are not expressed. A new method has been tried, intra-cytoplasmic injection of detergent-disrupted sperm and sperm-heads.[146] Up to 20% of transgenic mice produced offspring that expressed transgenes. The sperm heads were "considered dead due to disrupted membranes" but still supported full development. This method still requires microinjection and thus cannot be considered a mass transfer procedure that is substantially different from standard microinjection.

Electroporation of sperm with transgenic DNA prior to fertilization has been reported.[147–150] However most of the DNA that is brought into the egg remains on the exterior surface, as evidenced by its susceptibility to DNases.[145,151] The efficiencies for integration into chromosomes of DNA constructs brought into fish in this way is not known. At this point, the difficulties in the procedure appear to be greater than the benefits. Gene transfer into zebrafish by sperm nuclear transplantation has been tried as well.[152]

3.1.4. *Biolistics*

"Shock and awe" is sometimes employed when subtle means fail. Bombardment of cells with DNA-coated particles, *bio*logical bal*listics*, has been tried with fertilized loach, zebrafish and rainbow trout eggs, resulting in high mortality and low expression rates.[153–155] As an alternative, electrospray delivery has been attempted in zebrafish. For this, a fine mist of water containing transgenic DNA was electrostatically propelled onto a plate containing hundreds of embryos. Expression of the transgenic constructs was never detected, even at voltages that killed the embryos.

3.1.5. *Lipofection*

Nucleic acids encapsulated in synthetic lipid vesicles can be taken up into cells and tissues of animals where they can be expressed (e.g. Kren *et al.*[157]). In fish, several cocktails of DNA and proteins have been mixed and used to deliver transgene constructs to dechorionated zygotes at the 2–16 cell stage.[158,159] As with the sperm methods, the DNA could get into the embryos, but it was soon lost and its expression was transitory.

Together, the results of all of the transgenic procedures emphasize that delivery of transgene DNA is only half of the problem of transgenesis. The second half is integration into chromosomes for passage to subsequent generations and stable expression of the genetic material. For this, the DNA must penetrate the nuclear membrane and insert into a chromosome for genetic studies. The *Sleeping Beauty* transposon is designed to be catalytically inserted into chromosomes and SB transposases with their nuclear localization motifs may assist in conveying transposons through the nuclear membrane.

3.1.6. *Remobilization of transposons*

A genetic method of introducing transposons into new loci is to remobilize transposons already occupying a position a genome. This is the way transposons naturally spread from a single invading species to multi-copy numbers in genomes. This is very apparent with the several thousand-fold amplification in numbers over millennia that have occurred for most *Tc1/mariner*-type transposons in vertebrate genomes. Remobilization has been successful in mice using the *Sleeping Beauty* transposon system discussed in later Section 4.4. Three laboratories have reported efficiencies that range from an average of 0.2 to 2.0 remobilizations per newborn mouse pup following a mating between mice with SB transposons and mice expressing a SB transposase gene.[100,160,161] The higher rates of remobilization were obtained from animals in which transposons were remobilized from concatemers of transposons that entered the genome by random recombination rather than transposition.[160,161] There is also evidence for remobilization in zebrafish. Using probes to the endogenous transposon *Tzf* (also named *Tdr2*[91,92]), Lam *et al.*[162] found evidence of movement of *Tzf* sequences based on two-dimensional gel electrophoresis analysis. Although they found evidence for excision of transposons, they did not detect any new integration events, which would have been expected. However, the rate of the observed excision events correlates with the background mutation rate of 8×10^{-5} per gene per generation in zebrafish.[163] Additional evidence for the potential of element remobilization comes from the observation that *Sleeping Beauty* transposons can be excised from one integration site in transgenic zebrafish following injection of a transposase source into embryos.[164] Together, these data support the possibility that remobilization of transposons could be an effective way of delivering sequences to new sites in fish genomes.

3.2. Analysis for Integration of Transgenes into Chromosomes

About 1 to 10 million plasmids are microinjected into a single fish embryo in most experiments. Yet at best only a very few genomes take up even a single copy of DNA, an efficiency of integration of less than 10^{-4} percent. There are two important consequences of so few transgenic DNAs making their way into chromosomes. First, the integrating copies of transgenic DNA are diluted more than a million-fold with unintegrated DNA of identical sequence, making analysis of the integrated DNA exceedingly difficult. The unintegrated DNA can recombine, at variable rates that depend on the input conformation of the transgenic DNA, to form concatemers (reviewed in Hackett,[27] Iyengar *et al.*[112]). Second, after microinjection virtually all transgenic F_0 fish are mosaic.[129] As a result, screening for transgenic fish is quite labor-intensive, requiring the raising of all embryos subjected to transgenic DNA until gametes can be accurately assayed for the presence of integrated foreign DNA. In the early years of fish transgenesis, several assays were used to indicate integration had occurred, including dot-blotting or PCR amplification, expression of the transgene after larval development, Southern blotting to show that the size of the transgenic DNA changed as a result of its integration into chromosomes, and detection of transgenic DNA or its expression in F_1 progeny. None of these assays was sufficient. Because of the enormous amounts of unintegrated DNA that can recombine with itself, assays such as Southern blotting and PCR amplification that examined transgenic DNA size often gave misleading results. Stuart *et al.*[165,166] showed that unintegrated transgenic DNA was passed in an episomal state into F_1 progeny. The problems associated with the persistent presence of unintegrated transgenic DNA in fish have been reviewed.[6]

Three methods are reliable indicators of integration, chromosome *in situ* hybridization of tagged probes to metaphase chromosomes,[167] mendelian segregation of the transgenes (and their expression) in F_2 and subsequent progeny, and linker-mediated PCR or inverse PCR[164,168] to determine the sequence of the locus into which the transgenes integrated. The last method is essential in order to verify transposition of specific sequences rather than uptake of the transgenic DNA by random recombination.[75] In tissue culture, the initial SB transposon system using pT transposons and the SB10 transposase is 30- to 40-fold more efficient in directing integration of genes into chromosomes of HeLa cells than by random recombination[75] and about 10-fold more active than *Tc1*, *Tc3*, and various *mariner*-type transposon systems from invertebrates.[100] Since then both the pT transposon, pT2,[111] and the SB transposase, SB11[122] have been improved to now deliver transposition rates in cells more than 100-fold above random integration. In the mouse, the complete SB transposon system delivers expressing genes to liver[117,169] and lungs[118] in adult mice. In all of these studies, transposition was assessed by isolation and sequencing of the transposon junctions to verify that the increased gene transfer was due to transposition.

3.3. Transposition of Transgenes into Chromosomes of Fish

Several transposon systems have been used for gene transfer into zebrafish and medaka fish chromosomes, including the *Drosophila P* element[89,170] and *mariner* transposon,[97] the nematode *Tc3* transposon[96] that resembles the *Sleeping Beauty* transposon in that it has long IR-DR inverted terminal repeats, the medaka *Tol2*[105,107] and the salmonid-based SB transposon.[164,171] The invertebrate transposons have little or no activity above injection of plasmids and so are not discussed further.

3.3.1. *Transposition in fish using the SB transposon system*

The Hackett and Ekker labs have done a number of studies of using the SB transposon system in zebrafish in which the initial pT transposons were supplied on plasmids and an mRNA encoding SB10 transposase synthesized *in vitro* were co-microinjected into 1- to 4-cell stage embryos. The plasmid pSBRNAX was designed to generate ^{7m}G-capped SB10 mRNA by T7 RNA polymerase using the mMessage (Ambion) Machine transcription kit.[164,171] The embryos were raised to adulthood and examined for gene expression (made easy by the use of fluorescent protein markers). Fish that expressed a marker gene contained within the transposon in at least one cell were crossed with wild-type to test for transmission of the transposon. In these fishes, we found 5–10% of the population had germlines that produced at least one transgenic gamete. This suggested that the germline preferentially took up transgenic DNA and that most somatic tissues took up the transgenic DNA at a late stage of development. In most cases, we found that expressing fish in the F_1 generation would pass on the gene in a mendelian manner.[171,172]

This low rate of integration was disappointing, especially in light of the success of the transposon in delivering genes to chromosomes of somatic tissues of mice (cited above). This led to the use of the more advanced pT2 transposons and SB11 transposase. Davidson *et al.*[171] found that SB transposase enhanced the transgenesis and expression rate six-fold, from about 5% to more than 30%. This doubled the total number of tagged chromosomes over standard, plasmid injection-based transgenesis methods. In their report, they demonstrated that ubiquitous tissue-specific promoters, such as the lens γ-crystallin promoter could be used for reproducible and multi-generational gene expression.[55] Figure 6 summarizes the procedure and efficiencies for integration of transgenes in zebrafish chromosomes using SB transposons. The figure shows that the transposase is supplied via injection of mRNA encoding the enzyme rather than delivery of the SB transposase gene or the protein itself. By injecting the mRNA, there is a brief, transient presence of the transposase that declines as both the mRNA and the translated products decay. There

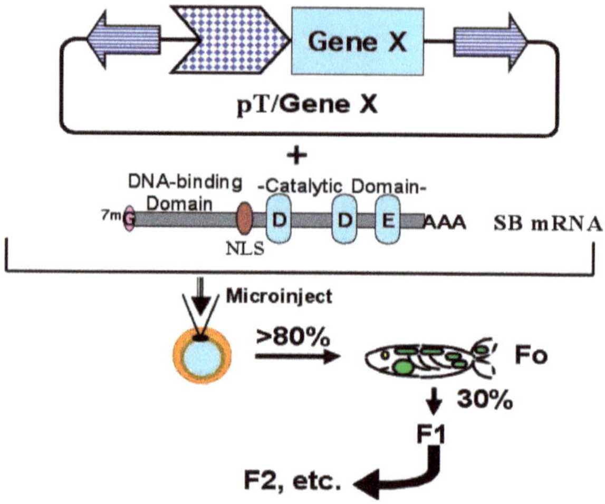

Fig. 6 Gene transfer of SB transposons into zebrafish embryos. The SB transposon comprises the IR-DRs (inverted arrows) and an expression cassette composed of a promoter (dotted arrowhead) and a gene (X). About 5×10^6 transposons and 10^8 mRNAs encoding SB transposase are injected per 1–4 cell embryo. About 80% of the injected embryos will express the transgene (filled "cells" in the fish). About 30% of the adults will have at least some gametes that are transgenic. When bred, these proceed to pass the transgene in a mendelian manner.

appears to be an optimal ratio of SB transposase protein to transposon DNA from studies done in tissue-cultured cells[122] that still needs to be determined for microinjection of zebrafish embryos. This may further lead to higher rates of insertion.

Although the rates of transposition look low following microinjection into fish embryos, the same is seen in other organisms. In insects, only about one plasmid per thousand injected will result in an excision event even though the transposition reaction can be mimicked *in vitro*.[173] Similarly, only one in a million transposons enters a mouse chromosome following microinjection of pronuclei.[161]

An important finding has been that generally expression of transgenes is maintained over several generations when introduced in a transposon[171] compared to that from plasmids where expression is often lost or altered.[175] This is probably due to the methylation of prokaryotic sequences in plasmids that accompany random integration.[6,176] Alternatively, injections of plasmids can lead to integration of concatemers and rearranged recombination products from which gene expression may be silenced or unstable.[127,165,166,177] For genetic studies, reliable, multi-generation expression is essential. Transposition removes all plasmid sequences and inserts single units in a given locus, which explains the reliable expression from transposon-mediated transgenic fish.

3.3.2. *Transposition of trap vectors in fish using the SB transposon system*

The various genome projects for vertebrates[123,124] have identified approximately 35,000 genes in vertebrates, most of which have functions whose physiological significance is unknown. Finding the "bottom line" functions of genes is important both for basic research as well as pharmaceutical development.[178] As noted earlier, finding the functions of genes can be accomplished by various mutagenesis screens using chemicals, but identifying the loci by positional mapping has been a problem — the procedure is extremely labor-intensive and slow (reviews by Driever *et al.*,[17] Haffter *et al.*,[18] Eisen[179]). While the total number of mutants recovered by insertional mutagenesis using retroviruses by the Hopkins lab is lower, these screens have been responsible for identifying as many of the corresponding mutated genes as in all the chemical mutagenesis screens in a large number of laboratories combined.[71] While the retroviruses served as little more than insertional "tags" that allowed interrupted genes to be identified in these screens, the large number of identified genes underscores the utility of an insertional mutagenesis approach. A more powerful insertional tag would be one that could be followed easily followed by its expression, e.g. a fluorescent protein. As demonstrated by Davidson *et al.*,[171] transposons harboring expressible fluorescent protein genes are eminently feasible.

Transposons have been used for insertional mutagenesis genome-scans of many species from viruses to mammals.[180–191] These transposons operate similar to the pseudotype retroviruses discussed earlier. A better way of investigating genes and their activities is to use "trap" vectors that express easily detectable reporter molecules when they insert in a transcriptional unit or in the vicinity of a transcriptional regulatory motif.[192] There are four basic types of trap vectors that differ in their requirements for activation of the reporter gene: enhancer traps,[193–195] promoter traps,[196–198] gene traps,[198–202] and poly(A) traps.[203] Gene traps and poly(A) traps can be designed as a single unit.[204] By scanning for activation of the traps, genes with specific responses to environmental, stress, or developmental cues can be discerned under various genetic conditions. For these reasons, insertional mutagenesis has become an important complement to the genome projects.[188,205–209]

With the above in mind, Clark and his colleagues initiated a project to produce panoply of trap vectors in *Sleeping Beauty* transposons. These vectors are illustrated in Fig. 7. Clark *et al.*[210] have compared the efficiencies of expression of SB transposons with a complete expression cassette, pT/SV40-Neo that has a neomycin phosphotransferase II gene (with poly(A) sequence) under the regulation of an SV40 promoter, with an equivalent SB transposon with a Neo gene trap in which the IRES-Neo gene cannot be expressed unless it integrates into an active transcriptional unit in the correct orientation, i.e. the gene is oriented in the same direction as the promoter behind which it inserts. Using tissue-cultured HeLa cells, they found that about 1 in 15 integrations resulted in activation of the gene trap, which corresponds to about 4 in 15 integrations into transcriptional units when orientation and percentage of the genes expressed is taken into account. This rate is close to the 1 in 4 rate expected for random integration into a mammalian genome.[210] Thus, the functional assay is consistent to other findings with the findings that SB transposons integrate nearly randomly in vertebrate genomes.[164,212–214]

One problem with trap vectors that use a reporter gene directly for activation is that most genes are expressed at relatively low levels.[215] Consequently, the intensity of a fluorescent protein trap might be below detection. This situation would be aggravated if short-half-life versions of fluorescent proteins were used to improve resolution of when and where trapped genes are expressed. Consequently, Clark *et al.*[210] replaced the reporter gene with one encoding a transcriptional enhancer-binding protein. When this gene is activated, it serves to direct expression of a fluorescent protein behind a strong promoter, thereby amplifying the signal up to 100-fold.

In summary, the best trap vectors are those that insert in a very random manner and allow assessment of a mutagenized locus. Hence,

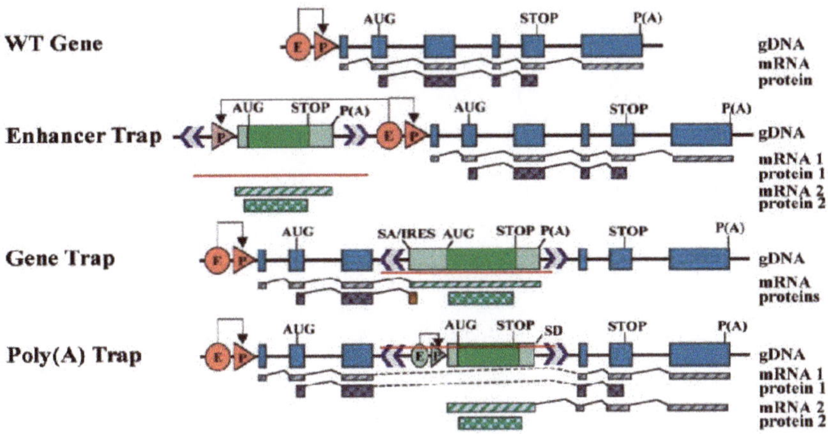

Fig. 7 SB transposon-based trap vectors for functional genomic studies in vertebrates. A genetic locus (WT Gene) is shown on the top line with exons as boxes, introns as lines, transcriptional enhancers as circles (E) that activates (arrow) a promoter (P). RNA transcripts with introns are indicated below the gene and the encoded proteins are shown at the bottom. The activities of enhancer, gene and poly(A) traps are shown in the three examples where the transposon is indicated by the blue, inverted, double arrowheads. Enhancer traps use the WT enhancer element to activate a minimal promoter (triangle with a "P" in the transposon) to activate a marker gene (e.g. GFP shown in green). When a transposon with a gene trap inserts in either a transcribed exon *or* intron, the marker gene is expressed from an Internal Ribosome Entry Site (IRES) that can activate protein synthesis at the normal initiation site for the marker protein. The splice acceptor (SA) site ensures that the marker sequences are spliced into the mRNA when the transposon integrates into an intron; the SA site may not be used if the transposon integrates into an exon. Poly(A) traps have both a splice acceptor and splice donor site. Poly(A) trap vectors must integrate into a transcriptional unit so that the marker gene can acquire a poly(A) sequence for stability. To ensure inclusion of the poly(A) sequence should the transposon integrate into an intron, the poly(A) trap vector has a splice-donor (SD) site at its 3' end. This vector has a strong enhancer/promoter driver so it does not need an SA site. Should the poly(A) trap integrate into an intron, the normal message may be formed (with splicing out of the poly(A) trap) along with the marker transcript that includes the same 3' poly(A) as the interrupted gene (adapted from Clark[211]).

it is likely that the SB transposon traps will integrate more randomly, and therefore detect more genes, than retrovirus vectors.

3.3.3. *Transposition in fish using the Tol2 transposon system*

The *Tol2* transposon is a member of the *hAT* family of transposons, so named because of the relatedness of the *hobo* element in flies, the *Activator* (*Ac*) transposon in maize, and the *Tam3* mobile element in snapdragons. The *Tol2* transposon was originally identified by its insertion into the tyrosinase gene that reduced coloration in certain medaka fish.[102] This element is active — it is found in different genomic positions, even in closed populations of fish.[216] The transposase of *Tol2* has four open reading frames that appear to be arranged in two spliced genes,[102,110] a more complicated arrangement than that of the *Sleeping Beauty* transposon. About 10 to 30 copies exist in medaka, and most appear to be autonomous, but *hAT* transposons have not been found in zebrafish.[105] The *Tol2* system is active in zebrafish[105–107] and in mammalian cells[109] and has been used as a gene transfer vector.[217] The integration site preferences of the *Tol2* transposase gene have not been identified, but *Tol2* transposons cause an 8-basepair duplication at the insertion site, compared with the TA duplication for SB transposons. The relative activities of the *Tol2* transposon and the SB transposon have not been compared side-by-side. These two transposon systems may have complementary features for functional genomic studies in fish. The identification of *Tol2*, *hAT*-type, and the SB, *Tc1/mariner*-type, transposons may just be the initial entries into synthetic and natural mobile elements that will be used for gene transfer in fish as well as other vertebrates. Further development of the *Tol2* system, along with more refinements in the available transposons and perhaps other natural and synthetic transposons, should lead to more versatility in precise gene-transfer in fish.

4. Future Directions

Transposons allow transfer of precise, single-copies of genetic material to chromosomes. At present, only transposons derived from vertebrates have high activities in vertebrates. The reasons for this are unclear and finding them is the focus of future directions. For transgenesis in fish, a major goal is to achieve integration of transgenes in the chromosomes of one-cell stage embryos. By doing this, effects of transgenesis can be elucidated in a single generation. Currently, the mosaicism in the founder populations of transgenic fish demands at least two generations before genetic studies can be conducted. Solutions to this problem may come from more complete understanding of the transposition process and the factors involved, or in the development of new delivery systems.

4.1. Parameters Affecting Transposition

Parameters that affect transposition are illustrated in Fig. 8. Several of these parameters are under investigation for the *Sleeping Beauty* transposon system, including sequences of the inverted terminal repeats and the direct repeats (DRs) that comprise each IR. Indeed, the DRs are not identical in the outer and inner positions[75] and it is crucial that DRs with weaker DNA-binding ability exist in the outer position.[111,217] The reason for this is unknown but probably reflects the dynamic nature of binding and release of transposase during the transposition reaction. A second component of the SB transposon system certainly can be improved — the transposase itself. Initial refinements have been reported[122] and more are sure to follow. The initial transposase and the modifications are largely based on phylogenetic consensus sequences.[90,92] However, DNA-based transposons exist in nature because they transpose very infrequently over evolutionary time. Hence, improvement in transposase activity under experimental conditions should be possible using appropriate screening following random mutagenesis. Other factors that affect transposition are the ratio of transposase to transposon and the lengths of the transposon.[220] The excision assay,[219,221] shown at the lower left corner of Fig. 8, is a relatively quick assay that has been developed to facilitate assays on the efficiency of transposition. The assay is based on the precise excision of the transposon from a plasmid and the subsequent repair of donor (Fig. 3; note the ambiguity in the central basepair of the canonical

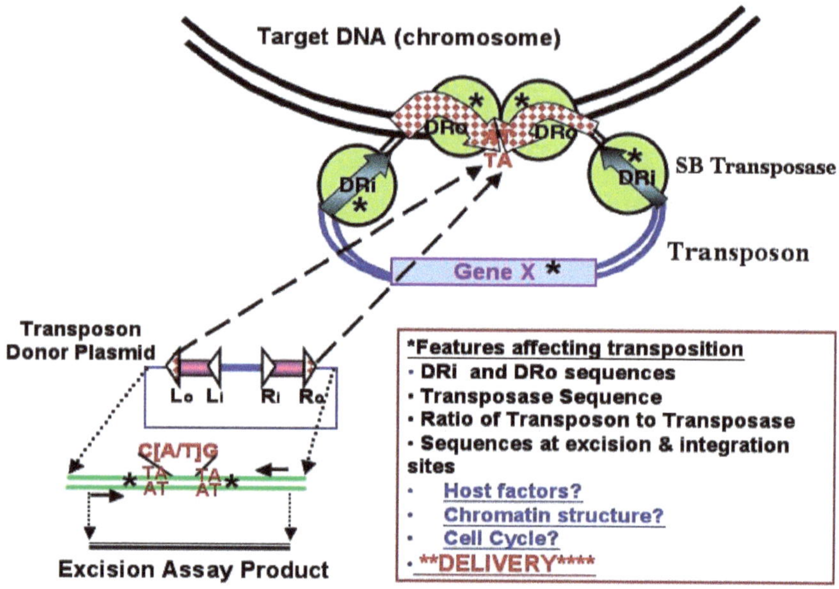

Fig. 8 Summary of parameters that can be altered to improve transposition. The schematic shows a transposon, represented by inverted terminal repeats composed of two DRs labeled Lo and Li for the left outer and inner DRs and Ri and Ro for the right inner and outer DRs on a plasmid (lower left). The efficiency of transposition (dashed lines) from a donor plasmid into a TA-target site on a chromosome is affected by the sequences surrounding the donor site (bottom left) as well as the conformation of the integration site. The differences in sequences of the outer and inner DRs are reflected by the different shadings. SB transposase is shown as a circle. Asterisks indicate components and motifs that affect transposition.

footprint) that defines a PCR product from primers on either side of the donor transposon (Fig. 8, bottom left corner). Several of these parameters that affect transposition are discussed in more detail in the following sections.

4.2. Identification of Host Factors Associated with Transposition

A major objective is to achieve integration in early-stage embryos at the one- to two-cell stage. With microinjection the amounts of DNA delivered to the average embryo can be better quantified than with the alternative procedures. As noted earlier, although about 1 to 10 million molecules are delivered to each embryo, the genomes of only a few cells incorporate a transposon. One physiological phenomenon in zebrafish and medaka that surely affects transposition is the speed of DNA replication. Owing to the rapidity of the early cleavage cycles, there must be at least 100 times as many DNA polymerases per unit length of fish chromatin as for mammalian chromatin, resulting in a situation where most of the chromatin is probably not in a form that can allow recombination.[6] DNA replication apparently does not depend on specific DNA sequences (*ori* sequences) that are required for DNA synthesis later on[222] because injected plasmids are replicated.[165] In zebrafish, transcription begins after the midblastula transition (ca. 1000-cell stage, Kane and Kimmel,[223] Kimmel *et al.*[224]). At this time replication of chromatin slows and the DNA apparently becomes available for enzymes that mediate integration. This is the basis for the late injection of pseudotype retroviral vectors for insertional mutagenesis in zebrafish. DNA conformation and chromatin binding factors also appear to influence transposition of SB transposons.[225–227] Presumably there are factors that, if co-injected into newly fertilized embryos with the transposon components, would lead to transient relaxation of chromatin at the one-cell stage to permit transposition. Identifying these factors is an important goal for fish transgenesis regardless of vector or method of introduction. It could be that introducing the transposon system into gametes prior to fertilization would allow access of transposons to chromatin before the onset of chromatin condensation and rapid rates of DNA replication.

4.3. Regulation of the Ratio of Transposase to Transposons

Transposition of SB transposons requires four transposase molecules per transposon. Binding of one to three transposase molecules might be able to facilitate nuclear import as a result of the nuclear localization motifs on transposase but not transposition. This hypothesis is supported by studies where one or more of the transposase binding sites were deleted in the IR/DR region of a transposon.[111,120] On the other hand, overexpression of transposase interferes with transposition.[122] Consequently, the initial source of transposase is important in transposition-mediated transgenesis. Studies in mice wherein plasmids harboring transposase genes have been injected indicate that the ratio of transposon plasmid to transposase has an effect.[117,118] A systematic investigation of the effects of different levels of DNA or mRNA encoding transposase has not been reported, nor has the injection of purified transposase with transposons.

Rapid achievement of optimal levels of transposase to transposon in early-stage embryos would facilitate early integration. The method of delivery of the components of the transposon system may have an effect on the ratio of transposase to transposon. For instance, if only a few molecules with a transposon and a few with a transposase gene are delivered to a cell, then the ratio of transposase to transposon will vary from cell to cell. Alternatively, if hundreds or more molecules are delivered to a cell, e.g. via microinjection, then the ratios of the two molecules will be similar from cell to cell.

4.4. Transposase-Expressing Lines of Fish

Functional genomics in mice using the SB transposon system has been facilitated by lines of mice that express SB transposase.[100,160,161,164] This allows remobilization of transposons in subsequent animals of up to an average of one to two new insertions per offspring. The SB-expressing lines of mice appear to have normal phenotypes with no indications of endogenous transposons being mobilized by the transposase.[230] Realizing the benefits of using fish that express fluorescent proteins when in various transposon trap vectors, attempts to develop lines of zebrafish that express SB transposase are being made (D. Balciunas, pers. comm.).

4.5. Site-Specific Integration

Transposition of the DNA transposons discussed in this review is characterized by relatively random integration throughout vertebrate genomes. This is the rule for transposons. Nevertheless, there is considerable interest in developing transposases with site-specific integration capability. The DNA-binding motif in the N-terminal third of the SB transposase is known to interact with the DR sequences of the transposon. However, the DNA-interacting motif for TA-dinucleotide basepairs at the target site resides somewhere in the catalytic domain in the carboxyl-terminal half of the transposase; it is not identified. As a result, directed mutagenesis to alter the specificity of integration has not been accomplished.

Site-specific integration has a broader implication than use in fish. The SB transposon system is an efficacious means of inserting defined sequences of DNA into mammalian chromosomes without using viruses. The SB system has been used to deliver genes for long-term expression to livers[117,231] and lungs[118] of mice. The system looks feasible for use in human gene therapy once delivery methods are better defined. The ability to direct specifically transposons to a given site in chromosomes would increase the safety of this method of delivery. Thus, methods that were initiated to improve transgenesis in fish are being adapted for use in humans. Truly, fish are an excellent model system for more than merely finding the functions of genes and their interactions.

Acknowledgments

We are grateful for all of the input from our laboratory colleagues over the years. Our academic laboratories are supported by grants from the Arnold and Mabel Beckman Center for Transposon Research and NIH to PBH and SCE.

References

1. Detrich HW, Westerfield M and Zon LI (eds.) (1999a). *The Zebrafish: Biology. Meth. Cell Biol.* vol 59. Academic Press, San Diego.
2. Detrich HW, Westerfield M and Zon LI (eds.) (1999b). *The Zebrafish: Genetics and Genomics. Meth. Cell Biol.* vol 60. Academic Press, San Diego.

3. Dodd A, Curtis PM, Williams LC and Love DR (2002). Zebrafish: bridging the gap between development and disease. *Hum. Mol. Genet.* **9**: 2443–2449.
4. Ekker SC (2000). Morphants: a new systematic vertebrate functional genomics approach. *Yeast* **17**: 302–306.
5. Drummond IA (2000). The zebrafish pronephros: a genetic system for studies of kidney development. *Pediatr. Nephrol.* **14**: 428–435.
6. Hackett PB and Alvarez MC (2000). The molecular genetics of transgenic fish. *Recent Adv. Mar. Biotech.* **4**: 77–145.
7. Fishman MC (2001). The genomic cosmos. *Nature* **410**: 1033.
8. Goldman D, Hankin M, Li Z, Dai X and Ding J (2001). Transgenic zebrafish for studying nervous system development and regeneration. *Transgen. Res.* **10**: 21–33.
9. Nasevicius A and Ekker SC (2000). Effective targeted gene 'knockdown' in zebrafish. *Nat. Genet.* **26**: 216–220.
10. Peterson RT, Link BA, Dowling JE and Schreiber SL (2000). Small molecule developmental screens reveal the logic and timing of vertebrate development. *Proc. Natl. Acad. Sci. USA* **97**: 12965–12969.
11. Shin JT and Fishman MC (2002). From zebrafish to human: modular medical models. *Ann. Rev. Genom. Hum. Genet.* **3**: 311–340.
12. Sehnert AJ and Stainier DYR (2002). A window to the heart: can zebrafish mutants help us understand heart disease in humans? *Trends Genet.* **18**: 491–494.
13. Streisinger G, Walker C, Dower N, Knauber D and Singer F (1981). Production of clones of homozygous diploid zebra fish (*Brachydanio rerio*). *Nature* **291**: 293–296.
14. Postlethwait J, Amores A, Force A and Yan Y-L (1999). The zebrafish genome. In: Detrich HW, Westerfield M and Zon LI (eds.), *The Zebrafish: Genetics and Genomics. Meth. Cell Biol.* **60**: 149–163.
15. Langheinrich U (2002). Zebrafish: a new model on the pharmaceutical catwalk. *BioEssays* **25**: 904–912.
16. Chakrabarti IS, Streisinger G, Singer F and Walker C (1983). Frequency of γ-ray induced specific locus and recessive lethal mutations in mature germ cells of the zebrafish, *Brachydanio rerio. Genetics* **103**: 109–122.
17. Driever W, Solnica-Krezel L, Schier AF, Neuhaus SCF, Malicki J, Stemple DL, Stanier DYR, Zwartkruis F, Abdelilah S, Rangini Z, Belak J

and Boggs C (1996). A genetic screen for mutation affecting embryogenesis in zebrafish. *Development* **123**: 37–46.

18. Haffter P, Granato M, Brand M, Mullins MC, Hammerschmidt M, Kane DA, Odenthal J, van Eeden FJM, Jiang Y-J, Heisenberg C-P, Kelsh RN, Furutani-Seiki M, Vogelsang E, Beuchle D, Schach U, Fabian C and Nüsslein-Volhard C (1996). The identification of genes with unique and essential functions in the development of the zebrafish, *Danio rerio*. *Development* **123**: 1–36.
19. Amsterdam A, Burgess S, Golling G, Chen W, Sun Z, Townsend K, Farrington S, Haldi M and Hopkins N (1999a). A large-scale insertional mutagenesis screen in zebrafish. *Genes Dev.* **13**: 2713–2724.
20. Amsterdam A and Hopkins N (1999b). Retrovirus-mediated insertional mutagenesis in zebrafish. In: Detrich HW, Westerfield M and Zon LI (eds.), *The Zebrafish: Genetics and Genomics. Meth. Cell Biol.* **60**: 87–98.
21. Beattie CE, Raible DW, Henion PD and Eisen JS (1999). Early pressure screens. In: Detrich HW, Westerfield M and Zon LI (eds.), *The Zebrafish: Genetics and Genomics. Meth. Cell Biol.* **60**: 71–86.
22. Pelegri F and Schulte–Merker S (1999). A gynogenesis-based screen for maternal-effect genes in the zebrafish. In: Detrich HW, Westerfield M and Zon LI (eds.), *The Zebrafish: Genetics and Genomics. Meth. Cell Biol.* **60**: 1–21.
23. van Eeden FJ, Granato M, Odenthal J and Haffter P (1999). Developmental mutant screens in the zebrafish. In: Detrich HW, Westerfield M and Zon LI (eds.), *The Zebrafish: Genetics and Genomics. Meth. Cell Biol.* **60**: 22–41.
24. Walker C (1999). Haploid screens and gamma-ray mutagenesis. In: Detrich HW, Westerfield M and Zon LI (eds.), *The Zebrafish: Genetics and Genomics. Meth. Cell Biol.* **60**: 44–70.
25. Darland T and Dowling JE (2001). Behavioral screening for cocaine sensitivity in mutagenized zebrafish. *Proc. Natl. Acad. Sci. USA* **98**: 116591–11696.
26. Crosier PS, Bardsley A, Horsfield JA, Krassowska AK, Lavallie ER, Collins–Racie LA, Postlethwait JH, Yan Y-L, McCoy JM and Crosier KE (2001). *In situ* hybridization screen in zebrafish for the selection of genes encoding secreted proteins. *Dev. Dynam.* **222**: 637–644.

27. Hackett PB (1993). The molecular biology of transgenic fish. In: Hochachka PW and Mommsen TM (eds.), *Biochemistry and Molecular Biology of Fishes*, Vol. 2. Elsevier, Amsterdam. pp. 207–240.
28. Gong Z and Hew CL (1995). Transgenic fish in aquaculture and developmental biology. *Curr. Top. Dev. Biol.* **30**: 177–214.
29. Maclean N (1988). Regulation and exploitation of transgenes in fish. *Mut. Res.* **399**: 255–266.
30. Hyatt TM and Ekker SC (1999). Vectors and techniques for ectopic gene expression in zebrafish. *Meth. Cell Biol.* **59**: 117–126.
31. Meng A, Jessen JR and Lin S (1999). Transgenesis. In: Detrich HW, Westerfield M and Zon LI (eds.), *The Zebrafish: Genetics and Genomics. Meth. Cell Biol.* **60**: 133–148.
32. Zbikowska HM (2003). Fish can be first — advances in fish transgenesis for commercial applications. *Transgen. Res.* **12**: 379–389.
33. Ishikawa Y, Hyodo–Taguchi Y and Tatsumi K (1997). Medaka fish for mutant screens. *Nature* **386**: 234.
34. Warren KS and Fishman MC (1998). "Physiological genomics": mutant screens in zebrafish. *Am. J. Physiol.* **275** (*Heart Circ. Physiol.* **44**): H1–H7.
35. Rubin EM and Tall A (2000). Perspectives for vascular genomics. *Nature* **407**: 265–269.
36. Talbot WS and Hopkins N (2002). Zebrafish mutations and functional analysis of the vertebrate genome. *Genes Dev.* **14**: 755–762.
37. Strehlow D, Heinrich G and Gilbert W (1994). The fates of the blastomeres of the 16-cell zebrafish embryo. *Development* **120**: 1791–1798.
38. Lee K-Y, Huang H, Ju B, Yang Z and Lin S (2002). Cloned zebrafish by nuclear transfer from long-term-cultured cells. *Nat. Biotech.* **20**: 795–799.
39. Chalfie M, Tu Y, Euskirchen G, Ward WW and Prasher DC (1994). Green fluorescent protein as a marker for gene expression. *Science* **263**: 802–805.
40. Amsterdam A, Lin S and Hopkins N (1995). The *Aequorea victoria* green fluorescent protein can be used as a reporter in live zebrafish embryos. *Dev. Biol.* **171**: 123–129.
41. Moss JB, Price AL, Raz E, Driever W and Rosenthal N (1999). Green fluorescent protein marks skeletal muscle in murine cell lines and zebrafish. *Gene* **173**: 89–98.

42. Long Q, Meng A, Wang H, Jessen JR, Farrell MJ and Lin S (1997). GATA-1 expression pattern can be recapitulated in living transgenic zebrafish using GFP reporter gene. *Development* **124**: 4105–4111.
43. Higashijima S, Okamoto H, Ueno N, Hotta Y and Eguchi G (1997). High-frequency generation of transgenic zebrafish which reliably express GFP in whole muscles or the whole body by using promoters of zebrafish origin. *Dev. Biol.* **192**: 289–299.
44. Ju B, Xu Y, He J, Liao J, Yan T, Hew CL, Lam TJ and Gong Z (1999). Faithful expression of green fluorescent protein (GFP) in transgenic zebrafish embryos under control of zebrafish gene promoters. *Dev. Genet.* **25**: 158–167.
45. Gibbs PDL and Scmale MC (1999). GFP as a genetic marker scorable through the lifecycle of transgenic zebrafish. *Mar. Biotech.* **2**: 107–125.
46. Fahrenkrug SC, Clark K and Hackett PB (1999). Dicistronic gene expression in developing zebrafish. *Mar. Biotech.* **1**: 552–561.
47. Halloran MC, Sato–Maeda M, Warren JT, Su F, Lele Z, Krone PH, Kuwada JY and Shoji W (2000). Laser-induced gene expression in specific cells of transgenic zebrafish. *Development* **127**: 1953–1960.
48. Finley K, Davidson A and Ekker SC (2001). Three color imaging using fluorescent proteins in living zebrafish embryos. *Biotechniques* **31**: 66–72.
49. Gong Z, Wan H, Tay TL, Wang H, Chen M and Yan T (2003). Development of transgenic fish for ornamental and bioreactor by strong expression of fluorescent proteins in the skeletal muscle. *Biochem. Biophys. Res. Commun.* **308**: 58–63.
50. Yoshizaki G, Takeuchi Y, Sakatani S and Takeuchi T (2000). Germ cell-specific expression of green fluorescent protein in transgenic rainbow trout under control of the rainbow trout vasa-like gene promoter. *Int. J. Dev. Biol.* **44**: 323–326.
51. Wakamatsu Y, Ju B, Pristyaznhyuk I, Niwa K, Ladygina T, Kinoshita M, Araki K and Ozato K (2001). Fertile and diploid nuclear transplants derived from embryonic cells of a small laboratory fish, medaka (*Oryzias latipes*). *Proc. Natl. Acad. Sci. USA* **98**: 1071–1076.
52. Wan H, He H, Ju B, Yan T, Lam YJ and Gong Z (2002). Generation of two-color transgenic zebrafish using the green and red fluorescent protein reporter genes *gfp* and *rfp*. *Mar. Biotech.* **4**: 146–154.

53. Peters KG, Ro PS, Bell BS and Kindman LA (1995). Green fluorescent fusion proteins: powerful tools for monitoring protein expression in live zebrafish embryos. *Dev. Biol.* **171**: 252–257.
54. Essner JJ, Laing JG, Beyer EC, Johnson RG and Hackett PB (1996). Expression of zebrafish *connexin43.4* in the notochord and tail bud of wild-type and *no tail* mutant embryos. *Dev. Biol.* **177**: 449–462.
55. Inoue K, Ozato K, Kondoh H, Iwamatsu T, Wakamatsu Y, Fujita T and Okada TS (1989). Stage-dependent expression of the chicken δ-crystallin gene in transgenic fish embryos. *Cell Diff. Dev.* **27**: 57–68.
56. Du SJ, Frenkel V, Kindschi G and Zohar Y (2001). Visualizing normal and defective bone development in zebrafish embryos using the fluorescent chromophore calcein. *Dev. Biol.* **238**: 239–246.
57. Farber SA, Pack M, Ho S-Y, Johnson ID, Wagner DS, Dorsch R, Mullins MC, Hendrickson HS, Hendrickson EK and Halpern ME (2001). Genetic analysis of digestive physiology using fluorescent phospholipid reporters. *Science* **292**: 1385–1388.
58. Vielkind J, Vielkind U, von Grotthuss E and Anders F (1971). Uptake of bacterial H3-DNA into fish embryos. *Experientia* **27**: 347–348.
59. Schwab M, Vielkind J and Anders F (1976). An approach to genetic transformation in the Xiphophorine fish. *Mol. Gen. Genet.* **144**: 151–158.
60. Vielkind J, Haas–Abdekam H, Vielkind U and Anders F (1982). The induction of a specific pigment cell type by total genomic DNA injected into the neural crest region of fish embryos of geneus *Xiphophorus. Mol. Gen. Genet.* **185**: 379–389.
61. Zhu Z, Xu Hm, Li G, Xie Y and He L (1986). Biological effects of human growth hormone gene microinjected into the fertilized eggs of loach *Misgurnus anguillicaudatus* (Cantor). *Kexue Tongbao* **31**: 988–990.
62. Burns JC, Friedmann T, Driever W, Burrascano M and Yee JK (1993). Vesicular stomatitis virus G glycoprotein pseudotyped retroviral vectors: concentration to very high titer and efficient gene transfer into mammalian and nonmammalian cells. *Proc. Natl. Acad. Sci. USA* **90:** 8033–8037.
63. Lundstrom K (2003). Latest development in viral vectors for gene therapy. *Trends Biotech.* **21**: 117–122.
64. Martineau D, Bowser PR, Renshaw RR and Casey JW (1992). Molecular characterization of a unique retrovirus associated with a fish tumor. *J. Virol.* **66**: 596–599.

65. Holzschu DL, Martineau D, Fodor SK, Vogt VM, Bowser P and Casey JW (1995). Nucleotide sequence and protein analysis of a complex retrovirus, walleye dermal sarcoma virus. *J. Virol.* **69**: 5320–5331.
66. Hart D, Frerichs GN, Rambaut A and Onions DE (1996). Complete nucleotide sequence and transcriptional analysis of the snakehead fish retrovirus. *J. Virol.* **70**: 3606–3616.
67. Lin S, Gaiano N, Culp P, Burns JC, Friedman T, Yee JK and Hopkins N (1994). Integration and germ-line transmission of a pseudotyped retroviral vector in zebrafish. *Science* **265**: 666–669.
68. Gaiano N, Allende M, Amsterdam A, Kawakami K and Hopkins N (1996a). Highly efficient germ-line transmission of proviral insertions in zebrafish. *Proc. Natl. Acad. Sci. USA* **93**: 7777–7782.
69. Gaiano N and Hopkins N (1996b). Insertional mutagenesis and rapid cloning of essential genes in zebrafish. *Nature* **383**: 829–832.
70. Allende ML, Amsterdam A, Becker T, Kawakami K, Gaiano N and Hopkins N (1996). Insertional mutagenesis in zebrafish identifies two novel genes, pescadillo and dead eye, essential for embryonic development. *Genes Dev.* **10**: 3141–3155.
71. Golling G, Amsterdam A, Sun Z, Antonelli M, Maldonado E, Chen W, Burgess S, Haldi M, Artzt K, Farrington S, Lin SY, Nissen RM and Hopkins N (2002). Insertional mutagenesis in zebrafish rapidly identifies genes essential for early vertebrate development. *Nat. Genet.* **31**: 135–140.
72. Linney E, Hardison NL, Lonze BE, Lyons S and DiNapoli L (1999). Transgene expression in zebrafish: a comparison of retroviral-vector and DNA-injection approaches. *Dev. Biol.* **213**: 207–216.
73. Wu X, Li Y, Crise B and Burgess SM (2003). Transcription start regions in human genome are favored targets for MLV integration. Science **300**: 1749–1751.
74. Smith KT, Shepard AJ, Boyd JE and Lees GM (1996). Gene delivery systems for use in gene therapy: an overview of quality assurance and safety issues. *Gene Therapy* **3**: 190–200.
75. Ivics Z, Izsvák Zs and Hackett PB (1997). Molecular reconstruction of *Sleeping Beauty*, a *Tc1*-like transposon from fish, and its transposition in human cells. *Cell* **91**: 501–510.
76. Izsvák Zs, Ivics Z and Hackett PB (1997). Repetitive elements and their genetic applications in zebrafish. *Biochem. Cell Biol.* **75**: 507–523.

77. Prack ETL and Kazian HH Jr (2002). Mobile elements and the human genome. *Nat. Genet. Rev.* **1**: 134–144.
78. Kazian HH Jr and Goodier JL (2002). LIKNE drive: retrotransposition and genome instability. *Cell* **110**: 277–280.
79. Izsvák Zs, Ivics Z, Garcia–Estefania D, Fahrenkrug SC and Hackett PB (1996). DANA elements: a family of composite, short interspersed, DNA elements associated with mutational activities in zebrafish (*Danio rerio*). *Proc. Natl. Acad. Sci. USA* **93**: 1044–1048.
80. Shimoda N, Chrevrette M, Ekker M, Kikkuuchi Y, Hotta Y and Okamoto H (1996). *Mermaid*: a family of short interspersed repetitive elements widespread in vertebrates. *Biochem. Biophys. Res. Commun.* **220**: 226–232.
81. San Miguel P *et al.* (1996). Nested retrotransposons in the intergenic regions of the maize genome. *Science* **274**: 765–768.
82. Takasaki N, Murata S, Saitoh M, Kobayashi T, Park L and Okada N (1994). Species-specific amplification of tRNA-derived short interspersed repetitive elements (SINES) by retroposition; a process of parasitization of entire genomes during the evolution of salmonids. *Proc. Natl. Acad. Sci. USA* **91**: 10153–10157.
83. Ivics Z, Izsvák Z and Hackett PB (1999). Genetic applications of transposons and other repetitive elements in zebrafish. In: Detrich HW, Westerfield M and Zon LI (eds.), *The Zebrafish: Genetics and Genomics. Meth. Cell Biol.* **60**: 99–131.
84. Plasterk RH (1993). Molecular mechanisms of transposition and its control. *Cell* **74**: 781–786.
85. Vos JC, De Baere I and Plasterk RH (1996). Transposase is the only nematode protein required for *in vitro* transposition of Tc1. *Genes Dev.* **10**: 755–761.
86. Lampe DJ, Churchill ME and Robertson HM (1996). A purified mariner transposase is sufficient to mediate transposition *in vitro. EMBO J.* **15**: 5470–5479.
87. Plasterk RHA, Izsvak Z and Ivics Z (1999). Resident aliens: the Tc1 mariner superfamily of transposable elements. *Trends Genet.* **15**: 326–332.
88. Ivics Z, Izsvák Z and Hackett PB unpub.
89. Gibbs PDL, Gray A and Thorgaard G (1994). Inheritance of *P* element and reporter gene sequences in zebrafish. *Mol. Mar. Biol. Biotech.* **3**: 317–326.

90. Radice AD, Bugaj B, Fitch D and Emmons SW (1994). Widespread occurrence of the Tc1 transposon family: Tc1-like transposons from teleost fish. *Mol. Gen. Genet.* **244**: 606–612.
91. Izsvák Zs, Ivics Z and Hackett PB (1995). Characterization of a Tc1-like trans-posable element in zebrafish (*Danio rerio*). *Mol. Gen. Genet.* **247**: 312–322.
92. Ivics Z, Izsvák Zs, Minter A and Hackett PB (1996). Identification of functional domains and evolution of Tc1 family of transposable elements. *Proc. Natl. Acad. Sci. USA* **93**: 5008–5013.
93. Reed KM (1999). Tc1-like transposable elements in the genome of lake trout (*Salvelinus namaycush*). *Mar. Biotech.* **1**: 60–67.
94. Liu Z, Li P, Kucuktas H and Dunham R (1999). Characterization of nonautonomous *Tc1*-like transposable elements of channel catfish (*Ictalurus punctatus*). *Fish Physiol. Biochem.* **21**: 65–72.
95. Leaver MJ (2001). A family of *Tc1*-like transposons from the genomes of fishes and frogs: evidence for horizontal transmission. *Gene* **271**: 203–214.
96. Raz E, van Luenen HGAM, Schaerringer B, Plasterk RHA and Driever W (1998). Transposition of the nematode *Caenorhabditis elegans Tc3* element in the zebrafish *Danio rerio*. *Curr. Biol.* **8**: 82–88.
97. Fadool JM, Hartl DL and Dowling JE (1998). Transposition of the *mariner* element from *Drosophila mauritiana* in zebrafish. *Proc. Natl. Acad. Sci. USA* **95**: 5182–5186.
98. Sherman A, Dawson A, Mather C, Gilhooley H, Li Y, Mitchell R, Finnegan D and Sang H (1998). Transposition of the *Drosophila* element *mariner* into the chicken germ line. *Nat. Biotech.* **16**: 1050–1053.
99. Zhang L, Sankar U, Lampe DJ, Robertson HM and Graham FL (1998). The *Himar1* mariner transposase cloned in a recombinant adenovirus vector is functional in mammalian cells. *Nucl. Acids Res.* **26**: 3687–3693.
100. Fischer SEJ, Wienholds E and Plasterk RHA (2001). Regulated transposition of a fish transposon in the mouse germ line. *Proc. Natl. Acad. Sci. USA* **98**: 6759–6754.
101. Inagaki H, Bessho Y, Koga A and Hori H (1994). Expression of the tyrosinase-encoding gene in a colorless melanophore mutant of the medaka fish (*Oryzias latipes*). *Gene* **150**: 319–324.
102. Koga A, Suzuki M, Inagaki H, Bessho Y and Hori H (1996). Transposable element in fish. *Nature* **383**: 30.

103. Koga A and Hori H (1997). Albinism due to transposable element insertion in fish. *Pigment Cell Res.* **10**: 377–381.
104. Hori H, Suzuki M, Inagaki H, Oshima T and Koga A (1998). An active Ac-like transposable element in teleost fish. *J. Mar. Biotech.* **6**: 206–207.
105. Kawakami K, Koga A, Hori H and Shima A (1998). Identification of a functional transposase of the *Tol2* element, an Ac-like element from the Japanese medaka fish, *Oryzias latipes*, in zebrafish, *Danio rerio*. *Gene* **225**: 17–22.
106. Kawakami K and Shima A (1999). Identification of the Tol2 transposase of the medaka fish *Oryzias latipes* that catalyzes excision of a nonautonomous *Tol2* element in zebrafish *Danio rerio*. *Gene* **240**: 239–244.
107. Kawakami K, Koga A, Hori H and Shima A (2000). Excision of the *Tol2* transposable element of the medaka fish and its transposition in the zebrafish germline. *Proc. Natl. Acad. Sci. USA* **97**: 11403–11408.
108. Koga A and Hori H (2000). Detection of *de novo* insertion of the medaka fish transposable element *Tol2*. *Genetics* **156**: 1243–1247.
109. Koga A, Iida A, Kamiya M, Hayashi R, Hori H, Ishikawa Y and Tachibana A (2003). The medaka fish *Tol2* transposable element can undergo excision in human and mouse cells. *J. Hum. Genet.* **48**: 231–235.
110. Koga A, Suzuki M, Maruyama Y, Tsutsumi M and Hori H (1999). Amino acid sequence of a putative transposase protein of the medaka fish transposable element *Tol2* deduced from mRNA nucleotide sequences. *FEBS Lett.* **461**: 295–298.
111. Cui Z, Geurts AM, Liu G, Kaufman CD and Hackett PB (2002). Structure-Function analysis of the inverted terminal repeats of the *Sleeping Beauty* transposon. *J. Mol. Biol.* **318**: 1221–1235.
112. Iyengar A, Müller F and Maclean N (1996). Expression and regulation of transgenes in fish. *Transgen. Res.* **5**: 147–165.
113. Caldovic L, Agalliu D and Hackett PB (1999). Position-independent expression of transgenes in zebrafish. *Transgen. Res.* **8**: 321–334.
114. Dorer DR (1997). Do transgene arrays form heterochromatin in vertebrates? *Transgen. Res.* **6**: 3–10.
115. Garrick D, Fiering S, Martin DIK and Whitlaw E (1998). Repeat-induced gene silencing in mammals. *Nat. Genet.* **18**: 56–59.

116. Henikoff S (1998). Conspiracy of silence among repeated transgenes. *BioEssays* **20**: 532–534.
117. Yant SR, Meuse L, Chiu W, Ivics Z, Izsvak Z and Kay MA (2000). Somatic integration and long-term transgene expression in normal and haemophilic mice using a DNA transposon system. *Nat. Genet.* **25**: 35–41.
118. Belur L, Frandsen J, Dupuy A, Largaespada DA, Hackett PB and McIvor RS (2003). Gene insertion and long-term expression in lung mediated by the *Sleeping Beauty* transposon system. *Mol. Ther.* **8**: 501–507.
119. Zanta AM, Belguise–Valladier P and Behr J-P (1999). Gene delivery: a single nuclear localization signal peptide is sufficient to carry DNA to the cell nucleus. *Proc. Natl. Acad. Sci. USA* **96**: 91–96.
120. Izsvak Z, Ivics Z and Plasterk RHA (2000). *Sleeping beauty*, a wide host-range transposon vector for genetic transformation in vertebrates. *J. Mol. Biol.* **302**: 93–102.
121. Karsi' A, Moav B, Hackett PB and Liu Z (2001). Effects of insert size on transposition efficiency of the synthetic transposon *Sleeping Beauty* in mouse cells. *Mar. Biotech.* **3**: 241–245.
122. Geurts AM, Yang Y, Clark KJ, Cui Z, Dupuy AJ, Largaespada DA and Hackett PB (2003). Gene transfer into genomes of human cells by the *Sleeping Beauty* transposon system. *Mol. Ther.* **8**: 108–117.
123. Lander ES *et al.* (2001). Initial sequencing and analysis of the human genome. *Nature* **409**: 860–921.
124. Venter JC *et al.* (2001). The sequence of the human genome. *Science* **291**: 1304–1351.
125. Inoue K (1992). Expression of reporter genes introduced by microinjection and electroporation in fish embryos and fry. *Mol. Mar. Biol. Biotech.* **1**: 266–270.
126. Zhao X, Zhang PJ and Wong TK (1993). Application of Baekonization: a new approach to produce transgenic fish. *Mol. Mar. Biol. Biotech.* **2**: 63–69.
127. Culp P, Nüsslein–Volhard C and Hopkins N (1991). High-frequency germ-line transmission of plasmid DNA sequences injected into fertilized zebrafish eggs. *Proc. Natl. Acad. Sci. USA* **88**: 7953–7957.
128. Müller F, Lele Z, Váradi L, Menczel L and Orbán L (1993). Efficient transient expression system based on square pulse electroporation and *in vivo* luciferase assay of fertilized fish eggs. *FEBS Lett.* **324**: 27–32.

129. Westerfield M, Wegner J, Jegalian BG, DeRobertis EM and Püschel AW (1992). Specific activation of mammalian *Hox* promoters in mosaic transgenic zebrafish. *Genes Dev.* **6**: 591–598.
130. Neumann E, Schaefer–Ridder M, Wang Y and Hofschneider PH (1982). Gene transfer into mouse myloma cells by electroporation in high electric fields. *EMBO J.* **1**: 841–845.
131. Xie Y, Liu D, Zou J, Lie G and Zhu Z (1989). Novel gene transfer in the fertilized eggs of loach via electroporation. *Acta Hydrobiologica Sinica* **13**: 387–389.
132. Inoue K, Yamashita S, Hata J, Kabeno S, Asada S, Nagahisa E and Fujita T (1990). Electroporation as a new technique for producing transgenic fish. *Cell Differ. Dev.* **29**: 123–128.
133. Buono RJ and Linzer PJ (1992). Transient expression of RSVCAT in transgenic zebrafish made by electroporation. *Mol. Mar. Biol. Biotech.* **1**: 271–275.
134. Lu J-K, Fu B-H, Wu J-L and Chen TT (2002). Production of transgenic silver sea bream (*Sparus sarba*) by different gene transfer methods. *Mar. Biotech.* **4**: 328–337.
135. Powers DA, Hereford L, Cole T, Chen TT, Lin CM, Kight K, Creech K and Dunham R (1992b). Electroporation: a method for transferring genes into genes of zebrafish (*Brachydanio rerio*), channel catfish (*Ictalurus punctatus*) and common carp (*Cyprinus carpio*). *Mol. Mar. Biol. Biotech.* **1**: 301–308.
136. Murakami, Y, Motoshashi K, Yano K, Ikebukuro K, Yokoyama K, Tamiya E and Karube I (1994). Micromachined electroporation system for transgenic fish. *J. Biotechnol.* **34**: 35–42.
137. Ono H, Hirose E, Miyazaki K, Yamamoto H and Matsumoto J (1997). Transgenic medaka fish bearing the mouse tyrosinase gene: expression and transmission of the transgene following electroporation of the orange-colored variant. *Pigment Cell Res.* **10**: 168–175.
138. Lavitrano M, Camaioni A, Fazio VM, Dolci S, Farace MG and Spadafora C (1989). Sperm cells as vectors for introducing foreign DNA into eggs: genetic transformation in mice. *Cell* **57**: 717–723.
139. Brinster RL, Sandgren EP, Behringer RR and Palmiter RD (1989). No simple solution for making transgenic mice. *Cell* **59**: 239–241.
140. Atkinson PW, Hines ER, Beaton S, Matthaei KI, Reed KC and Bradley MP (1991). Association of exogenous DNA with cattle and insect spermatozoa *in vitro*. *Mol. Reprod. Dev.* **29**: 1–5.

141. Arezzo F (1989). Sea urchin sperm as a vector of foreign genetic information. *Cell. Biol. Int. Rep.* **13**: 391–404.
142. Castro FO, Hernandez O, Uliver C, Solano R, Milanes C, Aguilar A, Perez A, de Armas R, Herrea L and de la Fuente J (1991). Introduction of foreign DNA into the spermatozoa of farm animals. *Theriogenology* **36**: 1099–1110.
143. Khoo H-W, Ang L-H, Lim H-B and Wong K-Y (1992). Sperm cells as vectors for introducing foreign DNA into zebrafish. *Aquaculture* **107**: 1–19.
144. Chourrout D and Perrot E (1992). No transgenic rainbow trout produced with sperm incubated with linear DNA. *Mol. Mar. Biol. Biotech.* **1**: 282–285.
145. Patil JG and Khoo HW (1996). Nuclear internalization of foreign DNA by zebrafish spermatozoa and its enhancement by electroporation. *J. Exp. Zool.* **274**: 121–129.
146. Perry ACF, Wakayama T, Kishikawa H, Kasai T, Okabe M, Toyada Y and Yanagimachi R (1999). Mammalian transgenesis by intracytoplasmic sperm injection. *Science* **284**: 1180–1183.
147. Müller F, Ivics Z, Erdélyi F, Papp T, Váradi L, Horváth L, Maclean N and Orban L (1992). Introducing foreign genes into fish eggs by electroporated sperm as a carrier. *Mol. Mar. Biol. Biotech.* **1**: 276–281.
148. Sin FYT, Bartley AL, Walker SP, Sin IL, Symonds JE, Hawke L and Hopkins CL (1993). Gene transfer in chinook salmon (*Oncorhynchus tshawytscha*) by electroporating sperm in the presence of pRSVlacZ DNA. *Aquaculture* **117**: 57–69.
149. Tsai H-J, Tseng FS and Liao IC (1995). Electroporation of sperm to introduce foreign DNA into genome of loach (*Misgurnus anguillicaudatus*). *Canad. J. Fish. Aquat. Sci.* **52**: 776–787.
150. Kang J-H, Yoshizaki G, Homma O, Strussmann CA and Takashima F (1999). Effect of an osmostic differential on the efficiency of gene transfer by electroporation of fish spermatozoa. *Aquaculture* **173**: 297–307.
151. Symonds JE, Walker SP and Sin FYT (1994). Electroporation of salmon sperm with plasmid DNA: evidence for enhanced sperm/DNA association. *Aquaculture* **119**: 313–327.
152. Jesuthasan S and Subburaju S (2002). Gene transfer into zebrafish by sperm nuclear transplantation. *Dev. Biol.* **242**: 88–95.
153. Klein TM, Wolf ED, Wu R and Sanford JC (1987). High-velocity microprojectiles for delivering nucleic acids into living cells. *Nature* **327**: 70–73.

154. Zelenin AV, Alimov AA, Barmintzev VA, Beniumov AO, Zelenina IA, Krasnov AM and Kolesnikov VA (1991). The delivery of foreign genes into fertilized fish eggs using high-velocity microprojectiles. *FEBS Lett.* **287**: 118–120.
155. Torgersen J, Collas P and Alestrom P (2000). Gene-gun-mediated transfer of reporter genes to somatic zebrafish (*Danio rerio*) tissues. *Mar. Biotech.* **2**: 293–300.
156. Montini E, Held PK, Noll M, Morcinek N, Al-Dhalimy M, Finegold M, Yant SR, Kay MA and Grompe M (2002). *In vivo* correction of murine tyrosinemia type I by DNA-mediated transposition. *Mol. Ther.* **6**: 759–769.
157. Kren BT, Bandyopadhyay P and Steer CJ (1998). *In vivo* site-directed mutagenesis of the *factor IX* gene by chimeric RNA/DNA oligonucleotides. *Nat. Med.* **4**: 285–290.
158. Szelei J and Duda E (1989). Entrapment of high-molecular-mass DNA molecules in liposomes for the genetic transformation of animal cells. *Biochem. J.* **259**: 249–553.
159. Szelei J, Váradi L, Müller F, Erdélyi F, Orbán L, Horváth L and Duda E (1994). Liposome mediated gene transfer in fish embryos. *Transgen. Res.* **3**: 116–119.
160. Horie K, Kuroiwa A, Ikawa M, Okabe M, Kondoh G, Matsuda Y *et al.* (2001). Efficient chromosomal transposition of a Tc1/mariner-like transposon Sleeping Beauty in mice. *Proc. Nat. Acad. Sci. USA* **98**: 9191–9196.
161. Dupuy AJ, Fritz S and Largaespada DA (2001). Transposition and gene disruption in the male germline of the mouse. *Genesis* **30**: 82–88.
162. Lam WL, Lee T-S and Gilbert W (1966). Active transposition in zebrafish. *Proc. Natl. Acad. Sci. USA* **93**: 10870–10875.
163. Mullins MC, Hammerschmidt M, Haffter P and Nüsslein–Volhardt C (1994). Large-scale mutagenesis in the zebrafish: in search of genes controlling development in a vertebrate. *Curr. Biol.* **4**: 189–202.
164. Dupuy AJ, Clark K, Carlson C, Fritz S, Finley K, Ekker S, Hackett PB, Horn S and Largaespada DA (2002). Mammalian germline transgenesis by transposition. *Proc. Natl. Acad. Sci. USA* **99**: 4495–4499.
165. Stuart GW, McMurray JV and Westerfield M (1988). Replication, integration and stable germ-line transmission of foreign sequences injected into early zebrafish embryos. *Development* **103**: 403–412.

166. Stuart GW, Vielkind JR, McMurray JV and Westerfield M (1990). Stable lines of transgenic zebrafish exhibit reproduction patterns of transgene expression. *Development* **109**: 577–584.
167. Tewari R, Michard-Vahee C, Parrot E and Chourrout D (1992). Mendelian transmission, structure and expression of transgenes following their injection into the cytoplasm of trout eggs. *Transgen. Res.* **1**: 250–260.
168. Ivics Z, Izsvák Zs and Hackett PB (1993). Enhanced incorporation of transgenic DNA into zebrafish chromosomes by a retroviral integration protein. *Mol. Mar. Biol. Biotech.* **2**: 162–173.
169. Score P, Frandsen JL, Jeske J, Hackett PB, Largaespada DA and McIvor RS (2003). Molecular evidence for *Sleeping Beauty*-mediated transposition and long-term expression *in vivo*. *Mol. Ther.* **7**: 59.
170. Ivics *et al.,* unpub.
171. Davidson AE, Balciunas D, Mohn D, Shaffer J, Hermanson S, Sivasubbu S, Hackett PB and Ekker SC (2003). Efficient gene delivery and expression in zebrafish using *Sleeping Beauty*. *Dev. Biol.* **263**: 191–202.
172. Clark KJ, Geurts AM, Bell J and Hackett PB (2004). Transposon vectors for gene-trap insertional mutagenesis in vertebrates. *Genesis* (in press).
173. Atkinson PW, Pinkerton AC and O'Brochta DA (2001). Genetic transformation systems in insects. *Ann. Rev. Entomol.* **46**: 317–346.
174. Koga A and Hori H (2001). The *Tol2* transposable element of the medaka fish: an active DNA-based element naturally occurring in a vertebrate genome. *Genes Genet. Syst.* **76**: 1–8.
175. Udvadia AJ and Linney E (2003). Windows into development: historic, current, and future perspectives on transgenic zebrafish. *Dev. Biol.* **256**: 1–17.
176. Chen Z-Y, He V-Y, Ehrhardt A and Kay MA (2003). Minicircle DNA vectors devoid of bacterial DNA result in persistent and high-level transgene expression *in vivo*. *Mol. Ther.* **8**: 495–500.
177. Cretekos CJ and Grunwald DJ (1999). Alyron, an insertional mutation affecting early neural crest development in zebrafish. *Dev. Biol.* **210**: 322–338.
178. Zambrowicz BP and Sands AT (2003). Knockouts model the 100 best-selling drugs — will they model the next 100? *Nat. Rev. Drug. Disc.* **2**: 38–51.
179. Eisen JS (1996). Zebrafish make a big splash. *Cell* **87**: 969–977.
180. Cooley L, Kelley R and Spradling A (1988). Insertional mutagenesis of the *Drosphila* genome with single P elements. *Science* **239**: 1121–1128.

181. Zwaal RR, Broeks A, van Meurs J, Groenen JTM and Plasterk RHA (1993). Target-selected gene inactivation in *Caenorhabditis elegans* by using a frozen transposon insertion mutant bank. *Proc. Natl. Acad. Sci. USA* **90**: 7431–7435.
182. Zhang P and Spradling AC (1994). Insertional mutagenesis of *Drosophila* heterochromatin with single P elements. *Proc. Natl. Acad. Sci. USA* **91**: 3539–3543.
183. Sundaresan V, Springer P, Volpe T, Haward S, Jones JDG, Dean C, Ma H and Martienssen R (1995). Patterns of gene action in plant development revealed by enhancer trap and gene trap transposable elements. *Genes Dev.* **9**: 1797–1810.
184. Korswagen HC, Durbin RM, Smits MT and Plasterk RHA (1996). Transposon Tc1-derived, sequence-tagged sites in *Caenorhabditis elegans* as markers for gene mapping. *Proc. Natl. Acad. Sci. USA* **93**: 14680–14685.
185. Martienssen R (1998). Functional genomics: Probing plant gene function and expression with transposons. *Proc. Natl. Acad. Sci. USA* **95**: 2021–2026.
186. Akerley BJ, Rubin EJ, Camilli A, Lampe DJ, Robertson HM and Mekalanos JJ (1998). Systematic identification of essential genes by *in vitro mariner* mutagenesis. *Proc. Natl. Acad. Sci. USA* **95**: 8927–8932.
187. Ross-McDonald P, Coelho PSR, Roemer T, Agarwal S, Kumar A, Jansen R, Cheung K-H, Sheehan A, Symoniatis D, Umansky L, Heidtman M, Nelson FK, Iwasaki H, Hager K, Gerstein M, Miller P, Roeder GS and Snyder M (1999). Large-scale analysis of the yeast genome by transposon tagging and gene disruption. *Nature* **402**: 413–418.
188. Brune W, Menard C, Hobom U, Odenbreit S, Messerle M and Koszinowski UH (1999). Rapid identification of essential and nonessential herpesvirus genes by direct transposon mutagenesis. *Nat. Biotechnol.* **17**: 360–364.
189. Hutchison III CA, Peterson SN, Gill SR, Cline RT, White O, Fraser CM, Smith HO and Venter JC (1999). Global transposon mutagenesis and a minimal mycoplasma genome. *Science* **286**: 2165–2169.
190. Klinakis AG, Zagoraiou L, Vassilatis DK and Savakis C (2000). Genome-wide insertional mutagenesis in human cells by the *Drosophila* mobile element *Minos*. *EMBO Rep.* **1**: 416–421.
191. Hoffman LM, Jendrisak JJ, Meis JRJ, Goryshin IY and Reznikof SW (1999). Transposon insertional mutagenesis and direct sequencing of microbial genomes. *Genetica* **108**: 19–24.

192. Evans MJ, Carlton MBL and Russ AP (1997). Gene trapping and functional genomics. *Trends Genet.* **13**: 370–374.
193. Hamada K, Tamaki K, Sasado T, Watai Y, Kani S, Wakamatsu Y, Ozato K, Kinoshitqa M, Kohno R, Takagi S and Kimura M (1998). Usefulness of the medaka β-actin promoter investigation using a mutant GFP reporter gene in transgenic medaka (*Oryzias latipes*). *Mol. Mar. Biol. Biotech.* 7: 173–180.
194. O'Kane CJ and Gehring WJ (1987). Detection *in situ* of genomic regulatory elements in *Drosophila. Proc. Natl. Acad. Sci. USA* **84**: 9123–9127.
195. Newfeld SJ and Takaesu NT (2002). An analysis using the *hobo* genetic system reveals that combinatorial signaling by the *Dpp* and *Wg* pathways regulates dpp expression in leading edge cells of the dorsal ectoderm in *Drosophila melanogaster. Genetics* **161**: 685–692.
196. von Melcher and Ruley HE (1989). Identification of cellular promoters by using a retrovirus promoter trap. *J. Virol.* **63**: 3227–3233.
197. von Melcher H, Reddy S and Ruley HE (1990). Isolation of cellular promoters by using a retrovirus promoter trap. *Proc. Natl. Acad. Sci. USA* **87**: 3733–3737.
198. Fridrich G and Soriano P (1991). Promoter traps in embryonic stem cells: a genetic screen to identify and mutate developmental genes in mice. *Genes Dev.* **5**: 1513–1523.
199. Gossler A, Joyner AL, Rossant J and Skarnes WC (1989). Mouse embryonic stem cells and reporter constructs to detect developmentally regulated genes. *Science* **244**: 463–465.
200. Skarnes WC, Auerbach BA and Joyner AL (1992). A gene trap approach in mouse embryonic stem cells: the *lacZ* reporter is activated by splicing, reflects endogenous gene expression, and is mutagenic in mice. *Genes Dev.* **6**: 903–917.
201. Niwa H, Araki K, Kimura S, Taniguchi S, Wakasugi S and Yamamura K (1993). An efficient gene-trap method using poly(A) trap vectors and characterization of gene-trap events. *J. Biochem.* (Tokyo) **113**: 343–349.
202. Mitchell KJ, Pinson KI, Kelly OG, Brennan J, Zupicich J, Scherz P, Leighton PA, Goodrich LV, Lu X, Avery BJ, Tate P, Dill K, Pangilinan E, Wakenight P, Tessier-Lavigne M and Skarnes WC (2001). Functional analysis of secreted and transmembrane proteins critical to mouse development. *Nat. Genet.* **28**: 241–249.

203. Yoshida M, Yagi T, Furuta Y, Takayanagi K, Kominami R, Takeda N, Tokunaga T, Chiba J, Ikawa Y and Aizawa S (1995). A new strategy of gene trapping in ES cells using 3'RACE. *Transgen. Res.* **4**: 277–287.
204. Zambrowicz BP, Friedrich GA, Buxton EC, Lilleberg SL, Person C and Sands AT (1998). Disruption and sequence identification of 2,000 genes in mouse embryonic stem cells. *Nature* **392**: 608–611.
205. Spradling AC, Stern DM, Kiss I, Roote J, Laverty T and Rubin GM (1995). Gene disruptions using *P* transposable elements: an integral component of the *Drosophila* genome project. *Proc. Natl. Acad. Sci. USA* **92**: 10824–10830.
206. Zhang JK, Prichett MA, Lampe DJ, Robertson HM and Metcalf WW (2000). *In vivo* transposon mutagenesis of the methanogenic archeon *Methanosarcina activorans* C2A using a modified version of the insect *mariner*-family transposable element *Himal*. *Proc. Natl. Acad. Sci. USA* **97**: 9665–9670.
207. Judson N and Mikalanos JJ (2000). Transposon-based approaches to identify essential bacterial genes. *Trends Microbiol.* **8**: 521–526.
208. Parimov S and Sundaresan V (2000). Functional genomics in aradbidopsis: large-scale insertional mutagenesis complements the genome sequencing project. *Curr. Opin. Biotechnol.* **11**: 157–161.
209. Vidan S and Snyder M (2001). Large-scale mutagenesis: yeast genetics in the genome era. *Curr. Opin. Biotechnol.* **12**: 28–34.
210. Clark KJ, Geurts AM, Bell JB and Hackett PB (2004). Identification of genes using gene traps in Sleeping Beauty transposons. *Genesis* (in press).
211. Clark KJ (2003). The development of Sleeping Beauty gene-trap transposons for insertional mutagenesis of vertebrates. Ph.D Dissertation, University of Minnesota.
212. Vigdal TJ, Kaufman CD, Izsvak Z, Voytas DF and Ivics Z (2002). Common physical properties of DNA affecting target site selection of *Sleeping Beauty* and other Tc1/mariner transposable elements. *J. Mol. Biol.* **323**: 441–452.
213. Roberg–Perez K, Carlson CM and Largaespada DA (2003). MTID: a database of Sleeping Beauty transposon insertions in mice. *Nucl. Acids Res.* **31**: 78–81.
214. Carlson C, Dupuy A, Fritz S, Roberg–Perez K, Fletcher CF and Largaespada DA (2003). Transposon mutagenesis of the mouse germline. *Genetics* (in press).

215. Hastie ND and Bishop JO (1976). The expression of three abundance classes of messenger RNA in mouse tissues. *Cell* **9**: 761–774.
216. Koga A and Hori H (1999). Homogeneity in the stucture of the medaka fish transposable element Tol2. *Genet. Res.* **73**: 7–14.
217. Koga A, Hori H and Sakaizumi M (2002). Gene transfer and cloning of flanking chromosomal regions using the medaka fish *Tol2* transposable element. *Mar. Biotech.* **4**: 6–11.
218. Hackett PB, Izsvák Zs, Ivics Z and Caldovic L (1999). Development of genetic tools for transgenic animals. In: Murray JD, Anderson GB, Oberbauer AM and McGlouoghlin MM (eds.), *Transgenic Animals in Agriculture.* CABI Publishing, Wallingford, UK, pp. 19–35.
219. Liu G, Cui Z, Aronovich EL, Whitley CB and Hackett PB (2003). Transposon-donor junction nucleotides are important for excision by *Sleeping Beauty* transposase. *J. Gene Med.* **6**: 574–583.
221. Koga A, Lida A, Kamiya M, Hayashi R, Hori H, Ishikawa Y and Tachibana A (2003). The medaka fish Tol2 transposable element can undergo excision in human and mouse cells. *J. Hum. Genet.* **48**: 231–235.
222. Hyrien O, Maric Ch and Mechali M (1995). Transition in specification of embryonic metazoan DNA replication origins. *Science* **70**: 994–997.
223. Kane DA and Kimmel CB (1993). The zebrafish midblastula transition. *Development* **119**: 447–456.
224. Kimmel CB, Ballard WW, Kimmel SR, Ullman B and Schilling TF (1995). Stages of embryonic development of the zebrafish. *Dev. Dyn.* **203**: 253–310.
225. Vigdal TJ, Kaufman CD, Izsvak Z, Voytas DF and Ivics Z (2002). Common physical properties of DNA affecting target site selection of *Sleeping Beauty* and other Tc1/*mariner* transposable elements. *J. Mol. Biol.* **323**: 441–452.
226. Zayed H, Izsvak Z, Khare D, Heinemann U and Ivics Z (2003). The DNA-bending protein HMGB1 is a cellular cofactor of *Sleeping Beauty* transposition. *Nucl. Acids Res.* **31**: 2313–2322.
227. Liu G, Geurts AM, Yae K, Srinivassan AR, Fahrenkrug SC, Olson WK, Takeda J, Horie K and Hackett PB. Target-site preference for *Sleeping Beauty* transposons. *Nucl. Acids Res.* (submitted).
228. Takeuchi Y, Yoshizaki G and Takeuchi T (1999). Green fluorescent protein as a cell-labeling tool and a reporter of gene expression in transgenic rainbow trout. *Mar. Biotech.* **1**: 448–457.

229. Largaespada D and Takeda J, personal communications (pg. 560).
230. Stainier DYR, Fouquet B, Chen J-N, Warren KS, Weinstein BM, Meiler SE, Mohideen M-APK, Neuhauss SCF, Solnica-Krezel L, Schier AF, Zwartkruis F, Stemple DL, Malicki J, Driever W and Fishman MC (1996). Mutations affecting the formation and function of the cardiovascular system in the zebrafish embryo. *Development* **123**: 285–292.
231. Amsterdam A, Lin S, Moss LG and Hopkins N (1996). Requirements for green fluorescent protein detection in transgenic zebrafish embryos. *Gene* **173**: 99–103.
232. Meng A, Tang H, Ong BA, Farrell MJ and Lin S (1997). Promoter analysis in living zebrafish embryos indentifies a cis-acting motif required for neuronal expression of GATA-2. *Proc. Natl. Acad. Sci. USA* **94**: 6267–6272.

Chapter 17

Evolution of the Zebrafish Genome

John H. Postlethwait
Institute of Neuroscience
University of Oregon
Eugene OR 97403, USA

Because the zebrafish lineage is basally diverging among Euteleost fish, analysis of its genome can complement information from other key species to understand the origin of genomes of teleost fish, the most species-rich group of vertebrates, and to infer some history of the human genome. Analysis of gene maps shows that zebrafish chromosomes have conserved syntenies with large segments of human chromosomes, suggesting that translocations have been rather infrequent during vertebrate evolution. Nevertheless, zebrafish chromosomes consist of mosaics of segments orthologous to several human chromosomes, in part, probably because ancestral chromosomes experienced fission in the human lineage. Because the zebrafish orthologs of small regions of human chromosomes appear to have exploded over much of the zebrafish chromosome, inversions have been fixed in populations much more frequently than translocations. Genetic maps show that segments of human chromosomes are generally present in zebrafish as two orthologous copies. These data suggest a genome duplication event in the zebrafish lineage. Comparative mapping of zebrafish, pufferfish and medaka genomes shows that this event occurred before the divergence of Euteleosts, suggesting the hypothesis that this genome duplication may have contributed to the evolutionary success of teleosts. The complete sequencing of the zebrafish genome will reveal the answer to many questions, such as the fraction of genes derived from this event that are retained in duplicate copy, and the evolutionary principles that cause gene duplicates to be retained. Because of the elegant functional analyses possible with zebrafish, the sequence of the zebrafish genome will facilitate exploration of conserved gene function in vertebrate development and physiology. In addition, functional analysis of duplicated zebrafish genes can reveal ancestral gene functions sometimes obscured in mammals by pleiotropy.

1. Introduction

An individual organism looks and acts the way it does because of biological change over two different time scales — the period of its development from egg to adult, and the stretch of evolutionary time from the ancient past to the present. Analysis of the zebrafish genome impacts broad problems in biology of each of these time scales. Over ontological scales, analysis of the zebrafish genome contributes to the ease of cloning of mutant genes that affect vertebrate embryology, development and physiological function of body organs, and models for human disease. Over evolutionary scales, comparative genomics contributes to our understanding of how vertebrate genomes arose, and how genome duplication influences the origin of evolutionary novelties and may have contributed to the spectacular radiation of teleost fishes. This review focuses first on the structural aspects of the zebrafish genome, then it investigates how genomic analysis affects analysis of vertebrate development and physiology, and finally it analyzes the role of zebrafish genomics on the understanding of evolutionary mechanisms.

2. Physical Aspects of the Zebrafish Genome

The haploid zebrafish genome has 25 chromosomes, most of which are metacentrics of similar size.[1,2] These chromosomes contain about 1.7×10^9 base pairs of DNA,[3] about half the size of the mammalian genome. Most teleost fish have about 24 chromosomes in the haploid state with very little variation, except in the case of relatively recent genome duplications in certain lineages such as salmonids, goldfish, and others.[4–6] This suggests that the original teleost probably had about 24 or 25 chromosomes. In contrast to teleosts, mammals have a more variable chromosome number.[6] Thus, in the 200 million years or so since the start of the teleost radiation,[7,8] teleost genomes may have been more stable than have mammalian genomes. This suggests that teleost karyotypes and the gene content of individual chromosomes may be more similar to the karyotype of the last common ancestor of ray-fin and lobe-fin fish than most mammalian genomes.

3. Gene Maps

Genetic mapping in zebrafish began with meiotic maps initially constructed mostly from anonymous DNA markers,[9–12] and microsatellite markers.[13,14] As sequences of more cloned genes and expressed sequence tags (ESTs) became available, they were added to the maps generally using Single Strand Conformation Polymorphisms (SSCP),[15–18] although a microarray-based system detecting Single Nucleotide Polymorphisms (SNPs) is promising.[19] An important addition to mapping tools came with the introduction of radiation hybrid mapping panels.[20–25] Meiotic maps generally have the advantage that the mechanisms of meiotic recombination are broadly understood, and so error-checking algorithms help identify anomalous data points that can be rechecked. Meiotic mapping panels have the disadvantage that limited amounts of DNA are generally available for analysis. In contrast, radiation hybrid mapping panels have the advantage that the cell lines that constitute the mapping panels can provide an essentially unlimited amount of DNA, so that the mapping panel DNAs can be widely distributed to the research community. Radiation hybrid panels, however, have the disadvantages that error checking is more difficult, and because the assay is the presence or absence of a band on a gel, false negatives can contribute to errors if conditions are not rigorously standardized.

The Zebrafish Information Network[26,27] (http://zfin.org/) currently lists on its maps 1588 genes, 1794 Sequence Tagged Sites (STS), 7539 Simple Sequence Length Polymorphisms (SSLPs) mostly CA-repeats, and 13,110 Expressed Sequence Tags (ESTs). These mapped loci help provide anchors for the contigs arising from the zebrafish sequencing project (http://www.ensembl.org/Danio_rerio/).

4. Genome Conservation

4.1. Levels of Conservation

The localization of coding sequences on meiotic and radiation hybrid mapping panels and the emerging zebrafish genome sequence provide comparative information to help understand the evolution of genomes and to improve connectivities among genomes of fish and mammals. These studies make comparisons at several different levels.

1. *Orthologs*

The first level is the identification of orthologs in different species. Orthologs are pairs of genetic elements, one in each of two different species, that are descended from a single gene in the last common ancestor of the two species. Because orthologs are not defined by function, but solely by evolutionary history, phylogenetic analysis is a key tool in ortholog identification. Rapid evolution or antiquity of origin can sometimes obscure the true history of a pair of genes in phylogenetic analyses, especially after gene duplication events.

2. *Conserved syntenies*

The second level of comparative genomics is the identification of conserved syntenies. Two loci are syntenic (syn, same; tene, thread) if they reside on the same thread of DNA, that is, on the same chromosome.[28] Two genes can be unlinked in the genetic sense, segregating in a meiotic cross independently as defined by Mendel, and still be syntenic because they are on the same chromosome. Recently, however, a new usage of the term has sprung up; where a gene in one species is described as being "syntenic" with a gene in another species. Clearly, a gene in one species cannot be on the same thread of DNA as a gene in another species, short of *in vitro* recombinant DNA technology, so this usage of the word is not consistent with the original meaning.

Shared syntenies are cases in which two genes are on the same chromosome in one species and the orthologs of those two genes occupy a single chromosome in a second species. An example would be *hoxb1a* and *dlx4a* on LG3 of zebrafish and *HOXB1* and *DLX4* on chromosome 17 (Hsa17) in human (Fig. 1). Because zebrafish and human each have a large number of chromosomes, it is unlikely that shared syntenies will occur by chance. Instead, shared syntenies will usually be conserved syntenies, syntenies that are shared because the ancestral copies of the two orthologous gene pairs were syntenic in the last common ancestor of the two species.

3. *Conserved chromosome segments*

The third level of genome comparison is the conserved chromosome segment. These are corresponding chunks of chromosomes in two different species

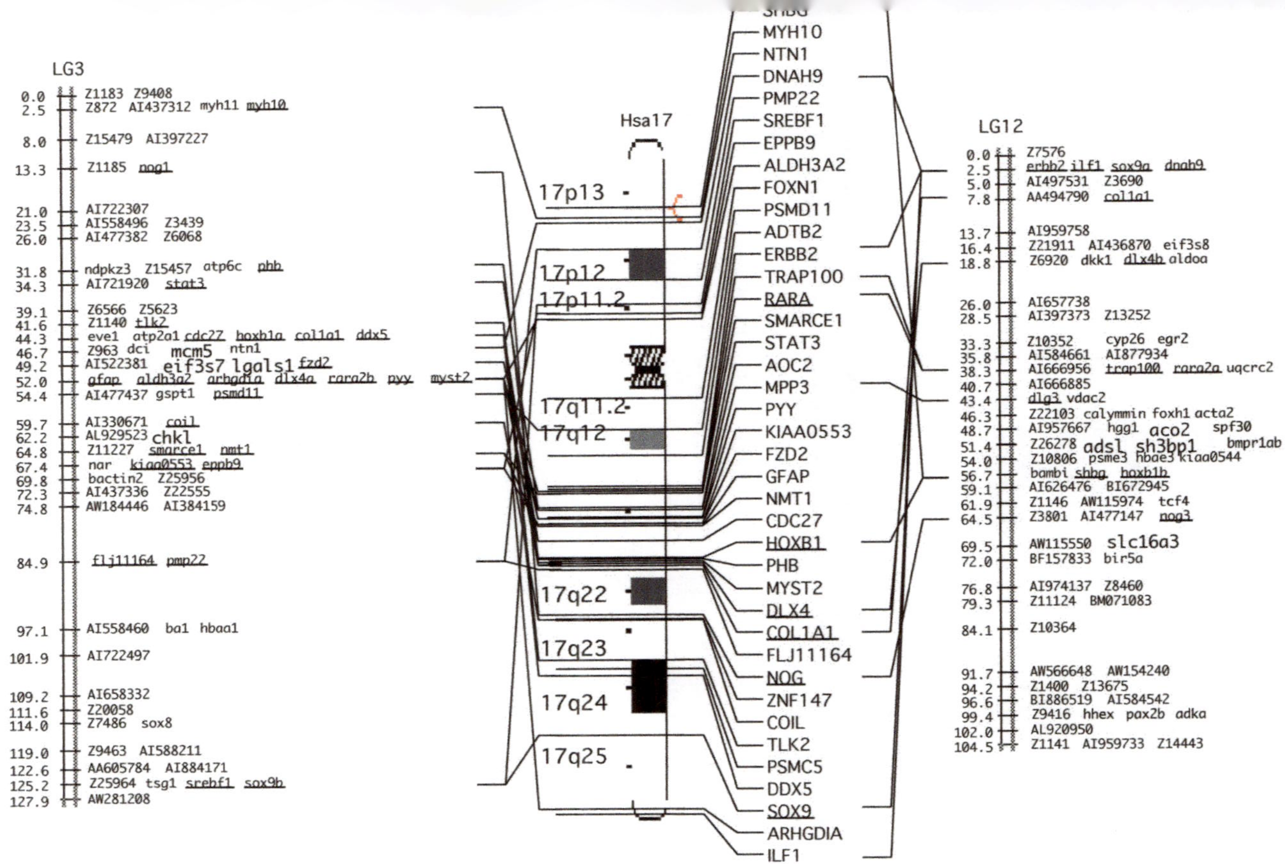

Fig. 1 Conserved syntenies for human chromosome 17 (Hsa17) and zebrafish LG3 and LG12. Orthologs of different human chromosomes are shown in different colors on zebrafish chromosomes. On Hsa17, genes found in duplicate in zebrafish are shown in red.

in which all genes in the two segments are orthologs. Some genes might be missing in one or the other chromosome segment, but as long as orthologs from other chromosome regions do not intrude, the chromosome segment will have been conserved. Inversions may have re-ordered corresponding genes within a conserved chromosome segment, but no genes from outside the segment will be present in the segment from either species.

4. *Conserved gene orders*
Finally, the fourth, and most rigorous level of genome conservation, is the conservation of gene orders within conserved chromosome segments. With conserved gene orders, three or more orthologs are in the same order and have the same transcription orientation in the two compared chromosomes. Such chromosome segments will have been retained without rearrangement from the last common ancestor of the two species. The question is: How frequently does the zebrafish genome conserve syntenies, chromosome segments, and gene orders with other vertebrates, especially other teleosts and humans?

4.2. Chromosome Evolution

Comparative mapping analysis suggests at least six general principles regarding the evolution of zebrafish and human chromosomes.

1. *Conserved syntenies*
Zebrafish chromosomes show conserved syntenies along large portions of human chromosomes.[15–18,24,29] For example, Fig. 1 shows 28 genes on LG3 whose orthologs are distributed throughout the length of Hsa17.[30] From such data, we infer that the last common ancestor of zebrafish and human had a chromosome that included this set of genes; after the human and zebrafish lineages diverged, large portions of this chromosome became parts of Hsa17 and LG3. Similar conclusions can be drawn for other human chromosome arms.

2. *Conserved gene orders*
The order of genes within long-range conserved syntenies is usually very different in the human and zebrafish chromosome. For example,

Fig. 1 shows that genes along LG3 are not co-linear with their orthologs along Hsa17. This observation can be explained if many inversions, probably in both lineages, rearranged gene orders with respect to the common ancestor and each other. Investigating the relative number of inversions in the two lineages will require chromosome level mapping data from an outgroup, such as a cartilaginous fish, or perhaps examination of basally diverging ray-fin and lobe-fin fish such as polypterus or coelacanth, respectively.

3. *Small blocks of conserved segments*

Although syntenies are often conserved over large distances, blocks of conserved chromosome segments are often rather small. This is consistent again with frequent inversions disrupting previously contiguous blocks of genes. As the color-coded loci in Fig.1 show, segments on LG3 with orthology to Hsa17 are interspersed among segments with orthologies to other parts of Hsa17 and indeed other human chromosomes. This leads to what looks like chromosome explosion, with the zebrafish orthologs of genes in small regions of a human chromosome littered along the length of a zebrafish chromosome. For example, a portion of Hsa1p appears to have exploded across LG19, and the distal tip of Hsa6q appears to have exploded across LG13 (Fig. 2). Examination of outgroups would be necessary to conclude that these two chromosome segments were syntenic in the last common ancestor of zebrafish and human and "exploded" by inversions only in the zebrafish lineage. It is also possible that the genes were separated in the last common ancestor and inversions in the human lineage brought them together.

4. *Chromosome mosaics*

Zebrafish chromosomes are mosaics of several human chromosomes. This is evident in Fig. 1, where LG3 and LG12 have orthologs not only from Hsa17, but also Hsa16, Hsa22, and other chromosomes. This situation would result either if these regions were syntenic in the last common ancestor of zebrafish and human but had separated in the human lineage, or if these regions were on different chromosomes in the last common ancestor and were joined by translocation in the zebrafish lineage.

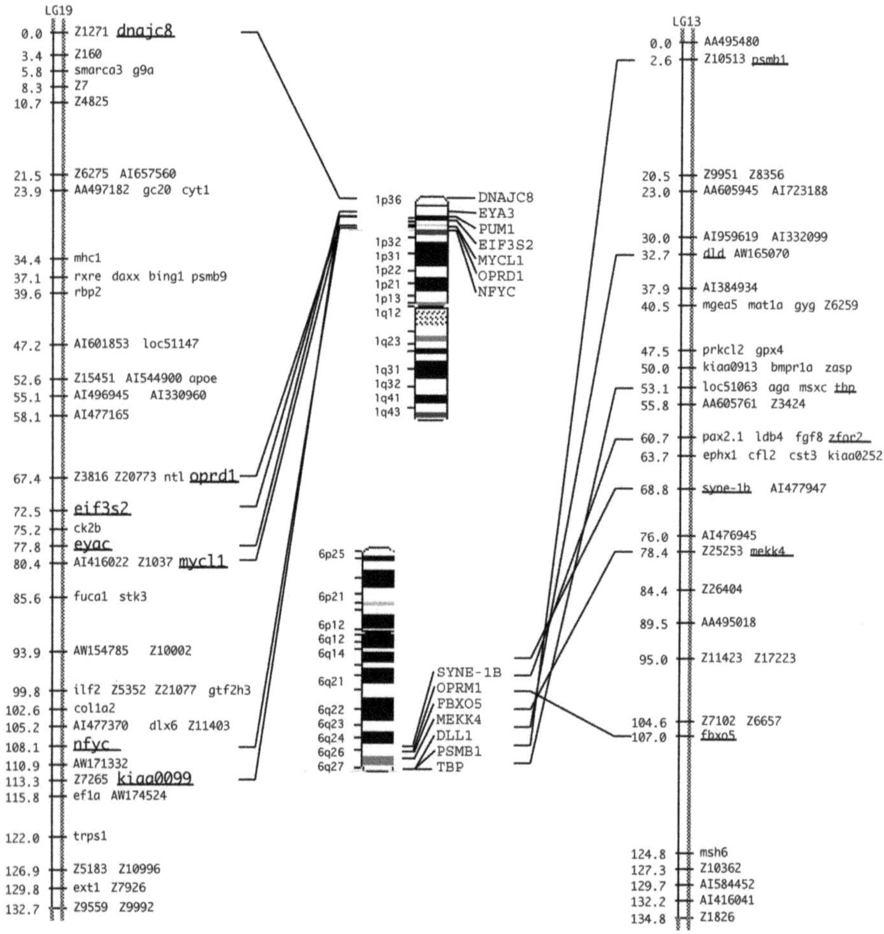

Fig. 2 Chromosome explosion. Orthologs of genes in small sections at the tips of Hsa1p and Hsa6q are scattered over large portions of zebrafish LG19 and LG13.

5. *Gene duplication*

Zebrafish has many duplicate copies, or co-orthologs, of single copy human genes and they often map in duplicated chromosome segments. Fig. 1 shows examples from Hsa17, including *rara2a* and *rara2b*, *hoxb1a* and *hoxb1b*, *dlx4a* and *dlx4b*, *col1a1a* and *col1a1b*, *nog1* and *nog3*, *sox9a* and *sox9b*, and *lhx1a* and *lhx1b*. Interestingly, all but *lhx1a/*

b are found on just two zebrafish chromosomes, linkage group (LG) 3 and LG12, while the *LHX1* co-orthologs are on LG5 and LG15. Note that LG12, LG5, and LG15 each have additional orthologs of Hsa17 genes that do not appear to be on LG3. These results and others like them suggest that many chromosome segments were duplicated in the zebrafish lineage.

6. Translocations

Although inversions appear to have been the predominant type of chromosome rearrangement, a few translocations are also evident. For example, although most orthologs of Hsa17 genes are on LG3 or LG12, at least 16 others from a few small regions of Hsa17 occur on LG5 and LG15. This includes an interval of about 2 Mb on Hsa17 that has four of these genes (*LIG3, AP2B1, TAF2N,* and *LHX1*), with *lhx1a* and *lhx1b* being zebrafish co-orthologs of human *LHX1* with one copy on each zebrafish chromosome. These results implicate translocation in either the zebrafish or human lineage. Clearly, these portions of Hsa17 were not with the LG3/LG12 portions before the chromosome duplication event, but without the examination of outgroups, it is not possible to know whether these two portions of Hsa17 material were syntenic in the last common ancestor of zebrafish and human and became separated in the zebrafish lineage by translocation, or if they were originally separate and were joined in the human lineage by translocation.

Another example of a translocation also involves LG12.[30] In addition to having orthologs of chromosome segments from Hsa17, 16, and 22 as does LG3, LG12 has orthologs from Hsa10 (Fig. 3). In addition, a portion of LG12 and LG13 appear to be largely co-orthologous to much of Hsa10, with orthologs spread out along much of the human chromosome and duplicated copies of the human genes *PAX2* and *BMPR1A*. This suggests that parts of LG12 and LG13 represent duplicated chromosome segments from a common ancestral chromosome that became much of Hsa10. These results can be interpreted with respect to two main alternative models, both of which involve translocations. Both models assume that LG3 and LG12 are largely duplicates of each other with respect to Hsa17, Hsa16, and Hsa22 material. In the first model, the

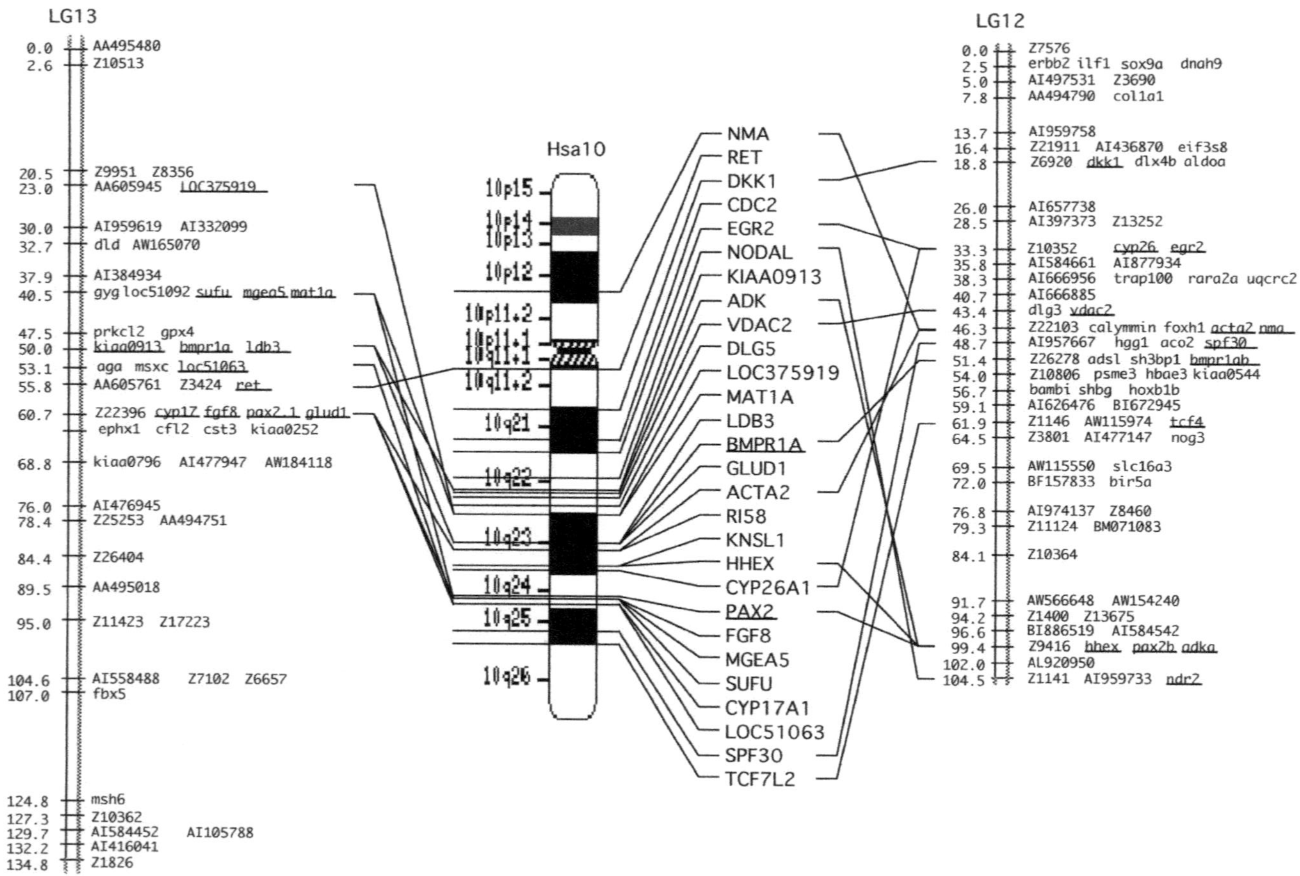

Fig. 3 Conserved syntenies of human chromosome 10q and zebrafish LG12 and LG13.

Hsa17, 16, and 22 material was on a single ancestral chromosome, which then doubled to become LG3 and LG12. Subsequently, a translocation brought Hsa10 material to LG12 after the duplication event. In the other model, the ancestral chromosome had material not only from Hsa17, 16, and 22, but also from Hsa10. After the chromosome duplication event, the model assumes that a translocation removed Hsa10 material from LG3.

5. Genome Duplication

Associated with the six general principles described above, is a seventh one regarding the origin of duplicated genes. Early experiments that systematically investigated entire gene families in zebrafish revealed frequent examples of two copies of many human genes.[31–33] A list of cases where two genes in zebrafish appear to be co-orthologous to a single gene in human is shown in Table 1. The question was, what mechanisms led to the duplicated zebrafish genes?

(1) *Models for genome amplification.* The genomic mapping of zebrafish co-orthologs of human genes can distinguish between three alternative hypotheses for the origin of these duplicates. In model 1, the duplicates arose from more frequent tandem duplication, or greater retention of tandem duplicates in the zebrafish lineage than in the human lineage. In model 2, the duplicates arose by extensive whole-chromosome duplications, as would happen for example in a genome duplication event. In model 3, the duplicates resulted from the persistence of genes that had been duplicated before the divergence of zebrafish and human lineages.

Mapping studies showed that zebrafish co-orthologs of human genes often occupy duplicated chromosome segments.[15,16,18,24,30,34–39] Examples are shown in Figs. 1 and 4. This observation shows that these chromosome segments arose as a duplication of an original single-copy chromosome segment.

Did these duplicated chromosome segments arise piecemeal, by the occasional duplication of parts of chromosomes? Or did it occur by whole chromosome duplication? Or by whole genome duplication?

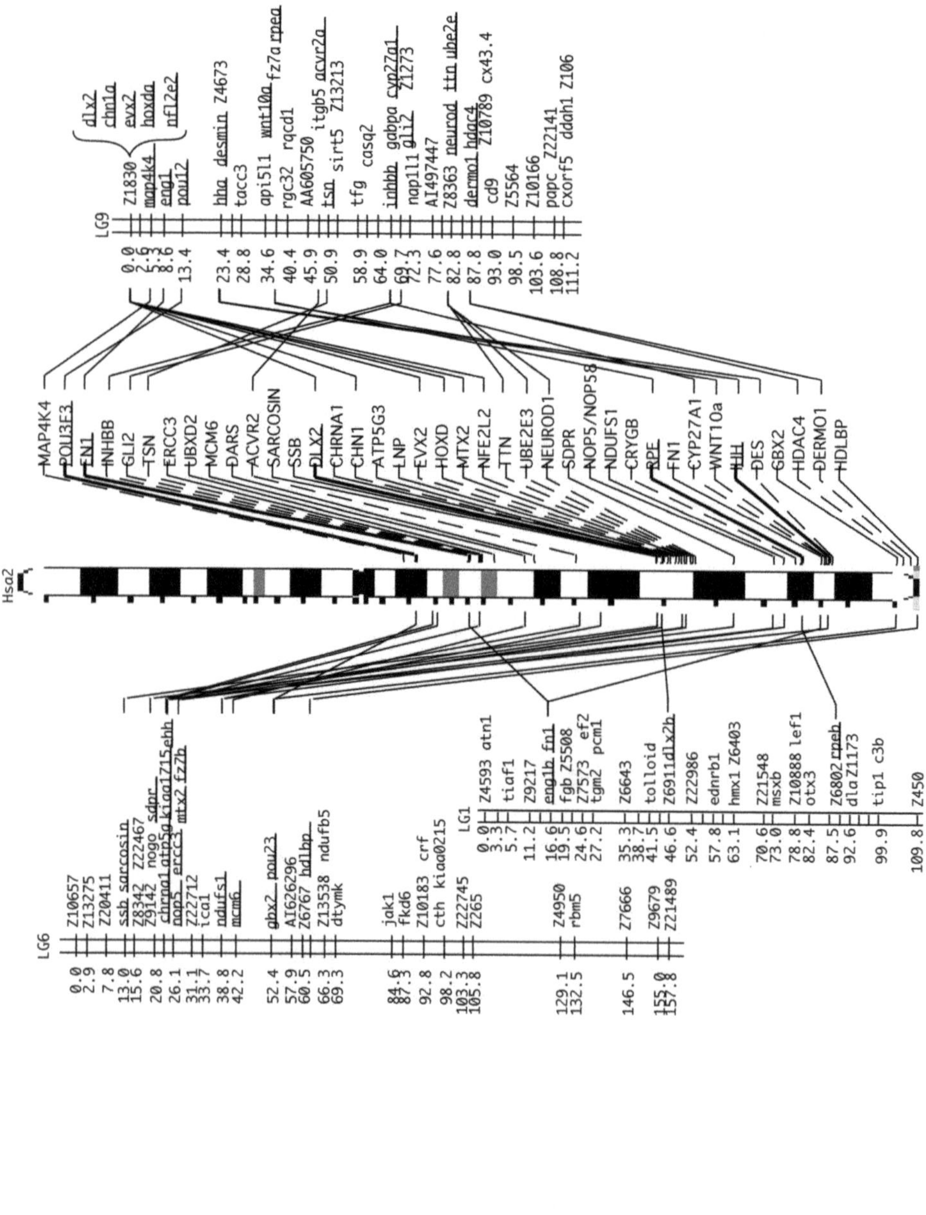
Hsa2
LG6
LG1
LG9
MAP4K4
POU3E3
FN1
INHBB
GLI2
TSN
ERCC3
UBXD2
MCM6
DARS
ACVR2
SARCOSIN
SSB
DLX2
CHRNA1
CHN1
ATP5G3
LNP
EVX2
HOXD
MTX2
NFE2L2
TTN
UBE2E3
NEUROD1
SDPR
NOP5/NOP58
NDUFS1
CRYGB
RPE
FN1
CYP27A1
WNT10a
IHH
DES
GBX2
HDAC4
DERMO1
HDLBP

Evidence suggests whole genome duplication for several reasons. First, random duplications of chromosome segments should produce some chromosome segments with three or four copies. The observation, however, is that chromosome segments generally occur in pairs.[38] Secondly, segments orthologous to long segments of human chromosomes or entire chromosome arms are sometimes observed, which argues against short, segmental duplications. Third, aneuploidy, as would occur by the duplication of a single chromosome, is often deleterious. In contrast, euploidy, as in triploids or tetraploids, often is consistent with near normal development and function.[40] Furthermore, tetraploidy has evolved several times independently in ray-fin fish, including salmonids, goldfish, carp, and suckers[41–43] and even in amphibians[44–46] and mammals.[47] We conclude that many zebrafish gene duplicates arose in a genome duplication event. We do not know if it was an allotetraploid event or an autotetraploid event.

(2) *Time of genome duplication.* When did the genome duplication event occur in ray fin-fish phylogeny? Re-analysis[34] of fugu *Hox* clusters[48] suggested that duplication of the *Hox* clusters occurred before the divergence of the fugu and zebrafish lineages, deep in the history of the teleost radiation. Subsequently, more complete analyses of pufferfish *Hox* clusters confirmed this conclusion.[49–51] Mapping of the medaka genome identified orthologs of zebrafish duplicated chromosomes, proving that the duplication is ancient in teleost fish.[6,52] Furthermore, phylogenetic analyses demonstrated that zebrafish duplicates are ancient,[35,37,38,53] and that without careful analysis, as conducted by Van de Peer and colleagues,[37,54] incorrect conclusions can be drawn.[55] All of the evidence best fits the occurrence of a genome duplication event at the base of the teleost radiation. Current work with deeply diverging teleosts, such as the eel, and with bowfin (*Amia calva*) representing the likely sister group of the teleosts, should pinpoint the timing more accurately.

(3) *Tetraploidy as a mechanism for genome amplification.* Although tetraploidy has occurred in multiple lines independently in vertebrates, one wonders how it could have occurred, and what selective forces would have allowed the tetraploid lineage to continue and become

successful. In the laboratory, tetraploid frogs and fish are manipulated readily by the application of high pressure or moderate heat shock at specific times in development.[56] It is not difficult to imagine that this could happen occasionally by chance in nature, and that entire clutches might be affected, providing several tetraploids at the same place and time. Resulting tetraploids will grow up rather normally, but problems arise with reproduction. One big problem is with meiosis — because chromosomes will pair in groups of four homologs rather than two as in diploids, segregation will give mostly aneuploid gametes. If two tetraploids mate, then there is a chance, though small, that the joining of two aneuploid gametes would result in a euploid zygote. Within a few generations, there would be strong selection for chromosome rearrangements that would re-diploidize chromosome pairing in meiosis. Even today, however, about 50 million years after a tetraploidization event in the salmonid lineage, some chromosomes in salmonids are not yet fully diploidized.[4] Despite these formidable problems — others exist with respect to sex determination and the finding of tetraploid mates in a population that is mostly diploids — tetraploidization does occur, and it seems like the best explanation for the results from the zebrafish genome.

6. Evolution of a Duplicated Chromosome Segment

What is the pattern of evolution within duplicated chromosome segments in zebrafish? After genome duplication, the duplicated chromosomes begin to rearrange and to lose genes, but over what scales of time do these events occur? Consider an approximately 2 Mb portion of the human genome in Hsa7q36 (Fig. 5), which illustrates in microcosm many of the principles listed above that were evident from chromosome-level analysis. This segment has 17 annotated genes, including the important developmental regulators *SHH* and *EN2*. Zebrafish has two co-orthologs of both of these genes, with *eng2a* and *shh* genes on LG7[31,57], and *eng2b* and *twhh* genes on LG2[31,58]. In both cases, the zebrafish genes had previously been mapped in the

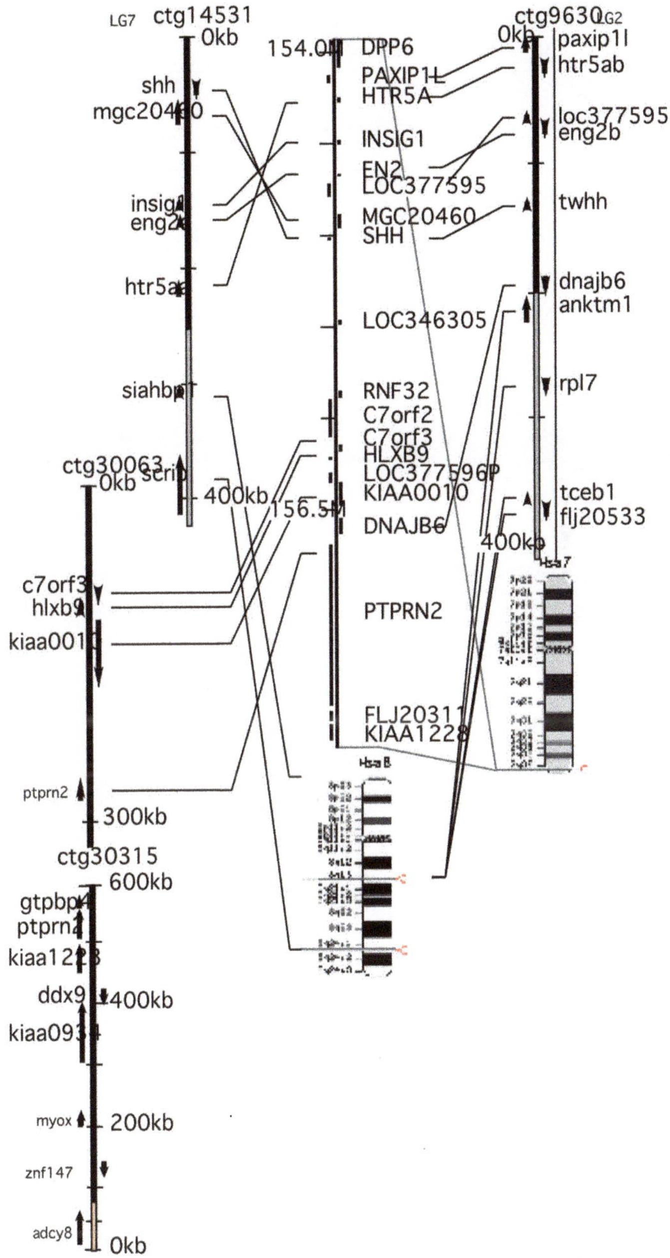

Fig. 5 Conserved syntenies for a portion of Hsa7 and zebrafish LG7 and LG2.

same bins, indicating they are quite close together.[15,34,59] (If the *twhh* gene had been called *shhb*, and *shh* subsequently renamed *shha*, their relationships to mammalian genes would have been clearer and its appreciation by mammalian biologists might have been greater.)

1. *Duplicate gene retention*

The zebrafish sequencing effort at the Sanger Institute currently shows two contigs that contain duplicated *shh* and *eng2* gene pairs and other genes from this portion of Hsa7q36. Consider first the seven genes between *PAXIP1L* and *SHH* (Fig. 5). Three of these genes are duplicated (*HTR5A, SHH,* and *EN2*) with one copy on each of the two zebrafish contigs. The other four genes are split, with two on one zebrafish contig, and the other two on the other. We infer that after the duplication, one copy of each of the fish orthologs of *PAXIP1L, INSIG1, LOC377595,* and *MCG20460* was lost on one of the two fish duplicated chromosome segments, and the loss was approximately equal in this case on each chromosome segment.

Of the remaining nine genes in the segment, one is a pseudogene (*LOC377596P*), and none of the rest appear to have duplicates in the current Sanger dataset. Orthologs of five of the four distal genes in the human segment are on zebrafish ctg30063 with the same order and transcription direction as in Hsa7q36; an ortholog of the other gene *DNAJB6* is on ctg9630 along with *twhh.* No zebrafish duplicates were detected for these genes, suggesting that all of these have been lost from the duplicated chromosome segment. Note that portions of *ptprn2* and *hlxb9* were on two other contigs, ctg30315 and ctg9427, respectively, so it is at this point unclear whether these are duplicates or as yet unassembled but contiguous regions of the genome. The *hlxb9* gene, however, does map to this location on the HS panel.[60]

2. *Local inversions*

Although taken together, these two duplicated segments preserved gene content compared to human, the gene order was apparently rearranged. The order of the duplicated genes *HTR5A, EN2,* and *SHH* was maintained in both contigs, but a single small rearrangement inverted the order of the central genes *INSIG1, EN2,* and *LOC377595.* The simplest explanation is that the pre-duplication chromosome in fish had the order *PAXIP1L — HTR5A — LOC377595 — EN2 — INSIG1 — MGC20460 — SHH.* A single inversion involving the three genes *LOC377595 — EN2 — INSIG1* and reciprocal loss of *LOC377595* and *INSIG1* would have resulted in the current situation. In addition, an inversion occurred that moved ortholog of *DNAJB6* in zebrafish relative to human. We conclude that inversions can occur over spaces of just a few genes. Whether these inversions occurred in the lineage of ray-fin fish or lobe-fin fish is not certain.

3. *Long-range inversions*

The other zebrafish orthologs of human genes in this part of Hsa7q36 are on other contigs and are apparently unduplicated, according to currently available data. The zebrafish ortholog of *DPP6* is on ctg12906 with an apparent ortholog of an Hsa10 gene; *LOC346305* does not appear to exist in the current dataset; *RNF32* is on ctg11508 with several genes each from a different human chromosome; and a portion of *C7orf2* is on ctg10623 with several genes each from a different human chromosome. We conclude that longe-range inversions also can affect small regions, as shown by the chromosome mapping studies.

4. *Non-conserved syntenies*

Both of the *SHH*-containing zebrafish gene contigs contain in addition to orthologs from Hsa7q36, orthologs of portions of Hsa8q. These include two genes that are nearest neighbors in both zebrafish and human on ctg14531 and Hsa8q24, and four genes in the same order in both species on ctg9630 and a 2 Mb region in Hsa8q13, although there are 14 genes annotated in human and just four in zebrafish ctg9630. We conclude that the pre-duplication ray-fin fish had portions of Hsa8q appended to this 2 Mb portion of Hsa7q36. Again, without appropriate outgroups, it is impossible to tell if the zebrafish or human arrangement was ancestral. Neither the mouse[61] nor the cat[62] chromosomes that contain this region of Hsa7q has conserved synteny with Hsa8.

7. An Estimate of the Rate of Duplicate Gene Retention

What fraction of duplicated genes has been retained in two copies in the zebrafish genome? A preliminary estimate can be obtained from Figs. 1, 4 and 5. Of the 38 Hsa17, 38 Hsa2q, and 16 Hsa7q36 genes investigated whose orthologs map to the duplicated linkage groups shown, at least 14 are present in duplicated copies on both zebrafish chromosomes. Bringing the whole *HOXB* cluster into the picture adds another 9 genes with 3 more duplicates for 103 genes and 17 duplicates, or 16.5%. Because lack of information is more likely to under-represent duplicated genes than to over-represent them, the retention rate is likely to be higher, in the realm of 20–30% or so. Full answers to the question of duplicate gene retention frequency await the complete sequencing and annotation of the zebrafish genome.

8. Chromosome Fission and Fusion

How can it be that most teleosts (including zebrafish) and human have about the same number of chromosomes, 24 or 25 and 23, respectively, but a genome duplication event preceded the teleost radiation? And how can we account for the fact that zebrafish chromosomes are mosaics of human chromosomes?

Two major hypotheses could account for these observations. In one hypothesis (the chromosome fusion hypothesis), the last common ancestor of zebrafish and human had about 24 chromosomes. According to this model, the ray-fin fish genome duplication would have resulted in 48 chromosomes in a haploid set. Subsequently, chromosomes would have fused together in the ray-fin lineage, which finally, on average, gave 24 or 25 chromosomes. This, coupled with frequent inversions and occasional translocations, would provide zebrafish chromosomes that are mosaics of sometimes small regions of human chromosomes.

In the other hypothesis (the chromosome fission hypothesis), the last common ancestor may have had only about twelve chromosomes, which then underwent fission sometime in the lineage leading to humans. If, on average, most of the original 12 chromosomes divided in two, it could produce the human set of 23 haploid chromosomes. For example, in the last common ancestor of zebrafish and human, material now on Hsa17 and Hsa16 may have been on the same chromosome. Chromosome fission in the human lineage may have then separated these segments into two different chromosomes, while they remained syntenic in zebrafish, although stirred by inversions.

The distribution of some genes are as expected from the fission hypothesis. For example, as Fig. 1 shows, the closely related genes *sox8, sox9b,* and *sox10* are all on LG3, but are on Hsa16, Hsa17, and Hsa22, respectively. Phylogenies show that these three genes form a monophyletic clade among vertebrate *SOX* genes.[63,64] If these three genes arose by tandem duplication sometime before the last common ancestor of human and zebrafish — a supposition supported by the very close linkage of *sox9b* and *sox8* in zebrafish — then inversions before and after lineage divergence could have spread them out on LG3, and chromosome fissions could have separated them to different chromosomes in human. Alternatively, these three *SOX* genes could have arisen from a single invertebrate chordate *SOX8/9/10* gene that then replicated in the genome duplication events that preceded the vertebrate radiation. Genome duplication would put the three genes on different chromosomes, as they are now in human, and subsequent translocations could have fortuitously brought the three genes together on a single chromosome in the zebrafish lineage. In the case of *sox9b* and *sox8*, the fortuitous translocation would have to have been quite surgical to place these two gene duplicates so close together. In addition, although analysis of gene duplicates in the human genome shows several paralog pairs between Hsa17 and Hsa22, it shows none between Hsa17 and Hsa16.[65] Thus, a working hypothesis is that *SOX10* and *SOX8/9* were duplicated in the vertebrate genome expansion,[66–71] and then tandem or segmental duplication gave separate *SOX8* and *SOX9* genes, which then dispersed to Hsa16 and Hsa17 in the human lineage and stayed nearby in the zebrafish lineage.

9. Retention of "Ohnologs"

A major problem for the annotation of the zebrafish genome will be lineage-specific retention of "ohnologs",[72,73] genes duplicated during a genome duplication, such as those hypothesized to have occurred early in the vertebrate radiation.[66–71] Examples of ohnologs in humans might include the four *HOX* clusters, the MHC region duplicates, the hedgehog genes *SHH, IHH,* and *DHH*, and many more,[70,74,75] (but see Hughes *et al.*[68]). The last common ancestor of zebrafish and human would have possessed many of these genes, and some may have been lost independently in the ray-fin and lobe-fin fish lineages. If an ohnolog has been lost in the human lineage, then the best match in a blast search of a zebrafish gene to the human genome will be a paralog, and a phylogenetic tree will be inconclusive about the human ortholog of the zebrafish gene. The key issue will be the genomic environs of the zebrafish gene — the nearest neighbors may share inappropriate conserved syntenies with the "wrong" ohnolog.

The first example of zebrafish retention of an ohnolog lost in the human lineage was the gene *eve1,* an *even-skipped* paralog in zebrafish with uncertain affinities to the human genes *EVX1* and *EVX2* in terms of sequence and expression.[76,77] The human *EVX1* and *EVX2* genes are located at the immediate 5' end of the *HOXA* and *HOXD* clusters, respectively, and the last common ancestor of vertebrates and non-vertebrate chordates likely had an *EVX* gene at the 5' end of its single *HOX* cluster.[78,79] The single ancestral *HOX* cluster duplicated twice to give four clusters, each with an *EVX* gene at its 5' end. Subsequent gene loss in the human lineage left an *EVX* gene next to only the *HOXA* and *HOXD* clusters. Mapping of the zebrafish *eve1* gene showed that it was at the immediate 5' end of the *hoxba* cluster, one of the two zebrafish *HOXB* cluster co-orthologs.[15,3] Thus, the last common ancestor of human and zebrafish likely had an *EVX* gene at the 5' end of its *HOXB* cluster, but this *EVX* gene, which logically would be called *EVX3*, was lost in the human lineage.

Another gene best explained as an ohnolog lost in the human lineage but retained in the zebrafish lineage is *adra2d*.[80] Humans have three alpha-2-adrenergic receptor genes *ADRA2A* (Hsa10q24-q26), *ADRA2B* (2p13-q13), and *ADRA2C* (4p16),[81–83] and these map in chromosome regions that appear to be anciently duplicated.[65,74,84,85] Zebrafish has orthologs of each of the three human genes that map in regions of conserved syntenies with the human orthologs, showing that the duplication events producing these genes occurred before the divergence of the human and zebrafish lineages. Amphibians and reptiles have an additional alpha-2-adrenergic receptor gene, and zebrafish has duplicates of this one, called *adra2da* and *adra2db*,[80] showing that the last common ancestor of human and zebrafish had four alpha-2-adrenergic receptor genes. In zebrafish, the fourth *adra2* genes map with conserved syntenies to Hsa5, suggesting that this would have been the location of the human gene, which would be called *ADRA2D* if it were not missing. The portion of Hsa4 that contains *ADRA2C* is clearly duplicated in *Hsa5*,[65,74,84,85] and this is where *ADRA2D* "should" map in human, if it existed. The identification of such missing ohnologs, and giving them a name that reflects their likely origin, will be a challenge for the annotation of the zebrafish genome sequence.

10. A Model for the Evolution of the Zebrafish Genome

The principles discussed above can provide an outline for a model of the origin of the zebrafish and human genomes. According to this model, the last common ancestor of human and zebrafish had about twelve chromosomes in the haploid set, only three of which are shown in Fig. 6. As the lineages of ray-fin and lobe-fin fish separated, inversions and translocations occurred in both lineages. In the ray-fin fish lineage, the model suggests that there was a whole genome duplication, which resulted in about 24 haploid chromosomes, approximating the karyotype of most teleost fish.[6] After the duplication, extensive inversions continued to occur, and a few translocations occurred as well. In addition, many duplicated genes degenerated to singletons because one member become

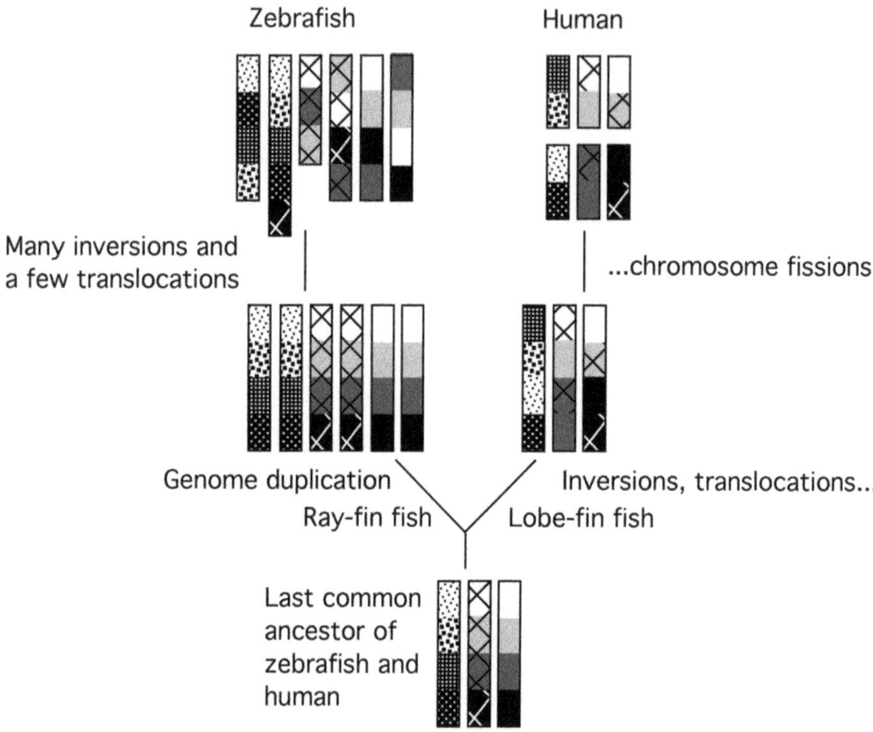

Fig. 6 A model for the evolution of zebrafish genome relative to human.

a pseudogene, the most common fate of gene duplicates.[86–91] In the lobe-fin lineage leading to humans, the model suggests that each chromosome, on average broke in half. The model accounts for the situation that humans and zebrafish have about the same number of chromosomes, 23 and 25 pairs. Furthermore, the model shows how syntenies could be broadly conserved between human and zebrafish, but that zebrafish chromosomes would be mosaics of more than one human chromosome, with orthologs of segments of human chromosomes distributed along much of the zebrafish chromosome. The sequence of the zebrafish genome will provide substantial data to test the model and refine specific aspects of the model.

Acknowledgements

Thanks are due to Catherine Wilson, Bryan Issa, Phuong Ngo–Hazlett and members of the Talbot lab for excellent work in generating mapping data. Thanks also to Drs. A. Andres, W. Cresko, S. Johnson, W. Talbot, T. Titus, and Y. Yan for their participation in this work. The work was supported by grants R01 RR10715 and P01 HD22486 from NIH.

References

1. Daga RR, Thode G and Amores A (1996). Chromosome complement, C-banding, Ag-NOR and replication banding in the zebrafish *Danio rerio*. *Chromosome Res.* **4**: 29–32.
2. Gornung E, Gabrielli I, Cataudella S and Sola L (1997). CMA3-banding pattern and fluorescence *in situ* hybridization with 18S rRNA genes in zebrafish chromosomes. *Chromosome Res.* **5**: 40–46.
3. Hinegardner R and Rosen DE (1972). Cellular DNA content and the evolution of teleostean fishes. *Am. Nat.* **106**: 621–644.
4. Allendorf F and Thorgaard G (1984). Tetraploidy and the evolution of salmonid fishes. In: Turner BJ and Turner BJs (eds.), *Evolutionary Genetics of Fishes.* Plenum Press, New York, pp. 1–46.
5. Risinger C and Larhammar D (1993). Multiple loci for synapse protein SNAP-25 in the tetraploid goldfish. *Proc. Natl. Acad. Sci. USA* **90**: 10598–10602.
6. Naruse K, Tanaka M, Mita K, Shima A, Postlethwait J and Mitani H (2004). A Medaka Gene Map: The Trace of Ancestral Vertebrate Proto-chromosomes Revealed by Comparative Gene Mapping Genome Research.
7. Hedges SB (2002). The origin and evolution of model organisms. *Nat. Rev. Genet.* **3**: 838–849.
8. Hedges SB and Kumar S (2002). Genomics. Vertebrate genomes compared. *Science* **297**: 1283–1285.
9. Johnson SL, Midson CN, Ballinger EW and Postlethwait JH (1994). Identification of RAPD primers that reveal extensive polymorphisms between laboratory strains of zebrafish. *Genomics* **19**: 152–156.

10. Johnson SL, Africa D, Horne S and Postlethwait JH (1995). Half-tetrad analysis in zebrafish: mapping the ros mutation and the centromere of linkage group I. *Genetics* **139**: 1727–1735.
11. Johnson SL, Gates MA, Johnson M, Talbot WS, Horne S, Baik K, Rude S, Wong JR and Postlethwait JH (1996). Centromere-linkage analysis and consolidation of the zebrafish genetic map. *Genetics* **142**: 1277–1288.
12. Postlethwait J, Johnson S, Midson CN, Talbot WS, Gates M, Ballenger EW, Africa D, Andrews R, Carl T, Eisen JS, Horne S, Kimmel CB, Hutchinson M, Johnson M and Rodriguez A (1994). A genetic linkage map for the zebrafish. *Science* **264**: 699–703.
13. Knapik EW, Goodman A, Atkinson OS, Roberts CT, Shiozawa M, Sim CU, Weksler–Zangen S, Trolliet MR, Futrell C, Innes BA, Koike G, McLaughlin MG, Pierre L, Simon JS, Vilallonga E, Roy M, Chiang PW, Fishman MC, Driever W and Jacob HJ (1996). A reference cross DNA panel for zebrafish (*Danio rerio*) anchored with simple sequence length polymorphisms. *Development* **123**: 451–460.
14. Knapik E, Goodman A, Ekker M, Chevrette M, Delgado J, Neuhauss S, Shimoda N, Driever W, Fishman M and Jacob H (1998). A microsatellite genetic linkage map for zebrafish (*Danio rerio*). *Nat. Genet.* **18**: 338–343.
15. Postlethwait J, Yan Y, Gates M, Horne S, Amores A, Brownlie A, Donovan A, Egan E, Force A, Gong Z, Goutel C, Fritz A, Kelsh R, Knapik E, Liao E, Paw B, Ransom D, Singer A, Thomson M, Abduljabbar T, Yelick P, Beier D, Joly J, Larhammar D and Talbot W *et al.* (1998). Vertebrate genome evolution and the zebrafish gene map. *Nat. Genet.* **18**: 345–349.
16. Gates MA, Kim L, Egan ES, Cardozo T, Sirotkin HI, Dougan ST, Lashkari D, Abagyan R, Schier AF and Talbot WS (1999). A genetic linkage map for zebrafish: comparative analysis and localization of genes and expressed sequences. *Genome Res.* **9**: 334–347.
17. Kelly PD, Chu F, Woods IG, Ngo–Hazelett P, Cardozo T, Huang H, Kimm F, Liao L, Yan Y-L, Zhou Y, Johnson SL, Abagyan R, Schier AF, Postlethwait JH and Talbot WS (2000). Genetic linkage mapping of zebrafish genes and ESTs. *Genome Res.* **10**: 558–567.
18. Woods IG, Kelly, PD, Chu F, Ngo–Hazelett P, Yan Y-L, Huang H, Postlethwait JH and Talbot WS (2000). A comparative map of the zebrafish genome. *Genome Res.* **10**: 1903–1914.

19. Stickney HL, Schmutz J, Woods IG, Holtzer CC, Dickson MC, Kelly PD, Myers RM and Talbot WS (2002). Rapid mapping of zebrafish mutations with SNPs and oligonucleotide microarrays. *Genome Res.* **12**: 1929–1934.
20. Chevrette M, Joly L, Tellis P and Ekker M (1997). Contribution of zebrafish-mouse cell hybrids to the mapping of the zebrafish genome. *Biochem. Cell Biol.* **75**: 641–649.
21. Geisler R, Rausch G-J, Baier H, van Bebber F, Brobeta L, Dekens MPS, Finger K, Fricke C, Gates MA, Geiger H, Geiger–Rudolph S, Gilmour D, Glaser S, Gnügge L, Habeck H, Hingst K, Holley S, Keenan J, Kirn A, Knaut H, Lashkari D, Maderspacher F, Martyn U, Neuhauss S, Neumann C, Nicolson T, Pelegri F, Ray R, Rick JM, Roehl H, Roeser T, Schauerte HE, Schier AF, Schönberger U, Schönthaler H-B, Schulte–Merker S, Seydler C, Talbot WS, Weiler C, Nüsslein–Volhard C, Haffter P and Geisler R (1999). A radiation hybrid map of the zebrafish genome. *Nat. Genet.* **23**: 86–89.
22. Hukriede NA, Joly L, Tsang M, Miles J, Tellis P, Epstein JA, Barbazuk WB, Li FN, Paw B, Postlethwait JH, Hudson TJ, Zon LI, McPherson JD, Chevrette M, Dawid IB, Johnson SL and Ekker M (1999). Radiation hybrid mapping of the zebrafish genome. *Proc. Natl. Acad. Sci. USA* **96**: 9745–9750.
23. Kwok C, Critcher R and Schmitt K (1999). Construction and characterization of zebrafish whole genome radiation hybrids. In: Detrich III HW, Westerfield M and Zon LI, (eds.), *The Zebrafish: Genetics and Genomics.* Academic Press, San Diego, CA, pp. 287–302.
24. Barbazuk WB, Korf I, Kadavi C, Heyen J, Tate S, Wun E, Bedell JA, McPherson JD and Johnson SL (2000). The syntenic relationship of the zebrafish and human genomes. *Genome Res.* **10**: 1351–1358.
25. Chevrette M, Joly L, Tellis P, Knapik EW, Miles J, Fishman M and Ekker M (2000). Characterization of a zebrafish/mouse somatic cell hybrid panel. *Genomics* **65**: 119–126.
26. Westerfield M, Doerry E, Kirkpatrick AE and Douglas SA (1999). Zebrafish informatics and the ZFIN database. *Methods Cell Biol.* **60**: 339–355.
27. Sprague J, Clements D, Conlin T, Edwards P, Frazer K, Schaper K, Segerdell E, Song P, Sprunger B and Westerfield M (2003). The Zebrafish Information Network (ZFIN): the zebrafish model organism database. *Nucl. Acids Res.* **31**: 241–243.

28. Kucherlapati RS and Ruddle FH (1975). Mammalian somatic hybrids and human gene mapping *Ann. Intern. Med.* **83**: 553–560.
29. Farber SA, De Rose RA, Olson ES and Halpern ME (2003). The zebrafish annexin gene family. *Genome Res.* **13**: 1082–1096.
30. Postlethwait JH, Woods IG, Ngo–Hazelett P, Yan Y-L, Kelly PD, Chu F, Huang H, Hill–Force A and Talbot WS (2000). Zebrafish comparative genomics and the origins of vertebrate chromosomes. *Genome Res.* **10**: 1890–1902.
31. Ekker M, Wegner J, Akimenko M-A and Westerfield M (1992). Coordinate embryonic expression of three zebrafish *engrailed* genes. *Development* **116**: 1001–1010.
32. Akimenko M-A, Ekker M, Wegner J, Lin W and Westerfield M (1994). Combinatorial expression of three zebrafish genes related to *Distal-less*: part of a homeobox gene code for the head *J. Neurosci.* **14**: 3475–3486.
33. Akimenko M-A, Johnson SL, Westerfield M and Ekker M (1995). Differential induction of four *msx* homeobox genes during fin development and regeneration in zebrafish. *Development* **121**: 347–357.
34. Amores A, Force A, Yan Y-L, Joly L, Amemiya C, Fritz A, Ho RK, Langeland J, Prince V, Wang Y-L, Westerfield M, Ekker M and Postlethwait JH (1998). Zebrafish *hox* clusters and vertebrate genome evolution. *Science* **282**: 1711–1714.
35. Taylor JS, Van de Peer Y, Braasch I and Meyer A (2001a). Comparative genomics provides evidence for an ancient genome duplication event in fish. *Philos. Trans. R. Soc. Lond. B* **356**: 1661–1679.
36. Postlethwait J, Amores A, Yan G and Austin CA (2002). Duplication of a portion of human chromosome 20q containing Topoisomerase (Top1) and Snail genes provides evidence on genome expansion and the radiation of teleost fish. In: Shimizu N, Aoki T, Hirono I, Takashima F, Shimizu N, Aoki T, Hirono I and Takashima FS (eds.), *Aquatic Genomics: Steps Toward a Great Future.* Springer–Verlag, Tokyo, pp. 20–31.
37. Van de Peer Y, Frickey T, Taylor J and Meyer A (2002). Dealing with saturation at the amino acid level: a case study based on anciently duplicated zebrafish genes. *Gene* **295**: 205–211.
38. Taylor J, Braasch I, Frickey T, Meyer A and Van de Peer Y (2003). Genome duplication, a trait shared by 22,000 species of ray-finned fish. *Genome Res.* **13**: 382–390.

39. Winkler C, Schafer M, Duschl J, Schartl M and Volff JN (2003). Functional divergence of two zebrafish midkine growth factors following fish-specific gene duplication. *Genome Res.* **13**: 1067–1081.
40. Wilkins NP, Cotter D and O'Maoileidigh N (2001). Ocean migration and recaptures of tagged, triploid, mixed-sex and all-female Atlantic salmon (Salmo salar L.) released from rivers in Ireland. *Genetica* **111**: 197–212.
41. Uyeno T and Smith GR (1972). Tetraploid origin of the karyotype of catostomid fishes. *Science* **175**: 644–646.
42. Larhammar D and Risinger C (1994). Molecular genetic aspects of tetraploidy in the common carp *Cyprinus carpio. Mol. Phylogenet. Evol.* **3**: 59–68.
43. Phillips R and Rab P (2001). Chromosome evolution in the Salmonidae (Pisces): an update. *Biol. Rev. Camb. Philos. Soc.* **76**: 1–25.
44. Becak W, Schwantes AR and Schwantes ML (1968). Polymorphism of albumin-like proteins in the South American tetraploid frog *Odontophrynus americanus* (Salientia: ceratophrydidae). *J. Exp. Zool.* **168**: 473–476.
45. Kamel S, Marsden JE and Pough FH (1985). Diploid and tetraploid grey treefrogs (*Hyla chrysoscelis* and *Hyla versicolor*) have similar metabolic rates. *Comp. Biochem. Physiol. A* **82**: 217–220.
46. Hughes MK and Hughes AL (1993). Evolution of duplicate genes in a tetraploid animal, *Xenopus laevis. Mol. Biol. Evol.* **10**: 1360–1369.
47. Gallardo MH, Bickham JW, Honeycutt RL, Ojeda RA and Köhler N (1999). Discovery of tetraploidy in a mammal: the red viscacha rat is unaffected by having double the usual number of chromosomes. *Nature* **401**: 341.
48. Aparicio S, Hawker K, Cottage A, Mikawa Y, Zuo L, Venkatesh B, Chen E, Krumlauf R and Brenner S (1997). Organization of the *Fugu rubripes* Hox clusters: evidence for continuing evolution of vertebrate Hox complexes. *Nat. Genet.* **16**: 79–83.
49. Aparicio S (2000). Vertebrate evolution: recent perspectives from fish. *Trends Genet.* **16**: 54–56.
50. Aparicio S, Chapman J, Stupk E, Putnam N, Chia JM, Dehal P, Christoffels A, Rash S, Hoon S, Smit A, Gelpke MD, Roach J, Oh T, Ho IY, Wong M, Detter C, Werhoef F, Predki P, Tay A, Lucas S, Richardson P, Smith SF, Clark MS, Edwards YJ, Doggett N, Zharkikh A, Tavtigian SV, Pruss D, Barnstead M, Evans C, Baden H, Powell J, Glusman G, Rowen L, Hood L, Tan YH, Elgar G, Hawkins T, Venkatesh B, Rokhsar D and

Brenner S (2002). Whole-genome shotgun assembly and analysis of the genome of *Fugu rubripes*. *Science* **297**: 1301–1310.

51. Amores A, Suzuki T, Yan YL, Pomeroy J, Singer A, Amemiya C and Postlethwait JH (2004). Developmental roles of pufferfish Hox clusters and genome evolution in ray-fin fish. *Genome Res.* **14**: 1–10.
52. Naruse K, Fukamachi S, Mitani H, Kondo M, Matsuoka T, Kondo S, Hanamura N, Morita Y, Hasegawa K, Nishigaki R, Shimada A, Wada H, Kusakabe T, Suzuki N, Kinoshita M, Kanamori A, Terado T, Kimura H, Nonaka M and Shima A (2000). A detailed linkage map of medaka, *Oryzias latipes*. Comparative genomics and genome evolution. *Genetics* **154**: 1773–1784.
53. Van de Peer Y, Taylor JS and Meyer A (2003). Are all fishes ancient polyploids? *J. Struct. Funct. Genomics* **3**: 65–73.
54. Taylor JS, Van de Peer Y and Meyer A (2001b). Revisiting recent challenges to the ancient fish-specific genome duplication hypothesis. *Curr. Biol.* **11**: R1005–R1008.
55. Robinson–Rechavi M, Marchand O, Escriva H and Laudet V (2001). An ancestral whole-genome duplication may not have been responsible for the abundance of duplicated fish genes. *Curr. Biol.* **11**: R458–R459.
56. Streisinger G, Walker C, Dower N, Knauber D and Singer F (1981). Production of clones of homozygous diploid zebrafish (*Brachydanio rerio*). *Nature* **291**: 293–296.
57. Krauss S, Concordet JP and Ingham PW (1993). A functionally conserved homolog of the Drosophila segment polarity gene hh is expressed in tissues with polarizing activity in zebrafish embryos. *Cell* **75**: 1431–1444.
58. Ekker SC, Ungar AR, Greenstein P, von Kessler DP, Porter JA, Moon RT and Beachy PA (1995). Patterning activities of vertebrate hedgehog proteins in the developing eye and brain. *Curr. Biol.* **5**: 944–955.
59. Force A, Lynch M, Pickett FB, Amores A, Yan Y-L and Postlethwait J (1999). Preservation of duplicate genes by complementary, degenerative mutations. *Genetics* **151**: 1531–1545.
60. Wilson C, Amores A and Postlethwait JH, unpub. observations (pg. 591).
61. NCBI (2003). Human-Mouse Homology Map.
62. Menotti–Raymond M, David VA, Chen ZQ, Menotti KA, Sun S, Schaffer AA, Agarwala R, Tomlin JF, O'Brien SJ and Murphy WJ (2003).

Second-generation integrated genetic linkage/radiation hybrid maps of the domestic cat (*Felis catus*) *J. Hered.* **94**: 95–106.
63. Bowles J, Schepers G and Koopman P (2000). Phylogeny of the SOX family of developmental transcription factors based on sequence and structural indicators. *Dev. Biol.* **227**: 239–255.
64. Schepers GE, Teasdale RD and Koopman P (2002). Twenty pairs of sox: extent, homology, and nomenclature of the mouse and human sox transcription factor gene families. *Dev. Cell.* **3**: 167–170.
65. Venter JC, Adams MD, Myers EW, Li PW, Mural RJ, Sutton GG, Smith HO, Yandell M, Evans CA, Holt RA et al. (2001). The sequence of the human genome. *Science* **291**: 1304–1351.
66. Ohno S (1970). *Evolution by Gene Duplication.* Springer–Verlag, New York.
67. Friedman R and Hughes AL (2001). Pattern and timing of gene duplication in animal genomes. *Genome Res.* **11**: 1842–1847.
68. Hughes AL, da Silva J and Friedman R (2001). Ancient genome duplications did not structure the human Hox-bearing chromosomes. *Genome Res.* **11**: 771–778.
69. Gu X, Wang Y and Gu J (2002). Age distribution of human gene families shows significant roles of both large- and small-scale duplications in vertebrate evolution. *Nat. Genet.* **31**: 205–209.
70. McLysaght A, Hokamp K and Wolfe KH (2002). Extensive genomic duplication during early chordate evolution. *Nat. Genet.* **31**: 200–204.
71. Spring J (2002). Genome duplication strikes back. *Nat. Genet.* **31**: 128–129.
72. Wolfe K (2000). Robustness — it's not where you think it is. *Nature Genetics* **25**: 3–4.
73. Leveugle M, Prat K, Perrier N, Birnbaum D and Coulier F (2003). ParaDB: a tool for paralogy mapping in vertebrate genomes. *Nucleic Acids Res.* **31**: 63–67.
74. Lundin LG (1993). Evolution of the vertebrate genome as relected in paralogous chromosomal regions in man and the house mouse. *Genomics* **16**: 1–19.
75. Ruddle FH, Amemiya C, Carr JL, Kim CB, Ledje C, Shashikant CS and Wagner GP (1999). Evolution of chordate hox gene clusters. *Ann. NY Acad. Sci.* **870**: 238–248.

76. Joly J-S, Joly C, Schulte–Merker S, Boulekbache H and Condamine H (1993). The ventral and posterior expression of the zebrafish homeobox gene eve1 is perturbed in dorsalized and mutant embryos. *Development* **119**: 1261–1275.
77. Avaron F, Thaeron–Antono C, Beck CW, Borday–Birraux V, Geraudie J, Casane D and Laurenti P (2003). Comparison of even-skipped related gene expression pattern in vertebrates shows an association between expression domain loss and modification of selective constraints on sequences. *Evol. Dev.* **5**: 145–156.
78. Pollard SL and Holland PW (2000). Evidence for 14 homeobox gene clusters in human genome ancestry. *Curr. Biol.* **10**: 1059–1062.
79. Minguillon C and Garcia–Fernandez J (2003). Genesis and evolution of the Evx and Mox genes and the extended Hox and ParaHox gene clusters. *Genome Biol.* **4**: R12.
80. Ruuskanen J, Xhaard H, Marjamaki A, Salaneck E, Salminen T, Yan YL, Postlethwait JH, Johnson MS, Larhammar D and Scheinin M (2003). Identification of duplicated fourth {alpha}2-adrenergic receptor subtype by cloning and mapping of five Receptor genes in zebrafish. *Mol. Biol. Evol.*
81. Kobilka BK, Dixon RA, Frielle T, Dohlman HG, Bolanowski MA, Sigal IS, Yang–Feng TL, Francke U, Caron MG and Lefkowitz RJ (1987). cDNA for the human beta 2-adrenergic receptor: a protein with multiple membrane-spanning domains and encoded by a gene whose chromosomal location is shared with that of the receptor for platelet-derived growth factor. *Proc. Natl. Acad. Sci. USA* **84**: 46–50.
82. Lomasney JW, Lorenz W, Allen LF, King K, Regan JW, Yang–Feng TL, Caron MG and Lefkowitz RJ (1990). Expansion of the alpha 2-adrenergic receptor family: cloning and characterization of a human alpha 2-adrenergic receptor subtype, the gene for which is located on chromosome 2. *Proc. Natl. Acad. Sci. USA* **87**: 5094–5098.
83. Bylund DB (1992). Subtypes of alpha 1- and alpha 2-adrenergic receptors. *FASEB J.* **6**: 832–839.
84. Pebusque MJ, Coulier F, Birnbaum D and Pontarotti P (1998). Ancient large-scale genome duplications: phylogenetic and linkage analyses shed light on chordate genome evolution. *Mol. Biol. Evol.* **15**: 1145–1159.
85. Wraith A, Tornsten A, Chardon P, Harbitz I, Chowdhary BP, Andersson L, Lundin LG and Larhammar D (2000). Evolution of the neuropeptide

Y receptor family: gene and chromosome duplications deduced from the cloning and mapping of the five receptor subtype genes in pig. *Genome Res.* **10**: 302–310.

86. Haldane JBS (1933). The part played by recurrent mutation in evolution. *Am. Nat.* **67**: 5–9.
87. Nei M and Roychoudhury AK (1973). Probability of fixation of nonfunctional genes at duplicate loci. *Am. Nat.* **107**: 362–372.
88. Bailey GS, Poulter RT and Stockwell PA (1978). Gene duplication in tetraploid fish: model for gene silencing at unlinked duplicated loci. *Proc. Natl. Acad. Sci. USA* **75**: 5575–5579.
89. Li W-H (1980). Rate of gene silencing at duplicate loci: a theoretical study and interpretation of data from tetraploid fishes. *Genetics* **95**: 237–258.
90. Watterson GA (1983). On the time for gene silencing at duplicate loci. *Genetics* **105**: 745–766.
91. Lynch M and Conery J (2000). The evolutionary fate and consequences of gene duplication. *Science* **290**: 1151–1155.

Chapter 18

Medaka Genome Mapping for Functional Genomics

Hiroshi Mitani and Akihiro Shima
Department of Integrated Biosciences
Graduate School of Frontier Sciences
University of Tokyo
Kashiwa, Chiba 277-8562, Japan

Kiyoshi Naruse
Department of Biological Sciences
Graduate School of Science
University of Tokyo
Hongo 7-3-1, Bunkyo-ku, Tokyo 113-0033, Japan

Minoru Tanaka
Center for Transgenic Animals and Plants
National Institute for Basic Biology
Nishigonaka 38, Myodaiji, Okazaki 444-8585 Aichi, Japan

Large-scale genome sequencing using several model organisms has provided a new pathway to identify new genes, the function of which can be discovered through comparative approaches. Pufferfish (*Fugu rubripes* and *Tetraodon nigroviridis*) has been sequenced by whole genome shotgun (WGS) assembly, and the zebrafish (*Danio rerio*) genome sequencing project, including whole genome shotgun sequencing and selective BAC shotgun sequencing, is now on going. Still, the number of model organisms is too small to understand fully how fish genomes evolved. Now, medaka (*Oryzias latipes*) is emerging as another important model fish that is phylogenically distant from zebrafish but closer to pufferfish. Recent additions to genetic toolkit of medaka, such as BAC resources, WGS sequences, and highly dense genome markers based on polymorphic inbred strains, can facilitate genome assembly with high quality (http://medaka.dsp.jst.go.jp/MGI/). About 1400 markers including 800 randomly selected EST markers were mapped, and all of them successfully assembled into 24 linkage groups that correspond to the medaka chromosome number. This genomic map is a powerful tool for positional cloning of mutated genes in medaka, and conserved syntenies of medaka, pufferfish and zebrafish genes to the human genome provide evidence for a whole genome duplication event that occurred after divergence of fish and tetrapods and before divergence of medaka and zebrafish. For gene regulation studies, polyploidy in model fish species might be advantageous, because regulatory elements and functional domains in each of the fish duplicates may have unique functional roles.

In this review we summarize the current status of medaka genome studies for functional genomics.

1. Introduction

Genome sequencing projects have established a new pathway to find the function of genes through comparative genomic approaches. Fugu (*Fugu rubripes*) has been sequenced to over 95% coverage, and more than 80% of the assembly is in multigene-sized scaffolds (http://genome.jgi-psf.org/fugu6/fugu6.home.html).[1] In February 2001, the Sanger Institute started sequencing the genome of the zebrafish following two strategies: clone mapping and sequencing from BAC (Bacterial Artificial Chromosome) and PAC (PI-derived Artificial Chromosome) libraries and whole genome shotgun sequencing with subsequent assembly. They aim at providing a finished genome sequence by the end of 2005 (http://www.sanger.ac.uk/Projects/D_rerio/faqs.shtml).

Fish provide more than 50% of vertebrate classes, with a particularly deep and broad phylogeny.[2] So using diverse fish species are particularly good to identify conserved, as well as species-specific, genetic and molecular mechanisms that underlie development and evolution, such as the mechanisms and processes that trigger adaptive radiation, the multiplication of a single ancestral species into a variety of functionally different species, and ecological adaptation related to speciation. However, the number of model organisms studied is still too low to understand how fish genomes evolved.

Medaka is a small freshwater fish native to Asia that is found primarily in Japan, but also in Korea and China. This fish has been used widely as an experimental animal because of its relatively short life cycle, high fecundity, transparent egg chorion, small size and so on. Medaka was first used to show Mendelian inheritance in fish.[3] It also provides an important test system for environmental research, cancer research, and developmental biology (for review, see Wittbrodt, Shima and Schartl[4] and see the Medakafish homepage http://biol1.bio.nagaya-u.ac.jp:8000/ for an overview of techniques). Furthermore, most experimental tools for gene function analysis can be applied to both zebrafish and medaka. In this review we summarize current topics in medaka genome mapping as a tool for functional genomics in vertebrates.

2. Medaka as a Model Organism for Functional Genomics

2.1. Phylogenic Position

Medaka, zebrafish, platyfish, pufferfish and rainbow trout are members of the Beloniformes, Cypriniformes, Cyprinodontiformes, Tetraodontiformes and Salmoniformes families, respectively, distributed in relatively different taxonomic groups in teleost phylogeny.[2] A large group of fishes of the superorder Acanthopterygii, over 1300 species have bony skeletons and spiny rays in the dorsal and anal fins and this includes medaka, pufferfish, bass, perch, mackerel, and swordfish. This large fish group inhabits any aquatic environments and show this huge diversity in their morphological and ecological characters.[2] Paleontological records suggests that many of perciform families radiated in the Paleocene or early Eocene after the Cretaceous/Tertiary boundary (K/T boundary) at about 65 Myr.[5] But Kumazawa *et al.*[6] suggested that the diversification of the perciforms was estimated to be substantially older than that deducible from the first occurrence evidence of the fossil records, and they suggested that perciformes (the most major group of Acanthopterygii) diversified at about 100–200 Myr ago. Although medaka and pufferfish diverged earlier than the Perciform fishes,[6,7] the evolutionary distances between the two model organisms may be comparable to those between human and rodents because those mammalian species also diverged at about 100 Myr ago.[8]

Medaka and zebrafish are more distant cousins that have evolved separately for at least 110 Myr since fossil records[5] or more than 250 Myr based on molecular clocks.[6] Thus if we want to compare the evolutionary differences in highly diverged Acanthopterygiian fishes for functional genomics, zebrafish may be too different from Acanthopterygii. On the other hand, medaka and pufferefish (both are Acanthopterygii and their body structures and ecological characters are highly differentiated) are ideal for comparative analysis to survey the relationships between adaptive evolution and genomic structure in fishes.

2.2. Genome Size and Chromosome Number

It has been known since the earliest days of genome size investigation that the amount of nuclear DNA content bears no relationship to intuitive notions of organismal complexity. And indeed, haploid genome sizes (C-values) do not correlate with the number of coding genes. Polyploidy is relatively common in certain orders of teleosts (Cypriniformes, Salmoniformes, and Siluriformes). Despite their otherwise incredible diversity, most teleosts have small genomes and 48 chromosomes as diploids, with only the ancient polyploids deviating substantially from this trend Gregory TR (2001) (Animal Genome Size Database http://www.genomesize.com/). On the other hand, there appears to be only one known case of polyploidy in mammals — red viscacha rat, *Tympanoctomys barrerae*.[9,10] The medaka genome consists of 24 pairs of chromosomes, and the genome size is estimated at 650–1000 Mb[11–13] in which is only one-third of the human genome size and less than half the size of the zebrafish genome. So, the medaka genome size and chromosome number is near average values among fish species (Fig. 1).

2.3. Polymorphisms and Inbred Strains

Egami[14] reported morphological polymorphisms in different wild populations of medaka in Japan, and allozyme analysis of these populations indicated the presence of four genetically distinct, highly polymorphic populations (northern and southern Japanese populations, China/West Korea and East Korea populations.[15,16] Sequence comparisons of orthologous loci showed single base pair polymorphisms at a frequency of ~3% in introns and ~1% in exons between northern and southern Japanese populations.[17]

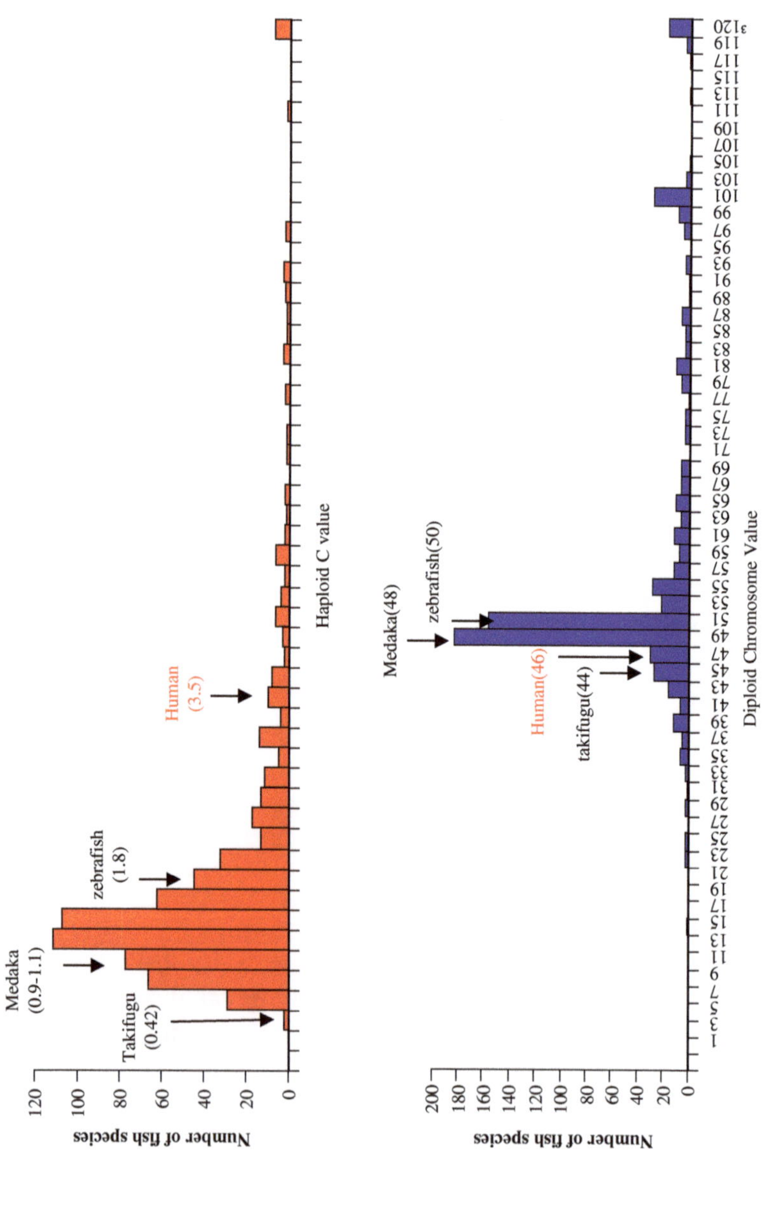

Fig. 1 Haploid C value in pg and diploid chromosome number of 681 fish species. Data is modified from Animal Genome Size Database: http://www.genomesize.com/ Gregory TR (2001). Last accessed August 2003. Values of human are imposed for comparison.

These polymorphisms are invaluable for genetic linkage mapping for any gene of interest. The establishment of inbred strains in medaka improves the ease and resolution of mapping and genome sequencing. The efforts to create a standard inbred strain of the medaka for laboratory use were started in 1974 by Hyodo–Taguchi and Egami, and in 1985 they published the first paper on inbred strains. Unlike in other lower vertebrate genetic systems, several inbred and highly fertile strains exist from northern, southern Japanese, and East-Korean populations of medaka, and these provide suitable materials for making mapping DNA panels for positional cloning studies.[18]

2.4. Spontaneous and Induced Mutants

Systematic screens for spontaneous medaka mutants in natural populations and breeding stocks were carried out by Tomita.[19,20] The spontaneous mutations, such as abnormal pigmentation, body size modification, and defects in fin or scale morphology, are still maintained at the medaka stock center at Nagoya University. Some mutants such as double anal fins (da), which has a mirror-image duplication of ventral body structures, and pectoral fin-less (pl), despite their severe morphological phenotypes, were not found in large-scale ENU mutagenesis screening of zebrafish, which indicates an only partially overlapping spectrum of embryonic mutant phenotypes between medaka and zebrafish. The see-through medaka strain was recently established by Wakamatsu *et al.*,[21] which is the first vertebrate model with a transparent body in the adult stage. In this fish model, most pigments are genetically removed from the entire body by a combination of spontaneous recessive alleles at four pigmentation loci; this strain is ideal for GFP transgenic analysis for gene regulation *in vivo*.[22]

Shima and Shimada[23] established a multi-locus tester strain that is homozygous at several recessive loci and is used to detect induced recessive mutations in the germline. The higher the number of marker loci, the higher the detection efficiency. The germ cell mutagenesis protocols they established turned out to be useful for mutagenesis not only in medaka but also in zebrafish. Loosli *et al.*[24] reported the first systematic mutagenesis approach to isolate embryonic-lethal developmental mutants in medaka, and recently, the ERATO project by Hisato Kondo and Makoto Furutani–Seiki started a large-scale genome-wide screening by ENU mutagenesis which is comparable to that done using zebrafish.[25] Surprisingly, many mutant phenotypes found by similar morphological screening are species-specific. So both spontaneous and induced medaka mutants should provide new insight into the function of genes during development.[25]

3. Genome Mapping

A medaka linkage map was first described by Aida.[26] He demonstrated that the male determining factor (Y) was linked with the gene that controls carotinoid deposition in xanthophores (R). Since Aida's study, over 60 visible mutants have been isolated and analyzed by allele sharing and allelic association.[19,20] The first multipoint linkage map including 170 loci and 28 linkage groups was reported using RAPD (Random Amplified Polymorphic DNA) fingerprints and allozyme analysis.[27,28] The segregation analysis was performed on the basis of genetic recombination during female meiosis using 134 random RAPD markers, 13 sequence-tagged sites (STSs), 15 polymorphic sequences from known genes, and the Da gene; these loci segregated into 26 linkage groups.[29] Later, 638 markers (489 AFLPs, 28 RAPDs, 34 IRSs, 78 ESTs, four STSs and four phenotypic markers) were mapped to 24 linkage groups corresponding to the haploid number of medaka chromosomes.[17,31] Because anonymous DNA markers, such as AFLP, RAPD, and IRSs markers, were expected to be randomly distributed throughout the genome, the number of these markers on each linkage group should reflect the physical size of each chromosome, and so linkage group number was assigned on that basis. Recently, some linkage groups were identified by chromosomal FISH analysis using mapped ESTs.[30,31]

The accumulation of mapping data using the same mapping panel is very important for eliminating linkage relationship ambiguities among markers. Because medaka is a small fish, the amount of DNA from each individual could be a limiting factor for extensive data accumulation. Most mammalian cells in culture are known to undergo growth crises, after which some may become established cell lines.[32] On the other hand, cell lines derived from the adult tissues of goldfish and medaka have never experienced growth crises since primary culture.[33,34] So 39 permanently growing cell lines, one each from 39 backcross progeny, were established for linkage analysis. No genetic alternation during cell subculture was detected using DNA extracted from the remaining carcasses of the backcross fish, or another typing panel.[17]

About 100,000 EST (expressed sequence tag) sequences of medaka have been sequenced and deposited in the public database. It was found that cultured cell lines could be a good source for cDNA libraries to increase the variation of cDNAs, since they could provide different cDNA species from various tissues. Furthermore, the gene expression pattern varied when cells were exposed to physical stress (Table 1). ESTs with high similarity to known vertebrate genes are a good source for the mapping. To assign about 800 loci encoding expressed genes to each linkage group, polymerase chain reaction-restriction fragment length polymorphism (PCR-RFLP) analysis is used. The current row mapping data and BLAST results for medaka ESTs can be viewed by our WEB site (http://mbase.bioweb.ne.jp/~dclust/ml_base.html). We found that the total map-length of all linkage groups (LGs) is about 1400 cM in male meiosis. If the total genome size of medaka is taken as 800 Mb, then the estimated physical length of each LG would range from 19 Mb to 59 Mb. The current map density is about one marker per Mb (one every 1.75 cM). This is sufficient to identify a DNA marker linked to any genetic traits within 0.9 cM. Comparisons of marker distribution for anonymous DNA markers, such as RAPD and AFLP markers, and for gene and EST markers suggest that distributions of genes are not uniform in each LG. For example, the gene density of LG2 is 4.3 times less than that of LG 22.

In many organisms, the rate of genetic recombination is not uniform along the length of chromosomes or between sexes. In mammals, many species show reduced recombination frequency in males. Averaged over the entire genome, human female-to-male recombination rates are 1.6:1.0.[35] Singer *et al.*[36] compared the relative recombination rates during meiosis in male and female zebrafish, and the recombination rate in male meiosis is dramatically suppressed relative to that of female

Table 1 Clustering results of ESTs derived from NHI strains.

Library name (source)	No. of clones	No. of clusters
OLa (adult fish)	1672	1063
OLb (cultured cell line OLHN1)	2204	1291
OLc (cultured cell line OLHN1 + UV)	1831	1081
OLf (cultured cell line OLHN1 + γ rays)	1480	1100
OLe (adult liver)	1271	570
OLa + OLb	3876	2226
OLb + OLc + OLf	5515	2890
OLa + OLb + OLc + OLf	7182	3768

(http://mbase.bioweb.ne.jp/~dclust/medaka_top.html)

meiosis, especially near the centromere. Summing the lengths of the greatest common interval for each linkage group on both the female and male zebrafish maps gave lengths of 2582.7 and 942.5 cM, respectively. This corresponds to a female-to-male ratio of 2.74:1.0.

In medaka, sex is determined chromosomally. The sex chromosomes differ from those of mammals in that the X and Y chromosomes are highly homologous in medaka.[30] We use mapping panels made from both male meiosis and female meioses to increase the resolution of map, since linkage distance could be different between male and female meioses depending on location along a chromosome. Using for linkage analyses backcross panels of XY male mated to XY sex-reversed females, Kondo *et al.* (2001)[37] mapped 21 sequence tagged site (STS) markers on the sex chromosomes (LG1). The genetic map of the sex chromosome was established using male and female meioses. The genetic length of the sex chromosome was shorter in male than in female meiosis. The region where male recombination is suppressed is the region close to the sex determining gene, while female recombination was suppressed in both of the telomere regions. We also compared the recombination rate on autosomal chromosomes in male and female meiosis, and found that male meiosis is also dramatically suppressed relative to that of female meiosis, especially near the centromere.[38]

These findings have practical applications for experimental design. The use of exclusively female meiosis in a positional cloning project maximizes the ratio of genetic map distance to physical distance. Alternatively, the use of exclusively male meiosis to localize a mutation initially to a linkage group or to maintain relationships of linked alleles minimizes recombination, thereby facilitating some types of analysis.

4. Positional Cloning of Mutated Genes

Koga *et al.*[39] reported the first example of gene responsible for a spontaneous mutant in medaka. Southern blot analysis of genomic DNA from the recessive albino mutant *i* with an authentic tyrosinase gene probe demonstrated that an extra 1.9 kb fragment, designated as a tol1 repetitive element is present inside the first exon.

Mapping data and other genome resources such as BAC clones now facilitate the positional cloning of mutated genes in medaka, and three mutated medaka genes and the sex determining gene have been identified. Many ENU-induced medaka mutants whose phenotypes are different from those so far discovered in zebrafish, have been isolated and they are good materials for identifying new gene functions.

4.1. *Colorless Melanophore*

Mammals have only one kind of chromatophore, the melanocyte. In contrast, the medaka is a suitable model of the lower vertebrates because it has all kinds of chromatophores, e.g. melanophores, xanthophores, leucophores and iridophores. Approximately 70 spontaneous pigmentation mutants have been isolated by Tomita. One of these, an orange-red variant, is a homozygote of a well known and common allele *b*, and has been bred for hundreds of years in Japan. Fukamach *et al.*,[40] reported that the gene AIM1 (MATP), which encodes a transporter that mediates melanin synthesis, is tightly linked to the *b*-locus. The protein is predicted to consist of 12 transmembrane domains and is 55% identical to a human EST of unknown function isolated from melanocytes and melanoma cells.

Analysis of AIM1 orthologs in mouse should provide new insights into the regulation of melanogenesis in both teleosts and mammals. The mouse *underwhite* locus, alleles of which manifest altered pigmentation of both the eye and fur, sometimes in an age-dependent fashion, was shown to be the mouse *Aim-1*.[41] Analysis of the three mutated alleles of mouse *Aim-1* alleles revealed that structure/function differences correlate with recessive versus dominant inheritance of *underwhite*. They also reported that MITF, a melanocyte-specific transcription factor essential to pigmentation and a clinical diagnostic marker in human melanoma, modulates AIM1 transcription.

4.2. Eyeless Mutation

The eyeless (*el*) mutation, initially isolated as a naturally occurring mutant from the southern population of medaka, was crossed to and kept in the Cab inbred strain, derived from southern Japanese population.[42] Crossing heterozygous *el* fish to the Kaga strain derived from the northern Japanese population revealed a close linkage of *el* and locus *b*. The Kaga strain is homozygous for the *b* allele that results in darkly pigmented melanophores. The Cab background carries the recessive *b* allele that leads to unpigmented melanophores. A genetic distance of 1.3 cM separates the *el* locus and the *b* locus on linkage group 12.[43] *In situ* hybridization showed that in *el* mutant embryo with defective optic vesicle evagination, the expression of Rx3, a member of the conserved vertebrate homeobox gene, Rx gene family, which is essential for normal eye development, was completely lost even in hypothalamus. A PCR amplified genomic Rx3 DNA fragment contains a polymorphic restriction site and the Cab-specific polymorphism always co-segregated with the *el*. In the DNA of homozygous mutant embryos (*el/el*) a larger PCR fragment is detected than in the wild-type (+/+), indicating an insertion larger than 13 kb in the mutant *el* locus. Furthermore, mutant embryos injected with either the Rx3 BAC or the Rx3 cosmid form eyes of wild-type phenotype. The temperature-sensitive expressivity of the mutant phenotype is tightly correlated with the expression levels of Rx3 in the presumptive retina. Medaka is very hardy and tolerates a wide range of temperatures (10–37°C).[44] Temperature-sensitive medaka mutants could provide important insights into the functional roles of gene.

4.3. Reduced Scales-3, *rs-3*

Fish scales are epithelial appendages that differentiate from the dermal mesenchyme. A mutation at the *rs-3* locus (reduced scales-3) was originally isolated from wild populations in a screen for spontaneous mutants.[20] It is a recessive mutation, and homozygous fish are viable and fertile but almost completely lack scales except for a few, which are larger in size and irregular in shape. Kondo *et al.*[45] reported that the EDAR (ectodysplasin-A receptor) gene is responsible for the *rs-3* phenotype. Preliminary linkage analysis revealed that the *rs-3* locus maps to medaka LG21 which contains the HOXDA cluster, whose orthologs in zebrafish and humans are located on zebrafish LG9 and the long arm of chromosome 2 (Chr2q), respectively. Studies of the zebrafish genome mapping had identified extensive conservation of between zebrafish LG9 and human Chromosome 2q.[46,47] Medaka orthologs of 14 human genes on chromosome 2q were cloned by degenerate PCR. Ten of them were mapped to LG21, whereas the remaining four were mapped to LG2. Although the conservation between human Chromosome 2q and medaka LG21 spans the entire chromosome arm, the gene orders are not conserved at all. A discrepancy in gene order reflects intrachromosomal inversions after the divergence of humans and medaka. This supports the hypothesis that inversions have been more frequently fixed than translocations during vertebrate genome evolution.

EDAR, located on medaka LG21 and human 2q, encodes a type I transmembrane protein that shows weak similarity to members of the TNF receptor superfamily, acting as a receptor for a TNF-like transmembrane ligand, ectodysplasin-A (EDA). Mutations in both the ligand and the receptor are known to result in the loss of certain types of hair in humans and mice. EDAR was closely linked to the *rs-3* phenotype and a novel transposon was inserted in the first intron. This insertion causes aberrant splicing. Although they are both skin appendages, hair and scales are not homologous organs; fish scales do not contain keratin. They are mineralized dermal elements that possibly contain dentine- and enamel-derived proteins. The identification of a common requirement for development or hair and fish scales thus reflects a common developmental mechanism for appendage formation involving epidermal-dermal interactions.

4.4. Sex-Determining Gene

The first artificial sex reversals were reported in 1958 by Yamamoto.[48] Treatment with steroid sex hormones during the larval period generated YY males, XY females, XX males and even YY females. The Y chromosome-specific region spans only about 280 kb, which contains a duplicated gene, has been found as a candidate gene from medaka that is functionally comparable to the mammalian male-determining gene, Sry by two groups independently. Masuda *et al.*[49] found a Y choromosome specific-dmY gene with a DM domain that was originally described as a DNA-binding motif shared between *doublesex* (*dsx*) in *Drosophila melanogaster* and *mab-3* in *Caenorhabditis elegans* found in other genes involved in sexual development in both vertebrates and invertebrates.[50] They also found that malfunctioning of dmY by spontaneous mutation causes feminization of XY fish in a wild population. Nanda *et al.*[51] sequenced 280 kb Y choromosome-specific region and they found that the dmrt1b same gene with dmY is a duplicate of the autosomal dmrt1a gene, which is the only functional gene in this chromosome segment, and maps precisely to the male sex-determining locus. At first, dmY/dmrt1b was thought to be a candidate for the primary sex-determining gene not only in the medaka, but also in other animal groups, however, it was suggested that the gene duplication generating dmrt1b occurred recently during the evolution of the genus Oryzias.[52–54] The gene is absent from all other fish species studied. Therefore, it may not be the male-sex-determining gene in all fishes. So new genetic strategies controlling sexual dimorphism in vertebrates should be revealed using other fish species.

5. Genome Duplication

In principle, the expansion of gene families in teleosts could be caused by tandem duplication or genome-wide duplication. Ohno[55] proposed that without duplicated genes, the creation of metazoans, vertebrates and mammals from unicellular organisms would have been impossible. Such big leaps in evolution, he argued, required the creation of new gene loci with previously non-existent functions. Because complete genome duplication increases gene number without upsetting gene dosage, it was advanced as the primary source of redundant genes.

Genome sequencing projects are now providing evidence that large-scale gene duplication and even complete genome duplication events have contributed significantly to gene family expansion and to genome evolution. The human genome has also been shaped by a diversity of duplication events including, perhaps, two complete genome duplication events very early during the evolution of vertebrates.[56,57]

The HOX genes are master genes for specifying posterior/anterior body axis in both vertebrates and invertebrates and the evolution of HOX clusters seem to be a typical case for whole genome duplication events in vertebrates. HOX genes cluster on 4 chromosomes in tetrapods and on one chromosome in invertebrates. The discovery that zebrafish, medaka and pufferfish possess seven Hox gene clusters, almost twice as many as human and mouse, led to the hypothesis that there was a whole-genome duplication after the divergence of ray-finned and lobe-finned fishes but before the teleost radiation.[17,58,61] Comparative genomics of zebrafish and humans have revealed that large conserved chromosome segments are retained between zebrafish and human, and that a genome-wide duplication may have happened in the ancestor of zebrafish.[46,47,59–61] Through phylogeny reconstruction, Taylor *et al.*[61] identified 49 genes with a single copy in man, mouse, and chicken, one or two copies in the tetraploid frog *Xenopus laevis*, and two copies in zebrafish. For 22 of these genes, both zebrafish duplicates had orthologs in the pufferfish. For another 20 of these genes, they found only one pufferfish ortholog but in each case it was more closely related to one of the zebrafish duplicates than to the other. These phylogeny and synteny data suggest that the common ancestor of zebrafish and pufferfish, a fish that gave rise to more than 20,000 species, experienced a large-scale gene or complete genome duplication event and that the pufferfish has lost many duplicates that the zebrafish has retained. Smith *et al.*[62] constructed cosmid and BAC contig maps across two pufferfish genomic regions containing the orthologs of human genes mapping to human Chromosome 20q and their data was also best explained by regional duplication, followed by substantial gene loss.

However, a complete pufferfish chromosomal map is not yet available because of the absence of highly polymorphic variants and the species' long life cycle, and zebrafish with a relatively large c-value and phylogenically distant from pufferfish, leave the possibility that those genes and extra-Hox clusters derived from independent gene duplication not from whole genome duplication. To clarify these questions, comparisons of distantly related fish species are essential.

To approach such questions, we mapped about 800 orthologous gene pairs between medaka and human to compare the syntenic relationships of medaka and human genomes.[31] A result of our genome-wide comparison of orthologous genes among medaka, zebrafish and human along with the evidence given above strongly indicates that the genome amplification is not partial, but involved the whole genome, and occurred before the last common ancestor of euteleosts. Figure 2 shows an Oxford grid comparing the medaka and human genomes. This pattern shows that, even though the distribution is scattered, it is not random and clusters of orthologous gene pairs are frequently observed as in zebrafish.[46] The degree of syntenic conservation between human and either medaka or zebrafish is almost the same. These results suggest that most of the fish genomes consist of paired chromosomes, each derived from single common proto-chromosomes, and the content of vertebrate proto-chromosomes can be identified by the comparison of conserved syntenic genes between distantly related teleost species like medaka and zebrafish and human (Table 2). An exception to this generalization is LG1 of medaka and LG1 of zebrafish, which appear to lack paired LGs. This suggests the deletion of an entire chromosome in the common ancestor of medaka and zebrafish. These results also

Human Chromosome

Medaka LG	1	2	3	4	5	6	7	8	9	10	11	12	13	14	15	16	17	18	19	20	21	22	X	Y
1		2		10	1		1		2	1	2			1		1	6		4			1	1	
2	1	7	2		1						1	2	1				1		1		2		2	
3	1	5	1		1	1				1	8	2			5	6							1	
4	9	3	4	1		1			1		1	1							6				1	
5	6	1	14	1	1	4		1	1		1	6				1	1		1	8			4	
6	1		1				3				4	4			8	3			2					
7	16	3	4				3	1			1	10		1		1				6			4	
8		1	1				3							1		4	16		8			8	2	
9		4	1		3		3	3	4	1		7					2	1	1	2		2		
10	1	1		2	12	1	1		1	1	3	1	2			1				1	1	1	8	
11	7	1	2	2		9	6	6	1			1			1	1		1	1	3				
12	1	1		2	9		1	2	9		1	3						3				3	1	
13		3	4		1	1	3				10	1	1	3		1	2		4					
14	1			2	8	1	2			2	3	1	1	1	1	1	4		1		3		2	
15	3	4			1	2		2		12	2	2		1										
16	8		2		1	3	4	7			3	5		1			1		8	1				
17	20	4	3				2	1		3	1							1	3				1	
18		1	1	4		2			2		3			1	2		2						1	
19	1	2	1							9	1					3	12		2	1		2		
20			3		2	2	5	6	1	2			1	1			1	1			1		2	
21	2	17	3	1			2	1					3		1	1	1				3		3	
22	8	3	1	2		3		1			1			16	1					3		1	1	
23			1				8					9			1			1						
24		3		1		12								4					1			1	1	

n=818

Fig. 2 An Oxford grid display is a matrix of cells demonstrating the number of orthologous genes. Each cell represents two chromosomes, one from each species. The chromosomes of human are arrayed as columns, with the chromosome numbers given along the top. Medaka chromosomes are shown in rows, with the chromosome numbers and species name appearing on the left side of the grid. (For details, see Naruse *et al.*[31])

suggest that the last common ancestor of ray-finned fish and lobe-finned fish (including tetrapods) may have had about 12 chromosomes in the haploid set, as previously suggested.[47]

GnRH plays pivotal roles in the regulation of vertebrate reproduction through binding to its specific membrane receptor. Two GnRH-Receptors, termed GnRH-R1 and GnRH-R2, have been identified in medaka.[63,64] Recently, a novel third member of GnRH-R, designated GnRH-R3, which

Table 2 Hypothetical proto-chromosomes in vertebrates and location of major segements. For details, see Naruse *et al.*[31]

Proto-chromosome	Major segments in Medaka LG	Major segments in Zebrafish LG	Major segments in Human Chromosome
1	16	16	1, 6, 7, 8, 19
1	11	19	1, 6, 7, 8
2	8	3	16, 17, 22, 19
2	19	12	16, 17, 22, 10
3	7	23	1, 3, 12, X
3	5	11	1, 3, 12, X
4	2	6	2, 3, X
4	21	9	2, 3, 13, 21, X
5	22	17	2, 6, 14
5	24	20	2, 6, 14
6	18	18	11, 15
6	3	7	11, 15, 16
6	6	25	11, 15, 16
7	10	14	5, X
7	12	5	5, 9, X
8	13	15	11, 17
8	14	10, 21	5, 11, 17
9	1	1	4, 17, 19
10	4	2	1, 3, 19
10	9	8	7, 8, 12
11	17	22	1, 3, 19
11	15	13	1, 6, 10
12	20	24	3, 7, 8
12	23	4	3, 7, 12

*Red characters indicate the pair of orthologous chromosomes with conserved synten between medaka and zebrafish.

seemed to be derived from whole genome duplication and acquired a new function, has been identified.[65] Phylogenetic analysis indicates that both GnRH-R1 and GnRH-R3 in the medaka are orthologous to the primate second GnRH-R, termed type 2 GnRH-R (GNRHR2). Genetic mapping revealed that the GnRH-R1 and GnRH-R3 genes are not clustered together — they are located on LGs 3 and 6, respectively. LGs 3 and

6 are structurally related to each other; both LGs 3 and 6 contain a number of genes whose human orthologs reside on chromosomes 1, 11, 15, and 16. The localization of GnRH-R3 expression is different from that of GnRH-R1; unlike GnRH-R1, GnRH-R3 exhibits an approximately equal selectivity for two of three native GnRH forms in the medaka, chicken-II-type GnRH (cGnRH-II) and salmon-type GnRH (sGnRH), and less sensitivity for the other form, medaka-type GnRH.

Because many genes in fish are present in single copy, proponents of the whole-genome duplication hypothesis postulate that many of the duplicated versions of genes have degenerated since the initial duplication event. But one would expect medaka and zebrafish to have considerably different sets of genes with differences due to subfunctionalization or neofunctionalization because this degeneration process would have affected different genes in the two fish lineages.

6. Future Aspects

The recent publications of the pufferfish genome gives us a new powerful tool for comparative genomics in vertebrates. As mentioned above, with their phylogenetic position, small genome size, and polymorphic inbred strains, medaka is one of the best model vertebrates for a genome project. An important genome resource, the gridded BAC (bacterial artificial chromosome) libraries, has been established from the southern and the northern inbred strains of medaka.[45,66] They are essential resources for both genome sequencing and fine mapping.[67] A first-generation physical map of the medaka genome in BAC clone was separated.[71]

As a research group with international collaboration, the Medaka Genome Initiative (MGI) is based on genetic and physical mapping resources. Laboratories that are part of this initiative collaborate to physically and computationally interconnect the resources (the MGI home page: http://www.dsp.jst.go.jp/MedGenIn/index.html). The shotgun libraries will be established from contig BACs for clone-by-clone sequencing project by Shimizu and his colleagues of Keio University and a whole-genome shotgun sequencing project by Kohara (NIG) has started from 2002. These international collaborations will provide an integrated genome database for medaka and zebrafish within a few years. It will provide us new insights how gene functions evolved in vertebrates, and how they can be experimentally examined in model systems.

Acknowledgements

We thank Dr. Takahiko Mukai for comments about phylogenic description of medaka. We also thank Dr. John Postlethwait for many valuable comments on an earlier version of the manuscript. This work was supported by a Grant-in-Aid for Scientific Research from the MEXT, Japan (Priority area #813) to HM and AS.

References

1. Aparicio S, Chapman J, Stupka E, Putnam N, Chia JM, Dehal P, Christoffels A, Rash S, Hoon S, Smit A, Gelpke MD, Roach J, Oh T, Ho IY, Wong M, Detter C, Verhoef F, Predki P, Tay A, Lucas S, Richardson P, Smith SF, Clark MS, Edwards YJ, Doggett N, Zharkikh A, Tavtigian SV, Pruss D, Barnstead M, Evans C, Baden H, Powell J, Glusman G, Rowen L, Hood L, Tan YH, Elgar G, Hawkins T, Venkatesh B, Rokhsar D and Brenner S (2002). Whole-genome shotgun assembly and analysis of the genome of *Fugu rubripes*. *Science* **297**: 1301–1310.
2. Nelson JS (1994). *Fishes of the World*. John Wiley and Sons Inc., New York.
3. Toyama K (1916). Some examples of Mendelian characters (in Japanese). *Rep. Jap. Breed. Soc.* **1**: 1–9.
4. Wittbrodt J, Shima A and Schartl M (2002). Medaka — a model organism from the Far East. *Nat. Rev. Genet.* **3**: 53–64.
5. Benton MJ (1993). *The Fossil record 2*, Chapman and Hall, London.
6. Kumazawa Y, Yamaguchi M and Nishida M (2000). Mitochondrial molecular clocks and the origin of euteleostean biodiversity: familial radiation of perciforms may have predated the Cretaceous/Tertiary boundary. In: M Kato (ed.), *The Biology of Biodiversity*. Springer–Verlag, Tokyo, pp. 35–52.
7. Miya M, Takeshima H, Endo H, Ishiguro NB, Inoue JG, Mukai T, Satoh TP, Yamaguchi M, Kawaguchi A, Mabuchi K, Shirai SM and Nishida M

(2003). Major patterns of higher teleostean phylogenies: a new perspective based on 100 complete mitochondrial DNA sequences. *Mol. Phyl. Evol.* **26**: 121–138.

8. Nei M, Xu P and Glazko G (2001). Estimation of divergence times from multiprotein sequences for a few mammalian species and several distantly related organisms. *Proc. Natl. Acad. Sci. USA* **98**: 2497–2502.
9. Gallardo MH, Bickham JW, Honeycutt RL, Ojeda RA and Kohler N (1999). Discovery of tetraploidy in a mammal. *Nature* **401**: 341.
10. Gallardo MH, Bickham JW, Kausel G, Kohler N and Honeycutt RL (2002). Gradual and quantum genome size shifts in the hystricognath rodents. *J. Evol. Biol.* **16**: 163–169.
11. Uwa H and Iwata A (1981). Karyotype and cellular DNA content of *Oryzias javanicus* (Oryziatidae, Pisces). *Chrom. Inf. Serv.* **31**: 24–26.
12. Hinegardner R and Rosen DE (1972). Cellular DNA content and the evolution of teleostean fishes. *Am. Nat.* **106**: 621–644.
13. Lamatsch DK, Steinlein C, Schmid M and Schartl M (2000). Noninvasive determination of genome size and ploidy level in fishes by flow cytometry: detection of triploid *Poecilia formosa*. *Cytometry* **39**: 91–95.
14. Egami N (1953). Studies on the variation of the number of the anal fin-rays in *Oryzias latipes.* I. Geographical variation. *J. Ichthyol.* **3**: 87–89.
15. Sakaizumi M, Moriwaki K and Egami N (1983). Allozymic variation and regional differentiation in wild populations of the fish *Oryzias latipes. Copeia* **2**: 311–318.
16. Sakaizumi M (1986). Genetic divergence in wild population of the fish *Oryzias latipes* (Pisces: Oryziatidae) from Japan and China. *Genetica* **69**: 119–125.
17. Naruse K, Fukamachi S, Mitani H, Kondo M, Matsuoka T, Kondo S, Hanamura N, Morita Y, Hasegawa K, Nishigaki R, Shimada A, Wada H. Kusakabe T, Suzuki N, Kinoshita M, Kanamori A, Terado T, Kimura H, Nonaka M and Shima A (2000). A Detailed Linkage Map of Medaka, *Oryzias latipes.* Comparative genomics and genome evolution. *Genetics* **154**: 1773–1784.
18. Hyodo–Taguchi Y and Egami N (1985). Establishment of inbred strains of the medaka *Oryzias latipes* and the usefulness of the strains for biomedical research. *Zool. Sci.* **2**: 305–316.
19. Tomita H (1975). Mutant gene in the Medaka. In: Yamamoto T (ed.), *Medaka (killifish) Biology and Strains* Keigaku, Tokyo, pp. 251–272.

20. Tomita H (1982). Gene analysis in the Medaka (*Oryzias latipes*). *Fish Biol. J. Medaka* **1**: 7–10.
21. Wakamatsu Y, Pristyazhnyuk S, Kinoshita M, Tanaka M and Ozato K (2001). The see-through medaka: a fish model that is transparent throughout life. *Proc. Natl. Acad. Sci. USA* **98**: 10046–10050.
22. Tanaka M, Kinoshita M, Kobayashi D and Nagahama Y (2001). Establishment of medaka (*Oryzias latipes*) transgenic lines with the expression of green fluorescent protein fluorescence exclusively in germ cells: a useful model to monitor germ cells in a live vertebrate. *Proc. Natl. Acad. Sci. USA* **98**: 2544–2549.
23. Shima A and Shimada A (1991). Development of a possible nonmammalian test system for radiation-induced germ-cell mutagenesis using a fish, the Japanese medaka (*Oryzias latipes*). *Proc. Natl Acad. Sci. USA* **88**: 2545–2549.
24. Loosli F, Koster RW, Carl MP, Henrich KR, Krone TMA and Wittbrodt J (2000). A genetic screen for mutations affecting embryonic development in medaka fish (*Oryzias latipes*). *Mech. Dev.* **97**: 133–139.
25. Furutani–Seiki M, Sasadao T, Morinaga C, Suwa H, Niwa K, Yoda H, Deguchi T, Hirose Y, Yasuoka A, Henrich T, Watanabe T, Iwanami N, Kitagawa D, Saito K, Asaka S, Osakada M, Kunimatsu M, Elmasri H, Winkler C, Ramialison M, Loosli F, Quiring R, Carl M, Grabher C, Winkler S, Del Bene F, Shinomiya A, Kota Y, Yamanaka T, Okamoto Y, Takahashi K, Todo T, Abe K, Takahama Y, Tanaka M, Mitani H, Katada T, Nishina H, Nakajima M, Wittbrodt J and Kondoh H (2004). A systematic genome-wide screen for mutations affecting organogenesis in Medaka, *Oryzias latipes Mechanisms of Development* **21**: 647–658.
26. Shima A, Himmelbauer H, Mitani H, Furutani–Seiki M, Wittbrodt J and Schartle M (2003). Fish genomes flying. *EMBO Rep.* **4**: 121–125.
27. Aida T (1921). On the inheritance of color in a fresh-water fish *Aplocheilus latipes* Temmick and Schlegel, with special reference to sex-linked inheritance. *Genetics* **6**: 554–573.
28. Kubota Y, Shimada A, Shima A (1995). DNA alterations detected in the progency of paternally irradiated Japanese medaka fish (Oryzias latipes). *Proc. Natl Acad. Sci.* USA **92**: 330–334.
29. Wada H, Naruse K, Shimada A and Shima A (1995). Genetic linkage map of a fish, the Japanese medaka *Oryzias latipes. Mol. Mar. Biol. Biotech.* **4**: 269–274.

30. Ohtsuka M, Makino S, Yoda K, Wada H, Naruse K, Mitani H, Shima A, Ozato K, Kimura M and Inoko H (1999). Construction of a linkage map of the medaka (*Oryzias latipes*) and mapping of the Da mutatant locus defective in dorsoventral patterning. *Genome Res.* **9**: 1277–1287.
31. Naruse K, Tanaka K, Mita K, Shima A. Postlethwait J and Mitani H (2004). A medaka gene map: the trace of ancestral vertebrate proto-chromosomes revelaed by comparative gene mapping. *Genome Res* **14**: 820–826.
32. Matsuda M, Matsuda C, Hamaguchi S and Sakaizumi M (1998). Identification of the sex chromosomes of the medaka, *Oryzias latipes*, by fluorescence *in situ* hybridization. *Cytogenet. Cell Genet.* **82**: 257–262.
33. Miyake Y *et al.* (unpub.).
34. Hayflick L and Moorhead PS (1961). The serial cultivation of human diploid cell strains. *Exp. Cell Res.* **253**: 585–621.
35. Shima A, Nikaido O, Shinohara S and Egami N (1980). Continued *in vitro* growth of fibroblast-like cells (RBCF-1) derived from the caudal fin of the fish, *Carassius auratus. Exp. Gerontol.* 305–314.
36. Komura J, Mitani H and Shima A (1988). Fish cell culture: establishment of two fibroblast-like cell lines (OL-17 and OL-32) from fin of the Medaka, *Oryzias latipes. In Vitro* **24**: 294–298.
37. Dib C, Faure S, Fizames C, Samson, D, Drouot N *et al.* (1996). A comprehensive genetic map of the human genome based on 5,264 microsatellites. *Nature* **380**: 152–154.
38. Singer A, Perlman H, Yan Y, Walker C, Corley–Smith G, Brandhorst B and Postlethwait J (2002). Sex-specific recombination rates in zebrafish (*Danio rerio*). *Genetics* **160**: 649–657.
39. Kondo M, Nagao E, Mitani H and Shima A (2001). Differences in recombination frequencies during female and male meioses of the sex chromosomes of the medaka, *Oryzias latipes. Genet. Res.* **78**: 23–30.
40. Mitani H *et al.* (unpub.).
41. Koga A, Inagaki H, Bessho Y and Hori H (1995). Insertion of a novel transposable element in the tyrosinase gene is responsible for an albino mutation in the medaka fish, *Oryzias latipes. Mol. Gen. Genet.* **249**: 400–405.
42. Fukamachi S, Shimada A and Shima A (2001). Mutations in the gene encoding B, a novel transporter protein, reduce melanin content in medaka. *Nat. Genet.* **28**: 381–385.

43. Du J and Fisher DE (2002). Identification of Aim-1 as the underwhite mouse mutant and its transcriptional regulation by MITF. *J Biol. Chem.* **277**: 402–406.
44. Winkler S, Loosli F, Henrich T, Wakamatsu Y and Wittbrodt J (2000). The conditional medaka mutation eyeless uncouples patterning and morphogenesis of the eye. *Development* **127**: 1911–1919.
45. Loosli F, Winkler S, Burgtorf C, Wurmbach E, Ansorge W, Henrich T, Grabher C, Arendt D, Carl M, Krone A, Grzebisz E and Wittbrodt J (2001). Medaka eyeless is the key factor linking retinal determination and eye growth. *Development* **128**: 4035–4044.
46. Oda S, Mitani H, Naruse K and Shima A (1991). Synthesis of heat shock proteins in the isolated fin of the Medaka, *Oryzias latipes*, acclimatized to various temperatures. *Comp. Biochem. Physiol. B* **98**: 587–592.
47. Kondo S, Kuwahara Y, Kondo M, Naruse K, Mitani H, Wakamatsu Y, Ozato K, Asakawa S, Shimizu N and Shima A (2001). The medaka rs-3 locus required for scale development encodes ectodysplasin-A receptor. *Curr. Biol.* **11**: 1202–1206.
48. Barbazuk WB, Korf I, Kadavi C, Heyen J, Tate S, Wun E, Bedell JA, McPherson JD and Johnson SL (2000). The syntenic relationship of the zebrafish and human genomes. *Genome Res.* **10**: 1351–1358.
49. Postlethwait JH, Woods IG, Ngo–Hazelett P, Yan YL, Kelly PD, Chu F, Huang H, Hill–Force A and Talbot WS (2000). Zebrafish comparative genomics and the origins of vertebrate chromosomes. *Genome Res.* **10**: 1890–1902.
50. Yamamoto T (1958). Artificial induction of functional sex-reversal in genotypic females of the medaka (*Oryzias latipes*). *J. Exp. Zool.* **137**: 227–264.
51. Matsuda M, Nagahama Y, Shinomiya A, Sato T, Matsuda C, Kobayashi T, Morrey CE, Shibata N, Asakawa S, Shimizu N, Hori H, Hamaguchi S and Sakaizumi M (2002). DMY is a Y-specific DM-domain gene required for male development in the medaka fish. *Nature* **417**: 559–563.
52. Nanda I, Zend–Ajusch E, Shan Z, Grutzner F, Schartl M, Burt DW, Koehler M, Fowler VM, Goodwin G, Schneider WJ, Mizuno S, Dechant G, Haaf T and Schmid M (1999). 300 million years of conserved synteny between chicken Z and human chromosome 9. *Nat. Genet.* **21**: 258–259.
53. Nanda I, Kondo M, Hornung U, Asakawa S, Winkler C, Shimizu A, Shan Z, Haaf T, Shimizu N, Shima A, Schmid M and Schartl M (2002).

A duplicated copy of DMRT1 in the sex-determining region of the Y chromosome of the medaka, *Oryzias latipes*. *Proc. Natl. Acad. Sci. USA* **99**: 11778–11783.

54. Kondo M, Nanda I, Hornung U, Asakawa S, Shimizu N, Mitani H, Schmid M, Shima A and Schartl M (2003). Absence of the candidate male sex-determining gene dmrt1b(Y) of Medaka from other fish species. *Curr. Biol.* **13**: 416–420.
55. Matsuda M, Sato T, Toyazaki Y, Nagahama Y, Hamaguchi S and Sakaizumi M (2003). *Oryzias curvinotus* has DMY, a gene that is required for male development in the Medaka, *O. latipes*. *Zoolog. Sci.* **20**: 159–161.
56. Volff JN, Kondo M and Schartl M (2003). Medaka dmY/dmrt1Y is not the universal primary sex-determining gene in fish. *Trends Genet.* **19**: 196–199.
57. Ohno S (1970). *Evolution by Gene Duplication*. Springer–Verlag, New York.
58. Spring J (1997). Vertebrate evolution by interspecific hybridization — are we polyploid? *FEBS Lett.* **400**: 2–8.
59. Lynch M (2001). The molecular natural history of the human genome. *Trends Ecol. Evol.* **16**: 420–422.
60. Amores A, Force A, Yan YL, Joly L, Amemiya C, Fritz A, Ho RK, Langeland J, Prince V, Wang YL, Westerfield M, Ekker M and Postlethwait JH (1998). Zebrafish hox clusters and vertebrate genome evolution. *Science* **282**: 1711–1714.
61. Amores A, Suzuki T, Yan Y, Pomeroy J, Singer A, Amemiya C and Postlethwait JH (2004). Developmental roles of pufferfish Hox clusters and genome evolution in ray-fin fish. *Genome Res.* **14**: 1–10.
62. Woods IG, Kelly PD, Chu F, Ngo–Hazelett P, Yan YL, Huang H, Postlethwait JH, Talbot WS (2000). A comparative map of the zebrafish genome. *Genome Res.* **10**: 1903–1914.
63. Taylor JS, Van de Peer Y, Braasch I and Meyer A (2001). Comparative genomics provides evidence for an ancient genome duplication event in fish. *Phil. Trans. R. Soc. Lond. B* **356**: 1661–1679.
64. Taylor JS, Braasch I, Frickey T, Meyer A and Van de Peer Y (2003). Genome duplication, a trait shared by 22,000 species of ray-finned fish. *Genome Res.* **13**: 382–390.
65. Smith SF, Snell P, Gruetzner F, Bench AJ, Haaf T, Metcalfe JA, Green AR and Elgar G (2002). Analyses of the extent of shared synteny and conserved gene orders between the genome of *Fugu rubripes* and human 20q. *Genome Res.* **12**: 776–784.

66. Okubo K, Nagata S, Ko R, Kataoka H, Yoshiura Y, Mitani H, Kondo M, Naruse K, Shima A and Aida K (2001). Identification and characterization of two distinct GnRH receptor subtypes in a teleost, the medaka *Oryzias latipes. Endocrinology* **142**: 4729–4739.
67. Okubo K, Mitani H, Naruse K, Kondo M, Shima A, Tanaka M and Aida K (2002). Conserved physical linkage of gnRH-R and RBM8 in the medaka and human genomes. *Biochem. Biophys. Res. Commun.* **293**: 327–331.
68. Okubo K, Ishii S, Ishida J, Mitani H, Naruse K, Kondo M, Shima A, Tanaka M, Asakawa S, Shimizu N and Aida K (2003). A novel third gonadotropin-releasing hormone receptor in the medaka Oryzias Latipes: evolutionary and functional implication. *Gene* **314**: 121–131.
69. Matsuda M, Kawato N, Asakawa S, Shimizu N, Nagahama Y, Hamaguchi S, Sakaizumi M and Hori H (2001). Construction of a BAC library derived from the inbred Hd-rR strain of the teleost fish, *Oryzias latipes. Genes Genet. Syst.* **76**: 61–63.
70. Matsuo Y, Asakawa S, Shimizu N, Kimura H and Nonaka M (2002). Nucleotide sequence of the MHC class I genomic region of a teleost, the medaka (*Oryzias latipes*). *Immunogenetics* **53**: 930–940.
71. Zadeh Khorasania M, Henniga S, Imrea G, Asakawa S, Palczewskia S, Bergera A, Hori H, Naruse K, Mitani H, Shima A, Lehrach H, Wittbrodt J, Kondoh H, Shimizu N, Himmelbauer H (2004). A first generation physical map of the medaka genome in BACs essential for positional cloning and clone-by-clone based genomic sequencing. *Mech. Dev.* **21**: 903–913.

Chapter 19

Medaka Embryonic Stem Cells

Hong Yunhan
Department of Biological Sciences
National University of Singapore
10 Kent Ridge Crescent, Singapore 119260
and
State Key Laboratory of Freshwater Ecology and Biotechnology
Institute of Hydrobiology
Chinese Academy of Sciences
Wuhan 430072, China

Manfred Schartl
Physiological Chemistry I
Biocenter of the University Würzburg
D-97074 Würzburg, Germany

Embryonic stem (ES) cells are undifferentiated cell cultures that are derived from early developing animal embryos. ES cells retain the potential of differentiation into all cell types including germ cells and therefore provide a unique bridge linking *in vitro* and *in vivo* genetic manipulations. ES cells have been widely used in the production knock-out mice. Attempts have been made to develop ES cells in fish. We used the medaka (*Oryzias latipes*) to develop the ES cell technology in a second vertebrate model. We have established feeder cell-free culture conditions and obtained several ES cell lines from midblastula embryos. These ES cells show all features of mouse ES cells including a diploid karyotype, the potential for differentiation into various cell types and chimera

*Correspondence: Yunhan Hong, Department of Biological Sciences, National University of Singapore, 10 Kent Ridge Crescent, Singapore 119260. Fax +65 6779 2486 Email: dbshyh@nus.edu.sg

Abbreviations: EC, embryonic carcinoma; ES, embryonic stem; GT, gene targetin; HR, homologous recombination; MES1, medaka ES cell line 1; neo, gene for neomycin resistance; NT, nuclear transfer; RI, random integration; tk, gene for thymidine kinase of the herpes simplex virus.

competence. This review is to use medaka ES cells to highlight the major advances and future prospects for obtaining and utilizing ES cells in model and aquaculture fish species.

Keywords: Chimera; ES; gene targeting; medaka; pluripotency.

Introduction

Palmitter *et al.*[1] reported dramatic growth in mice that had developed from eggs microinjected with a metallothionein-growth hormone fusion gene. This showed the enormous potential of genetic engineering by gene transfer. Since then, germline transgenesis for improving animal production has been extensively performed in numerous animal species including fish.[2] In fish, a transgene is usually delivered directly into oocytes or early embryos by microinjection. In most cases, desired integration of a transgene and its proper expression have rarely been achieved, and the animals derived from microinjected embryos are highly mosaic for the introduced gene (for reviews see Gong and Hew,[3] Iyengar *et al.*[4]). In this classical transgenesis, a foreign gene is randomly added to a genome and the fate of the transgene is neither controllable nor selectable. This elusive nature of transgenes makes this method fairly inefficient. Thus, interest in novel approaches is steadily increasing (for review see Hackett and Alvarez[5]). In particular, because they are extremely rare and unrealistic to achieve by large-scale screening of genomes at the individual level, highly demanding genetic alterations such as site-specific gene addition, deletion, replacement and precise modifications cannot be achieved by classical transgenesis. Fortunately, cultured cells provide an unlimited number of genomes which can easily be manipulated and screened by drug selection for such rare events. Cultured cell genomes containing the desired modifications can generate normal animals by passing through the germline of a host embryo. This technology has been widely used in the mouse. This is the embryonic stem (ES) cell technology that relies on the derivation of ES cell lines, their use in gene targeting via homologous recombination and their germline transmission through chimera formation.

Work towards establishing embryo-derived cell lines and, ultimately, ES cell lines in fish began twelve years ago. Various methods used to develop mouse ES cell cultures have been adapted to various fish species.

However, as in mammalian species other than the mouse, the results obtained with fish have generally been unsatisfactory. An exception is a small aquarium fish, the Japanese medaka (*Oryzias latipes*), which we have used as a model for establishing ES cell technology in fish. We are now nearly at the point where we can use the full range of ES cell technology in this species. In this paper, we summarize the present status and perspectives of ES cell cultures and gene targeting in fish. The focus is on the work conducted in our laboratories on the medaka. We also review experiments on other fish species when these are directly relevant. Data from fish are compared to those from other vertebrate species when available. The results obtained so far indicate that it is possible to obtain ES cell lines and to use them for genetic engineering in teleost fish.

ES Cells

ES cell lines are undifferentiated cells that after long-term cultivation *in vitro* retain totipotency — the ability to give rise to all cell types of a whole organism. Upon reintroduction into early embryos they can enter various cell lineages, and most importantly, colonize the germline. This ability is maintained after gene transfer and selection for the desired, but rare, events of transgene integration and expression. Therefore, ES cells provide a bridge between *in vitro* manipulations of vertebrate genomes and germline transmission of transgenes. Furthermore, ES cells offer a unique opportunity to identify novel genes of particular interest and to study gene functions by gene targeting. In the past two decades, the enormous potential of ES cell lines has caused an exponential increase in the interest in model animals for basic biomedical research and in genetically engineering farm animals for higher productivity.

The key requirement for the development of ES cell-based germline transgenesis is the availability of ES cell lines. ES cell lines have so far been limited to the mouse because they are elusive to obtain in non-murine species. The success in the mouse has benefited much from the experience gathered during a long period of working with embryonic carcinoma (EC) cells. Similar to ES cells, EC cells are also derived from early developing embryos and have the pluripotency required for differentiation into various cell types. Unlike ES cells, EC cells are often aneuploid and of tumor origin. In 1975, Mintz and Illmensee produced genetically normal mosaic mice from malignant teratocarcinoma cells that had experienced almost 200 transplant generations *in vivo* as a highly malignant tumor.[6] In 1981, Stewart and Mintz generated mice from an established cell line of euploid teratocarcinoma cells.[7] In the same year, first ES cell lines from mouse embryos by using a feeder layer[8] or EC cell-conditioned medium.[9] One of the main advantages of ES over EC is efficient germline transmission.[10]

The last 20 years have seen two stages of investigation on ES cells. The first stage covered the initial 15 years and involved two main lines of research. Those working on mice specialized in making full use of ES cells to produce knock-out mice in order to analyze the physiological functions of candidate genes involved in important developmental or pathological processes. This resulted in the establishment of ES cell technology.[11,12] This technology has reached an extremely high level and has currently become a routine, invaluable tool for identifying and unraveling the biological function of novel genes in mice.[13,14] During the same time period, much work was done towards ES cell derivation in many other mammalian species, and in chickens and zebrafish. This line of research achieved little success since, in these non-murine species, the pluripotency of embryo-derived cells could be maintained only for a limited period or was lost after a few subcultures. The inability to obtain ES cell lines in farm animals has at least partly forced a search for alternatives, such as nuclear transfer. A good example was the birth of cloned sheep[15] and pigs.[16]

Two major concerns arose from these early unsuccessful attempts. The first was whether the ability to obtain totipotent ES cell lines was entirely restricted to the mouse. The second was that the lack of non-murine germline-competent ES cell lines may be due to the conditions that were used in different heterologous systems. To answer these questions, it is now widely recognized that understanding the biology of ES cells is crucial for their reliable derivation and full utilization. Therefore, during the second stage that spans the last 5 years, much emphasis has been placed on the investigation of ES cell biology at the molecular and cellular levels. Several genes and signaling pathways have been shown to be involved in regulation of ES cell self-renewal and differentiation. For example, the octamer-binding transcription factor Oct4 plays a key role in mouse ES cell renewal and differentiation.[17] The differentiation inhibiting activity of a feeder cell layer or conditioned medium can be mimicked by LIF, the leukemia inhibitory factor.[18] LIF exerts its effect through binding to its cell surface receptor and activates the gp130 signaling pathway.[19] Noticeably, downstream of gp130, Stat3 signaling is necessary for ES cell renewal.[19,20] Most recently, Nanog encoding a homeodomain protein of 305 amino acids has been reported to be necessary for maintenance of pluripotentcy in mouse epiblast and ES cells.[21,22] Expression of Oct4 is necessary but not sufficient for ES cell self-renewal, whereas overexpression of either Nanog or Stat3 is sufficient for ES cell self-renewal. Oct4 and Nanog are expressed exclusively in pluripotent stem cells and show a significant sequence divergence, making the isolation of their homologs in non-mammalian species difficult. Stat3 is expressed highly in ES but also in non-ES cells, it shows high sequence conservation in vertebrates. A model is proposed to explain the action and roles of Oct4, Nanog and Stat3 in ES cell self-renewal and differentiation (Fig. 1).

A better understanding of ES cell biology has promoted ES cell derivation. A milestone was the derivation of monkey and human ES cell lines.[23–25] The totipotency of these primate ES cell lines has not, however, been demonstrated, because of the obvious difficulty of testing their germline transmission. The availability of human ES cell lines has tremendous clinical potential. They may provide a universal source for particular cell types required for cell replacement therapy. This potential is causing increasing interest in the full understanding of stem cell renewal and differentiation.

ES Cells in Fishes

The major challenge in the establishment of ES cell lines is to inhibit their spontaneous differentiation. In the mouse this is achieved by cultivating the inner cell mass cells of blastocysts on a layer of feeder cells[8] or in a conditioned medium (CM).[9] Although application of both

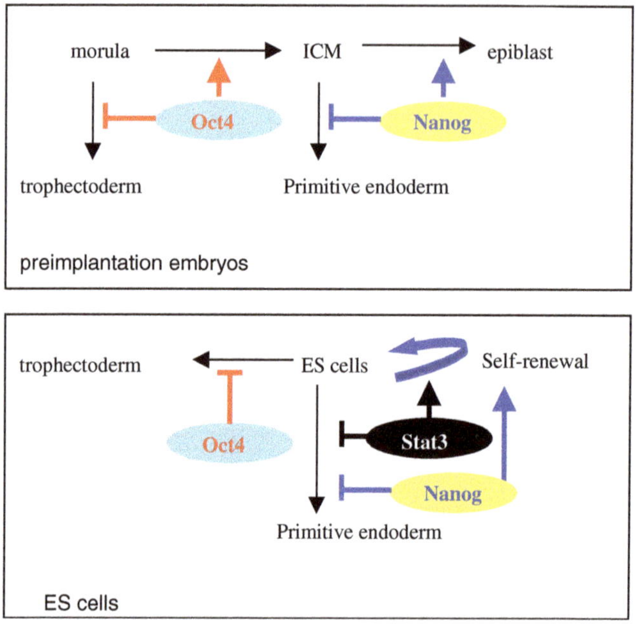

Fig. 1 Proposed functions of Oct4, Nanog and Stat3 in preimplantion embryos and ES cells. Oct4 is crucial for the first embryonic lineage specification, and Nanog is essential for the second. In the ES cells, Oct4, Nanog and Stat3 are essential for self-renewal.

techniques to non-murine mammals has achieved only partial success,[26] similar approaches were initially adopted in fish. Small model fish species, such as the zebrafish (*Danio rerio*) and the medaka, are best suited for the development of ES cells.

In zebrafish, Barnes' group[27] attempted the CM approach. They formulated a medium (LDF; a mixture of L15, DMEM and F12 media) containing embryo extract and serum from trout (*Oncorhynchus mykiss*) and produced CM by using the Buffalo rat liver (BRL) cell line which had proven capable of maintaining the ES phenotype of mouse blastocyst-derived cell cultures. The combination of BRL-CM and LDF supported the cultivation of midblastula embryo (MBE) cells for more than 40 population doublings in a seemingly undifferentiated state. Growth of these zebrafish cells was stable during this culture period and had a doubling time of 72 hours. Later, the same group tested the feeder layer procedure using the BRL line or zebrafish embryo fibroblasts (ZEF) prepared from early gastrulae and obtained MBE cell cultures that showed the ability to differentiate into neurons and astrocytes.[28,29] Both ZEF and BRL feeder cells were similarly effective in enhancing proliferation of MBE cells and suppressing their differentiation into melanocytes. This effect of both feeder cell types was found to be mimicked by basic fibroblast growth factor (bFGF) alone.[30] Recently, Ma *et al.*[31] cultured zebrafish embryo cells on a feeder layer of rainbow trout spleen cells line (RTS34st) and obtained germline chimeras.

In the medaka, Wakamatsu *et al.*[32] independently applied the feeder layer technique. They used primary cultures from blastula- and gastrula-stage embryos of the medaka strain HNI as feeder cells and developed a medium formulation which included bFGF and also fish serum from carp (*Cyprinus carpio*), besides other major supplements common to media for mouse ES cell cultures. These culture conditions enabled them to establish a pluripotent cell line, OLES1, from a blastula embryo of strain HNI. OLES1 exhibited stable growth, ES-like morphology, high alkaline phosphatase (AP) activity, and, most importantly, the potential to be induced by retinoic acid to differentiate into several cell types. Although the *in vivo* pluripotency of these cultured MBE cells from both species remains to be determined by chimera formation, the work has been valuable with respect to ES cell derivation in fish.

In order to establish conditions for deriving fish ES cell lines, we have used the medaka as a model system. This fish is, in many respects, an ideal model for vertebrate development.[33] In particular, it produces daily a number of eggs that can be used for cell culture initiation and for chimera experiments. The cells used for initiating medaka ES cell lines were dissociated MBE cells. MBE cells are easy to use for cell culture initiation, whereas cells from younger embryos are not. MBE cells are indeterminant cells,[34] have the ability to form germline chimeras,[35,36] and thus should possess the potential to give rise to cultured ES cells. Indeed, all cells of medaka midblastulae are strongly positive for AP staining,[37] a marker for undifferentiated ES cells in the mouse.

Nichols *et al.*[18] successfully derived mouse ES cell lines in the presence of LIF without using feeder cells. Since no suitable feeder cells have conclusively been established for maintaining fish ES cells in an undifferentiated state, we developed feeder cell-free culture conditions under which MBE cells can be grown on a gelatin-coated surface.[37] A medium has been formulated for the initiation and maintenance of MBE cells. The basic medium is DMEM (Dulbecco's Modified Eagle Medium) containing various supplements. The first medium ESM1 contains LIF, a higher concentration of growth factors (bFGF) and crude fish embryo extract (FEE). Growth response assays revealed that several supplements besides fetal calf serum are essential for stable growth. These include fish serum, 2-mercaptoethanol, bFGF and most importantly, FEE. Interestingly, recombinant human LIF has no effect on the proliferation and differentiation of medaka or zebrafish MBE cells[37] as well as those of marine species, the gilthead seabream (*Sparus aurata*[38]) and sea perch (*Lateolabrax japonicus*[39]). It is likely that LIF is not necessary for fish MBE cell cultures, or that LIF activity is contained in, or mimicked by, e.g. the heterogeneous embryo extract. Alternatively, the mammalian LIF sequence is too specialized to elicit its effect in fish cells. The latter appears to be supported by the fact that no LIF has so far been identified in fish. Mitogenic activity has been documented for FEE from various fish species including trout,[27] zebrafish,[29] medaka,[37,40] and gilthead seabream.[38] It is worth noting that although FEE from one species may be mitogenic to cells from other species, FEE from the same species or close relatives appears to be more mitogenic. An even greater species-specific difference was found for bFGF: besides its strong mitogenic activity in both species, bFGF irreversibly inhibits the differentiation of MBE cell cultures into melanocytes in zebrafish[32] but not in medaka.[37,40]

Several medaka ES-like cell lines have been obtained under feeder-free conditions from MBE cultures. One of them, MES1, has been extensively characterized *in vitro* and *in vivo*. *In vitro* this line shows all the known features of mouse ES cells: stable growth, ES cell morphology (small size with relatively large nuclei and prominent nucleoli, and round or polygonal shape) and high AP activity (Fig. 2), a normal diploid karyotype, the ability to form compacted colonies when seeded at low density, and to form embryoid body-like structures in suspension culture. MES1 cells have the potential to be induced to differentiate under defined conditions into various cell types including melanin-synthesizing pigment

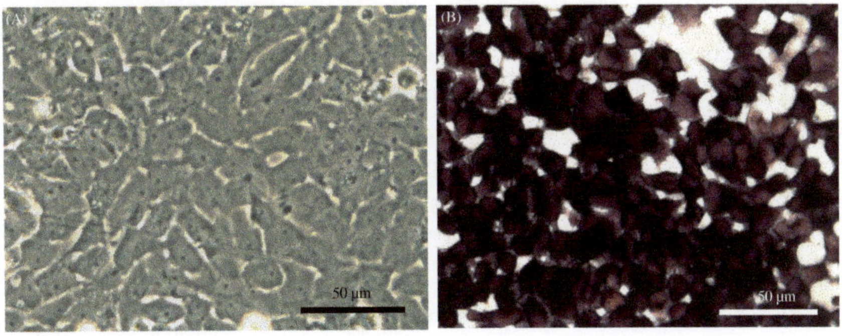

Fig. 2 Morphology of medaka ES cells (MES1 cells at passage 60 and 337 days of culture). (A) Phase contrast micrograph of MES1 cells grown at confluence on gelatin-coated surface. MES1 cells show a small size, a round or polygonal shape, compacted morphology, large nuclei and prominent nucleoli. (B) Cells showing strong staining for alkaline phosphatase, a widely used marker to monitor the undifferentiated state of ES cells. Cells were fixed in cold methanol:acetone (1:1) for 30 min at room temperature and stained for 2 hours in NBT/BCIP at 37°C.

cells, contracting muscle cells, nerve cells and fibroblasts. MES1 cells can easily be trypsinized into single cells for subculture, and, more importantly, for clonal growth. Thus a procedure was established for subclonal culture experiments which provided compelling evidence that the MES1 line is pluripotent: all descendants from a single colony of cells uniformly displaying the ES cell morphology were able to give rise not only to ES-like cells but also to the same spectrum of differentiated derivatives as the parental line.[40] This property is retained after long-term cultivation (>140 passages during >2 years of culture) and is not abolished by cryostorage. The derivation of the MES1 line was not a single lucky event. In fact, stable medaka MBE cell cultures can be obtained repeatedly using the feeder-free technique.

The successful derivation of ES cell lines is strongly dependent not only on proper culture media and conditions such as cell density[40] but also on the genetic background. This latter point has been well documented in mice. Most of the currently available mouse ES cell lines have been derived from a limited number of permissive strains.[41,42] One such strain is 129/Sv, which is characterized by a high incidence of spontaneous teratomas and teratocarcinomas, and has served as the source of embryonic carcinoma cell lines. The genetic background also has an apparent effect on the derivation of medaka MBE cell cultures. For example, of 12 medaka strains tested, only certain strains such as HNI, HB32C and HB12A reliably give rise to chimera-competent ES-like cell cultures, whereas other strains, such as Sakura, Kaga and Yokote, do not.[43] In refractory genotypes, a first problem is that MBE cells adhere poorly to substrata or not at all, e.g. medaka strain HB11C[43] and trout.[27] A second difficulty is differentiation during early days of culture, as is the case for the medaka strains Kaga and Yokote[43] as well as certain zebrafish sources. It follows that different strains or populations should be tested for their suitability for MBE cell derivation when ES cell lines are to be established in a particular fish species.

Is the feeder-free condition we developed for the medaka widely applicable for derivation of pluripotent cells from fish blastula embryos? We have so far tested four additional species of the genus *Oryzias*. Three species (*O. minutillus, O. curvinotus* and *O. mekongenesis*) allow for easy initiation and maintenance of MBE cell cultures, whereas *O. celebensis* does not. We also obtained one MBE cell line with the same conditions from a particular source of zebrafish. This feeder-free system has been successfully used for the derivation of ES-like cells in fish, even from several marine species including the gilthead seabream (*Sparus aurata*[38]) and sea perch (*Lateolabrax japonicus*[39]). Because the medaka, zebrafish, and the marine species are distantly related and live in either freshwater or seawater habitats, it appears that the ease with which MBE cell cultures can be obtained in fish has little to do with the phyletic relatedness. This is analogous to the situation in mammals. For instance, the rat, rabbit and hamster are very closely related to the mouse. They are, however, no better than other species for ES cell derivation using the techniques established in the mouse. On the other hand, primates, including human, that are distant relatives of the mouse allow for the establishment of ES cell lines. It is noteworthy that this feeder-free system is also suitable for germ cell cultures in medaka.

Mouse ES cells display a unique gene expression pattern. For example, expression of Oct4, Sox2 and Nanog is high in undifferentiated ES cells and down-regulated upon differentiation.[17,21,22,44] Medaka ES cells appear to share this feature as they retain the ability to activate gene expression from the mouse Oct4 promoter by reporter assay and express the Sox2 homolog by RT-PCR. Interestingly, medaka ES cells also express several germline-specific genes such as the DEAD family RNA helicase Vasa which is not expressed in mouse ES cells. Therefore, medaka ES cells will provide an *in vitro* system to study expression and regulation of genes important for stem cell renewal and differentiation.

Gene Targeting in Medaka ES-Like Cells

One major application of ES cells is gene targeting (GT), the replacement of an endogenous genomic sequence by a genetically engineered version via homologous recombination (HR). In the mouse, this is currently the most powerful tool for introducing defined genetic modifications into specific sites of the genome, thus allowing the elucidation of the physiological functions of the genes under study. However, this approach is so far limited to this organism.

HR events are very rare compared to random integrations (RI). Thus, in GT experiments suitable selectable markers are required in order to distinguish HR from IR and to enrich targeted ES cell colonies. A positive selectable marker permits the selection for cells containing an integrated HR sequence, while a negative selectable marker allows for the killing of cells harboring one or more copies of randomly integrated HR sequence, leaving cells from HR alive. This dual selection system, the positive-negative selection (PNS) procedure, was firstly devised in mice as a universal tool for targeting genes that are either expressed or not expressed in ES cells.[12] However, it is also possible in several cell lines of other mammals. Usually, the bacterial neomycin (*neo*) gene for resistance to G418 is inserted within one exon or used to replace an exon-flanked fragment of the gene, while the Herpes simplex virus thymidine kinase (*tk*) gene for sensitivity to gancyclovir (Gc) is attached to the end(s) of the genomic sequence. The insertion of the positive marker *neo* will result in frame-shift mutations and/or truncated translation of the gene of interest, and will therefore disrupt the function of that gene. The expression of *neo* and *tk* will confer on the cells resistance to G418 selection and sensitivity to Gc, respectively. Thus, a cell with randomly integrated HR vector DNA will, on the one hand, contain both *neo* and *tk* gene. This cell will survive G418 but die from Gc selection. On the other hand, during HR only the *neo* gene is co-integrated into the specific site on a chromosome, but the *tk* gene is lost. Consequently, a cell with a targeted sequence survives a double selection of G418 and Gc.

MES1 cells were tested for gene transfer. MES1 cells can be transfected at moderate efficiency by a simple calcium phosphate co-precipitation procedure. By electroporation, a transfection efficiency of up to 40% can be reproducibly achieved. Furthermore, the relative activities of various promoter/enhancer sequences were analyzed in order to choose regulatory sequences suitable for the expression of selectable markers.

Prior to the development of suitable GT constructs for fish cells, sensitivity assays were performed. It was found that fish cells, including MES1 cells, did not survive G418 or puromycin selection. On the other hand, Gc has no effect on cell growth. Based on these data, we constructed several cassettes expressing *neo*, *tk* and a bicistronic plasmid, consisting of both the *neo* and *tk* cassettes for PNS. Transfection and drug selection in the carp cell line EPC demonstrated that these selectable marker genes worked as well as in the mouse.[45]

The key to the gene replacement approach in fish is the presence of HR activity in fish ES-like cells. Recently, Hagmann *et al.*[46] have shown that introduced plasmids were able to undergo intramolecular HR events in zebrafish. In parallel, MES1 cells display HR activity, as evidenced by the appearance and frequency of sister-chromatid exchanges following sister-chromatid differential staining. The medaka *p53* gene was isolated and found to be a single copy gene in the medaka genome.[47] It was used for construction of HR vectors.[48] MBE cell cultures, including the MES1 line, were transfected by electroporation with the vectors. Five independent electroporated cell pools, before and after dual drug selections, were subjected to screening by PCR. All the five pools yielded a PCR product with a size expected for the correct HR event. Furthermore, drug-selected pools gave rise to a significantly stronger signal than non-selected ones, indicating the effectiveness of PNS in medaka ES-like cells. Most importantly, Southern blot analysis revealed that four of the five drug-selected pools produced a detectable band of the expected size. Clonal expansion of these pools under PNS conditions led to the formation of single cell colonies which were isolated and expanded to cell clones. Identification of targeted cell clones is in progress.

Chimera Formation in Fish

Chimera formation is a stringent criterion for testing the pluripotency of putative ES cells. To establish conditions for chimera production, the procedure for transplanting non-cultivated medaka blastomeres obtained from the deep layer[36,49] was modified for blastomeres previously dispersed from whole blastulae. This led to the efficient production of chimeras (> 90%) displaying donor-derived wild-type melanocytes in as early as two days of embryonic development.[43] In this case of uncultivated donor blastomeres, transplantation of as many cells as possible had little effect on the survival and development of host embryos, which should be due to full physiological compatibility between the donor blastomeres and those of the recipient blastulae. Interestingly, although different donor strains gave rise to a similar frequency of pigmented chimeras, they showed various patterns of chimeric pigmentation in terms of compartmental distribution of donor-derived melanocytes.

A procedure to transplant medaka MBE cell cultures was established. Introduction of too many cultured MBE cells (> 100) severely affected the viability and chimera frequency in surviving embryos. This phenomenon is common to all cultures that have been maintained over three days *in vitro*, regardless of the donor strain. When up to 100 cultured MBE cells were transferred into each recipient, pigmented chimeras were obtained. The frequency of chimeras and the degree of chimerism were high when early MBE cultures were used as donors. Although there was a stepwise decline in the efficiency of chimera formation after prolonged cultivation of donor MBE cells, in general the chimera-competence was retained in MBE cells following cultivation up to 70 days.[43] In zebrafish, Barnes' group reported chimera formation from MBE cell cultures: 37% for pigmented chimeras from 2-day-old MBE cell cultures[30] and 15% for PCR-detectable chimeras from 14-day-old MBE cell cultures.[29]

When MES1 cells after 27–66 passages (205–397 days) were transplanted into albino blastulae, 90% of the recipients developed into chimeric fry as revealed by genotype-specific PCR-assays.[50] Moreover, pigmented chimeras (5–6%) were repeatedly obtained in numerous independent experiments. To investigate whether MES1 cells were able to contribute not only to the pigment cell lineage but also to other cell lineages, they were labeled by transfection with a construct expressing the green fluorescent protein (GFP) from a strong constitutive, ubiquitously active promoter. Two days after transfection, 7% of the cells were GFP-positive. Such transfected cultures were used for transplantation to host blastulae and the resulting embryos were examined up to the hatching stage. More than 90% of these blastulae developed into GFP-positive fry. These chimeras contained from one to more than 50 GFP-expressing MES1 cells that were found in one to several different areas. The GFP-positive cells contributed to all major organ systems of all three germ layers, for example, epithelial cells in undulating fins, contracting muscles in hearts and extraembryonic cells in the yolk sac.[50] Furthermore, when uniform populations of cells stably expressing GFP were obtained by long-term drug selection and used for transplantation, we obtained 100% chimera formation. Stable GFP-expressing transfectants also allowed following the behavior of MES1 cells in more detail. It turned out that the fate of MES1 cells depend largely on their distribution. When in regions where the future heart, blood or fins will form, MES1 cells differentiated into corresponding cell types and participated in these organs. This may reflect that MES1 cells are totipotent *in vivo* and that they express this totipotency by responding in a correct way to various local signals of the developing embryo.

The degree of chimerism, the donor contribution in the MES1-derived chimeras, was generally low (2–10%). This phenomenon was most obvious from the small number (6%) of pigmented chimeras and the degree of chimerism in pigmented organs, which is significantly lower compared to mouse ES cells[42] or to chimeras formed from non-cultivated medaka blastomeres.[36,43] A series of transplantation experiments using a large number of independent MBE cell cultures has proven that this is not specific to the MES1 line but common to cultured fish MBE cells.[43] The reduced degree of chimerism obtained with MBE cell cultures compared to blastomeres may be ascribed to a collection of several possible barriers to the physiological and genetic compatibility between the donor and host.

On average, a total of 100 MES1 donor cells were introduced into each host blastula that consists of 1000 cells. Thus, providing the transferred donors could behave like normal blastomeres, a degree of chimerism of approximately 10% should be observed. Based on pigmentation that detects the genuine contribution of donors in a physiologically functional cell lineage, we generally obtained a degree of 1–2% (1–2 melanocytes in chimeras versus 84 melanocytes in an embryo of the wild-type donor strain). Considering a frequency of 5–6% for pigmented chimeras, the degree of chimerism for all MES1-transplanted embryos will be around 0.1%. Thus, there is a 100-fold difference between expected and observed degrees of chimerism. This indicates that cultivated donors are weaker than endogenous blastomeres in terms of propagation. In parallel, cultured cells differ remarkably from blastomeres of host embryos in cell cycle length: 33–48 hours for MES1 under different culture conditions versus 30 minutes for mid-blastomeres. The difference is approximately 70–100 fold, which is comparable to the 100-fold difference in the degree of chimerism. The lengthy cell cycling time may be a major reason for the low degree of chimerism. We reasoned that use of weaker host blastulae could improve chimera production. Joly *et al.*[51] has enhanced germline transmission of non-cultivated medaka blastomeres using gamma-irradiated host embryos. Similar conditions were adopted for chimera formation from MES1. Extensive transplantation experiments showed a dramatic enhancement in both chimera frequency and degree of chimerism.

In mice, the combination of donor and host strains strongly affects chimera frequency and degree of chimerism.[41,42,52] A similar phenomenon was observed in fish, as medaka ES-like cells from different donor strains showed highly variable efficiencies in chimera formation.[43]

Of the two albino strains we have used as hosts, i^1 is superior to i^3 in the formation of pigmented chimeras from MES1 and other MBE cell cultures: $> 5\%$ in i^1 but $< 1\%$ in i^3 and a wide distribution of donor-derived melanocytes in i^1 but distribution restricted to the yolk sac (extraembryonic structure) in i^3. Interestingly, when gamma-irradiated blastulae were used as the hosts, i^1 became unable to produce pigmented chimeras, whilst i^3 gave rise to a considerably enhanced formation of pigmented chimeras in which the donor-derived melanocytes were found predominantly in the embryonic body. This suggests that the donor-host genetic compatibility can be experimentally modulated. Although we do not know how the donor-host compatibility operates and is modulated by irradiation, this observation provides a clue to adjust this compatibility in the future. Of particular importance, germline transmission depends heavily not only on totipotency of ES cells, but also on the host genetic background. For example, some C57BL/6 ES lines produce germline chimeras in embryos of inbred BalbC but not C3H.[41,42] Availability of many different medaka strains and populations allows extensive examination of various combinations. It is possible that intensive investigation will reveal an optimal donor-host combination. Ma *et al.*[31] reported the production of zebrafish germline chimeras from embryo cell cultures.

Alternatives to ES Cells

Primordial germ cells (PGCs) are totipotent and efficient for germline formation. In the mouse, PGCs can form totipotent cell lines, the embryonic germ (EG) cell lines. PGCs have attracted considerable interest as an alternative to ES cells. Chicken PGCs after an extended culture period retained the ability to produce germline chimeras.[53,54]

In order to obtain pure PGCs for cell line initiation, a reliable diagnostic marker for PGCs is required. Among the known cytoplasmic germline components of the well-studied organisms *Drosophila* and *C. elegans*, Vasa protein shows a high conservation and allows for cloning using the homology approach. The cDNA or gene has recently been isolated from mammals,[55] *Xenopus*,[56] zebrafish,[57,58] medaka[59] and trout.[60] In mice, Vasa is expressed exclusively in germ cells by immunohistochemical analysis[61] and the phenotypic analysis of gene knock-outs indicates that it is required for the development of male germ cells.[62] *In situ* hybridization and/or transgenic reporter assays have also demonstrated germline-specific expression of Vasa in fish. More importantly, stable transgenic lines specifically expressing GFP from the Vasa promoter in the germline have been made available in the trout[63] and medaka.[64,65] This permits the isolation and characterization of candidate PGCs and facilitates the cultivation of PGCs in these species. For example, Takeuchi *et al.*[66] has reported the mass isolation of PGCs from the transgenic trout carrying the GFP gene driven by the *Vasa* promoter. Germline-specific expression from the Vasa promoter also makes it possible to prescreen germline transgenics at embryonic stages.[64] Recently, *ziwi*, the zebrafish homolog of *piwi*, another important component of the *Drosophila* germ plasm, has been shown to express I specifically in the gonads, and very probably in germ cells. It is noteworthy that the culture procedure for mouse EG cells is very similar to that for ES cells: essential components like a feeder layer, LIF and bFGF are common to both sources of cells. In this context, it is interesting to know if the feeder-free conditions we used for medaka ES-like cells can support the initiation and maintenance of germ cell cultures in medaka and other fish.

Alternatives to Germline Chimeras

The advent of Dolly, a sheep cloned by nuclear transfer or transplantation (NT) from a somatic nucleus,[15] surprised the world. Dolly grew to adulthood and reproduced normally.[67] This success encouraged mammalian cloning. The increasing interest in NT in mammals is generally possibility of using NT as an alternative to ES cell transplantation, necessary because of the lack of ES cell lines in these farm species such as sheep[15] and cattle. Screening a large number of animals for a rare, but highly desirable genetic modification, such as targeted transgenesis, is time- and cost-consuming in these large farm animals. In cattle, ES-like cells are available, but elusive, for single cell preparations and clonal growth. Therefore, Cibelli *et al.*[68] used NT and ES-chimera formation for the production of transgenic bovine chimeras. They first got transgenic clones of fibroblasts that are quite robust for trypsinization and gene transfection. The nuclei of the transgenic fibroblasts were transferred into bovine oocytes and the resulting blastocysts were used for ES-like cell derivation. These resulting ES-like cells were finally transplanted into blastocysts, leading to chimeric calves containing the transgenic ES-like donor cells in various somatic tissues. NT is a highly demanding experiment, requiring skilled operators and its efficacy generally is extremely low. After Dolly, only NT work with the aim of investigating parameters to improve the efficiency of germline transmission from somatic nuclei has been made in species with well-established ES cell lines like mice.[69] Clearly, even in species where the NT procedure is well-established and ES-like cells are available, ES-chimera formation remains the most efficient method for germline transmission.

Historically, NT has been most successful in fish[70] and frogs.[71] Advanced stage embryos have been obtained in *Xenopus* from the nuclei of transgenic cultured cells.[72] In fact, fish are the first vertebrate in which NT was attempted as early as half a century ago.[73] By the early 1960s, NT was already well-established in fish.[74] From then on, NT has extensively been carried out in China using various cyprinid fishes. As early as in 1986, exactly ten years before the birth of Dolly, Chen *et al.*[75] at the Institute of Hydrobiology generated adult fish from cultured cell nuclei. The donor nuclei were prepared from short-term cultured kidney cells and long-term cultured blastula cells. Initially, they obtained 41–61% blastulae from cultured kidney cell nuclei and 83% blastulae from long-term cultured embryonic cell nuclei, but none of these blastulae developed to fry. They continued this experiment by performing serial NTs in the hope that cultured cell nuclei could have more time for reprogramming in the embryonic environment during two successive rounds of NT experiments. They chose well-developed blastulae produced in the initial NT experiments to prepare donor nuclei for a second generation of NT and did obtain six fry, two of which normally developed to adulthood. Although both fish were aneuploid and sterile, this work has clearly proven both the possibility and basic procedure for fish cloning from cultured somatic cell nuclei.

Wakamatsu's group[76,77] reported the transplantation of blastula nuclei to non-enucleated eggs in the medaka. They obtained adult NT fish. Because the host eggs were not enucleated, the resulting fish were triploid and abnormal in fertility. Wakamatsu *et al.*[78] established a protocol of enucleation by gamma-irradiation and obtained fertile and diploid nuclear transplants from non-cultivated embryonic cells. Recently, Lin's group[79] has succeeded in the production of cloned zebrafish by nuclear transfer from long-term cultured cells. In this context, MES1 represents an ideal candidate cell line for generating NT fish because of its normal karyotype and confirmed pluripotency.

So far, the major interest in ES cells and NT focuses on their use for germline transgenesis and, in the case of human ES cells, for the production of large amounts of particular cell types for cell replacement therapy. Eight years ago, we proposed a novel approach based on ES cells to cryopreserve genetic sources in the form of cell banks, and to recover whole animals from these frozen cells by germline chimera formation or NT.[40] This strategy would be applicable to endangered species such as the Chinese paddle fish, which now has a population size of only approximately 100. This approach has been already tested by the genetic rescue of an endangered mammal by cross-species nuclear transfer using post-mortem somatic cells.[80] Our preliminary test showed that MBE cells and their cultures from *O. minutillus* were able to produce chimeras in heterologous host embryos of *O. latipes*. Again, cryostorage did not affect chimera-competence. Also blastomeres of *O. celebensis* produced normal pigmented chimeric fry in the albino host of *O. latipes*. These two species show obvious incompatibility, as their hybrids are sterile. Apparently, there is no serious interspecies barrier chimera formation between these three medaka species, indicating that it was possible to rescue whole animals by passing ES cells through the germline of a related species. Fertile NT fish have been obtained from various combinations with respect to the phylogenetic relationships between the nuclear donor and oocyte host species.[70] This includes one interfamilial combination. Even if species of different orders are used as the donor and host, such as tilapia nuclei into common carp eggs, young fish can be obtained, as well as normally developing embryos that should be suited for ES cell initiation. It appears to be possible to reconstitute whole fish from their cryopreserved ES cells by NT into eggs of a related species.

Most recently, Scholer's group has reported their success in the derivation of oocytes from mouse ES cells.[81] Therefore, differentiation of ES cells into germ cells and their use for gametogenesis *in vitro* will be the focus of future investigation of ES cells for germline transmission.

Perspectives

ES cells have great potential in basic and applied sciences such as functional genomics, cell replacement therapy of human diseases, preservation and recovery of biodiversity and genetic engineering. The major current application is the production of gene knockout animals through germline chimera formation. Our ability to derive stable euploid ES-like cell lines and to use them to generate fertile chimeras, the ability to transfect and select stable transgenic ES-like cell clones and the initial success in gene targeting in fish ES-like cells, are major steps towards fully establishing ES cell technology in fish. We have recognized a collection of potential problems and, more importantly, a range of potential solutions or alternatives to the problems.[82] In zebrafish, embryo cells after 4 weeks of culture have been shown to be able to form germline chimeras.[84] All this has pointed out the direction to the ultimate destination. The key in the near future is to further enhance the degree of chimerism of fish ES-like cell cultures for efficient germline transmission and to study the molecular signature of stem cell self-renewal and differentiation for our understanding of mechanisms underlying germ stem cell development. In this regards, medaka ES cells have shown the excellence for directed differentiation into a uniform population of functional cells.[85] At the molecular level, medaka ES cells share many features with mammalian ES cells as shown by the activation of the pluripotency-specific mouse Oct4 promoter.[86] The ability to modulate genetic and physiological compatibility between the ES donor and host embryo will be of critical importance. Most recently, medaka has successfully given rise to the first cell line of normal spermatogonia capable of meiosis sperm production *in vitro*.[87] It will be intriguing to determine whether this spermatogonial cell line could offer a potential approach for germline transmission through sperm production followed by artificial insemination. With the increasing interest in ES cells from various species, and in particular human ES cells, progress in stem cell research will greatly be speeded up. Provided more fish laboratories become involved and more fish species are investigated, it is not far away from a new era of fully establishing and exploiting ES cell technology in the new millennium.

Acknowledgment

We thank Dr. H. Zhao for unpublished data and J. Deng for fish breeding. This work was supported by grants to YH from the Chinese Academy of Sciences (KSCX2-SW-303), the National University of Singapore (R-154-000-147-101, R-154-000-156-112 & R-154-000-153-720), and to MS from the Deutsche Forchungsgemeinschaft (SFB 172 and SFB465), the Commission of the European Community (FAIR-PL98-3482).

References

1. Palmiter RD, Brinster RL, Hammer RE, Trumbauer ME, Rosenfeld MG, Birnberg NC and Evans RM (1982). Dramatic growth of mice that developed from eggs microinjected with metallothionein-growth hormone fusion gene. *Nature* **300**: 611–615.
2. Zhu Z, Li G, He L and Chen L (1985). Novel gene transfer into the fertilized eggs of goldfish (*Carassius auratus* L. 1785). *Z. Angew. Ichthyol.* **1**: 31–34.

3. Gong Z and Hew CL (1995). Transgenic fish in aquaculture and developmental biology. *Curr. Top. Dev. Biol.* **30**: 177–214.
4. Iyengar A, Müller F and Maclean N (1996). Regulation and expression of transgenes in fish — a review. *Transgen. Res.* **5**: 147–166.
5. Hackett PB and Alvarez MC (2000). Molecular genetics of transgenic fish. In: Fingerman M and Nagabhushanam R (eds.), *Recent Advances in Marine Biotechnology-Aquaculture Fishes,* Vol 4. Enfield, USA, pp. 77–145.
6. Mintz B and Illmensee K (1975). Normal genetically mosaic mice produced from malignant teratocarcinoma cells. *Proc. Natl. Acad. Sci. USA* **72**(9): 3585–3589.
7. Stewart TA and Mintz B (1981). Successive generations of mice produced from an established culture line of euploid teratocarcinoma cells. *Proc. Natl. Acad. Sci. USA*: **78**(10): 6314–6318.
8. Evans MJ and Kaufman MH (1981). Establishment in culture of pluripotential cells from mouse embryos. *Nature* **292**: 154–156.
9. Martin GR (1981). Isolation of a pluripotent cell line from mouse embryo cultures in medium conditioned by teratocarcinoma stem cells. *Proc. Natl. Acad. Sci. USA* **78**: 7634–7638.
10. Bradley A, Evans M, Kaufman MH and Robertson E (1984). Formation of germ-line chimaeras from embryo-derived teratocarcinoma cell lines. *Nature* **309**: 255–256.
11. Thomas KR and Capecchi MR (1987). Site-directed mutagenesis by gene targeting in mouse embryo-derived stem cells. *Cell* **51**: 501–512.
12. Mansour SL, Thomas KR and Capecchi MR (1988). Disruption of the proto-oncogene *int-2* in mouse embryo-derived stem cells: a general strategy for targeting mutations to non-selectable genes. *Nature* **336**: 348–352.
13. Müller U (1999). Ten years of gene targeting: targeted mouse mutants from vector design to phenotype analysis. *Mech. Dev.* **82**: 3–21.
14. Cecconi F and Meyer BI (2000). Gene trap: a way to identify novel gene and unravel their biological function. *FEBS Lett.* **480**: 63–71.
15. Campbell KHS, McWhir J, Ritchie WA and Wilmut I (1996). Sheep cloned by nuclear transfer from a cultured cell line. *Nature* **380**: 64–66.
16. Polejaeva IA, Chen SH, Vaught TD, Page RL, Mullins J, Ball S, Dai Y, Boone J, Walker S, Ayares DL, Colman A and Campbell KHS (2000). Cloned pigs produced by nuclear transfer from adult somatic cells. *Nature* **407**: 86–90.

17. Pesce M, Gross MK and Schöler HR (1998). In line with our ancestors: Oct-4 and the mammalian germ. *BioEssays* **20**: 722–732.
18. Nichols J, Evans EP and Smith AG (1990). Establishment of germ-line-competent embryonic stem (ES) cells using differentiation inhibiting activity. *Development* **110**: 1341–1348.
19. Niwa H, Burdon T, Chambers I and Smith A (1998). Self-renewal of pluripotent embryonic stem cells is mediated via activation of STAT3. *Genes Dev.* **12**(13): 2048–2060.
20. Matsuda T, Nakamura T, Nakao K, Arai T, Katsuki MM, Heike T and Yokota T (1999). STAT3 activation is sufficient to maintain an undifferentiated state of mouse embryonic stem cells. *EMBO J.* **18**: 4261–4269.
21. Mitsui K, Tokuzawa Y, Itoh H, Segawa K, Murakami M, Takahashi K, Maruyama M, Maeda M and Yamanaka S (2003). The homeoprotein Nanog is required for maintenance of pluripotency in mouse epiblast and ES cells. *Cell* **113**(5): 631–642.
22. Chambers I, Colby D, Robertson M, Nichols J, Lee S, Tweedie S and Smith A (2003). Functional expression cloning of Nanog, a pluripotency sustaining factor in embryonic stem cells. *Cell* **113**: 643–655.
23. Thomson JA, Kalishman JK, Golos T, Urning M, Harris CP, Becker RA and Hearn JP (1995). Isolation of a primate embryonic stem cell line. *Proc. Natl. Acad. Sci. USA* **92**: 7844–7848.
24. Thomson JA, Itskovitz–Eldor J, Shapiro SS, Waknitz MA, Swiergiel JJ, Marshall VS and Jones JM (1998). Embryonic stem cell lines derived from human blastocysts. *Science* **282**: 1145–1147.
25. Reubinoff BE, Pera MF, Fong CY, Trounson A and Bongso A (2000). Embryonic stem cell lines from human blastocysts: somatic differentiation *in vitro*. *Nat. Biotechnol.* **18**: 399–404.
26. Anderson GB (1992). Isolation and use of embryonic stem cells from livestock species. *Anim. Biotechnol.* **3**: 165–175.
27. Collodi P, Kamei Y, Sharps A, Weber D and Barnes D (1992). Fish embryo cell cultures for derivation of stem cells and transgenic chimeras. *Mol. Mar. Biol. Biotechnol.* **11**: 257–265.
28. Sun L, Bradford CS and Barnes DW (1995a). Feeder cell cultures for zebrafish embryonal cells *in vitro*. *Mol. Mar. Biol. Biotechnol.* **4**: 43–50.
29. Sun L, Bradford CS, Ghosh C, Collodi P and Barnes DW (1995b). ES-like cell cultures derived from early zebrafish embryos. *Mol. Mar. Biol. Biotechnol.* **4**: 193–199.

30. Bradford CS, Sun L and Barnes WD (1994). Basic fibroblast growth factor stimulates proliferation and suppresses melanogenesis in cell cultures derived from early zebrafish embryos. *Mol. Mar. Biol. Biotechnol.* **3**: 78–86.
31. Ma C, Fan L, Ganassin R, Bols R and Collodi P (2001). Production of zebrafish germ-line chimeras from embryo cell cultures. *Proc. Natl. Acad. Sci. USA* **98**: 2461–2466.
32. Wakamatsu Y, Ozato K and Sasado T (1994). Establishment of a pluripotent cell line derived from a medaka (*Oryzias latipes*) blastula embryo. *Mol. Mar. Biol. Biotechnol.* **3**: 185–191.
33. Wittbrodt J, Shima A and Schartl M (2002). Medaka — a model organism from the Far East. *Nat. Rev. Genet.* **3**: 53–64.
34. Kimmel CB and Warga RM (1986). Tissue-specific cell lineages originate in the gastrula of the zebrafish. *Science* **231**: 365–368.
35. Lin S, Long W, Chen J and Hopkins N (1992). Production of germ-line chimearas in zebrafish by cell transplants from genetically pigmented to albino embryos. *Proc. Natl. Acad. Sci. USA* **89**: 4519–4523.
36. Wakamatsu Y, Ozato K, Hashimoto H, Kinoshita M, Sakaguchi M, Iwamatsu T, Hyodo–Taguchi Y and Tomita H (1993). Generation of germ-line chimeras in medaka (*Oryzias latipes*). *Mol. Mar. Biol. Biotechnol.* **2**: 325–332.
37. Hong Y and Schartl M (1996). Establishment and growth responses of early medakafish (*Oryzias latipes*) embryonic cells in feeder layer-free cultures. *Mol. Mar. Biol. Biotechnol.* **5**: 93–104.
38. Bejar J, Hong Y and Alvarez MC (2002). An ES-like cell line from the marine fish *Sparus aurata*: characterization and chimaera production. *Transgen. Res.* **11**(3): 279–289.
39. Chen S, Sha Z and Ye H (2003). Establishment of a pluripotent embryonic cell line from sea perch (*Lateolabrax japonicus*) embryos. *Aquaculture* **218**: 141–151.
40. Hong Y, Winkler C and Schartl M (1996). Pluripotency and differentiation of embryonic stem cell lines from the medakafish (*Oryzias latipes*). *Mech. Dev.* **60**: 33–44.
41. Lerdermann B and Burki K (1991). Establishment of a germ-line competent C57BL/6 embryonic stem cell line. *Exp. Cell Res.* **197**: 254–258.
42. Kawase E, Suemori H, Takahashi H, Okazaki K and Nakatsuji N (1994). Strain difference in establishment of mouse embryonic stem (ES) cell lines. *Int. J. Dev. Biol.* **38**: 385–390.

43. Hong Y, Winkler C and Schartl M (1998b). Efficiency of cell culture derivation from blastula embryos and of chimera formation in the medaka (*Oryzias latipes*) depends on donor genotype and passage number. *Dev. Genes Evol.* **208**: 595–602.
44. Avilion AA, Nicolis SK, Pevny LH, Perez L, Vivian N and Lovell–Badge R (2003). Multipotent cell lineages in early mouse development depend on SOX2 function. *Genes Dev.* **17**: 126–140.
45. Chen S, Hong Y and Schartl M (2002a). Development of a positive-negative selection procedure for gene targeting in fish cells. *Aquaculture* **214**: 67–79.
46. Hagmann M, Bruggmann R, Xue L, Georgiev O, Rungger D, Schaffner W, Spaniol P and Gerster T (1998). Homologous recombination and DNA-end joining reactions in zygotes and early embryos of zebrafish (*Danio rerio*) and *Drosophila melanogaster. Biol. Chem.* **379**: 673–681.
47. Chen S, Hong Y, Scherer S and Schartl M (2001). Lack of UV light inducibility of the medakafish (*Oryzias latipes*) tumor suppressor gene p53. *Gene* **264**: 197–203.
48. Chen S, Hong Y and Schartl M (2002b). Cloning, structural analysis and construction of homologous recombination vector of p53 gene in medaka fish (*Oryzias latipes*). *Acta Zool. Sin.* **48**: 519–526.
49. Ando S and Wakamatsu Y (1995). Production of chimeric medaka (*Oryzias latipes*). *Fish Biol. J. Medaka* 7: 65–68.
50. Hong Y, Winkler C and Schartl M (1998a). Production of medakafish chimeras from a stable embryonic stem cell line. *Proc. Natl. Acad. Sci. USA* **95**: 3679–3684.
51. Joly J-S, Kress C, Vandeputte M, Bourrat F and Chourrout D (1999). Irradiation of fish embryos prior to blastomere transfer boosts the colonisation of their gonads by donor-derived gametes. *Mol. Reprod. Dev.* **53**: 394–397.
52. Schwartzberg PL, Goff SP and Robertson EJ (1989). Germline transmission of a *c-abl* mutation produced by targeted gene disruption in ES cells. *Science* **246**: 799–803.
53. Park TS and Han JY (2000). Derivation and characterization of pluripotent embryonic germ cells in chicken. *Theriogenology* **56**: 475–482.
54. Han JY, Park TS, Hong YH, Jeong DK, Kim JN, Kim KD and Lim JM (2002). Production of germline chimeras by transfer of chicken gonadal primordial germ cells maintained *in vitro* for an extended period. *Theriogenology* **58**: 1531–1539.

55. Komiya T and Tanigawa Y (1995). Cloning of a gene of the DEAD box protein family which is specifically expressed in germ cells in rats. *Biochim. Biophys. Res. Commun.* **207**: 405–410.
56. Komiya T, Itoh K, Ikenishi K and Furusawa M (1994). Isolation and characterization of a novel gene of the DEAD box protein family which is specifically expressed in germ cells of *Xenopus laevis*. *Dev. Biol.* **162**: 354–363.
57. Olsen LC, Aasland R and Fjose A (1997). A *vasa*-like gene in zebrafish identifies putative primordial germ cells. *Mech. Dev.* **66**: 95–105.
58. Yoon C, Kawakami K and Hopkins N (1997). Zebrafish vasa homologue RNA is located to the cleavage planes of 2- and 4-cell-stage embryos and its expression in the primordial germ cells. *Development* **124**: 3157–3166.
59. Shinomiya A, Tanaka M, Kobayashi T, Nagahama Y and Hamaguchi S (2000). The *vasa*-like gene, *olvas*, identifies the migration path of primordial germ cells during embryonic body formation stage in the medaka, *Oryzias latipes*. *Dev. Growth Differ.* **42**: 317–326.
60. Yoshizaki G, Sakatani S, Tominaga H and Takeuchi T (2000a). Cloning and characterization of a *vasa*-like gene in rainbow trout and its expression in the germ cell lineage. *Mol. Reprod. Dev.* **55**: 364–371.
61. Toyooka Y, Tsunekawa N, Takahashi Y, Matsui Y, Satoh M and Noce T (2000). Expression and intracellular localization of mouse *vasa*-homologue protein during germ cell development. *Mech. Dev.* **93**: 139–149.
62. Tanaka S, Toyooka Y, Akasu Y, Katoh–Fukui Y, Nakahara Y, Suzuki R, Yokoyama M and Noce T (2000). The mouse homolog of *Drosophila* Vasa is required for the development of male germ cells. *Genes Dev.* **14**: 841–853.
63. Yoshizaki G, Takeuchi A, Sakatani S and Takeuchi T (2000b). Germ cell specific expression of green fluorescent protein in transgenic rainbow trout under control of the rainbow trout *vasa*-like gene promoter. *Int. J. Dev. Biol.* **44**: 323–326.
64. Tanaka M, Kinoshita M, Kobayashi D and Nagahama Y (2001). Establishment of medaka (*Oryzias latipes*) transgenic lines with the expression of green fluorescent protein fluorescence exclusively in germ cells: a useful model to monitor germ cells in a live vertebrate. *Proc. Natl. Acad. Sci. USA* **98**: 2544–2549.

65. Zhao and Hong, unpublished.
66. Takeuchi Y, Yoshizaki G, Kobayashi T and Takeuchi T (2002). Mass isolation of primordial germ cells from transgenic rainbow carrying the green fluorescent protein gene driven by the *vasa* gene promoter. *Biol. Reprod.* **67**: 1087–1092.
67. Wilmut I, Schnieke AE, McWhir J, Kind AJ and Campbell KHS (1997). Viable offspring derived from fetal and adult mammalian cells. *Nature* **385**: 810–813.
68. Cibelli JB, Stice SL, Golueke PJ, Kane JJ, Jerry J, Blackwell C, Ponce de Leon FA and Robl JM (1998). Transgenic bovine chimeric offspring produced from somatic cell-derived stem-like cells. *Nat. Biotechnol.* **16**: 642–646.
69. Wakayama T, Perry ACF, Zuccotti M, Johnson KR and Yanagimachi R (1998). Full-term development of mice from enucleated oocytes injected with cumulus cell nuclei. *Nature* **394**: 369–374.
70. Yan S (1998). *Cloning in Fish: Nuclearcytoplasmic Hybrids.* IUBS Educational and Cultural Press Ltd., Hong Kong.
71. DiBerardino MA and Orr NH (1992). Genomic potential of erythroid and leucocytic cells of *Rana pipiens* analysed by nuclear transfer into diplotene and maturing oocytes. *Differentiation* **50**: 1–13.
72. Kroll KL and Gerhart C (1994). Transgenic *X. laevis* embryos from eggs transplanted with nuclei of transfected cultured cells. *Science* **266**: 650–652.
73. Tung TC, Chang CY and Tung YFY (1945). Experiments on the development potencies of blastoderm and fragment of teleostean egg separated latitudinally. *Proc. Zool. Soc. Lond.* **115**: 175–188.
74. Tung TC, Wu SC, Tung YFY, Yan SY, Tu M and Lu TY (1963). Nuclear transplantation in fishes. *Scientia* (Peking) **14**: 1244–1245.
75. Chen H, Yi Y, Chen M and Yang X (1986). Studies on the developmental potentiality of cultured cell nuclei of fish. *Acta Hydrobiol.* **10**: 1–7.
76. Niwa K, Ladygina T, Kinoshita M, Ozato M and Wakamatsu Y (1999). Transplantation of blastula nuclei to non-enucleated eggs in the medaka, *Oryzias latipes. Dev. Growth Differ.* **41**: 163–172.
77. Niwa K, Kani S, Kinoshita M, Ozato K and Wakamatsu Y (2000). Expression of GFP in nuclear transplants generated by transplantation of embryonic cell nuclei from GFP-transgenic fish into non-enucleated eggs of medaka, *Oryzias latipes. Cloning* **2**: 23–34.

78. Wakamatsu Y, Ju B, Pristyaznhyuk I, Niwa K, Ladygina T, Kinoshita M, Araki K and Ozato K (2001). Fertile and diploid nuclear transplants derived from embryonic cells of a small laboratory fish, medaka (*Oryzias latipes*). *Proc. Natl. Acad. Sci. USA* **98**: 1071–1076.
79. Lee K-Y, Huang H, Ju B, Yang Z and Lin S (2002). Cloned zebrafish by nuclear transfer from long-term-cultured cells. *Nat. Biotechnol.* **20**: 795–799.
80. Loi P, Ptak G, Barboni B, Fulka J, Cappai JrP and Clinton M (2001). Genetic rescue of an endangered mammal by cross-species nuclear transfer using post-mortem somatic cells. *Nat. Biotechnol.* **19**: 962–964.
81. Hubner K, Fuhrmann G, Christenson LK, Kehler J, Reinhold R, De la Fuente R, Wood J, Strauss JF, Boiani M and Scholer HR. Derivation of oocytes from mouse embryonic stem cells. *Science* **300**(5623): 1251–1256.
82. Hong Y, Chen S and Schartl M (2000). Embryonic stem cells in fish: current status and perspectives. *Fish Physiol. Biochem.* **22**: 165–170.
83. Hong Y, Chen S, Gui J and Schartl S (2004). Retention of the developmental pluripotency in medaka embryonic stem cells after gene transfer and long-term drug selection for gene targeting in fish. *Transgenic Res* **13**: 41–50.
84. Fan L, Crodian J, Liu X, Aleströme A, Aleströme P, Collodi P (2004). Zebrafish embryo cells remain pluripotent and germ-line competent for multipassages in culture. *Zebrafish* **1**: 21–626.
85. Bejar J, Hong Y, Schartl M (2003). Mitf expression is sufficient to direct differentiation of medaka blastula derived stem cells to melanocytes. *Development* **130**: 6545–6553.
86. Hong Y, Winkler C, Liu T, Chai G, Schartl M (2004). Activation of the mouse Oct4 promoter in medaka embryonic stem cells and its use for ablation of spontaneous differentiation. *Mech. Dev.* (in press).
87. Hong Y, Liu T, Zhao H, Xu H, Wang W, Liu R, Chen T, Deng J, Gui J (2004). Establishment of a normal medakafish spermatogonial cell line capable of sperm production *in vitro. Proc. Natl. Acad. Sci. USA* **101**: 8011–8016.

Subject Index

www.ingramcontent.com/pod-product-compliance
Lightning Source LLC
LaVergne TN
LVHW070907300926
849162LV00010B/1841